I0055046

Biodiversity in
HORTICULTURAL CROPS

Prof. K V. Peter, basically a horticulturist, a plant breeder and a University Professor is an acknowledged research manager. He associated with development of several varieties in tomato, brinjal, chilli, bittergourd, snake gourd, oriental pickling melon, amaranth and cowpea. Breeding for resistance to bacterial wilt initiated in 1979 led to release of popular varieties like Pant C-1 and Pant C-2 in chilli, Sakthi in tomato and Surya in brinjal. Sources of resistance to aphids in cowpea and viral leaf curl in chilli were located. A vegetable seed production complex established by him at Kerala Agricultural University in 1980 continues to supply quality seeds to farmers even today. Prof. Peter established a spices biotechnology facility at Indian Institute of Spices Research, Calicut and nurtured an active team of scientists who worked out protocols for rapid multiplication of 32 spices including black pepper and cardamom. *In vitro* conservation of spices germplasm consisting 36 species and 315 accessions was accomplished, along with know-how for minimal growth methods. Prof. Peter provided effective managerial support to the Indian Institute of Spices Research (IISR), Calicut to possess the world's largest collection of black pepper and cardamom germplasm. The technology package for protected cultivation of bush black pepper would make available green pepper throughout the year. Prof. Peter authored/edited 64 books and is author/co-author of 16 bulletins and 110 full research papers. Prof. Peter edited 3 volumes of 'Handbook of Herbs and Spices' published by Woodhead Publishing Company, UK and CRC, USA now getting revised to two volumes to be published in 2012. He co-authored Handbook of Industrial Crops published by Haworth Press, USA. The National Book Trust of India published the book 'Plantation Crops' and the book 'Tuber Crops' authored by him. The book 'Plantation Crops' is available in three languages-English, Hindi, Malayalam-.The New India Publishing Agency (NIPA) New Delhi published a 12 volumes Horticultural Science Series, 5a volumes Underutilized and Underexploited Horticultural Crops, one volume Basics of Horticulture and two volumes Science of Horticulture. The Daya Publishing House, New Delhi published his edited 4 volumes of Biodiversity of Horticultural Crops. He co-edited with Prof. P.A. Sebastian "Spiders of India" published by Orient and Blakswan (University Press, Hyderabad). The University Press also published his co-edited book "Plant Biotechnology: Methods of Tissue culture and Gene Transfer". The Studium Press (USA) India published "Cashew: A Monograph", co-authored with Dr. Abdul Salam. He is a Fellow of National Academy of Agricultural Sciences, New Delhi, Fellow of National Academy of Sciences, Allahabad and Fellow of National Academy of Biological Sciences, Chennai. He was awarded the "Rafi Ahmed Kidwai Award" by Indian Council of Agricultural Research, New Delhi. The National Academy of Agricultural Sciences, New Delhi conferred 'Recognition Award 2000' and the 'Dr. K. Ramiah Memorial Award "for contributions to Plant Improvement-2008-2009 . The University of Agricultural Sciences, Bangalore bestowed 'Dr. M.H. Marigowda Award' to him for outstanding contributions to Horticultural Sciences. He is the recipient of Silver Jubilee Medal for outstanding contributions in vegetable research, instituted by the Indian Society of Vegetable Science, Varanasi, UP. He was conferred the Shivasakthi-HSI Award for Life time achievement in Horticulture-2008. Dr. Prem Nath Agricultural Science Foundation conferred the National PNASF Gold Medal Award for the outstanding scientist of 2009.The IIHR, Bangalore appreciated his contributions to Vegetable Research by conferring a plaque on 18 January, 2010. During 1991-2000, he got the then National Research Centre of Spices elevated into Indian Institute of Spices Research. Indian Council of Agricultural Research New Delhi conferred The Best Institute Award on IISR for its research contributions to spices. Dr. Peter was Member, Standing Committee, Association of Indian Universities, New Delhi (2004-2005) and Member, Spices Board of India (1991-1998, 2007-2009). He was Member, Governing Body, ICAR (2002-2005) and Member, Board of Management IARI, New Delhi and IIVR, Izatnagar (2003-2006). Prof. Peter was Member, Task Force (Plant Biotechnology) DBT and National Task Force on Women in Science DST. He was member, Consultative Expert Committee of Gi and IPR Authority, Chennai, Chairman, Tea Research and Extension Council, Valparai, India and Member RAB, KSSF, PDS Peerumedu. He was, Member, Academic Council, IGNOU New Delhi, Member, School Board of Agriculture, IGNOU New Delhi and Chairman, CAC, NAIP Project-'A value chain on cashew'. He is Trustee Federal Ashwas Trust promoted by Federal Bank, Aluva. He is a member of Governing Body of Bishop Jerome Memorial Educational Trust Quilon; Member Academic Committee, Shrehrdaya Educational Trust, Kodakara; Nominated Member Kerala Regional Latin Catholic Council, Aluva, currently Chairman, Research Advisory Committee, Indian Vegetable Research Institute, Varanasi 2011-2014 and Member Board of Management IARI, New Delhi 2012-2014 and Member, School Board School of Agriculture IGNOU New Delhi (2010-2014).

Born in 17th May, 1948 at the Island village of Kumbalangi, he had secondary School Education at St. Peters High School, Kumbalangi, pre-degree at Maharajas College, Ernakulam, B.Sc. (Ag) at Agricultural College and Research Institute, Vellayani and Mastoral and Doctoral degrees at G.B. Pant University of Agriculture and Technology, Pantnagar and post doctoral training at USDA, BARC-W, Maryland. He attended scientists meet at Bangladesh, Taiwan, Goudeloupe (FWI) and USA. He was co-ordinator of two network projects on plant biotechnology, funded by Department of Biotechnology, Government of India. The Research Team led by him received the National Award For Processing and Product Development given by The Department of Biotechnology. During his Vice-Chancellorship at Kerala Agricultural University (June, 2001 to June 2006), the University received 1st Rank in 2001-2002, 2002-2003 and 2003-2004 and third rank in 2004-2005 in academic performance on all India basis from ICAR, New Delhi. The ICAR New Delhi conferred The Sardar Patil Award for the Best ICAR Institution 2003 to Kerala Agricultural University for all round performance. As Vegetable Breeder at G.B. Pant University of Agriculture and Technology from 1976-1979 and as Professor of Horticulture at KAU from 1979 to 2008 he guided several students for M.Sc. and Ph.D. programmes in the Universities of Calicut, Kerala Agricultural University and GBPUAT, Pantnagar. He is now Director, World Noni Research Foundation, Chennai, Contact E-mail: kvptr@yahoo.com

Biodiversity in
HORTICULTURAL CROPS
— Volume 4 —

Editor

Professor K.V. Peter

2013

Daya Publishing House®

A Division of

Astral International Pvt. Ltd.

New Delhi – 110 002

© 2013 K.V. PETER (b. 1948–)
ISBN 9788170359692

Despite every effort, there may still be chances for some errors and omissions to have crept in inadvertently. No part of this publication may be reproduced in any form or by any means, electronically, mechanically, by photocopying, recording or otherwise, without the prior permission of the copyright owners.

The views expressed in various articles are those of the authors and not of editor or publisher of the book.

Published by ·: **Daya Publishing House®**
A Division of
Astral International Pvt. Ltd. –
ISO 9001:2008 Certified Company –
4760-61/23, Ansari Road, Darya Ganj
New Delhi-110 002
Ph. 011-43549197, 23278134
E-mail: info@astralint.com
Website: www.astralint.com

Laser Typesetting : **Classic Computer Services**
Delhi - 110 035

Printed at : **Salasar Imaging Systems**
Delhi - 110 035

PRINTED IN INDIA

Devotion

I devote the 4ᵗʰ Volume "Biodiversity in Horticultural Crops" to Prof. V.L. Chopra Former Member Planning Commission Government of India, National Professor ICAR, Director General ICAR and above all a wonderful teacher who motivated me to be a good scientist and Science writer. He co-authored with me a book "Handbook of Industrial crops" published by Taylor and Francis, USA (Haworth Press, USA).

Prof. K.V. Peter

Acknowledgement

I acknowledge with gratitude Prof. P.I. Peter Chairman NoniBiotech, Chennai for the encouragement and facilities provided for my ventures in science journalism. He gave me free time and provided academic environment to edit the several series. I profusely thank my wife Vimala and sons Anvar and Ajay and daughter in law Cynara for the encouragement. I also acknowledge the scientists who contributed the 21 chapters.

My special gratitude to Dr. P.N. Mathur, Co-ordinator, Biodiversity International, Rome, and Dr. Brahma Singh, Former Director, DRDO, New Delhi for contributing FOREWORD and PREFACE to the book respectively. I also express gratitude to the publisher Daya Publishing House, New Delhi for meticulous printing.

Prof. K.V. Peter

Foreword

Biodiversity in flora and fauna is the wealth of any nation. Billions of genes with considerable value are habitated in the bioreserves designated. The shift in emphasis on biodiversity as property in global public domain to actual national, state and gram panchayat domain has decentralized the distribution of wealth to actual custodian of plants. This shift led to maintenance of People's biodiversity registry, state biodiversity registry and national biodiversity registry as documents of immense value. National Biodiversity Authority, State Biodiversity Board and local bodies Biodiversity Management Committees are established in India by the Biological Diversity Act, 2002 and Biological Diversity Rules, 2004 to ascertain richness of India in biological diversity and associated traditional and contemporary knowledge system. India is a founding partner to the UN Convention on Biological Diversity (UN CBD) signed at Rio de Janeiro on 5th June, 1992. The convention reaffirms the sovereign right of states over their biological resources. Conservation of biological diversity, sustainable use of its components and fair and equitable sharing of the benefits arising out of the utilization are objectives of the UN CBD.

Biodiversity boosts ecosystem productivity where each species, no matter how small, all have an important role to play. With dwindling land under crops, irrigation water scarcer and plundering forests for human habitation and human greed, conservation of biodiversity is a high priority item for survival of flora and fauna. Plant Genetic Resource (PGR) is fundamental to crop improvement programme and the key to establish future food and nutritional security. In India National Bureau of Plant Genetic Resources, New Delhi has now come out with the Vision-2025 focusing on strengthening national net working on acquisition, conservation, evaluation, strengthening data base on PGR information, utilization of crop germplasm and monitoring national and international regulations in PGR policy to suitably ensure harmonized PGR management. Emphasis is given to introduction of trait specific germplasm of different field and horticultural crops encompassing industrial, ornamental, vegetable and new crops with trade potential. At international level Biodiversity International, Rome a constituent partner of CGIAR is in the forefront of assisting member countries in identifying genetic resources for nutritional and food security without hampering the fragile eco-system.

The present series of books "BIODIVERSITY OF HORTICULTURAL CROPS" edited by Professor K.V. Peter Former Vice-Chancellor, Kerala Agricultural University are in 4th volume covering vegetables, fruits, spices, plantation crops and ornamentals especially orchids. Thirty nine scientists from thirty eight Research Institutes and Universities have contributed to the 4th volume. I congratulate all of them for their valuable contributions. Daya Publishing House, New Delhi is to be complimented for taking up publication of the series which add to the wealth of knowledge in this area of science.

Dr. P.N. Mathur

Co-ordinator, Biodiversity International (Erstwhile IPGRI)
NASC, DPSM, New Delhi

Preface

Horticultural crops contribute to nutrition and livelihood security. Mankind depends on about 5000 plant species worldwide for food, shelter and clothing. This is in addition to the several industrial uses to which plants are put into. This number is just a fraction of half million species of mosses, ferns, conifers and flowering plants. More than 50000 plant species are yet to be documented. Horticultural crops encompass medicinal plants, tubers, vegetables, fruits, plantation crops, ornamentals, spices and aromatic plants and mushrooms. Horticultural crops provide essentially the much needed nutrition, colour, flavour, taste and digestability. Biodiversity in horticultural crops demands attention for survey, collection, characterization, documentation, bioprospecting and in registration under the Plant Protection of Variety and Farmers Right Act once distinctiveness, uniformity and stability (DUS) of plant varieties are documented by evaluation at selected research centres. The Biological Diversity Act, 2002 and Biological Diversity Rules, 2004 provide legal framework for registry in three levels-village, state and national-. The series BIODIVERSITY IN HORTICULTURAL CROPS focus on conservation, characterization, bioprospecting and documenting variability in horticultural crops. Volume I carries 18 chapters-conservation and use of tropical fruit species, temperate fruit crops, tropical fruits, banana and plantains, temperate and subtropical vegetables, tropical vegetable crops, tuber crops, orchids of western ghats, conservation of spices genetic resources through *in vitro* conservation through cryopreservation, biodiversity of black pepper; ginger; tree spices; cardamom; large cardamom; kokum; seed spices; cashew and rubber-. Volume II carries 14 chapters-centres of origin and diversity of horticultural crops, convention on biodiversity, sacred groves for biodiversity conservation, biodiversity of amaranth, drumstick, radish, tropical cauliflower, jackfruit, grape, cocoa, tamarind, Indian gooseberry, orchids of medicinal value and plumeria. Volume III covers biodiversity in ash gourd, satputia, watermelon, cranberry, jamun, litchi, mangoes, noni, pomegranate, heliconia, marigold, zinnia, bydagi chilli and bambara groundnut. The current volume IV covers 21 chapters – management of genetic resources of horticultural crops in the emerging IPR regime, horticultural biodiversity heritage sites, impact of climate change on biodiversity, role of sacred groves in the perspective of biodiversity conservation with special reference to Kumaun Himalaya, biodiversity in anthurium, aonla, bael, capsicums and paprikas, chrysanthemums, citrus, custard apple, jujube, okra, orchids in Arunachal Pradesh and Himalayas, papaya, potato, sponge gourd, tamarind, temperate and sub-tropical vegetables and tomato-.

The whole series target consolidation of information on biodiversity of horticultural crops-characterisation, documentation, bioprospecting, molecular characterization and conservation. I appreciate the efforts of Dr. K.V. Peter, Director World Noni Research Foundation for consolidation. I also compliment the publisher Daya Publishing Hosue, New Delhi for undertaking publication of the series.

Brahma Singh
Former Secretary, Life Science Research Board
Former Director, Life Sciences
Defence Research and Development Organization (DRDO)
New Delhi.

Contents

List of Contributors

Chapter 1

J.C. Rana
National Bureau of Plant Genetic Resources, Regional Station, Shimla – 171 004, H.P.
E-mail: ranajc2003@yahoo.com; pgripr@gmail.com

Sanjeev Saxena
IPR&P (ICAR), Krishi Anusandhan Bhawan–II, Pusa Campus, New Delhi – 110 012

Chapter 2

Anurudh K. Singh and J.P. Yadav
Department of Genetics, M.D. University, Rohtak – 124 001, Haryana
E-mail: anurudhksing@gmail.com

Chapter 3

Anita Srivastava and Om Kumar
Biodiversity Climate Change (BCC) Division,
Indian Council of Forestry Research and Education, Dehra Dun
E-mail: srivastavaa@icfre.org; kumarom@icfre.org

Chapter 4

Tariq Husain, Priyanka Agnihotri and Harsh Singh
Biodiversity and Angiosperm Taxonomy,
CSIR-National Botanical Research Institute, Rana Pratap Marg, Lucknow – 226 001
E-mail: hustar_2000@yahoo.co.uk

Chapter 5

Manjunath S. Patil and Anil R. Karale
College of Horticulture,
Mahatma Phule Krishi Vidyapeeth, Pune – 411 005, Maharashtra
E-mail: patilmsflori@rediffmail.com

Chapter 6

K. Madhavi Reddy
Principal Scientist,
Indian Institute of Horticultural Research, Hessaraghatta Lake Post, Bangalore – 560 089
E-mail: kmr14@iihr.ernet.in

Chapter 7

P. Narayanaswamy
Dean, College of Horticulture,
University of Horticultural Science, Bagalkot – 587 102 Karnataka
E-mail: deancoh.hiriyur@gmail.com

Chapter 8

S.S. Hiwale
Central Horticulture Experiment Station,
Godhra-Baroda Highway, Vejalpur – 389 340, Dist. Panchmahals, Gujarat
E-mail: sshiwale@yahoo.com

Chapters 9 and 10

Sunil Kumar Sharma
IARI Regional Station,
Agricultural College Estate, Shivajinagar, Pune – 411 005, M.S.
E-mail: sunilksharma1959@yahoo.co.in

Chapter 11

K.M. Indiresh
Professor and Head,
Department of Vegetable Crops, Post Graduate Center, UHS Campus, GKVK, Bangalore – 65
E-mail: indiresh_kabali@yahoomail.com

H.M. Santhosha
Ph.D Research Scholar
University of Horticultural Sciences, Bagalkot, Karnataka

Chapter 12

D.K. Singh and Mangaldeep Sarkar
Department of Vegetable Science,
College of Agriculture, GBPUA&T, Pantnagar – 263 145, U.S. Nagar, Uttarakhand
E-mail: dks1233@gmail.com

Chapter 13

V. Ponnuswami, M. Prabhu, S.P. Thamaraiselvi and J. Rajangam
Horticultural College and Research Institute, Periyakulam
E-mail: swamyvp200259@gmail.com

Chapter 14

Ravindra Mulge
I/C Professor and Head,
Department of Vegetable Science, K.R.C. College of Horticulture, Arabhavi,
University of Horticultural Sciences, Bagalkot, Karnataka
E-mail: ravindramulge@yahoo.com

Chapter 15

Sant Ram
Scientist,
Central Institute for Arid Horticulture,
Beechwal, Bikaner – 334 006, Rajasthan, India
E-mail: santramiari@gmail.com

S.K. Sharma
Director
Central Institute for Arid Horticulture,
Beechwal, Bikaner – 334 006, Rajasthan, India
E-mail: ciah@nic.in

Chapter 16

S.K. Sharma, R.S. Singh* and A.K. Singh
Central Institute for Arid Horticulture,
Bikaner – 334 006, Rajasthan
*E-mail: rssingh1@yahoo.com

Chapter 17

T.K. Hazarika
Department of Horticulture, Aromatic and Medicinal Plants
School of Earth Sciences and Natural Resources Management,
Mizoram University, Aizwal – 796 004, Mizoram
E-mail: tridip28@gmail.com

Chapter 18

Amnuai Adthalungrong
Senior Agricultural Researcher,
Horticulture Research Institute, Department of Agriculture, Thailand
E-mail: amnuai.th@gmail.com

Chapter 19

Krishna S. Tomar and Sunil Kumar
College of Horticulture and Forestry,
Central Agricultural University, Pasighat, Arunachal Pradesh – 791 102, India
E-mail: tomarhorti@rediffmail.com

Chapter 20

R.B. Ram*, Rubee Lata and M.L. Meena
Department of Applied Plant Science (Horticulture)
Babasaheb Bhimrao Ambedkar University (A Central University),
Vidya Vihar, Rae Bareli Road, Lucknow – 226 025 (U.P.), India
*E-mail: rbram@rediffmail.com

Chapter 21

D. Ram, Mathura Rai and Major Singh
Indian Institute of Vegetable Research,
Varanasi – 221 005, India
E-mail: singhvmo@gmail.com

Introduction

Volume 4 of "*Biodiversity in Horticultural Crops*" is continuation of previously published three volumes. The Indian gene centre is one of the 12 centres of diversity of crop plants having linkages and contiguity with Indo-Chinese, Chinese and Central Asian regions. The Horticultural Genetic Resources (HGR) constitute an important component of Indian Agriculture. The HGR includes cultivated species, their varieties and wild relatives in a heterogeneous range of fruit and nut crops, vegetables, food legumes, root and tuber crops, plantation crops, ornamentals, spices, condiments and medicinal and aromatic plants needed to enhance horticultural production and productivity in India. India is the centre of diversity for Mango having the largest number of commercial cultivars grown in different regions. Truely wild mangoes are recorded in North-Eastern India, sub-Himalayan tracts and deep gorges of the Bahraich and Gonda hills in Uttar Pradesh. Rich genetic diversity of wild *Musa* spp. exists in the North-Eastern region and of cultivated types of genomic groups AA, AB, AAB and ABB exist in the Southern region. About 30 species of Citrus occur in India. Guava, although originated in tropical America has been naturalized in parts of Eastern India where sufficient variability is available. Indian Jujube, pomegranate, jamun, litchi, sapota and coconut have rich genetic diversity in India. Among the 30 species of *Malus* worldwide only two occur wild in India-*M. baccata* and *M. sikkimensis*. The genera *Pyrus* and *Prunus* have 14 and 22 species respectively in India. The Indian region also has unique diversity for minor fruits-tropical and temperate-like bael, Indian gooseberry, papaya, jackfruit, custard apple, karonda, lasoda, phalsa, fig, kokam, mangostean, keir, seabuckthorn, palu, kaphal and ghai-. Among vegetables *Solanum melongena* has three cultivated species-*S. torvum, S. incanum* and *S. melongena var. insanum* –which are widely distributed in South India, Shivalik hills and North Eastern region. Among spices, black pepper and cardamom originated in the tropical evergreen forests of the Western Ghats of South India while for ginger and turmeric, the South Asian region is the centre of origin; possibly India in view of the fact that maximum diversity for these crops occur in India. Indian region has enormous diversity for ornamental plants which are grown commercially and in home gardens. The important ornamentals native to India are orchids, rhododendrons, primulas, musk rose, begonias, lily, foxtail lily, chameli, lotus, water lily and wild tulip. In India, there is rich diversity in medicinal and aromatic plants and traditional uses associated with their uses. The Western Ghats and North-Eastern region are particularly rich in this flora. The indigenous germplasm has been enriched by introducing about 40,000 accessions of various HGR. Safe movement of germplasm to protect the native genetic resources is an important management aspect as there are glaring examples of enormous losses caused by the introduced pests. The Directorate of Plant Protection, Quarantine and Storage is the nodal organization for implementation of plant quarantine regulations and also to deal with commercial exchange. The National Bureau of Plant Genetic Resources (NBPGR), New Delhi has national responsibility to carry out plant quarantine for exchange of germplasm meant for research. Pest Risk Analysis (PRA) standards are being worked out for 48 crops of horticultural importance. The import of transgenic material is regulated by the DBT and NBPGR is the authorized agency for carrying out the quarantine test. A National Facility for evaluation of transgenics has been established at NBPGR. National Plant Biosecurity System is being developed by harmonizing national legislations with international agreements and protocols. The National Gene Bank housed at NBPGR is primarily responsible for conservation of all kinds of HGR on long term basis. Nearly 40, 000 accessions have been conserved in the National Gene Bank for long term storage in the form of seed, while around 87, 000 accessions of perennial crops are conserved in the field gene banks at 10 regional stations of NBPGR and 59 National Active Germplasm Sites (NAGS) located in different agro-climatic and phyto-geographical regions of the country. Besides 2000 accessions of 158 vegetative propagated crops *in vitro* and 8, 981 accessions of 726 species have been cryo preserved. There is a national net work for evaluation and conservation of HGR which includes NBPGR, Horticultural Institutes, National Research Centres, AICRPs of ICAR, Department of Biotechnology, Department of Environment and Forests, State Agricultural Universities, State Departments of Agriculture/Horticulture/Forestry. The documentation of PGR becomes important at the national, regional and global levels for effective conservation of rapidly disappearing genetic stocks for their immediate and future uses. The Agricultural Research and Information System (ARIS) cell at NBPGR which is mainly responsible for the data base activities at NBPGR has developed many on line databases related to PGR and plant varieties, which are in use by researchers in India. The genetic resources are considered as common heritage of humankind. The Convention on Biological Diversity (CDB) has emerged out of the global concern for conserving biodiversity and was signed by 150 countries at the United Nations Conference on Environment and Development in 1992 in Rio de Janeiro (the Earth Summit) and entered into force on 29th December, 1993. There are currently 191 parties and the treaty has been ratified by 168 countries. As the national response to CBD, the Biological Diversity Act, 2002 was enacted on December 11, 2002 and came into force since April 15, 2003 with the notification of the Biological Diversity Rules 2003. The implementation of the Act is coordinated by three functional bodies namely the National Biodiversity Authority (NBA), The State Biodiversity Boards (SBB) and the Biodiversity Management Committee (BMC). Protection of Plant Varieties and Farmers Rights Act-2005 is the second Indian IPR legislation which India adopted to comply TRIPS requirements. Geographical Indications (GI) is the third important IPR law covered under the TRIPS and concerned to agriculture. Indian law on GI of goods (Registration and Protection) Act was enacted in 1999 and enforced from 15th September, 2003 with the GI Registry located at the Patent Office, Chennai. The ICAR has framed a detailed guideline which deals with IPR awareness and literacy, enhances the work environment for higher innovativeness, ensures that the scientists/innovators are duly rewarded with their share of benefits accrued and guide the manner of technology transfer which would be competitive and better serve the interest of agriculture and farmers.

Chapter II deals with horticultural heritage sites-history, traditions and qualities-. The World Heritage Convention (WHC) is a unique international instrument for conserving cultural and natural heritage of outstanding value. By virtue of physiognomy and climatic diversity, India harbours a rich flora and is considered one of the 12 mega diversity centres, housing an estimated 12 per cent of the world flora. India contains three of the 34 biodiversity hotspots, the Himalayas, the Indo-Burma region and the Western Ghats with thousands of endemic flora. India possesses 141 endemic genera belonging to over 47 families. As per Botanical Survey of India (BSI), India has 46,214 plant species. Of these about 17, 500 represent flowering plants (7 per cent of the World Flora); 37 per cent of them are endemic. Of the endemic species (4, 950), the largest number (about 2532) species are located in Himalayas followed by Peninsular region (1, 788 species) and Andaman and Nicobar Islands (185 species). Around 583 species are cultivated of which 417 belong to horticultural crops. About 27 species in fruits and nuts;23 in vegetables;15 in plantation and tuber crops; and 16 in spices and condiments are cultivated on a regular basis. Vavilov (1935) recognizes 8 primary centres of origin of which two are associated with India namely the India Centre and Indo-Malayan Centre. The Hindustani Centre is placed next to Chinese in richness for crop species. Based on the geographical association with various life forms, India is divided into 10 biogeographical and 21 agroecological regions. Based on the indices associated with agrobiodiversity, PPV and FR authority identified 22 agro-biodiversity hotspots. These are the Western Himalayan region, Eastern Himalayan Region, Brahmaputra Valley Region, Garo, Khasi and Jainthia Hill Region, Northeastern Hills of Nagaland, Manipur, Mizoram and Tripura; Arid Western Region and Semi-arid Kathiawar peninsula; Malwa Plateau Region, The Bundelkhand Region, Upper Gangetic Plains, Middle Gangetic Plain Region, Lower Gangetic Plain or Delta Region, Chotanagpur Plateau Region, Bastar Region, Koraput Region, South Eastern Ghats, Kaveri Region, Northwestern Deccan Plateau, Konkan Region and Malabar Region.

Chapter III deals with impact of climate change on biodiversity. It is established that Earth's climate is warmer approximately by 0.7 degree centigrade over past 100years due to the green house gases (GHGs). These gases trap heat from the sun and thus lead to climate change. Climate change and global warming are major issues affecting production and productivity in horticultural crops. Early flowering and early fruit set may reduce yield in tropical and sub-tropical fruits, vegetables and plantation and spices. Temperate horticultural crops may have reduced life span. Climate induced disasters can have devastating effects on the economy, cause huge human losses and can set back development efforts.

Chapter IV refers role of sacred groves in the perspective of biodiversity conservation with special reference to Kumaon Himalayas. Sacred groves are special sites of areas having one or more attributes which distinguish them as somehow extraordinary, usually in a religious or spiritual sense. Sacred groves are the small or large patches of near natural vegetation dedicated to local ancestral spirits or deities and are preserved on the basis of religious and cultural beliefs represented the climax vegetation of the region. Significance of sacred groves has assumed importance from the point of view of anthropological, taxonomical and ecological considerations. Such a grove may consist of a multi-species, multi-tier primary forest, a clump of trees belonging to one species or even a single old tree, depending on the history of the vegetation and local culture of the region or area. These groves are protected by local communities usually through customary taboos and sanctions with traditional, cultural and ecological implications. Sacred groves are called Law lyngdhoh in Meghalaya, Kovil kadu in Tamil Nadu, Dev bhumi in Uttarakhand, Kavu in Kerala, Sarna and Deori in MP, oran in Rajasthan, Jaherthan and Garamthan in West Bengal, Deovan in HP, Ummanglai in Manipur etc.

Existence of such undisturbed pockets is mostly due to certain taboos, strong beliefs supplemented by mystic folklores. The groves exhibit diversity in species of trees, huge climbers, epiphytes and other shade loving plants. Sacred groves were the most prominent in Ancient East and prehistoric Europe and also feature in various cultures throughout the world. They play significant role in biodiversity conservation. There is urgent need to protect these invaluable heritages and treasure houses of plants and animals.

Biodiversity of selected ornamentals, vegetables, fruits, spices, aromatic crops and plantation crops are elaborated from Chapter Vth onwards. Anthurium is the largest and the most complex genus in the monocotyledonous family Araceae. The genus Anthurium is adopted to a number of different tropical wet forest environments as well as coastal forest with sandy soil and consists of about 1100 neotropical species with the greatest concentration of species occurring between 0-1500 m elevations. Most of the species originated from Central or South America especially from Northern Andes, from Colombia and Venezuela through Equador. Anthuriums are used as potted plants, cut flowers and for bouquet making. There are flowering, foliage and fragrant groups. Diversity with respect to spread, height, colour of spathe, leaf and petiole are amazing. Anthurium cultivation is a promising source of income to women starting from nursery for seedling production, cut flower production and flower placement in hotels and restaurants.

Capsicums and paprika's are vegetables and spices respectively. Grown both under field conditions and in protected cultivation, the above crops are assuming economic significance. The genus *Capsicum* consists of at least 25 wild species of which five species *viz., annuum, baccatum, chinense, frutescens* and *pubescens* are domesticated. There are more than 400 varieties of Capsicum available all over the world. They differ in pungency, size, shape and colours. Peppers are divided into pungent (hot) and non-pungent (sweet) groups. A few of the peppers are Aji, Anaheim, Ancho, Bell peppers, Bolita, Cascabel, Cayenne, Cherry, Chilepin, Piquin, de Arbol, Habanero, Jalapeno, Jamaican Hot, Mirasol, Mulato, New Mexican, Paprika, Pasilla, Pimento, Poblano, Scot's Bonnet, Serrano, Squash/ Tomato/Cheese, Tabasco and Yellow Wax. Indian pepper (chilli) varieties are Naga Jolokia, Bird Eye, Byadagi, Ellachipur Sannam, Guntur Sannam, Hindupur Chillies, Jwala, Kandhari, Kashmiri Mirch, Mathania Chilli, Mundu, Nalcheti, Tomato chilli or Warangal Chilli and Sankeswar chilli.

Chrysanthemums are the second largest cut flower group after rose among the ornamental plants traded in the global flower market. It is cultivated for both potted and cut flower. Chrysanthemum means Gold flower. Originated in China, the flowers are integral part of royal ceremonies in China. A few of the cultivar names are Emperor's Garment, Golden Dish, Embroidered Hibiscus, Star of the sky, Jade Ball, Agate Rings, Full Moon etc. The flower became the symbol of the sacred ruling dynasty of Japan. Breynius, a native of Holland wrote the first book about Chrysanthemum in 1689. The first Chrysanthemum show was organized in Vienna in 1831. Chrysanthemum was introduced to America in 1798 by John Stevens of New Jersey. Interspecific hybridization was resorted to generate colourfull variants. There are many diverse forms in Chrysanthemum-single or daisy type, semi-double, button or baby type, pompon, anemone, decorative, cactus, incurved, reflexed, spoon, quill, thread, tube, spidery and exhibition-. The National Botanical Garden and Research Institute, Lucknow has an ambitious programme on Chrysanthemum. Many cultivars were introduced to India from Australia- J. S. Lioyd-, France-Gloria Deo-, Japan-Ajina Purple-, New Zealand-Gusman Red-, UK-Alfred Wilson- and USA-Cassa Grande-. In India there are mainly two types-Garland purpose, Cut sprays-. Chrysanthemums are propagated through vegetative means. Hence genetic diversity is limited.

Custard apple is a tropical fruit grown in Andhra Pradesh, Maharashtra, Assam, Karnataka, MP, Odissa, Rajasthan and Tamil Nadu. Of late commercial orchards have come up in Maharashtra,

Gujarat and AP. The fruit is unique for its taste, flavour, colour and texture. Uses are as dessert, in ice creams and other milk products, jam and jelly, in Ayurveda medicine and in seed oil industry. Ideotypes of custard apple are proposed-prolific bearing, low seed content, better keeping quality, large fruit, sweet, pleasant aroma, resistance to drought and salinity-. The edible species other than *A. squamosa* are *A. reticulata, A. cherimola, A. muricata* and *A. diversifolia.*. Seeds are traditionally used to propagate most of the annonas. Grafting (soft wood) and budding are successful methods of propagation. The above species are grafted on their own seedlings or each other.

The Indian jujube (ber) is cultivated for almost 4000 years in India. There are two species *Ziziphus mouritiana* (Indian jujube) Lam. and *Z. jujube* (Chinese jujube). The Indian jujube is cultivated in Southern Asia, Africa and Europe (a likely introduction). The tree has a wide range of morphological variations and is widespread in Africa and Southern Africa while in Africa it has naturalized and so-called 'wild' types are found. It adapts to warm to hot tropical climates with low to relatively high rain fall, tolerating poor soils. Chinese jujube is a small shrub having a small canopy (3.5-4.5m), trunks short or long depending on genotype. Chinese jujube is spread west wards to the Mediterranean, throughout the Near East and South West Asia and eastwards to Korea and Japan. Genetic biodiversity of *Ziziphus* is high in India and about 20 species are found between 8.5-32.5 degree N and 69-84 degree E. Economically important species are *Z. nummularia, Z. oenoplia, Z. rugosa, Z. sativa, Z. vulgaris* and *Z. xylopyrus*. A large number of cultivars are available for both Indian and Chinese jujubes. A modern assessment of the genus *Ziziphus* is needed in terms of better characterization using molecular fingerprinting of accessions in all existing germplasm collections including both wild and cultivated taxa. The cultivated jujubes represent polyploidy series and not enough is known about the distribution of ploidy levels. Being rich in minerals, vitamins and antioxidants, jujube is an emerging fruit for India.

Papaya (*Carica papaya*) is grown in tropics for fresh fruits and the latex, papain having many industrial uses. India contributes about 36 per cent of world production from one forth of the world area under papaya (98, 000 ha). *Carica papaya* is likely to have originated in Tropical Central America. The family Caricaceae has six genera comprising 35 species. *Vasconcellea* is the largest genus with 21 species, genus *Jacaratia* with 7 species, followed by genus *Jarilla* with three species. Genus *Cylicomorpha* has two species while genera *Carica* and *Horovitzia* have one species each. *Carica papaya* produces edible fruits, other species in the remaining genera are wild except a few species in the genera *Jacaratia* and *Vasconcellea*. Papaya varieties in India show great biodiversity based on their height, vigour, stem and petiole colour, sex expression etc. Old cultivars from which modern cultivars emerged are Washington, Honey Dew, Ranchi, Ceylon and Bangalore. The modern cultivars from old cultivars are Pusa Giant, Pusa Delicious, Pusa Majesty, Pusa Nanha, Pusa Dwarf, Co-1, Co-2, Co-3, Co-4, Co-5, Co-6 and Co-7 from Tamil Nadu Agricultural University. Pant Papaya-1, Pant Papaya-2 and Pant Papaya-3 are from G B Pant University of Agriculture and Technology, Pant Nagar. IIHR Bangalore released Coorg Honey Dew, Pink Flesh Sweet and Surya. Red Lady 786 (Taiwan) and Sinta (Philippines) are cultivars imported to India. Biodiversity was created unintentionally by developing two transgenic (TG) papaya cultivars, SunUp and Rainbow resistant to Papaya ringspot virus (PRSV) at Hawaii. Papaya is a nutritious dessert fruit ideally suited to home stead growing.

Potato is grown in more than 148 countries and is one of the world's major non-cereal food crops. Till 16th century, it was unknown to Europe, Asia and North America. Potato has it's origin in South America where it grows wild. There is great deal of genetic biodiversity in potatoes. Central Potato Research Institute Shimla HP possesses a great deal of biodiversity in potato. At early days, potato varieties were introduced to India-Up-to-date, Atlantic, FTL-1533, Criags Defiance and Alley. Varieties developed by clonal selection are Kufri Red and Kufri Safed. Hybrids are Kufri Chandramukhi (S-4485 x Kufri Kuber), Kufri Badshah (Kufri Jyothi x Kufri Alankar), Kufri Sheethman (Phulwa x Craigs Defiance) and Kufri Jyothi (Phulwa x CP 1787). Other methods of crop improvement followed are interspecific hybridization, mutation breeding, molecular breeding and recombinant DNA breeding. In addition to use as food, potatoes are used to brew alcoholic beverages like Vodka, potcheen or akvavit. Potato starch has varied uses as thickners, binders of soups and sauces, in the textile industry, as adhesives and for manufacture of papers and boards. Many companies are exploring possibilities of using waste potatoes to obtain polylactic acid for use in plastic products. Research is on the way to use the starch as a base for biodegradable packaging.

Sponge gourd is a cucurbit fruit vegetable commonly grown in tropics and sub-tropics of Asia, Europe and America. An annual climber, the fruit has a lot of fibre adding to the value of the vegetable in a nutrition perspective. The genus *Luffa t*o which sponge gourd belongs has 5-7 species, out of which *L. cylindrica* and *L. acutangula* are domesticated. The wild species *L. graveolens* is the progenitor of the cultivated smooth gourd and ridge gourd. Sizeable germplasm of smooth gourd is maintained at National Bureau of Plant Genetic Resources, New Delhi and at Project Directorate of Vegetable Research attached to Indian Institute of Vegetable Research, Varanasi UP. Rich diversity occurs in Indo-Gangetic plains, Tarai Region and North Eastern Plains. Sponge gourd is grown in Sunderbans (West Bengal)

Tamarind also called 'Indian date' is a multipurpose tree known for drought tolerance and used primarily for its fruits eaten fresh or processed, used as seasoning or spice or the fruits and seeds processed for non-food uses. Tamarind has wider geographical distribution-Africa, Asia and Americas-. It is naturalized in Indonesia, Malaysia, Philippines and the Pacific Islands. Thailand has the largest plantations of the Asian nations, followed by Indonesia, Myanmar and the Philippines. Tamarind is often used as a roadside or avenue tree grown along canals, particularly in the North and South dry zones. Two major types-sweet and sour-are recognized based on sweetness of the fruit pulp. Cultivars are differentiated by pulp colour-red and brown. Tamarind leaf, pulp and seeds are extensively used in traditional Indian and African medicine. Many delicacies especially Ready to Serve (RTS) drinks are made from tamarind in Thailand.

The tomato is a relatively recent addition to the world's important food crops. Leading tomato growing countries are China, USA, India and Turkey. Cultivated tomato is native to South America although it is domesticated in Mexico. The first recorded mention of the tomato in North America was in 1710. The Asian Vegetable Research and Development Centre Taiwan (AVRDC) renamed "World Vegetable Centre" maintains one of the largest collections of tomato germplasm. A tomato genetics co-operative based in University of California documents tomato genes for possible uses any where. Among vegetables, tomato is the most researched crop for its easiness to grow, high seed content, suitability to be grown under protected cultivation and genetic vulnerability.

Aonla (Indian gooseberry) is a well known fruit tree finding mention in Charaka Samhita, Susrutha Samhita and in Ayurvedic texts for its medicinal nutritional and wellness values. Wild aonla fruits are ingredients of Chavanprash recommended to children and pregnant women. A rich source of Vitamin A, aonla has high content of minerals-calcium, iron, boron- and anti-oxidants. Being rich in fibre, aonla forms an ingredient in pickles. Banarsi, Francis and Chakaiya are traditional varieties.

Bael-Shri Phal, Bael patra, Bengal quince, beel-is a fruit of India. Bael is a religious tree as well planted near temples devoted to Lord Shiva. Bael fruit is mentioned in the Indian Pharmacopoea. Importance of bael fruit lies in its curative properties mentioned in Charaka Samhita. All the parts of the tree-stem, bark, root, leaf, flower, seed, fruit of all stages-are used in Ayurvedic medicines. Bael fruits are one of the most nutritious fruits-protein 1.8g, fat 0.39g, carbohydrates 31.8g, minerals 1.7g, carotene 0.13mg, thiamine 0.13mg, riboflavin 1.19mg, niacin 1.1mg and vitamin C 8.0 mg.per 100 g of edible portion. Bael tree grows naturally in UP, Bihar, Jharkhand, M.P., Odissa, West Bengal and Chattisgarh. Development of low seeded variety is a tree improvement goal. Bael is the fruit of arid climate with less water demanding. The fruit is mentioned in Indian Pharmacopoeia. Tree improvement resulted in varieties like Narendra Bael-5, Narendra Bael-7, Narendra Bael-9, Pant Aparna, Pant Shivani, Pant Sujata, Pant Urvasi, CISH Bael-1, CISH Bael-2 and Goma Yashi.

Collectively called Citrus, it comprises of sweet oranges, mandarins, acid lime-lemons, pumello, grape fruit, kinnow, sour orange and an array of interspecific hybrids. High Vitamin C is the common nutritional content of citrus fruits. Citrus is now a major food industry crop for squashes, ready to serve drinks, marmalades, concentrates and dehydrated peeled fruits and peels for oils and volatiles. India enjoys a remarkable position in the "citrus belt of the world" due to her rich wealth of Citrus genetic resources, both wild and cultivated. The species *Citrus reticulata* (loose skinned orange), *C. unshiu* (cold hardy mandarin), *C. deliciosa* (Mediterranean mandarin), *C. nobilis* (King orange), *C. reshni* (Cleopatra mandarin), *C. medurensis* (calamondin), *C. madaraspatana* (Kitchli sour orange) and *C. tangarina* (tangerine) belong to mandarin group.Orange group consists of *C. sinensis* (tight skin orange) and *C. aurantium* (Sour orange).Pummelo-grape fruit group consists of *C. grandis* (pummelo) and *C. paradisi* (a natural cross between *C. sinensis* and *C. grandis*). Acid group has *C. limon* (lemon), *C. jambhiri* (rough lemon), *C. aurantifolia* (Acid lime) and *C. limettioides* (sweet lime), *C. medica* (citron), *C. karna* (karma-khatta), *C. latifolia* (Persian lime), *C. limetta* (Sweet lime) and *C. limonia* (Rangpur lime). The citrus genus has been extensively studied by Swingle and Tanaka with their merits and demerits of academic value. With the development of fractional chemistry supported by organic chemists, a large number of flavourants have been identified in citrus and used industrially. Several value added products like dehydrated fruits, volatiles from rind, seed oil, leaf oil and dried flowers are made. Lime juice-salted/sweetened-is a hydrant in case of loss of body water due to digestive disorders.

Okra (Ladies finger, bhindi) is a major tropical fruit vegetable rich in fibres and minerals. Tender pods are taken in as salad or cooked partially alone or along with other vegetables. Belonging to cotton family, okra has the same pests and diseases especially the yellow vein mosaic. Broadcasted okra yields small fruits as being heavily populated. Line sown okra yields long fruits as plants thinly populated. There are varietal differences also. Pod colour ranges white to green to red and their shades. Dehydrated okra pods are available. Okra seed oil is edible and has value in lubricating industry. Abysenia (present Ethiopea) is reported as the centre of origin. Southeast Asia especially Thailand, Malaysia, Philippines and Indonesia is a major grower and consumer of okra. Okra soup, okra fry, okra with fish and meat are delicacies.

Orchids are highly evolved group of plants with diverse habits, bizarre and curious flower structure with brilliant colours adding beauty to the land and landscape. The very fact that there are about 20, 000 species distributed all over the world speaks of their diversity and varied nature having a realm of their own. In India, there are about 1150 species distributed in various states. Arunachal Pradesh alone accounts for 601 species, almost 52% of the total species known in India. Arunachal's rich and colourful orchids find a place of pride. Arunachal Pradesh is called the Orchid Paradise of India. The entire orchid family with about 601 species in Arunachal Pradesh is classified into six sub families, 17 tribes, 24 sub-tribes and 111 genera. Orchids in Arunachal Pradesh can also be classified into 140 species of terrestrial orchids with 15 saprophytes and about 340 epiphytes found in different forest types. Orchids have ornamental, medicinal and spiritual uses.

Orchids of Himalayas provide two very recognizable end points-highly restricted endemics and a set of widespread species-. The largest family of flowering plants in the hot spot of Himalayas is Orchidaceae with nearly 800 species. The word orchid is derived from the Greek word "Orchids" meaning testicle because of the appearance of subterranean tubers of the genus Orchids. Orchids are perennial herbs. They may be either terrestrial (growing in soils), epiphytes (growing on plants but not parasitizing on them), lithophytes (growing on rocks and sand grains) or saprophytes and show a great diversity in their floral structure, developed mainly as a result of their adaptation to pollination by a wide variety of insects. The North East India Himalayas being the cradle, comprises eight states-Arunachal Pradesh, Assam, Manipur, Meghalaya, Mizoram, Nagaland, Sikkim and Tripura – is the abode of unique orchids. Many are now endangered due to man's invasion and "burn and grow" type of shifting cultivation.

Temperate and sub-tropical vegetables were listed as distinct from tropical, arid and aquatic vegetables. Tropicalisation of temperate vegetables is a significant contribution of vegetable breeders. Now temperate vegetables like carrot, radish, cabbage, cauliflower and broccoli can be grown in nontraditional areas in tropics. There are now tropical carrots, tropical cauliflower and tropical radishes suited for cultivation throughout India in all seasons except monsoons. A large number of vegetables–leaf, fruit, stem, inflorescence, whole plant-are grown in all the eight agro-ecological regions of India. Majority are well known vegetables like amaranth, cucurbits, solanaceous vegetables, okra, peas and beans, potato, onions and garlic. There are a vast number of underexploited and underutilized vegetables like curry leaf, drum stick, chow chow, basil, Ceylon spinach, basella, chekkurmanis, clove bean, agathi, lotus root, water cheshnut and so on. Scope of aquatic plants as possible vegetables is being looked into.

The Biodiversity in Horticultural Crops Vol. 4 has 21 chapters of much scientific delightment to students, nature lovers and botanists.

Prof. K.V. Peter

2013, Biodiversity in Horticultural Crops Vol. 4   *Pages 1–23*
Editor: Professor K.V. Peter
Published by: DAYA PUBLISHING HOUSE, NEW DELHI

Chapter 1

Management of Genetic Resources of Horticultural Crops in the Emerging IPR Regime

J.C. Rana[1] and Sanjeev Saxena[2]
[1]National Bureau of Plant Genetic Resources Regional Station, Shimla – 171 004, H.P.
[2]IPR&P (ICAR), Krishi Anusandhan Bhawan–II, Pusa Campus, New Delhi – 110 012
E-mail: ranajc2003@yahoo.com; pgripr@gmail.com

Indian gene centre is credited for having the most valuable and diverse genetic resources required for enhancing productivity and ensuring food security. Besides, having linkages and contiguity with Indo-Chinese, Chinese-Japanese and Central-Asian region, it is one of the twelve centers of diversity of crop plants. The current matrix of Plant Genetic Resources (PGR) consists of the gene pool of 166 indigenous crops, 320 wild relatives including many introduced crops. Horticultural Genetic Resources (HGR) constitute an important component of Indian agriculture. The important genera comprising HGR for which wide range of genetic diversity exist include *Artocarpus, Citrus, Grewia, Mangifera, Musa, Prunus, Pyrus, Rubus, Syzygium, Vitis, Ziziphus* in fruits, *Abelmoschus, Amorphophallus, Amaranthus, Cucumis, Cucurbita, Dioscorea, Luffa, Momordica, Solanum, Trichosanthes* in vegetables, *Allium, Amomum, Curcuma, Elettaria, Cinnamomum, Garcinia, Piper, Trigonella, Zingiber* in spices and condiments and *Coffea* and *Camellia* in plantation crops, *Rosa,* and *Jasminum* in ornamentals. It is also a secondary centre of diversity for crops such as cowpea, cluster bean, okra, niger, safflower, chayote, *Cucurbita* spp., chillies and amaranths. Genetic diversity does exist for introduced crops like *Prunus, Pyrus, Rubus, Punica granatum, Psidium guajava, Cocos nucifera, Capsicum annuum, Lablab purpureus,* and *Camellia sinensis*. Apart from cultivated crop species, there are >1500 wild edible plant species mainly comprised of horticultural plants, widely used by various tribes and rural people.

Nonetheless, the increasing human population pressure coupled with several other factors has adversely impacted HGR and their productivity. The pressure on HGR is likely to increase further in the emerging threats of climate change and changing IPR regimes. The genetic resources are no more a common heritage of humankind. Greater awareness has been created on the conservation and sustainable utilization of genetic resources, the legal regimes have changed and the principal of free availability of genetic resources has become unacceptable to many countries. As a result, the issue of managing and sharing genetic resources and benefits accrued from their use got driving seat in all TRIPS and WTO related negotiations. The stakeholders across the globe agreed to make sustainable use and conservation of genetic resources and also to develop *sui generis* protection instruments harmonizing with TRIPS and WTO regulations. In the subsequent pages, we have discussed the scenario of HGR in India, their management and various national and international legal instruments available for their management and protection in the emerging IPR regimes.

Status of HGR

The wide ranges of HGR which includes cultivated species, their varieties and wild relatives in a heterogeneous range of fruit and nut crops, vegetables, food legumes, root and tuber crops, plantation crops, ornamentals, spices, condiments, and medicinal and aromatic plants are needed to enhance horticultural production and productivity in India.

Fruit Crops

Among fruits, tropical fruits constitute a major proportion of the spectrum of fruit diversity. India is centre of diversity for *Mangifera indica* having the largest number of commercial cultivars grown in different regions, for instance, Dashehari, Langra, Chausa, Bombay Green and Fazli in North India; Banganapalli, Totapuri (Banglora), Neelum, Prairi, Suvarnarekha, Mulgoa, Kalapadi and Rumani in South India; Alphonso, Kesar, Mankurad, Fernandin and Vanraj in Western India, and Langra, Fazli, Chausa, Zardalu, Himsagar and Malda in eastern India (Karihaloo *et al.*, 2005). Truly wild mango trees have been recorded in North-eastern India, sub-Himalayan tracts and deep gorges of the Bahraich and Gonda hills in Uttar Pradesh (Brandis, 1874; Kanjilal *et al.*, 1934). Wild species such as *M. andamanica, M. khasiana, M. sylvatica* and *M. camptosperma* have been reported from India (Mukherjee, 1985; Yadav and Rajan, 1993). Rich genetic diversity of wild *Musa* spp. in the North-eastern region and of cultivated types especially those belonging to the genomic groups AA, AB, AAB and ABB exist in the southern region. Maximum genetic variability of *M. acuminata* and *M. balbisiana* occur in South-east Asia. Wild species like *M. acuminata* ssp. *burmannica, M. balbisiana, M. nagensium, M. sikkimensis, M. cheemanii, M. ornata, M. laterita, M. velutina* and *M. sanguinea* have been recorded from southern and western regions while *M. flaviflora* is localised to Manipur and Meghalaya. In Citrus, *Citrus reticulata, C. sinensis, C. aurantifolia* and *C. limon* are the major cultivated species whereas *C. latifolia, C. grandis, C. paradisi* and *C. maderaspatana* are cultivated to a lesser extent. About 30 species of *Citrus* are reported to occur in India (Anonymous, 1950) while 17 *Citrus* species, their 52 cultivars and 7 probable natural hybrids are endemic to the north-eastern region (Bhattacharya and Dutta, 1956). *Citrus limettioides, C. paratangerina, C. rugulosa* and *Glycosmis pentaphylla* (Limonia) are occurring in sub-Himalayan tract of North-west India. Guava, although originated in tropical America has been naturalized in parts of Eastern India where sufficient variability exists. Similarly, *Ziziphus mauritiana* (Indian jujube), *Punica granatum* (pomegranate), *Syzygium cuminii* (Jamun), *Litchi chinensis* (litchi), (*Achras zapota* (sapota), and *Cocos nucifera* (coconut) have equally rich genetic diversity.

Among temperate fruits, genus *Malus* has 30 species worldwide and only two *Malus baccata* and *M. sikkimensis* occur wild in India. Two varieties of *M. baccata* have also been identified as *M. baccata* var. *himalaica* in North-western region and Meghalaya and *M. baccata* var. *dirangensis* in Arunachal Pradesh (Joshi and Rana, 1994, Dhillon and Rana, 2005). The genus *Pyrus* and *Prunus* have 14 and 22 species, respectively, in India and constitute an important component of the temperate fruit diversity.

Table 1.1: Diversity of Cultivated and Wild Species of Horticultural Crops Genera in India

Genus	Species (No.)		Genus	Species (No.)	
	Cultivated	Wild		Cultivated	Wild
Fruit and nut crops					
Malus	0	2	Mangifera	1	4
Pyrus	1	13	Musa	2	11
Prunus	5	17	Citrus	4	26
Sorbus	0	10	Vitis	1	9
Rubus	1	33	Ziziphus	2	15
Ribes	0	7	Phoenix	1	13
Fragaria	0	4	Ficus	1	11
Diospyros	1	40	Morus	1	3
Juglans	1	0	Grewia	1	39
Corylus	1	2	Syzygium	3	72
Vegetable crops					
Abelmoschus	1	9	Amaranthus	3	22
Solanum	1	38	Chenopodium	1	7
Cucumis	2	4	Fagopyrum	2	4
Citrullus	1	2	Malva	1	4
Momordica	1	6	Rumex	1	12
Luffa	2	3	Polygonum	1	79
Trichosanthes	1	19	Canavalia	2	2
Coccinia	1	0	Dolichos	1	3
Morniga	1	1			
Plantation and Tuber Crops					
Colocasia	1	2	Camellia	1	5
Alocasia	1	1	Coffea	2	3
Amorphophallus	1	19	Dioscorea	6	44
Xanthosoma	1	30	Ipomoea	1	49
Areca	1	1			
Spices and condiments					
Allium	7	28	Garcinia	1	35
Alpina	1	14	Zingiber	1	18
Amomum	3	26	Curcuma	2	38
Carum	1	2			

Source: Dhillon and Rana, 2005.

Species like *Pyrus pyrifolia*, *Pyrus pashia* and *Prunus armeniaca* have both cultivated and wild types/variants used for edible and rootstock purposes (Sharma *et al.*, 2005, Rana *et al.*, 2008). Considerable genetic variability exists in *Juglans regia* and *Prunus dulcis* for shell thickness, nut size, nut-kernel ratio in Western Himalaya (Rana *et al.,* 2007). The cultivated hazelnut (*Corylus avellana*) is essentially an introduced nut but another species *Corylus colurna* popularly known as *bhotia badam* or *thangee* produces equally good quality nuts endemic to cold region (Dhillon and Rana, 2004). In soft fruits, *Rubus* is the most variable genus and has 57 species in India (Chandel and Shrama, 1996) while *Fragaria nubicola, F. daltoniana, F. nilgerrensis* and *F. vesca* are predominant in the genus *Fragaria*.

The Indian region also has unique diversity for minor fruits both the tropical and temperate regions, for example, bael (*Aegle marmelos*), Indian gooseberry (*Emblica offcinalis*), papaya (*Carica papaya*), jackfruit (*Artocarpus heterophyllus, A. integrifolia*), custard apple (*Annona squamosa*), karonda (*Carissa carandus*), lasoda (*Cordia myxa*), phalsa (*Grewia subinaequalis*), fig (*Ficus carica*), kokam (*Garcimia indica*), mangosteen (*Garcinia mangostana*), ker (*Capparis decidua*), currant (*Ribes* sp.), sea buckthorn (*Hippophae* sp.), palu (*Sorbus lanata*), kaphal (*Myrica nagi*), and ghai (*Elaeagnus angustifolia*) are cultivated in the arid regions of India and also occurring naturally in other parts of India (Rana *et al.,* 2008).

Vegetable Crops

Among vegetables, *Solanum melongena* complex has three cultivated species, namely, *S. melongena, S. incanum* and *S. melongena* var. *insanum* and also wild relatives like *S. torvum, S. indicum, S. insanum, S. surattense, S. pubescens, S. gilo,* and *S. khasianum* which are widely distributed in South India, Shivalik hills and North-eastern region (Kalloo *et al.,* 2005). There are about 100 species of Cucurbitaceous fruits including 34 endemic species majoring *Cucurbita moschata, C. pepo, C. maxima, Cucumis hardwickii, C. trigonus, Luffa graveolens, L. acutangula* var. *amara, L. cylindrica, Trichosanthes anguina, Momordica cochinchinensis, M. dioica, Neoluffa sikkimensis* and *Citrullus colocynthis* (Chakravarty, 1982; Arora and Nayar, 1994). In *Abelmoschus* only *A. esculentus* is cultivated while *A. angulosus, A. tuberculatus, A. manihot, A. moschatus, A. ficulneus, A. tetraphyllus* var. *tetraphyllus, A. tetraphyllus* var. *pungens* and *A. crinitus,* are wild taxa (Dhankar *et al.,* 2005). Intra-specific diversity is also high in *Capsicum annuum, Coccinia cordifolia* and *Moringa oleifera.* Among root and tuber crops, *Amorphophallus paeoniifolius, Manihot esculenta, Ipomoea batatas, Dioscorea alata, Dioscorea rotundata, Dioscorea esculenta, Dioscorea bulbifera* var. *sativa, Colocasia esculenta, Alocasia macrorrhiza* are important cultivated types while *Maranta arundinacea, Solenostemon rotundifolius, Moghania vestita, Canna edulis, Psophocarpus tetragonolobus* and *Pachyrrhizus erosus* are considered as minor and used as food in some parts of western peninsular region. Genus *Allium* represents a major group and about 30 species are found in Indian region (Babu, 1977, Negi and Pant 1992). The cultivated Alliums are *Allium cepa* var. *cepa, A. cepa* var. *aggregatum, A. cepa* var. *viviparum, A. fistulosum, A. tuberosum, A. sativum, A. ampeloprasum* var. *porrum, A. schoenoprasum,* whereas *A. carolinianum, A. chinensis, A. consanguineum, A. humile, A. przewalskianum, A. stolczkii, A. stracheyi, A. victorialis* and *A. wallichii* are important wild species occurring in Himalayan region (Pandy *et al.,* 2008).

Plantation Crops, Spices and Condiments

Plantation crops constitute high value commercial crops *viz. Cocos nucifera, Areca catechu, Anacardium occidentale, Coffea arabica* and *Camellia sinensis.* The genetic diversity of *Camellia sinensis* consists of cultivated species, wild and weedy relatives, old seedling populations, polyclonal seed stocks and clones. Besides tea producing taxa, several other species such as, *C. caudata, C. irrawadiensis, C. taliensis, C. kissi* and *C. drupifera,* and other related genera, such as *Eurya, Pyrenaria, Schima* and *Gordonia* are also found in the forests of North-east India (Singh, 1999). The genus *Areca* has 76 species. *A. catechu* along with five botanical varieties, namely, *communis, silvatica, batanensis, deliciosa* and *longicarpa* are cultivated and have variability for size and shape of fruits and kernels (Nampoothiri *et al.,* 1999). Among spices, black pepper (*Piper nigrum* L.) and cardamom (*Elettaria cardamomum* (L.)

Maton) have originated in the tropical evergreen forests of the Western Ghats of South India while for ginger (*Zingiber* spp.) and turmeric (*Curcuma* spp.), the South Asian region is believed to be the centre of origin; possibly India, in view of the fact that maximum diversity for these crops occurs in India. Baker (1886) described 24 species from the Indo-Malayan region while Sabu (1991) described 18 species of *Zingiber* from Western Ghats and adjacent areas. Velayudhan *et al.* (1999) described 32 species of *Curcuma,* of which 9 are endemic to India. In addition, the other important spices relevant in the Indian context are various species of *Coriandrum, Trigonella, Foeniculum, Cuminum, Murraya, Cinnamomum, Myristica, Syzigium, Tamarindus, Garcinia* and *Vanilla* in the tropical while *Carum, Bunium, Arnebia, Angelica* and *Thymus* in the temperate region (Rana and Sharma 2000).

Ornamental Crops

Indian region has enormous diversity for ornamental plants which are grown commercially and in home gardens. The important ornamentals native to India are orchids, rhododendrons, primulas, musk rose (*Rosa moschata*), begonias (*Begonia* ssp.), lily (*Lilium bulbiferum*), foxtail lily (*Eremurus himalaicus*), chameli (*Michelia champaka*), lotus (*Nelumbo nucifera*), water lily (*Nymphaea* spp.), and wild tulip (*Tulipa stellata* and *T. aitchisonii*). Orchids like *Arachnis, Ascocentrum, Cymbidium, Dendrobium, Paphiopedilum, Phalaenopsis, Phaius, Renanthera, Rhynchostylis* and *Vanda* have exceptionally beautiful and long-lasting flowers of myriad shapes, sizes and colours (Sharma and Rana, 2000). The important native ornamental trees are *Bauhinia variegata, Cassia fistula, Butea frondosa, Lagerstroemia flosreginae, Pongamia glabra, Tecomella undulata, Saraca indica, Ficus benghalensis, F. religiosa* and *Anthocephalus cadamba* while *Jasminum* spp., *Hiptage madablota, Mussaenda frondosa, Clerodendron inerme* and *Passiflora leschenaultii* are important shrubs and climbers.

Medicinal and Aromatic Plants

In India, there is rich diversity in medicinal and aromatic plants and traditional knowledge associated with their uses. The Western Ghats and North-eastern region are particularly rich in this flora. There are a number of medicinal plants, native to India, such as ashwagandha (*Withania somnifera*), rauvolfia (*Rauvolfia serpentina*), safed musli (*Chlorophytum* spp.), babchi (*Psoralea corylifolia*), belladona (*Atropa acuminata),* atees (*Aconitum heterophyllum*), Indian berberry (*Berberis asiatica*), Indian gentian (*Gentiana kurroo*), kutki (*Picrorrhiza kurroa*), Indian henbane (*Hyoscyamus niger*), jatamansi (*Nardostachys jatamansi*) and many more. Many of the medicinal plants, such as aloe (*Aloe barbadensis*) indigenous to African continent and Mediterranean region, periwinkle (*Catharanthus roseus*) occurring mainly in West Indies and Madagascar have widely naturalized and some of them are now commercially cultivated, for example, psyllium (*Plantago ovata*), native of Persia, senna (*Cassia angustifolia*), native of Saudi Arabia and opium poppy (*Papaver somniferum*) with center of origin in western Mediterranean region.

HGR Management

HGR are managed primarily by National Bureau of Plant Genetic Resources as its national mandate in collaboration with other crop based institutes designated as National Active Germplasm Sites (NAGS) and State Agricultural Universities. The different activities related to HGR management are briefly discussed here:

Germplasm Collection

The germplasm collecting priorities are based on availability of genetic diversity, collections already made, threat of genetic erosion, economic importance of the crop and breeders' requirements. Since inception, NBPGR in collaboration with other institutes, has undertaken 2415 germplasm explorations and assembled >250000 accessions of agri-horticultural crops and their wild relatives. Out of these, HGR constitute about 75000 accessions of which fruits (7000), vegetables (35000), spices and condiments (14000), plantation and tuber crops (10000), medicinal and aromatic plants (5000) and ornamentals number (4000). The number is relatively low than field crops which primarily due to perennial and vegetative propagated nature of majority of HGR, nevertheless, in the 11th plan (2007-2012), the HGR was given major focus and priority in our collection programmes. In order to collect trait specific germplasm during explorations, most desirable traits and the traits lacking in our existing collections of many horticultural crops have been identified (Dhillon and Rana, 2005) and targeted in collection programmes. Wild relatives, which are extremely good resources of genes for drought, water logging, heat, cold, and many biotic stresses as the case may be, have been taken on priority.

Germplasm Introduction

The indigenous germplasm has been enriched by introducing about 40000 accessions of various HGR. A few of them have been utilized both as breeding material and variety *per se* and include Red and Golden Delicious group of cultivars in apple, Barlett and Maxred Bartlett pear, Elberta peach, Santa Rosa plum, Loose Perlette and Thompson Seedless grapes, Red Blood Malta, Valancia Orange, Jaffa and Kinnow oranges and Sugar Baby water melon in fruits; Copenhagen Market and Golden Acre cabbage, Erfut-Alfa and Snowball groups of cauliflower, Green Sprouting broccoli, Early Grano onion, Bragg, Lee, and Hardee soybean, Contender and Kentucky Wonder french bean, Sioux tomato, Bonneville, Lincoln, and Sylvia garden pea, Poinsette cucumber and All Green spinach. Many introductions have been used as parents to develop new cultivars and also as rootstocks in fruits. There are many popular introduced ornamentals which include rose, gladiolus, cacti, chrysanthemum, bougainvillea, carnation etc. Some new crops like kiwi fruit, tree tomato, *Atriplex* sp., rambuttan, mangosteen, pawpaw, durian, pecan nut, persimmon, Pine apple, guava, and hops have also been introduced and popularized.

Plant Quarantine

Safe movement of germplasm to protect the native genetic resources is an important management aspect as there are glaring examples of enormous losses caused by the introduced pests. To prevent the spread of pests associated with PGR into India, a strict plant quarantine protocol, proposed through legislative measures, is enforced. The Directorate of Plant Protection, Quarantine and Storage is the nodal organization for implementation of plant quarantine regulations and also to deal with commercial exchange. The NBPGR has national responsibility to carry out plant quarantine for exchange of germplasm meant for research. So far, NBPGR has screened more than 40000 samples of introduced HGR, of which about 4000 were found infected but also salvaged successfully (99.5 per cent). During the course of investigation, 18 pests not reported earlier in India and 5 new host species have been intercepted. Pest Risk Analysis (PRA) standards are being worked out for 48 crops of horticultural importance. The import of transgenic material is regulated by the DBT and NBPGR is the authorized agency for carrying out the quarantine tests. A National Containment Facility for evaluation of transgenics has been established at NBPGR. National Plant Biosecurity System is being developed by harmonizing national legislatives with International agreements and protocols.

Characterization, Evaluation and Utilization

The germplasm has been characterized and evaluated for agronomic, quality, biotic and abiotic stresses. The characterization data of each year are published and sent to users. Crop catalogues containing data of 22179 accessions of 26 crops have also been published. Until now, >30000 accessions of HGR are characterized and evaluated which led to identification of > 800 potential donors and development of about 25 varieties of national repute. A few to mention are Pusa Purple Long and Pusa Purple Round in brinjal, and Pusa Sawani, Pusa Makhmali of okra, Harbjan of pea, Rituraj of cowpea, Pusa Kesar of carrot, Pusa Red of onion, Annapurna and Durga of amaranth, Dashehari, Langra and Chausa in mango, Sharbati in peach, Karan and Gobind in walnut and Subhakara and Sreekara in black pepper. The potential donors have also been identified and used as breeding material. Some worth mentioning donors are Elaichi strain of mango - source of resistance against mango malformation; Mangala for early bearing, Thirthahalli and Sagar for more nuts/bunch, in arecanut; Pournami against nematodes, Panniyur-5 and OP-Karimunda for drought tolerance in pepper; K61, K62 for *Fusarium* wilt resistance, SM81, SM56, SM71, SM72, SM74, Kopek for bacterial wilt resistance,; MM392, MM450 for nematodes (*M. incognita* and *M. aremeriai*) and H165, H407, H408 for Shoot and fruit borer resistance in brinjal. Germplasm comprising > 30000 samples of promising accessions and donors have been supplied to users. Wild relatives, although have not been used to the extent it would have been, yet enjoyed a great success in a few crops such as *Lycopersicon pennellii* - tomato fruit borer, leaf worm, aphid and spider mites; *Solanum incanum* - Fusarium wilt, *S. khasianum, S. gilo, S. indicum* - Shoot and fruit borer, *S. sisymbriifolium* – resistant to *Meloidogyne* spp, *M. incognita* and *M. javanica* nematodes; *S. macrocarpon* and *S. macrosperma* for drought tolerance in brinjal; *Lycoperscium pimpinellifolium* and breeding lines, Poter Narcariang, Early North, Puck, Fireball, Red Cloud, Cold Set, Avalanche, Illalihin, Severianin for cold tolerances, *L. pennelli*, IIHR 14-1, 146-2, 383, 553, 555, EC-130042, 65992 for drought tolerance, *L. cheesmani*, EC–41824, 7764, 164855, 169308, 164653, 122063, 37311, 31515, 126955,130042 35220, 163709 for heat tolerance,. *cheesmani*, Sabour Suphala, EC-251706, 151628 for salt tolerance in tomato, *Abelmoschus crinitus* - YVMV, *A. moschatus, A. crinitus* – jassids, *A. tuberculatus, A. moschatus* for fruit borer, *A. tetraphyllus, A. angulosus, A. manihot* and *A. pungens* for fruit borer in okra; *Manihot pseudoglaziovii* - drought tolerance, *Cucumis hardwickii* and *C. callosus* for downy mildew and fruit fly in cucumber, *Allium roylei* - downy mildew, anthracnose, *Vitis parviflora* and *V. lanata* - downy mildew, powdery mildew and anthracnose, *Psidium molle* - guava wilt, *Phylanthus acidus* - rust, *Musa acuminata* subsp. b*urmannica* - leaf spot and wilt, *Piper attenuatum* - pollu beetle. While *Malus baccata* in apple, *Pyrus pashia* in pear, *Prunus cerasoides* in cherry, *Prunus mira* in peach, *Citrus jambhiri* and *C. limonia* in citrus, *Diospyros lotus* in persimmon, *Piper colubrinum* in black pepper and *Myristica prainii* and *M. malabarica* in nutmeg have been exploited as rootstocks resistant to soil borne diseases, nematodes and salinity.

Germplasm Conservation

The National Gene Bank housed at NBPGR is primarily responsible for conservation of all kinds of HGR on long-term basis. Nearly 40,000 accessions have been conserved in the National Gene Bank for long-term storage in the form of seed, while around 87,000 accessions of perennial crops are conserved in the field gene banks at 10 regional stations of NBPGR and 59 NAGS (29 for horticultural crops) located in different agro-climatic and phyto-geographical regions of the country (Table 1.2). Besides, 2000 accessions of 158 vegetative propagated crops under *in vitro* and 8,981 accession of 726 species have been *cryo* preserved. Pollen grains of 279 accessions of *Mangifera indica* and *Garcinia* spp and dormant buds of 197 accessions of *Morus* spp. are conserved. New protocols have been developed for micro-propagation and *in-vitro* conservation of vegetative propagated species like wild yams, *Alocasia indica, Tylophora indica, Coleus forskohlii, Gentiana kuroo, Picrorhiza kurroa, Rauvolfia serpentina* and cryopreservation of vegetative propagated (*Picrorhiza kurroa, Allium tuberosum, Dioscorea* spp., *Morus* spp., *Rubus* spp.) and non-orthodox seed species (tea, black pepper, almond, neem, trifoliate orange).

Table 1.2: Germplasm of Important Horticultural Crops being Maintained as Active Collections in Various NAGS

Temperate fruits and nuts

Apple	840	Plum	160	Strawberry	189
Pear	275	Cherry	64	*Rubus* spp.	97
Apricot	186	Walnut	426	*Ribes* spp.	39
Almond	205	Pecan nut	89	Persimmon	56
Peach	166	Hazelnut	26	Minor fruits	246
Total					**3064**

Tropical and Subtropical fruits

Mango	2942	Jamun	227	Khirini	30
Grape	1512	Data palm	352	Phalsa	28
Banana	1603	Tamarind	136	Sapota	36
Cashew	1631	Litchi	210	Fig	48
Citrus	728	Bael	59	Jackfruit	69
Ber	398	Aonla	88	Lasoda	79
Papaya	198	Karonda	59	Mulberry	42
Guava	176	Chironji	30	Pineapple	38
Pomegranate	359	Custard apple	64	Wood apple	18
Total					**11160**

Vegetable crops

Tomato	3498	French bean	4156	Faba bean	548
Brinjal	3026	Pea	2469	Winged bean	507
Chilli	1589	Cowpea	2698	Cole crops	965
Okra	3469	Leafy vegetables	1489	Root vegetables	526
Cucurbits	2652	Lablab bean	1100	Cucumber	456
Total					**29148**

Plantation and Tuber crops

Taro	929	Cassava	1824	Betel vine	65
Elephant foot yam	224	Cashew	1189	Coleus	21
Yams	1224	Coconut	489	Maranta	18
Sweet potato	1678	Tannia	115	Canna	16
Potato	3924	Cocoa	296	Giant taro	77
Tea	2972	Arecanut	218	Under-utilized tuber crops	289
Total					**15568**

Contd...

Table 1.2–*Contd...*

		Spices and Condiments			
Piper	3695	Cumin	580	*Syzygium*	215
Elettaria	1284	Methi	470	Dil	103
Coriander	1960	Cinnamon	508	*Garcinia*	71
Curcuma	2668	Nutmeg	571	Nigella	53
Ginger	1115	Fennel	324	Ajwain	95
Allium	1627	Clove	314	Celery	56
Fenugreek	917	Vanilla	215	Anise	42
Total					**16483**
Medicinal and Aromatic Plants					**7589**
Ornamentals					**4221**
Grand total					**87233**

Source: Dhillon and Rana 2005 (updated 2011).

Network for Evaluation and Conservation of HGR

The management of HGR especially because of their perennial and vegetative nature and also of large number of accessions is a gigantic task. Therefore, enormity and diversity of the task required a networking of all the stakeholders. The national network should include National Bureau of Plant Genetic Resources (NBPGR), horticultural institutes, National Research Centers (NRCs) and All India Coordinated Research Projects (AICRP) of ICAR, Department of Biotechnology, Department of Environment and Forests, State Agriculture Universities (SAUs), State Departments of Agriculture, Horticulture and Forests, and other organizations (Rathore, 2001). Already, there are 29 National Active Germplasm Sites (NAGS) entrusted with the responsibility of crop specific collection, multiplication, evaluation, maintenance and conservation of active collections and their distribution of respective crops to users for research purposes.

Table 1.3: National Active Germplasm Sites (NAGS) for Various Horticultural Crops

Sl.No.	*Crop(s)*	*Institute/AICRP/NRC/Directorate*
1.	Agroforestry crops	NRC Agroforestry, Pahuj Dam, Jhansi, Uttar Pradesh
2.	Arid fruits	Central Institute for Arid Horticulture, Sadulganj, Bikaner, Rajasthan
3.	Banana	NRC Banana, Thayanur Santhai, Tiruchirapalli, Tamil Nadu
4.	Cashew	Directorate of Cashew, Puttur, Dakshin Kanada, Karnataka
5.	Citrus species	NRC Citrus, Shankar Nagar, Nagpur, Maharashtra
6.	Floriculture	AICRP Floriculture, Division of Floriculture and Landscaping, Indian Agricultural Research Institute, New Delhi
7.	Grapes	NRC Grapes, Sholapur Road, Pune, Maharastra
8.	Litchi, bael, aonla and jackfruit	NRC, Litchi, Bihar

Contd...

Table 1.3–*Contd...*

Sl.No.	Crop(s)	Institute/AICRP/NRC/Directorate
9.	Medicinal and aromatic plants	Directorate of Medicinal and Aromatic Plants, Boriavi Taluka, Anand, Gujarat
10.	Mango	Central Institute for Subtropical Horticulture, Rehmankhera, Lucknow, Uttar Pradesh
11.	Sub-tropical fruits	AICRP Sub-tropical Fruits, Lucknow, Uttar Pradesh
12.	Mulberry	Silkworm and Mulberry Germplasm Station, Hosur
13.	Oil palm	Directorate of Oil Palm Research, Pedavegi, West Godavari Distt, Andhra Pradesh
14.	Onion and garlic	Directorate of Onion and Garlic, Rajgurunagar, Pune, Maharashtra
15.	Orchids	NRC Orchids, Pakyong, Sikkim
16.	Ornamentals and non-traditional crops	National Botanical Research Institute, Lucknow, Uttar Pradesh
17.	Plantation crops	Central Plantation Crops Research Institute, Kasargod, Kerala
18.	Potato	Central Potato Research Institute, Shimla, Himachal Pradesh
19.	Spices	Indian Institute of Spices Research, Marikunnu, Kerala
20.	Spices	AICRP Spices, Marikunnu, Calicut, Kerala
21.	Tea	UPASI Tea Research Foundation, Valparai, Coimbatore, Tamil Nadu
22.	Temperate horticulture crops	Central Institute of Temperate Horticulture, Sanatnagar, Srinagar (J&K)
23.	Temperate horticulture crops	NBPGR Regional Station, Phagli, Shimla
24.	Tropical fruits	Indian Institute of Horticulture Research, Hassarghatta Lake, Bangalore, Karnataka
25.	Tropical fruits	AICRP Tropical Fruits, Hassaraghatta Lake Post, Bangalore, Karnataka
26.	Tuber crops	Central Tuber Crops Research Institute, Thiruvananthapuram, Kerala
27.	Tuber crops	AICRP Tuber Crops, Thiruvananthapuram, Kerala
28.	Vegetables	Indian Institute of Vegetable Research,, Gandhi Nagar, Naria, Varanasi, Uttar Pradesh

DNA Fingerprinting and Diagnostics

DNA fingerprinting protocols using different molecular marker techniques like STMS (microsatellites), AFLP, ISSR and RAPD have been developed and DNA finger printing of 1177 varieties comprising vegetables (tomato -27, chillies-42, brinjal-38, bittergourd-38, cucumber-47); fruits (banana-243, mango-250, cashew-140, *Citrus*-34) and medicinal plants (andrographis-25, neem-69, vetiver-86, plantago-48, palmarosa-34, chlorophytum-21, saffron-13 and giloe-22) are completed while for others it is being done. A total of 1593 lines of 60 imported transgenic planting materials comprising 12 different crops have been tested for absence of embryogenesis deactivator terminator gene sequence. PCR and multiplex-PCR methods have also been developed for detection of more than 30 trans-genes comprising promoters, structural and terminator gene sequences in all imported transgenic planting material using primers designed in the laboratory.

Genetic Resources Knowledge Management System

The documentation of PGR becomes very important at the national, regional and global levels for effective conservation of rapidly disappearing genetic stocks for their immediate and future uses. The

Agricultural Research and Information System (ARIS) Cell at NBPGR, which is mainly responsible for the databases activities at NBPGR, has developed many on-line databases related to PGR and plant varieties, which are in use by researchers in India. Two databases namely IINDUS (Indian Information System as per the DUS Guidelines) and NORV (Notified and Released Varieties of India) are in use at the Protection of Plant Varieties and Farmers Rights Authority for the purpose of registration of extant and new varieties. It has also established a "National Information Sharing Mechanism" (NISM) to implement the Global Plan of Action for conservation and sustainable utilization of PGRFA in India. NBPGR is making earnest efforts to collate the PGR related information from all the partners of NAGS of India and incorporate the geographical location with the passport information in the form of GIS parameters in the existing databases.

The rapid developments in cell and molecular biology and subsequent sequencing of genomes is helping to unravel the genetic mechanisms underlying regulation of many important traits. The enormous number of gene and protein sequences, structure and expression data have been deposited in public online databases such as Genbank, PDB, TIGR, and Unigene. Ever increasing quantities of this deposited genomic data have started aiding discovery and annotation of new genes of agricultural importance. A National Agricultural Bioinformatics Grid (NABG) has recently been initiated to provide computational framework and support to carry out biotechnological research. NABG has a two tier architecture with the centre head quartered in the campus of Indian Agricultural Statistics Research Institute, New Delhi and linked to five Domain Bioinfomatics Centres (DBC) ie. NBPGR (crops), NBAGR (animals), NBFGR (fisheries), NBAII (Insects) and NBAIM (Agricultural Microbes). The DBCs are equipped with High Performance Computational infrastructure, connected through high speed Wide Area of Network (WAN) for seamless information flow with 32 Mbps bandwidths and provided with advanced and customized statistical and bioinformatics software packages to analyze genomic resources to derive knowledge. The centers will be responsible for information generation and knowledge delivery from NABG to respective domain institutions for conducting experiments. The DBC's will maintain and verify data quality of genomic information from all sources in their respective domains and derived genomics data will be function annotated for applications in agriculture.

Protecting HGR in the IPR Regime

The importance of genetic resources for providing food, feed and fodder and numerous medicines, industrial products and other uses cannot be over-emphasized. However, the advancement in new technologies situated at the interface of biology and disciplines such as material sciences, physical chemistry, and engineering and particularly biotechnology including molecular biology and genetic engineering/metabolic engineering (genomics, proteomics, metabolomics) have profoundly changed the conception of agriculture research and development. Creation of new knowledge utilising living materials is now perceived as a means of generating commercial goods by converting it into useful value added products, production process and services. The shift in the perception has been aided by the unprecedented enhancement in the last two decades in the level, scope, territorial extent and role of intellectual property rights (IPR) which were traditionally applied to mechanical inventions of one kind or another, or to artistic creations.

The complexities witnessed in the subject related to management of genetic resources have largely emanated from two global developments namely the Convention on Biological Diversity (CBD) and the Agreement on Trade-Related Aspects of Intellectual Property Rights (TRIPS) administered by the World Trade Organisation. These two major developments have confronted the plant scientists with new dimensions of obligations particularly related to access, utilisation and benefit sharing. There are now many statutary provisions and regulatory obligations which make access and utilisation of genetic resources not as unrestricted as it used to be and also provide an opportunity to establish ownership on the technologies, products and processes developed by scientific organisations. In addition, to the legal issues there are others which are political, ethical and related to environment and food safety. However, before dealing with these issues, it is pertinent to first understand the various provisions and developments consequent to the adoption of these treaties which impinge upon research in genetic resources.

The Commission on Genetic Resources for Food and Agriculture

The genetic resources were considered as common heritage of humankind and there was no organized international effort for their management. In 1983, FAO established an intergovernmental forum: the Commission on Plant Genetic Resources (now the Commission on Genetic Resources for Food and Agriculture). The Commission has developed a global system for the Conservation and Sustainable Utilization of PGRFA, of which the Report on the State of the World's Plant Genetic Resources and the Global Plan of Action (GPA) are key elements (http://www.fao.org/ag/cgrfa/en). The GPA is intended as a framework, guide and catalyst for action at community, national, regional and international levels. It made significant contribution to the successful implementation of the CBD and seeks to create an efficient system to:

☆ Ensure the conservation of plant genetic resources for food and agriculture as the basis of food security

☆ Promote sustainable use of plant genetic resources to foster development and reduce hunger and poverty

☆ Promote the fair and equitable sharing of the benefits arising from the use of plant genetic resources

☆ Assist countries and institutions to identify priorities for action

☆ Strengthen existing programmes and enhance institutional capacity.

Convention on Biological Diversity

The CBD had emerged out of the global concern for conserving biodiversity and was signed by 150 countries at the United Nations Conference on Environment and Development in 1992 in Rio de Janeiro (the 'Earth Summit'), and entered into force on 29 December, 1993. It recognised the sovereign rights of nations over their biological diversity and has three main objectives; (i) conservation of biological diversity, (ii) the sustainable use of its components, and (iii) fair and equitable sharing of the benefits arising from the use of genetic resources, including by appropriate access to genetic resources and by appropriate transfer of relevant technologies. There are currently 191 parties and the treaty has been ratified by 168 countries.

The CBD negotiations witnessed unprecedented debates among the nations with developed and developing economies on issues that included *inter alia* the question of access to genetic resources even beyond the scope of plant genetic resources for food and agriculture. Since much of the developing countries were rich in genetic resources it was argued that they did not necessarily benefit equally from industrial, medical, agricultural, and other uses of genetic resources that take place with technologically advanced developed countries, which were equipped or were equipping themselves with legal instruments that expanded and reinforced the implementation of IPR regimes to life forms. Thus the commitment of the Parties to "fair and equitable sharing of the benefits arising out of the utilization of genetic resources" was incorporated in the CBD objective.

There are a range of Articles incorporated in the treaty that reflect the concern of Parties for the issues of access to genetic resources, access to and transfer of technology, handling of biotechnology and distribution of its benefits (Box 1.1). It can be seen from these provisions that CBD reflects not only the commitment of global community for conservation and sustainable use of biodiversity, but also makes the facilitated access to genetic resources by developing countries a two-way process by effectively linking it with equitable sharing of benefits. The CBD incorporates two instruments for providing access namely "Mutually Agreed Terms" (MAT) and "Prior Informed Consent" (PIC). The concept of MAT establishes a new participatory relationship among country of origin, provider and user of genetic resources. It is a contractual arrangement, executed on a bilateral basis that provides an opportunity for the providing country to negotiate a share of the benefits derived from the use of the genetic resources. It also provides an opportunity to the provider of the genetic resource to specify the permitted or prohibited uses of the genetic resources provided, including whether or not it may be commercialised, any IPR rights which may or may not be taken over the resource or its derivatives and benefits that are to be shared. The PIC empowers the provider of genetic resource to establish an authority that can grant or refuse access following a request from the applicant as per the national legislation. The applicant is required to provide information concerning the genetic resources required, purpose for which required and any proposal for benefit sharing thereby providing the party an opportunity to become equal partner and monitoring the use. As part of the PIC procedure, the authority may consult indigenous and local communities or other stakeholders concerned before granting the access.

Access and Benefit Sharing

Implementation of the CBD is a dynamic process and the Conference of the Parties (CoP) is the governing body of the Convention that takes decisions on outstanding issues at its periodic meetings. In order to facilitate better implementation of the CBD obligations with respect to access and benefit-sharing Parties have undertaken several initiatives. The CBD constituted two Ad Hoc Open-ended Working Groups one on Traditional Knowledge (COP IV/9/1998) and other on Access and Benefit Sharing (ABS) issues {COP V/6/2000); and one panel of experts on ABS (CBD/COP IV/8, 1998) with representation from diverse interest groups. The terms of references of these Working Groups and Expert Panel included mainly the development of a framework that would help assist Parties to develop national and regional regulations or guidelines on ABS, with special focus on evolving standards and principal elements of PIC, MAT, Material Transfer Agreements (MTA) and Monetary and Non – monetary Benefit Sharing Agreements.

One of the important outcomes of Sixth Meeting of the Conference of Parties held in April 2002 (COP VI/24/2002) in The Hague was adoption of the Bonn Guidelines on Access to Genetic Resources and Fair and Equitable Sharing of the Benefits Arising out of their Utilisation.

(*i*) The Bonn Guidelines

The Bonn Guidelines can be construed as the first affirmative step in operationalising the relevant CBD provisions with respect to ABS. These Guidelines are voluntary in nature and indicate procedures intended to provide the Parties and Stakeholders with a transparent framework to facilitate access to genetic resources through the practices and procedures of PIC of the country of origin as well as MAT, MTA, and other relevant agreements. The Guidelines assist in establishing and developing national access and benefit sharing regimes while promoting capacity building, transfer of technology and the provision of financial resources. They also seek to promote sustainable use of genetic resources by

Box 1.1: Important Provisions Under CBD Related to Conservation, Access and Benefit Sharing

General Measures (Art 6)

☆ Develop national strategies, plans or programmes for conservation and sustainable use of biological diversity

☆ Integrate, as far as possible and as appropriate, the conservation and sustainable use of biological diversity into relevant sectoral or cross-sectoral plans, programmes and policies.

Conservation of biological diversity (Art 8 and 9)

☆ Establish a system of protected areas or areas where special measures need to be taken to conserve biological diversity

☆ Develop guidelines for the selection, establishment and management of protected areas

☆ Promote the protection of ecosystems, natural habitats and the maintenance of viable populations of species in natural surroundings

☆ Rehabilitate and restore degraded ecosystems and promote the recovery of threatened species

☆ Regulate, manage or control the risks associated with the use and release of living modified organisms resulting from biotechnology

☆ Prevent the introduction of, control or eradicate those alien species which threaten ecosystems, habitats or species

☆ Respect, preserve and maintain knowledge, innovations and practices of indigenous and local communities and with the approval and involvement promote their wider application and encourage the equitable sharing of the benefits arising from the utilization of such knowledge, innovations and practices

☆ Cooperate in providing financial and other support for *in situ and ex-situ* conservation particularly to developing countries.

☆ Adopt measures for the *ex situ* conservation of components of biological diversity

☆ Establish and maintain facilities for *ex situ* conservation of and research on plants, animals and micro-organisms

☆ Adopt measures for the recovery and rehabilitation of threatened species and for their reintroduction into their natural habitats

☆ Regulate and manage collection of biological resources so as not to threaten ecosystems and *in situ* populations of species

Access and benefit sharing (Art 15)

☆ Authority to determine access to genetic resources rests with the national governments and is subject to national legislation

☆ Create conditions to facilitate access for environmentally sound uses

☆ Access subject to "prior informed consent" of resource provider and on "mutually agreed terms"

☆ Endeavour to develop and carry out scientific research based in genetic resources provided by other Contracting Parties with full participation of, and where possible in, such Contracting Parties

☆ Take legislative, administrative or policy measures, as appropriate, with the aim of sharing in a fair and equitable way the results of research and developments and the benefits arising from the commercial or other utilization of genetic resources.

Access to and transfer of technology (Art 16)

☆ Endeavour to develop and carry out scientific research based on genetic resources provided by other Contracting Parties with the full participation of, and where possible in, such Contracting Parties. Access to and transfer of technology using genetic resources to countries providing the genetic resources.

☆ Take legislative, administrative or policy measures, as appropriate, with the aim that countries, which provide genetic resources are provided access to and transfer of technology on mutually agreed terms

☆ Ensure that IPRs are supportive of and do not run counter to the implementation of the Convention objectives.

promising to improve users' (commercial and non-commercial) access to valuable genetic resources in return for sharing the benefits with the countries of origin and with local and indigenous communities.

(*ii*) International Regime on Access and Benefit-Sharing

The World Summit on Sustainable Development held at Johannesburg, in September 2002, invited Parties to take appropriate steps *'to negotiate within the framework of the Convention on Biological Diversity [CBD], bearing in mind the Bonn Guidelines, an international regime to promote and safeguard the fair and equitable sharing of benefits arising out of the utilisation of genetic resources'*. Following it the Convention's Conference of the Parties at its seventh meeting, in 2004, (COP VII/19/2004) mandated its Ad Hoc Open-ended Working Group on ABS to elaborate and negotiate, together with the Working Group on Article 8(j), an international regime on access to genetic resources and benefit-sharing in order to effectively implement Article 15 and Article 8(j) of the CBD. Subsequently, discussions were held by the Working Group in the fifth and sixth meeting held in meeting in Montreal, Canada (8-12 October 2007) and in Geneva, Switzerland (21-25 January 2008) respectively. Thereafter, recommendations on international regime on ABS were consolidated in Cali, Colombia by COP 9 (COP IX/12/2008). Thus after six years of negotiations, the 'Nagoya Protocol on Access to Genetic Resources and the Fair and Equitable Sharing of Benefits Arising from their Utilization to the Convention on Biological Diversity' was adopted at the tenth meeting of the Conference of the Parties on 29 October 2010, in Nagoya. The Nagoya Protocol will be open for signature by Parties to the Convention from 2 February 2011 until 1 February 2012 at the United Nations Headquarters in New York.

CBD and Plant Genetic Resources for Food and Agriculture

The CBD provisions encompass all biological resources, however, the plant genetic resources for food and agriculture (PGRFA) received special attention during the negotiations. The International Undertaking on Plant Genetic Resources (IUPGR) at the Food and Agriculture Organisation (FAO) was the first intergovernmental undertaking establishing the principle of PGR as the 'heritage of mankind' and consequently these resources were available without restriction. The recognition of the sovereign right of nations under the CBD, *i.e.*, "to exploit their own resources pursuant to their own environmental policies" raised two issues (i) the facilitated access to the *ex situ* collections held world over prior to the CBD and (ii) the realization of farmers' rights enshrined in the revised agreed interpretation of the IUPGR. Resolution 3 of the Final Act to the CBD referred these outstanding issues to be dealt with by the FAO's Global System on Plant Genetic Resources, of which the International Undertaking was the corner stone. The 1993 FAO Conference provided the framework for the revision of the IUPGR on these lines and the negotiations were initiated for the revision of the IUPGR in harmony with the CBD. The negotiations culminated with the adoption of the International Treaty on Plant Genetic Resources for Food and Agriculture (ITPGRFA) on 3 November 2001. It became effective on 29 June 2004; presently the Treaty is having 115 Contracting Parties.

International Treaty on Plant Genetic Resources for Food and Agriculture

The ITPGRFA provides a legal framework that not only recognizes the need for conservation, sustainable agriculture and food security for the PGRFA but also defines a regime for ABS linking it with IPRs (Box 1.2). It provides broad guidelines but entrusts the responsibility for realizing farmers' rights to member states. The specific obligations include the need to keep an inventory of PGRFA, promote their collection, promote farmers and local communities' efforts to manage and conserve on-farm their PGRFA, promote *in situ* conservation of wild crop relatives and wild plants for food production, and cooperate to promote the development of an efficient and sustainable system of *ex situ* conservation.

Box 1.2: Important Provisions Under ITPRFA Related to Conservation, Access and Benefit Sharing

Farmers rights (Article 9)

☆ To save, use, exchange and sell farm saved seed/propagating material

☆ Protection of traditional knowledge relevant to PGRs

☆ Right to equitably participate in sharing benefits arising from the utilisation of PGRs

☆ Right to participate in making decisions on matters related to conservation and sustainable use of PGRs.

Establishment of Multilateral system (MLS) (Article 10)

☆ Recognizes sovereign rights of states over their PGRFA

☆ Authority to determine access rests with states and is subject to national legislation

Coverage of MLS (Article 11)

☆ List of 35 crops/genepools and forages - Annex. I

☆ PGRFA under "the management and control of the Contracting Parties and in the public domain"

☆ Includes Annex I genetic resources held by CGIAR and other international institutions

☆ Other holders of Annex I PGRFA (*e.g.* private sector) invited to include these in MLS

Guidelines for Facilitated Access to PGRFA (Article 12)

☆ Solely for the purpose of utilisation and conservation for research, breeding and training for food and agriculture

☆ Accorded expeditiously, with a minimal or no charge at all

☆ All available non-confidential descriptive information shall be made available

☆ No IPR or other rights so as to limit access shall be claimed over the PGRs or their genetic parts or components in the form received from the multilateral system

☆ Access to PGRs protected by IPR shall be consistent with international and national arrangements;

☆ PGRs accessed under the multilateral system shall continue to be made available

☆ Access to PGRs in *in situ* conditions shall be provided subject to national legislation.

Benefit Sharing Mechanism (Article 13)

☆ Establish a Global Information System to facilitate exchange of information

☆ Facilitated access to technologies for the conservation, characterization, evaluation and use of PGRFA, subject to applicable IPRs.

☆ Technologies protected by IPRs particularly for use in conservation as well as for the benefit of farmers in developing countries should be transferred under 'fair and most favorable terms'.

☆ Establishing/strengthening/developing scientific and technical education and training, facilities for conservation and sustainable use of PGRs and carry out scientific research in the developing countries.

☆ Standard Material Transfer Agreement, for facilitated access to include a requirement that an equitable share of the benefits arising from the commercialisation of product that incorporates material accessed through the Multilateral System will have to be paid to the Trust Account set up under the Treaty.

☆ The benefits that arise under the benefit sharing arrangements must be primarily directed to farmers who conserve and sustainably use PGRFA.

In exercise of sovereign rights of nations over their PGRFA, the Treaty provides for a multilateral system (MS) covering a list of 35 food crops/group of crops and 29 forage crops/group of crops (provided as Annex. 1 of the ITPGRFA) selected on the criteria of food security and interdependence include Breadfruit, asparagus, oat, beet, brassica complex, pigeon pea, chickpea, citrus, coconut, major aroids, carrot, yams, finger millet, strawberry, sunflower, barley, sweet potato, grass pea, lentil, apple, cassava, banana/plantain, rice, pearl millet, beans, pea, rye, potato, eggplant, sorghum, triticale, wheat, faba bean/vetch, cowpea and maize. The legume forages include: *Astragalus, Canavalia, Coronilla, Hedysarum, Lathyrus, Lespedeza, Lotus, Lupinus, Medicago, Melilotus, Onobrychis, Ornithopus, Prosopis, Pueraria* and *Trifolium.* Grass forages include *Adropogon, Agropyron, Agrostis, Alopecurus, Arrhenatherum, Dactylis, Festuca, Lolium, Phalaris, Phleum, Poa* and *Tripsacum.* The other forages include *Atriplex* and *Salsola.*

The facilitated access to PGRFA is to be provided in pursuance of Standard Material Transfer Agreement (SMTA) adopted on 16 June 2006 by the Governing Body of ITPGRFA in their first meeting held at Madrid, Spain. The SMTA has 9 Articles including Rights and obligations of the provider and the recipient under Articles 5 and 6 and contains the provisions of the Articles 12 and 13 and other relevant provisions of ITPGRFA as referred in the Box 2. The conditions of SMTA shall apply to the transfer of PGRFA among the Contracting Parties and also to another person or entity, as well as to any subsequent transfers of these PGRFA. The ITPGRFA includes ground-breaking, innovative provisions for monetary benefit-sharing as reflected in the SMTA. If a product that incorporates material from the MS is commercialised in such a way that is not "available without restriction to others for further research and breeding" a mandatory payment of 1.1 per cent of the sales of the product(s) less 30 per cent or alternatively at 0.5 per cent of the sales of any product and of sales of other products that are PGRFA belonging to the same crop, into the mechanism established for this purpose. However, payment shall be voluntary if product(s) are available without restriction for further research and breeding.

The Biodiversity Act 2002

As the national response to CBD, The Biological Diversity Act, 2002 was enacted on December 11, 2002 and came in force since April 15,2003 with the notification of the Biological Diversity Rules 2003. The Act is primarily aimed to regulate, in the territorial jurisdiction of India, access to genetic resources and associated knowledge with the purpose of realizing equitable sharing of benefits arising out of the use of these resources with the local people. The implementation of the Act is coordinated by three functional bodies namely the National Biodiversity Authority (NBA), the State Biodiversity Boards (SBB), and the Biodiversity Management Committee (BMC). The NBA is the national competent authority to discharge all decisions pertaining to ABS to persons who are not Indian citizens, Indian citizens who are non-resident and a body corporate, association or organization not incorporated or registered in India or incorporated or registered in India under any law which has any non-Indian participation in its share capital or management. Also the prior approval of the NBA is mandatory before applying for IPRs based on biological resources or traditional knowledge obtained from India.

Further, the Authority, while granting the approval shall ensure equitable sharing of benefits between the applicant and local bodies concerned or the benefit claimers, on mutually agreed terms and conditions. It may develop guidelines to impose and fix criteria for benefit sharing fee or royalty or both or impress conditions including the sharing of financial benefits arising out of the commercial utilization of such resources; and shall notify the specific details of benefit sharing formula in an official gazette on a case-to-case basis. The National Biodiversity Authority is also empowered to take necessary measures to oppose the grant of IPRs in any other country on any biological resource or associated knowledge derived/obtained from India.

The National Biodiversity Authority is empowered to advise the Central Government on matters relating to conservation and sustainable use of biological diversity and advise the State Governments in the selection of areas to be notified as heritage sites. Share of monetary benefits will be deposited into the National Biodiversity Fund except in cases where biological resources and knowledge are accessed from specific individual or group of individuals or organizations, in which case the monetary benefit may be directly made to the providers on the direction of the Authority in accordance with the terms of any agreement and in such manner as it deems fit.

The Indian citizens or a body corporate, association or organization which is registered in India, shall obtain any biological resource for commercial utilization, or bio-survey and bio-utilization only after giving prior intimation to the concerned State Biodiversity Board. The State Biodiversity Board, on receipt of such intimation, may, in consultation with the local bodies concerned, prohibit any such activity if it is considered detrimental or contrary to the objectives of conservation and sustainable use of biodiversity or allow use on the basis of equitable sharing of benefits. Exception to this has been provided to the local people and communities of the area, including growers and cultivators of biodiversity, and vaids and hakims.

CBD and Bio-safety Issues

Biotechnology has opened up multitude of opportunities for the scientists to alter the genetic structure of any organism, thereby developing a number of genetically modified organisms and products from them. Recognising the potential risk arising from such living modified organisms (LMOs), the CBD, in Article 19.3 provided for the Parties to consider the need for and modalities of a protocol to ensure safe transfer, handling and use of LMOs. In November 1995, at the second meeting of CoP, an open ended Ad-hoc Working Group was established to develop a protocol on biosafety through a negotiation process. After six meetings of the Working Group held from June 1996 to February 1999, the Cartagena Protocol on Biosafety was adopted in Montreal, Canada on June 29, 2000 and entered into force on September 11, 2003.

The Protocol is a legally binding instrument governing transfer of LMOs from one nation to another. It deals primarily with GMOs that are to be intentionally introduced into the environment (such as seeds, trees or fish) and with genetically modified farm commodities (such as corn and grain used for food, animal feed or processing). The objective of the protocol is to contribute to adequate level of protection in the field of the safe transfer, handling and use of LMOs resulting from modern biotechnology that may have adverse effects on the conservation and sustainable use of biological diversity, taking also into account risks to human health, and specifically focussing on transboundary movements (Article 1 of the Protocol, SCBD 2000). An important feature of the Protocol is the procedure of "Advance Informed Agreement" (AIA) that applies to the first intentional transboundary movement of LMOs for intentional introduction into the environment of the Party of import. It obliges the exporter, to notify the competent national authority of the importing country and provide the necessary relevant information to provide the importing countries both the opportunity and the capacity to assess risks that may be associated with the LMO before agreeing to its import.

Further to fulfil the commitment set forth in Article 27 of the Cartagena Protocol, the Nagoya-Kuala Lumpur Supplementary Protocol on Liability and Redress to the Cartagena Protocol was adopted by the Parties to the Cartagena Protocol on Biosafety on 15 October 2010, in Nagoya, Japan. The protocol specifically addresses that Parties may use criteria set out in their domestic law to address damage that occurs within the limits of their national jurisdiction. Therefore, this agreement will enable countries to adopt and implement their own liability provisions and redress legislation and financial security.

Protection of Plant Varieties and Farmers' Rights Act

This is the second Indian IPR legislation which India adopted to comply TRIPS requirements. TRIPS provide option to member countries for protecting them by patents, or by an effective *sui generis* system or by a combination of patent and *sui generis* system. In the case of plant varieties, India chose not to give patent to plants and to protect plant variety by the *sui generis* system. It can also assume a wider scope to cover those aspects of IP system embodying the rights of communities, which are not protectable under the patent system. It grants an exclusive right to the innovator of a plant variety for producing, processing and stocking, commercializing, importing or exporting the propagating material of the protected variety. This *sui generis* system of law to protect plant varieties called 'Protection of Plant Varieties and Farmers Rights Act, 2001 (PPVFR Act)'. This Act provides for three rights- the Plant Breeder's Right (PBR), the Farmers' Rights (FRs) and the Researcher's Rights (RRs). The process of granting *sui generis* right or the PBR is called registration.

The FRs in the PPVFR Act includes the right on seeds, the right to equitable benefit sharing, the right to register farmers' varieties, the right for reward and recognition for conservation and improvement of plant genetic resources, the right for compensation when a registered variety does not perform to the promised level, etc. RRs refer to the free and unrestricted right to researchers in accessing any plant material including the varieties registered under this Act for the purpose of research, including breeding new varieties. Also commercialization of such varieties bred by using a registered variety does not require prior approval from the PBR - holder of the registered variety or payment of royalty. This right is important to promote research leading to the development of more and more superior varieties, which would benefit farmers and promote the cause of national agriculture. PPVFR Act partially came into force from November 2005 with the establishment of the PPVFR Authority at New Delhi to implement this law.

A variety will become eligible for registration on qualifying for the distinctness, uniformity and stability test (DUS). Registration granted is valid for 18 years for vines and trees and 15 years for varieties of annual species. The process is on to register extant varieties (which are in commerce or public domain), new varieties (those bred by individuals including farmers and institutions and possessing novelty) and farmers' varieties (traditional varieties developed and conserved by farmer communities) in 30 crops species. Out of 30 crops for which registration has been started seven *viz.* black pepper, pea, french bean, small cardamom, turmeric, ginger and chrysanthemum are horticultural crops. The DUS guidelines have been prepared for another ten horticultural crops *viz.* potato, garlic, onion, tomato, brinjal, cabbage, cauliflower, okra, rose and mango. So far, 228 varieties were registered and four *viz.* Kashi Nandini, Kashi Shakti, Kashi Mukti of garden pea and Kashi Puram of French bean are of horticultural crops

Geographical Indications

Geographical Indications (GI) is the third important IPR law covered under the TRIPS and concerned to agriculture. Indian law on GI of Goods (Registration and Protection) Act was enacted in 1999 and enforced from 15th September 2003 with the GI Registry located at the Regional Patent Office in Chennai. Unlike other IPRs, which are created and owned by individuals or institutions, the GI is a collective IPR of many people belonging to a territory of a country or the country itself. For this reason GI is not transferable or licensable. GI refers to an essentially attributable association between a given quality, reputation or other characteristics of any goods such as agricultural, natural or manufactured, which is valued in commerce, with its either origin or production or manufacture or any act of processing or preparation in a definable territory, or a region of a country. For this exclusive relationship between the region/territory/country and the quality and distinctiveness of given goods, it commands a commercial reputation conferring premium market space in demand and price. Examples of horticultural products having distinctiveness for GI are Coorg Orange, Nanjanagud Banana, Mysore Jasmine, Udupi Jasmine, Hadagali Jasmine, Malabar Pepper, Allahabad Surkha, Monsooned Malabar Arabica Coffee, Monsooned Malabar Robusta Coffee, Alleppey Green Cardamom, Coorg Green Cardamom, Eathomozhy Tall Coconut, Laxman Bhog Mango, Khirsapati (Himsagar) Mango, Fazli Mango grown in the district of Malda, Naga Mircha, Nilgiri orthodox tea, Assam orthodox tea, Virupakshi Hill Banana, Sirumalai Hill Banana, Mango Malihabadi Dusseheri, Vazhakulam Pineapple, Devanahalli Pomello, Appemidi Mango, Kamalapur Red Banana, Guntur Sannam Chilli, Mahabaleshwar Strawberry, Nashik Grapes, Byadagi Chilli, Darjeeling tea, and Kangra tea etc. which are famous for their uniqueness in quality, reputation or other characteristics and produced within a definite territory in the country. According to Indian GI Act, any association of persons or producers or any organization or authority established under any law representing the interest of producers of the concerned goods are eligible to apply for the GI. Using the flexibility of this provision, efforts are underway in the country to appropriate the collective IP of farmers associated with some reputed agricultural commodities by the organized processors and traders.

ICAR Initiatives

The Indian Council of Agricultural Research (ICAR) recognizes the need of becoming competitive in the intellectual property rights (IPR) regime so that we can ultimately bring the Indian farmers away from subsistence with the transfer of our IPR enabled technologies through commercial, cooperative and public routes. In the early 1990s, ICAR had taken initial steps for the pro-active management of IPRs generated by our scientists. This was in line with the upcoming global developments on trade and IPR but also within the limited scope of our Patents Act. As a follow up of the World Trade Organization (WTO) Agreements, the legislative scenario concerning IPRs changed fast and by 2005 the country had revised or enacted various IPR laws to meet the national requirements vis-à-vis the Agreement on Trade Related Aspects of Intellectual Property. The ICAR framed a detailed guideline (http://www.icar.org.in) which deals with IPR awareness and literacy, enhance the work environment for higher innovativeness, ensure that the scientists/innovators are duly rewarded with their share of benefits accrued, and guide the manner of technology transfer which would be competitive and better serve the interests of agriculture and farmers. In order to have to provide a protection to the genetic resources of actual or potential value, the ICAR instituted a mechanism called "Registration of Plant Germplasm" at the National Bureau of Plant Genetic Resources (NBPGR), New Delhi for registration of experimentally developed genetic resources. It also accorded recognition to those associated with the development of improved/unique germplasm and genetic stocks, such as plant breeders and/or farmer-breeders or other developers/innovators. Out of 942 potentially valuable germplasm belonging to 160 crops which have been registered so far, 194 belong to 80 crop species of horticultural importance (59 in vegetables (brinjal, tomato, chilli, snap melon, cucumber, sword bean, potato, bottle gourd, pointed gourd, fenugreek), 28 in fruits (mango, aonla, ber, cashew, kokam, watermelon), 45 in medicinal and aromatic plants and spices (betel vine, blackpepper, turmeric, malabar tamarind), 21 in tubers and 41 in ornamentals (balsam, gladiolus, trumphet bush, orchid). This venture is expected to enhance the utilization of genetic resources and its availability to researchers across the country in the light of emerging IPR regime.

Looking Ahead

There is need to strengthen planning and conducting collaborative explorations involving different stakeholders. We also need to prepare more exhaustive and informative inventory of collections maintained at various research stations to facilitate their use and eliminate duplicates. GIS based mapping will further help to increase germplasm collection and utilization. Greater research and development realizing the importance of horticultural crops in food and nutritional security, diversification and sustainability, are required to be undertaken on areas like adaptability, breeding for specific trait to promote specific uses including export, valuation of HGR as sources of desirable traits and rootstocks, exploitation of wild relative for incorporation of resistance to biotic and abiotic stresses, nutritional quality and yield.

Since majority of HGR produce recalcitrant or intermediate seeds and are vegetatively propagated therefore, cryo preservation and *in vitro* conservation activities have to be strengthened. The crops which are difficult-to-propagate, have high economic importance and wild germplasm need to be considered at priority. The national networking on HGR management which includes acquisition, conservation, evaluation and utilization of PGR, strengthening database on information and monitoring national and international regulations in policy to suitably ensure efficient PGR management are required and for this purpose greater role of NAGS for evaluation and maintenance of germplasm has been advocated in many forums. Bilateral programmes need to be developed with resource countries for the introduction of germplasm of desired traits. The threat of biopiracy and infringement of plant variety related IPR necessitates thorough characterization, including DNA fingerprinting of the elite germplasm. There is also a need to have integrated conservation of genomic resources to collect, validate and facilitate the use of useful genes and gene constructs generated in the country. The efforts are under way for creating genomic resources centre especially for those species, which have reached at the verge of extinction. Plant variety protection, geographical appellations, patents and trademarks would help to protect and utilize PGR more efficiently. It also requires a significant commitment from governments, industry sectors and the wider community to ensure a long-term balance between sustainable resources use and conservation.

References

Annonymous, 1950. *Wealth of India: Raw materials.* CSIR, New Delhi, 2: 188–209.

Arora, R.K. and Nayar, E.R., 1994. Wild relatives of crop plants in India. *Science Monograph,* National Bureau of Plant Genetic Resources, New Delhi, 7: 90.

Babu, C.R., 1977. *Herbaceous Flora of Dehradun.* Publication and Information Department, Council of Scientific and Industrial Research, New Delhi, India, p. 721.

Baker, J.G., 1886. Scitamineae. In: *Flora of British India,* (Ed.) J.D. Hooker. 6: 198–264.

Bhattacharya, S.C. and Dutta, S., 1956. Classification of *Citrus* fruits of Assam. *Sc. Monogr.,* ICAR, New Delhi, 20: 110.

Brandis, D.D., 1874. *Forest Flora of North West and Central India.* Allen, London.

Chakravarty, H.L., 1982. *Fascicles of Flora of India, Fascicle II, Cucurbitaceae.* Botanical Survey of India, Howrah, India.

Dhankar, B.S., Mishra, J.P. and Bisht, I.S., 2005. Okra. In: *Plant Genetic Resources: Horticultural Crops,* (Eds.) B.S. Dhillon, S. Saxena, A. Agrawal and R.K. Tyagi. Narosa Publ. House, New Delhi, India, pp. 59–74.

Dhillon, B.S., Rana, J.C. and Singh, A.K., 2005. Collection, conservation and utilization of indigenous horticultural crops germplasm. In: *Proc: First Indian Horticulture Congress,* (Eds.) K.L. Chadha and S.P. Singh. IARI, New Delhi, India, pp. 11–35.

Dhillon, B.S. and Rana, J.C., 2004 .Temperate fruits genetic resources management in India: Issues and strategies. *Acta Hort.,* 662: 139–146.

Joshi, B.D. and Rana, J.C., 1994. Collecting temperate fruits and their wild relatives in north-west Himalayas. *IBPGR Newsletter,* pp. 15.

Kanjilal, U.N., Kanjilal, P.C. and Das, A., 1934. *Mangifera indica* L. *Flora of Assam,* 1: 335–336.

Karihaloo, J.L., Malik, S.K., Rajan, S., Pathak, R.K. and Gangopadhyay, K.K., 2005. Tropical fruits. In: *Plant Genetic Resources: Horticultural Crops,* (Eds.) B.S. Dhillon, S. Saxena, A. Agrawal, and R.K. Tyagi. Narosa Publ. House, New Delhi, India, pp. 121–145.

Kostermans, A.J.G.H., 1983. The South Indian species of *Cinnamomum* Schaeffer (Lauraceae). *Bull. Bot. Survey India,* 25: 90–133.

Mukherjee, S.K., 1985. Systematic and ecogeographic studies on crop genepools. 1. *Mangifera* L. IBPGR, Rome.

Nayar, E.R., Pandey, A., Venkateswaran, K., Gupta, R. and Dhillon, B.S., 2003. Crop plant of India: A check list of scientific names. *NBPGR Monograph on Agro-biodiversity* No. 26.

Negi, K.S. and Pant, K.C., 1992. Less known wild species of *Allium* L.(Amaryllidaceae) from mountainous region, India. *Eco. Bot.,* 46(1): 112–114.

Pandey, A., Pandey, Ruchira, Negi, K.S. and Radhamani, J., 2008. Realizing value of genetic resources of *Allium* in India. *Genet. Resour. Crop. Evol.,* 55(7): 985–994.

Rana, J.C. and Shrama, B.D., 2000. Naturally occurring spices and aromatic plants of North-western Indian Himalayas and their domestication potential. In: *Proc. Centennial Conference on Spices and Aromatic Plants,* IISR, Calicut, pp. 35–42.

Rana, J.C., Singh, Dinesh, Yadav, S.K., Verma, M.K., Kumar, K. and Pradheep, K., 2007. Genetic diversity collected and observed in Persian Walnut (*Juglans regia*) in Western Himalaya region of India. *Plant Genetic Resources Newsletter (Biodiversity International),* 151: 68–73.

Rana, J.C., Pradheep, K. and Verma, V.D., 2007. Naturally occurring wild relatives of temperate fruits in Western Himalayan region of India: An analysis. *Biodiversity and Conservation,* 16(14): 3963–3991.

Rathore, D.S., 2001. Networking for genetic resources management of horticultural crops. Paper presented at *Symposium on Plant Genetic Resources Management Advances and Challenges,* National Bureau of Plant Genetic Resources, New Delhi.

Sabu, M., 1991. A taxonomic and phylogenetic study of South Indian Zingiberaceae. *Ph.D. Thesis,* University Calicut, Kerala, India.

Sharma, B.D. and Rana, J.C., 2000. Survey, collection and scope of floriculture germplasm. In: *Horticulture Technology, Vol. 1: Production,* (Eds.) V.K. Sharma and K.C. Azad. Deep and Deep Publication, New Delhi, pp. 381–398.

Sharma, S.D., Kumar, K., Gupta, S., Rana, J.C., Sharma, B.D. and Rathore, D.S., 2005. Temperate fruits. In: *Plant Genetic Resources: Horticultural Crops,* (Eds.) B.S. Dhillon, S. Saxena, A. Agrawal and R.K. Tyagi. Narosa Publ. House, New Delhi, India, pp. 146–167.

Sharma, B.D. and Chandel, K.P.S., 1996. Occurrence, distribution and diversity of soft fruits in N-W and W-Himalayas and prospects of their conservation and utilization. *Indian J. Pl. Genet. Resources,* 9(2): 237–246.

Singh, I.D., 1999. Plant Improvement. In: *Global Advances in Tea Science*, (Ed.) N.K. Jain. Aravalli Books International Pvt. Ltd., New Delhi, India, pp. 427–448.

Singh, N.P., 1993. Clusiaceae. In: *Flora of India, Vol 3: Portulacaceae–Ixonanthaceae*, (Eds.) B.D. Sharma and M. Sanjappa. Botanical Survey of India, Calcutta, pp. 86–151.

Srivastava, S.K., 1994. *Garcinia dhanikhariensis* (Clusiaceae), a new species from the Andaman Islands, India. *Nordic J. Bot.,* 14: 51–53.

Velayudhan, K.C., Asha, K.I., Mithal, S.K. and Gautam, P.L., 1999. Genetic resource of turmeric and its relatives in India. In: *Biodiversity, Conservation and Utilization of Spices, Medicinal and Aromatic Plants*, (Eds.) B. Sasikumar, B. Krishnamoorty, J. Rema, P.N. Ravindran and K.V. Peter. Indian Institute of Spices Research, Kerala, India, pp. 101–109.

Yadav, I.S. and Rajan, S., 1993. Genetic resources of *Mangifera*. In: *Advances in Horticulture Vol 1, Part 1*, (Eds.) K.L. Chadha and O.P. Pareek. Malhotra Publishing House, New Delhi, pp. 77–93.

2013, Biodiversity in Horticultural Crops Vol. 4
Editor: Professor K.V. Peter
Published by: DAYA PUBLISHING HOUSE, NEW DELHI

Pages 25–59

Chapter 2

Horticultural Biodiversity Heritage Sites

Anurudh K. Singh and J.P. Yadav
Department of Genetics, M.D. University, Rohtak – 124 001, Haryana
E-mail: anurudhksing@gmail.com

Heritage refers to history, traditions and qualities that a country/place and society had over generations and considered important features. The World Heritage Convention (WHC), 1972 is a unique international instrument for conserving cultural and natural heritage of outstanding value. Convention has recognised three categories of cultural landscapes as heritage sites, namely, a) clearly defined landscapes designed and created by humans; b) organically evolved landscapes and c) the landscapes associated with the virtues of religious, artistic or cultural practices that are intangible. It is very much true with agriculture, because of the basic fact that economically important crop plant could not have originated without richness in biodiversity and ingenuity of local people. For this reason, Singh and Varaprasad (2008) proposed Agricultural Biodiversity Heritage Sites, as the sites where agriculture has evolved with domestication of landscape with crop species, initially by random planting of economically important plants in the forest areas, referred as 'domiculture', followed by shifting agriculture, with cultivation of economically selected crop species and crop cultivars, in concentrated agriculture plots, as a part of multi-species complex agro-ecosystem, and overall landscape organisation (Ramakrishnan, 1992). Further, cultural diversification and settlement lead to more intensely managed multi-species complex agro-ecosystem, maintaining the overall integrity of landscape (Swift *et al.,* 1996). Therefore, heritage sites account for all significant changes which evolved in time and space in the biotic composition of component of domesticated or economically important plant and animal species with evolution of suitable genetic diversity and agricultural practices for their eco-friendly management and sustainable cultivation, responding to local needs and goals and to climatic and marketing scenario. Consequently, these sites are reservoirs of genetic diversity for domesticated plant and animal species, including horticultural species, and for knowledge regarding their effective eco-friendly management based on background information about the surrounding landscape, resources, their recycling, which offered sustainability without much use of energy.

India: A Centre of Rich Biodiversity

By the virtue of physiognomy and climatic diversity, India harbours a rich flora and is one of the 12 "mega-biodiversity centres", housing an estimated 12 percent of the world flora. It contains three of the 34 biodiversity "hotspots", the Himalayas, the Indo-Burma region and the Western Ghats, with thousands of endemic flora (Conservation International, 2005). The floristically rich India has about 141 endemic genera belonging to over 47 families. As per Botanical Survey of India (BSI), India has 46,214 plants species. Of these, about 17,500 (7,000 species in north east region alone) represent flowering plants (7 per cent of the world flora); 37 percent of them are endemic. Of the endemic species (4,950), the largest number of (about 2532) species are located in Himalayas followed by peninsular region (1,788 species) and Andaman Nicobar Islands (185 species). India is a rich centre of floristic and economic plant diversity (Arora *et al.,* 2006), particularly for horticulture species like fruits, vegetables, spices, ornamentals, medicinal plants etc. both at species and intra-specfic levels. Consequently around 583 species are cultivated, of which around 417 belong to horticultural crops (Dhillon *et al.,* 2005). Around 27 species in fruits and nuts, 23 in vegetables, 15 in plantation and tuber crops and 16 in spices and condiments are cultivated on regular basis, which has a large number of wild relatives, around 331 in fruits and nuts, 215 in vegetables, 154 in plantation and tuber crops and 161 in spices and condiments (Singh 2010). Thus the amount of species diversity available offers substantial genetic diversity to meet the future needs, particularly for the present scenario of climate change.

India: A Centre of Origin of Crops and Agriculture

India offers great diversity both in ecological and socio-political term, because of being large land mass with all possible ecological variation, from tropical rain forest of Western Ghats, to temperate high altitude regions of Himalayas, to Thar desert of north western plains. It offers most diverse ecologies for adaptation of plants. For example, the Gangetic plains, blessed with fertile soil and hospitable climate attracted settlers, because of these attributes, fundamental to agricultural productivity, evolving it into a primary centre of plant and animal domestication and agriculture (Tripathi, 2008). Therefore, throughout the history of Indian civilization, the human ingenuity, interest, intervention and support attempting to domesticate the indigenous/endemic plants and animals or evolving agriculture with introduction of useful domesticated species from various sources have been an important feature. The level of interaction of human population with natural resources, particularly biodiversity, has been quite high in Indian sub-continent, resulting in domestication of a large number of species starting from Indo-Gangetic plains in north, identified with rice grains of earliest cultivation in the world (Sharma, 1980), to South India, where as per recent archeobotanical evidences, domestication of crops like minor millets, *Vigna* L. spp. (used as vegetable), fruits, etc. took place (Fuller *et al.,* 2004), which were indigenous to the Indian peninsula. Vavilov (1935) recognized eight primary centres of origin, of which two are associated with India, namely The Indian Centre (including the core of subcontinent) - based originally on rice, millets, legumes and a number of vegetables and fruits, with a total of 117 (168) species, and The Indo-Malayan Centre (including northeast region of India and southeast Asian countries like Myanmar, Thailand, Malaysia, Indonesia, Philippines, etc. - with a number of root crops (*Dioscorea* L. spp., *Tacca* Frost., etc.) and preponderant fruit crops, sugarcane, spices, etc., of around 55 species. For these reasons, Vavilov (1934) kept Hindustani Centre second after Chinese in richness for crop species.

The Concept of Agricultural Biodiversity Heritage Sites

Generations of farmers and herbalists on one hand and travellers/traders on the other, for more than 10,000 years have helped developing ingenious farming/production systems. These efforts of generations have resulted in establishment of significant reservoirs of agro-biodiversity heritage sites, associated with supportive landscape, environment and ingenious practices and products (landraces/varieties/breeds) as part of cultural inheritance. They are mostly confined to river valley's, which ensure availability of physical natural resources (water, fertile enriched soil with nutrition) for the growth of cultigens and domesticated animals and have great aesthetic beauty for human settlement. Such sites have been referred to "Agro-biodiversity Hotspots" by the task forces of "Protection of Plant Varieties and Farmers' Rights Authority" (Nayar *et al.,* 2009) and National Agro-biodiversity Heritage Sites (NAHS) by Singh and Varaprasad (2008). The traditional societies of most NAHS have maintained/conserved agro-biodiversity in the form of multi-species (including multi-cultigens) complex agro-ecosystems, presently managed casually or at low intensities as an integral part of a cultural landscape, based on local value system. They have strong socio-cultural, socio-economical interconnections with the landscape and biodiversity and are the products of eco-cultural interactions occurring in space and time and may still be evolving.

The changed scenario of intellectual property rights on biological diversity from heritage of mankind to sovereign ownership to nations in the post Convention on Biological Diversity (CBD) era demands identification, revisit to these sites for validation, value assessment, revitalization and conservation or further improvement of traditional products and agriculture systems to facilitate their use by designing sustainable models. This would need documentation, as the traditional agricultural systems have two important perspectives- a) conserving and generating genetic diversity responding to prevailing environments ensuring its availability to modern agriculture and b) conservation of crop biodiversity to ensure the livelihood of local people realizing that they are dependent upon some specific species and crop level diversity contained within these traditional systems. In addition to richness in agro-biodiversity, which includes floristic diversity, number of endemic species, crop species and varieties cultivated and wild relatives of crop naturally habitat to the region, Singh and Varaprasad (2008) suggested following additional indices to be associated with agricultural biodiversity heritage sites:

1. The agriculture should be geographically and/or socially important providing the livelihood to most population,
2. Should be functioning as the custodian of natural resources,
3. Providing opportunities for continued evolution, development and adaptation,
4. Agricultural practices in the site shall have intangible religious, artistic or cultural associations,
5. Community ingenuity should be visible through innovative practices evolved,
6. Might have external inputs from other civilizations, accepting good from all sources,
7. Consequent *farming/production systems* may have integration of other systems in term of accounting Traditional Ecological Knowledge (TEK) for cultural values, participation, sustainability and complex management of resources by different members with visible ingenuity at any level of complexity or diversity.

To identify the actual site or epicentres of domestication and evolution of genetic diversity in crop species is difficult because on domestication, the early settlers carried the evolved genotypes/genetic diversity to new settlements and then dispersed it to a wide area (including foreign lands) through traditional systems of seed dispersal or barter trade and further evolved it, making vast areas as centres of genetic diversity and conserving it through generations. Therefore, the identified heritage sites are regions of adaptation of crop species, with genetic diversity captured and conserved by locals in form of local landraces, farmer's varieties and ancestral wild relatives.

Agricultural Biodiversity Heritage Sites

Based on the geographical association with various life forms, India is divided into 10 biogeographical, 9 phytogeographical (Chatterjee 1939) and 21 agroecological regions (Sehgal *et al.,* 1992). While based on the indices associated with agrobiodiversity, PPV and FR, identified 22 agro-biodiversity hotspots (Nayar *et al.,* 2009), and whereas the author has been writing on possible National Agricultural Biodiversity Heritage sites, numbering 21. These sites are important from agricultural heritage point of view and most are also important from horticultural biodiversity point of view, as most farming/production systems are integrated with diverse horticulture crops for sustainability, conservation of natural resources, cultural value etc. by the same or different member of local communities with visible ingenuity at different levels. Table 2.1 lists these sites along with location and extent and salient horticultural features and subsequently discusses them in brief.

Table 2.1: Possible Horticultural Biodiversity Heritage Sites, their Location and Unique Features

Sl.No. Possible Horticultural Biodiversity Heritage Sites	Location and Extent	Unique Features
1. The Western Himalayan Region	Srinagar, Anantnag, Udhampur, Riasi, Kathua in Jammu and Kashmir, Himachal Pradesh, Siwalik hills of Punjab, and Uttrakhand	Secondary centre of origin/diversity of *Prunus, Pyrus, Sorbus, Rubus and primary centres* of *Allium* spp.
2. The Eastern Himalayan Region	Most of Arunachal Pradesh and Sikkim, and northern tip of West Bengal (Darjeeling and Kalimpong)	Centre of diversity of *Rhododendron, Primula, Pedicularis,* orchids, spices, *Prunus rufa*
3. Brahmaputra valley	Most parts of Assam	Tea, cucumber, *Musa,* bamboo
4. Khasi, Jaintia and Garo Hills	Meghalaya	Centre of diversity of *Rhododendron, Schima, Orchidaceae, Zingiberaceae* spices, *Citrus, Garcinia*
5. North-eastern Hills	North-eastern states of Nagaland, Manipur, Mizoram and Tripura, and district Cachar of Assam	Cucumber, brinjal, *Citrus,* mango, tropical and subtropical minor fruits
6. Arid Western and semi-arid Kathiawar Region	Western part of Rajasthan, parts of the south-western Haryana and Kthaiawar peninsula of Gujarat	Centre of diversity of *Citrullus, Cucumis,* arid fruits (*Ziziphus*), seed spices
7. Malwa Plateau	Western part of Madhya Pradesh and parts of southeastern Rajasthan	Centre of diversity of *Papaver somniferum, Abelmoschus, Cucumis* spp.
8. Bundelkhand	South of Yamuna, between fertile Gangetic plains stretching across northern Uttar Pradesh and southern highlands of Madhya Pradesh.	*Brinjal, Ziziphus, Aegle marmelos, Buchanania lanzan*
9. Upper Gangetic plains	Northern Punjab, most of Haryana and Western, central and parts of eastern Uttar Pradesh	Centre of diversity of *Benincasa hispida, Citrullus, Abelmoschus*

Contd...

Table 2.1–*Contd...*

Sl.No. Possible Horticultural Biodiversity Heritage Sites	Location and Extent	Unique Features
10. Middle Gangetic Plains	Eastern Uttar Pradesh and parts Bihar, lying on either side of Ganga and Saryu (Ghaghara)	Centre of diversity of Cucurbits, *Trichosanthes dioica*, *Luffa hermaphrodita*,
11. Lower Gangetic Plains or Delta region	Parts of West Bengal and Bangladesh where Ganges, Brahmaputra and Meghna meet ending into Bay of Bengal	*Ipomoea aquatica*, Okra, *Momordica*, *Trichosanthes*, Musa, ornamentals
12. Chotanagpur Plateau region	South-eastern plateau of Jharkhand	Litchi, jackfruit, Cucurbits, *Dendrobium* (orchid)
13. Baster region	Chattisgarh state	Root and tuber crops, *Diospyros melanoxylon*
14. Koraput region	Southern-eastern Orissa and some districts of north-eastern Andhra Pradesh	Root and tuber crops, Cucurbits-*Luffa*, *Momordica*
15. Southern-eastern Ghats	Dry areas of southern Andhra Pradesh and Karnataka	Root and tuber crops
16. Kaveri region	Kaveri delta of Tamilnadu	*Trichosanthes*, Root and tuber crops, *Syzygium*, banana
17. Northwest Deccan Plateau	Maharashtra and parts of north-west Andhra Pradesh	*Vigna* spp., *Citrus*, *Annona squamosa*, *Vitis vinifera*, *Tamarindus indica*
18. Konkan region	Hot humid-perhumid region of Western Ghats, coastal plains of Maharashtra, Goa, and Uttar Karnataka	Centre of diversity of *Vigna* spp., spices, Mango, *Artocarpus heterophyllus*, *Garcinia*
19. Malabar region	Hot humid, perhumid, southern part of Western Ghats and coastal plains	Primary centre of diversity of spices, Syzygium, *Garcinia*, *Artocarpus heterophyllus*, coconut

In addition, cold arid region of Himalayas (*Vicia*, *Trigonella* and Medicinal and Aromatic Plants) and Andaman and Nicobar Islands offer diversity for wild relatives of a number of horticultural crop species

1. The Western Himalayan Region

This heritage site extends from Srinagar, Anantnag, Udhampur, Riasi, Kathua in Jammu and Kashmir, Himachal Pradesh, Siwaliks hills of Punjab to Uttrakhand. It is characterized by high mountains and narrow valleys with large variation in latitude and availability of moisture promoting evolution and perpetuation of rich natural bio-diversity. It is cool-humid to warm sub-humid eco-region with mild summers and cool to cold winters. The mean annual rainfall varies from 1600 to 2000 mm ensuring soil moisture availability between 150 to 210 days for cultivation. Being part of one of the global Biodiversity Hot Spots, the Himalayas, it is very rich in floristic diversity. The natural vegetation comprises Himalayan moist temperate, subtropical pine and sub-alpine forests.

Rain-fed farming is the traditional practice in the valleys and on terraces. Horticulture based farming systems are predominant. Being part of mountainous terrain, it has evolved unique horticulture-based multi-crop farming/production systems. The traditional mixed cropping system accounts for 66 percent of total cropped area. This site falls in one of the three centres recognised for the origin of temperate fruits, the Central Asiatic Centre, extending from Tien-Shen South, the Hindu

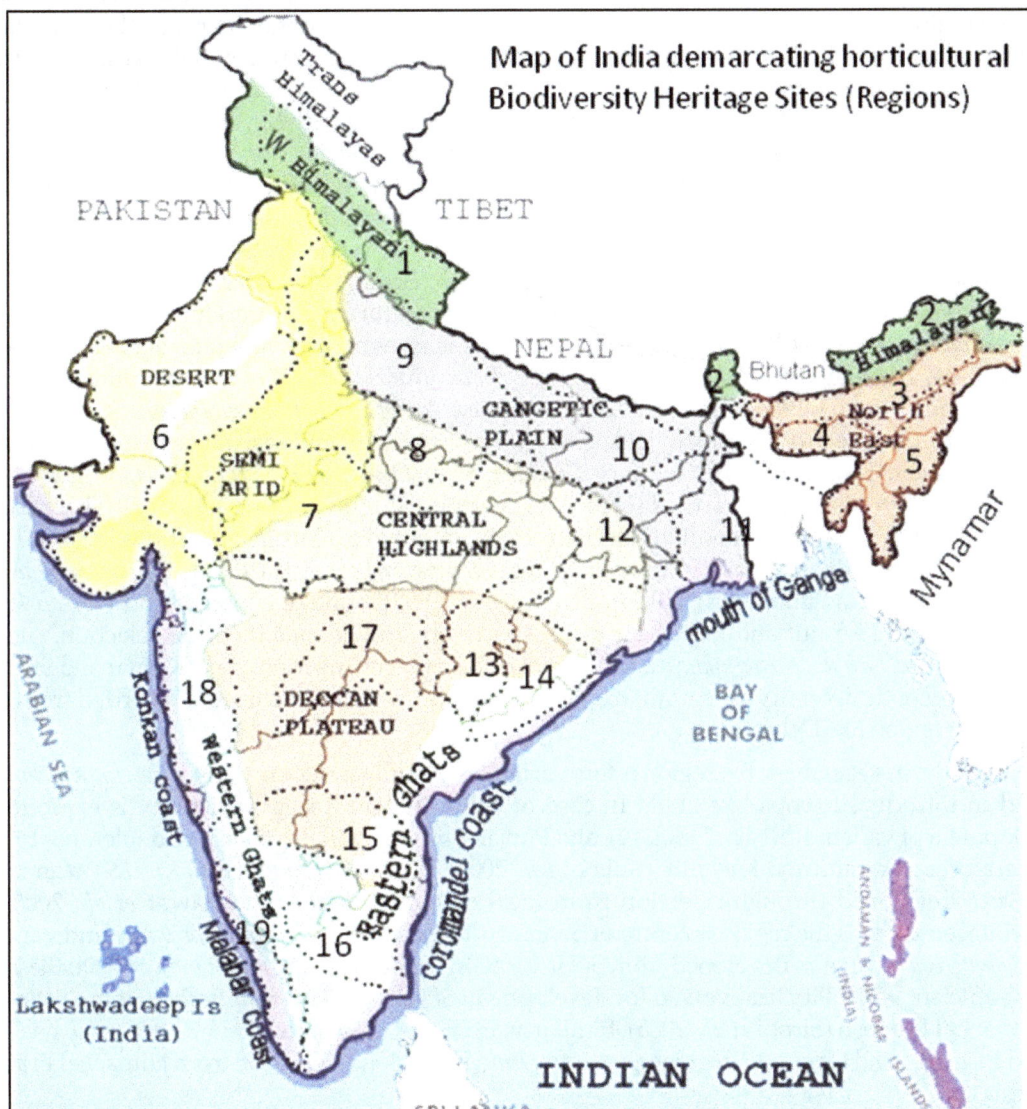

Figure 2.1: Location and Extent of Various Horticultural Biodiversity Heritage Sites (Regions)

Kush to Kashmir. It is believed that some species of the temperate fruit's genera such as *Malus* L., *Pyrus* L., *Prunus* L., *Rubus* L., etc. escaped to the Himalayas and over a period of a few thousand years, they not only survived but got established in the hills, leading to evolution of new species and varieties. The cultivation of temperate fruits extends from Jammu and Kashmir to sub-tropical plains. The fruits grown traditionally are Apple (*Malus domestica* Borkh.), Pear (*Pyrus communis* L.), Peach (*Prunus persica* (L.) Batsch.], Plum (*Prunus domestica* L.), Almond (*Prunus amygdalus* Batsch.), Apricot (*Prunus armeniaca* L.), Cherry (*Prunus avium* L.) and Walnut (*Juglans regia* L.) making the region, a reservoir of species and genetic diversity for temperate fruits, vegetables including leafy and root and tuber, ornamentals and medicinal plants, the main horticultural crops and their wild relatives (Dhillon *et al.*, 2005).

The region is inhabitated by an admixture of Indo-Aryan or Mongolian races, though original inhabitants were the Kinnars, Kilinds, and Kiratas. The region has native variability in the form of local stocks and clones in sand pears, wild apricots, *Prunus* species, almond, walnut, pecan nut, hazelnut and many minor fruits. In Kashmir, high-density planting with MM 106 rootstock of superior varieties, such as *Starkrimson, Oregon spur, Red Chief, Vance Delicious, Ambri Selection* with suitable pollinizers, like *Gold Spur, Red Gold, Golden Hornet* is doing well, whereas, in Uttarakhand, *Anupam, Chaubattia, Agrim, Chaubattia Swarnima, Red June, Early Shanburry* is doing well (Prakash *et al.,* 1997). In pear, *Red Blush* and *Punjab Sunehri* are examples of superior strains selected from plantations of Baggugosha (Uppal *et al.,* 1993). In peach, *Strak Earliglo, Early White Giant, Starking Delicious* and *Candor* are suitable for cultivation in mid hills due to early maturing feature. Cultivars, *Prairie Dawn, Prairie Ramber* and *Prairie Rose* are suitable for high altitude areas of dry temperate region (Sharma and Kumar, 1994). In plum, *Green Gage, Frontier, Kanto5, Kubio, Red Ace* and *Tarrol* were promising for mid hills (Bisht and Sharma, 1996). In sweet cherry, cultivars like *Triumph Domini, Pietro Nigra, Bella Italia* and *Foaya Tardiva* are promising for yield, quality, early maturity and resistance to insect pests and diseases (Gautam *et al.,* 1992). In strawberry, *Chandler, Selva, Douglas, Confictura, Dana, Belrubi, Gorella, Addie, Pajaro, Fern* and *Tioga* are large fruited, ever bearing and high yielding selections (Sharma *et al.,* 2005). In almond, Kaul (1990) reported selections HS8, HS9 and HS10 from plantations in Jammu and Kashmir, whereas Tripathi *et al.* (1992) reported that *Supernova,* Ferragnes, *Genco, California Papershell* and IXL doing well in Pithoragarh. In Walnut, Bhat *et al.* (1992) found selection P3 and *Wassan 4* from Kashmir, suitable for nut and kernel export, respectively. Indigenous seedling selection, such as *Solding Selection, Gobind, Roopa Akhrot* and *Kotkhai* are also promising for export (Kumar and Sharma, 1995). Rich genetic diversity for certain tropical fruits like *Citrus* and lichi were recorded from sub-Himalayan region like Dehradun.

Similarly, in vegetables, the region offers variability in indigenous crops, like *Solanum melongena* L. and in introduced crops like chilli. In case of chilli, the landraces' variability is exploited in development of varieties, Nishat 1 (Sel. 12), and Punjab Mirch 27 with high yield and tolerance to high temperatures in Jammu and Kashmir, (Kalloo *et al.,* 2005). In potato, *Chamba Red, Kufri Safed* and *Kufri Red* were developed through selection from local cultivar *Phulwa* (Shekhawat *et al.,* 2005). In cucurbitaceous crops the region is centre of diversity for cucumber, *Cucumis sativus* L. and varieties like *Solan green* have been developed from Solan local. In introduced *Cucurbita moschata* Duchesne ex Poir., sufficient variability has evolved for development of variety like *Sadan Badami* from a collection of Himachal Pradesh (Sirohi *et al.,* 2005). Similarly, in case of spices, rich genetic diversity has been found in ginger and turmeric. In ginger, variety *Himgiri* is a clonal selection from Himachal Pradesh germplasm.

2. The Eastern Himalayan Region

The Eastern Himalayan Region consists of Arunachal Pradesh, Sikkim and the hills of West Bengal. The landscape consists of high mountains, glaciers, passes and valleys. The general climate is characterized by moderate-to-severe winters and mild summers with an annual rainfall of around 2,000 mm with growing period of more than 270 days. Arunachal Pradesh receives a comparatively higher rainfall of 2,000 to 4,000 mm. The region has shallow-to-medium loamy, brown forest soil and deep, organic-matter-rich, *Tarai* soils. It is rich in floristic diversity being part of Himalayas, one of the hotspots of global biodiversity and falling in the meeting ground of the Indo-Malayan and the Indo-Chinese biogeographical realms. Arunachal Pradesh and Sikkim contain 5,000 and 4,500 species respectively, 60 per cent of which are endemic (Chatterjee *et al.,* 2006). This area has been identified as centre of diversity for taxa, such as rhododendron, *Primula* Linn. fide Kuntze, and *Pedicularis* Linn.

The region has vast variation in altitude (ranging from 300 m to 8,585 m) within very short distance and this has played a prime role in designing the ecoregional diversity, which broadly has four zones of vegetation: tropical, subtropical, temperate and alpine and trans-Himalayan.

The region is predominantly inhabitated by Mongoloids with the influence of Tibetan culture. The original inhabitants of Sikkim are said to be the *Lepchas*. The other major tribes include the *Adi, Galo, Nishi, Khamti, Monpa, Apatani*, and *Hill Miris* (Gangwar and Ramakrishnan, 1989). As per one estimate, a total of 171 plant species are used by the Nishis, the Hill Miris, the *Sulungs* and the Apatanis of the Lower Subansiri district of Arunachal Pradesh. Of these, 38 per cent were leafy vegetables and 28 per cent were edible fruits. Most farming is rainfed. The *jhum* system of shifting cultivation has been the traditional way of farming, including horticultural species. The dominating traditional agro-forestry system involves cultivation of large cardamom in Sikkim. The other principal horticultural crops include ginger, turmeric, orange, apple, pear, and off-season vegetables and flowers, such as gladiolus, orchids, lilies, gerbera, carnation and *Anthurium* Schott. The ginger available in Sikkim is less fibrous with high moisture content and suitable for the manufacture of ginger products, like preserves, candy, crystallized ginger, ginger biscuits, etc. Sikkim grows a special chilli locally known as '*Dalle Khorsani*'. It has good aroma with considerable pungency. Consequently the region is an important centre of diversity of spices and ornamentals in addition to vegetables (including tuberous crops), and fruits, along with large number of wild relatives. Significant genetic variability was recorded in potato (Darjeeling Red Round), for fruit and plant types in okra and chilli with good aroma and considerable pungency. Among spices, the region is known for variability in *Piper nigrum* L., cardamom, large cardamom, crisp ginger with high moisture content (Sikkim) and turmeric (Arunachal Pradesh), besides tea and coffee. The locals have generated information about the medicinal value of local plants, including orchids, for example, the *Chakma* community in Arunachal Pradesh use *Achyranthes aspera* L. against urinary disorders; *Cataranthus roseus* (L) G.Don, known for anti-cancer drug, is being used against diabetes; and *Centella asiatica* (L) Urban is used against stomach disorders by different tribes. The wild relatives of horticultural crops are represented by *Luffa graveolens* Roxb. in vegetables (Sirohi *et al.,* 2005); *Actinidia callosa* Lindl. and *A. strigosa* Hook.f. and Thomson (kiwi fruit), *Citrus reticulata* Blanco. (wild forms), *Mangifera indica* L., *M. sylvetica* Roxb. in fruits, making it a reservoir of gene in continuum of Western Himalayas. The region inhabits a large number of orchid and rhododendron species presenting useful variability for attractive flowers in many colors and shapes.

3. The Brahmaputra Valley Region

The northeastern region of India can be physiographically divided into three distinct divisions – the Meghalaya Plateau, the Northeastern Hills, and the Brahmaputra Valley. The State of Assam is predominantly made up of the Brahmaputra Valley. The region has mountain slopes, rivers, coldwater streams; floodplain wetlands, reservoirs, lakes and ponds. However, most part of it consists of the plains around the Brahmaputra and Barak rivers. The area is characterized by hot summers and mild to moderately cool winters. The mean annual rainfall ranges from 1,600–2,000 mm with 270 days long growing period. The region as a whole is very rich in floristic diversity, with around 2,833 dicotyledonous and 1,072 monocotyledonous angiosperms, and has been referred as "Biological Gateway" of north-eastern India. The species diversity is so spectacular that it often becomes difficult to clearly assign separate niches to existing plant formations. The Brahmaputra Valley was naturally associated with Tropical Wet Evergreen Forests, Tropical Semi-Evergreen Forests, Tropical Moist Deciduous Forests, Littoral and Swamp Forests, and Grassland and Savannahs. However, most of the above forests have been converted into grasslands by centuries of forest fires and other human interference, and now only small patches are left, scattered along the Indo-Bhutan border and along the border of Assam and Meghalaya.

The region is predominantly inhabited by Indo-Mongoloids, Indo-Aryans and Austrics races. The Bhutia and Bodo are the main tribes associated with agriculture. About 75 per cent of the population is directly or indirectly dependent on agriculture. But the progress from traditional shifting *Jhum* cultivation to permanent plough cultivation has remained very slow. Ponds called *Dongs* are constructed by the Bodo tribes to harvest water for irrigation during the lean period. The agro-climatic condition with high rainfall has favoured cultivation of agri-horticultural crops, like plantation crops, fruits, vegetables, flowers, spices, medicinal and aromatic plants, nuts, and tuber crops, under diverse farming systems. The biggest contribution of the region is tea produced from the indigenous variety *Camellia sinensis* (L.) Kuntze var. *assamica* (J.W.Mast.) Kitam. and floriculture, as the region has over 600 varieties of orchids, of which around 200 varieties are unique to the region, of which 60 per cent are ornamental in nature, offering tremendous opportunities to the local farmers for increasing their income. Muga silk, *Antheraea assama* is raised outdoors primarily on two trees – som (*Machilus bombycina* King ex Hook.f.) and soalu (*Litsaea polyantha* Juss.).

The region is primary or secondary centre of diversity of several horticultural crops, such as cucumber and several other cucurbitaceous vegetables, banana, citrus, mango, tea, etc., and has evolved reservoir of rich genetic diversity in these crops. Brinjal, *Solanum melongena* is represented by the variability for fruit shape, size, and colour (Kalloo *et al.,* 2005); okra has both five-edged and multi-edged fruit types and extended distribution of *Abelmoschus pungens* Wall. Welsh onion, *Allium fistulosum* L. cultivated in China, also extends to the region. Natural hybridization in chillies between *Capsicum chinensis* Jacq., and *C. frutescens* L., followed by natural or farmer's selection, evolved local landraces such as *Bhut jolokia* (Borgohain *et al.,* 2008) with high capsaicin content. In case of spices such as ginger and turmeric, many local cultivars with desirable features are known (Ravindran *et al.,* 2005). Among fruits, the region is known for variability in jackfruit (Karihaloo *et al.,* 2005), and in *Citrus* for the mandarin, *Citrus reticulata* Blanco. Locals have been using plants for medicinal value, for example, Sonowal Kacharis tribe uses *Allium sativum* L., *Oryza sativa* L., *Cassia sophera* L., *Ricinus communis* L. and *Ananas comosus* (L.) Merr. First-hand information on the uses of 16 wild and 11 cultivated plant species are reported for revalidation (Kalita and Deb, 2006).

4. The Garo, Khasi and Jainthia Hill Region

Physiographically, this is the second region of northeastern India characterized by high rainfall, heavy cloud cover. It consists of Garo Hills, Khasi Hills and Jaintia Hills, representing the Western, Central and Eastern parts of the State of Meghalaya. These hills have significant variation in latitude. The landscape of the region is mostly rolling plateau with extremely steep south-facing slopes. The region is a fairyland of great scenic beauty, a panorama of lush, undulating hills, fertile valleys, meandering rivers, waterfalls, sparkling mountain streams, lakes, etc. It is a warm perhumid ecoregion. The climate is characterized by warm summers and cold winters. This is one of the world's most wet regions, with annual rainfall varying from 1,600 to 2,600 mm and growing period up to 270 days. The soils are reddish and lateritic in origin, and vary from shallow to very deep sandy loam, red and yellow loam to clayey loam soils. Being part of the biodiversity hotspot "Indo-Burma Region", it is very rich in floristic diversity. One of the unique features of the landscape is the occurrence of numerous 'Sacred Groves' developed by locals, comprising of mixed evergreen forest with oak, rhododendrons, *Schima* Aucl. ex Steud., cinnamon, and a number of orchids and epiphytes. This has led to conservation of rich biodiversity and establishment of sanctuaries in the Tura range of the Garo Hills for conservation of *Citrus* L. and *Musa* L. species diversity.

The region is inhabited by Mongoloid and Austric races, consisting of Khasi, the Jaintia and the Garo tribal communities associated with agriculture. Because of richness in horticulture biodiversity, the forestry, horticulture, and animal husbandry are more common. Most traditional farming practices, besides agroforestry, revolve around silviculture, the science and art of cultivating forest crops, based on the knowledge of the life history and general characteristics of forest trees and horticulture, resulting in the development of systems, such as agri-silviculture, agri-horticulture, silvi-horticulture, pastoral-silviculture, and pastoral-horticulture. It includes mixed farming, which may include even medicinal plants along with crop or tree species. Shifting (*Jhum*) and terrace (*bun*) cultivation are the two major farming systems. *Jhum* has progressed to tree-based organized farming with cultivation of crops in association of trees like alder (*Aquilaria* Lam.), arecanut, coconut, bamboo, *Khasi* pine, etc. In bamboos and canes the region has evoved significant variability. The important fruit crops are oranges (Khasi mandarins), pineapple, lemon, guava, litchi, jackfruit, whereas banana, *tezpatta*, arecanut, betel leaf and black pepper are the chief plantation crops. Even in recently introduced crops, like Potato (*Solanum tuberosum* L.), planted under special planting system, called '*Nur Bun*' has evolved significant variability

This region offers significant genetic diversity in *Solanum melongena* for fruit and plant characteristics, besides *S. khasianum* with resistance to stem and fruit borer (Kalloo, 1993) and *Solanum kurzii* Br. used as vegetable and medicine, and is endemic to Garo hills. In chilli, genotype with the highest pungency has been reported from the region (Kalloo *et al.,* 2005). In Cucurbits, genetic diversity has been collected in *Momordica charantia* L., *Benincasa hispida* (Thunb) Cong., *Lagenaria siceraria* (Molina) Standley, *Sechium edule* (Jacq.) Swartz., etc. (Sirohi *et al.,* 2005), besides the *Cucumis hardwickii* Royle, the likely progenitor of cucumber. In okra, local landraces have variability for plant and fruit type with varying degree of tolerance to stresses. Among fruits, in mango, *Mangifera khasiana* Pierre., is endemic, distinguishable from *M. sylvatica* Roxb., in having smaller and narrower leaves, small flowers and inflorescence, with fasciculate divergent branching, oval to lanceolate petals and disc glandular or with distinct lobes (Mukherjee, 1985). In cultivated mango, dwarf and poly-embryonic types are known. In *Citrus*, 8 of the 17 species reported from the northeast region, like *C. indica* Tanaka, *C. macroptera* Montrouz., *C. latipes* Tanaka etc. naturally occur, presenting rich genetic diversity. In *Musa* L., *Musa flaviflora* Simmonds is localized with four additional species. Other fruits with significant variability are *Artocarpus heterophyllus* Lam., *Litchi chinensis* Sonn. Mill., etc. Khasi hills particularly harbour genetic variability for several temperate fruits, such as *Malus* spp., *Pyrus* spp., *Prunus* spp., *Prunus persica*, *Rubus* spp., *Sorbus* spp., *Corylus* spp. and *Castanea sativa* Mill. (Sharma *et al.,* 2005). The Shillong plateau of Khasi hills has *Prunus napaulensis* (Ser.) Steud. *P. undulate* Buch.-Ham. ex D. Don and *P. cerasoides* C.Don., while *Pyrus pyrifolia* Nakai var. *cubha makai* (*P. serotina* Rehder) is grown semi commercially.

North-eastern and South-western regions of India are recognized as two independent centres of diversity in spices. This region is known for diversity in black pepper, ginger and turmeric. In genetic diversity, it is second to Malabar in pepper, maximum in ginger and significant in case of turmeric. Cultivar, *Megha Turmeric 1* has been produced through selection from local Lakadong type with high curcumin content and bold rhizome (Ravindran *et al.,* 2005). Rich diversity is also recorded in aroids and minor tuber crops. The locals have tremendous knowledge about the medicinal plants of the region. Sajem and Gosai (2006) reported medicinal plant species used by Jaintias of North Cachar Hills district of Assam, belonging to 27 families and 37 genera. In total 39 medicinal plant species are used in curing 30 types of ailments, of which the highest number (20 species) are used for the treatment of gastrointestinal disorders, such as indigestion and constipation. The root powder of *Asparagus racemosus* Willd, known as Shatavari is effective in chronic peptic ulcer, the Jaintias use it for urinary disorders, as well as stomach ache that could be due to high peptic juice secretion.

5. The North-eastern Hills of Nagaland, Manipur, Mizoram and Tripura

This is the third physiographic region of northeastern India, occupying 65 percent of the total land area. It consists of states of Nagaland, Manipur, Mizoram and Tripura, and the adjoining districts of Cachar of Assam. The landscape of the region is dominated by the diverse range of hills, valleys, gorges, lakes and a network of streams and rivers running across the region. It is a warm perhumid ecoregion with red and lateritic soil. The climate of the region in general is characterized by warm summers and cold winters. It is extremely humid, with mean annual precipitation varying from 1,600 to 2,600 mm. The length of the growing season exceeds 270 days. Tripura has a tropical climate, hot and humid. The major soil of the region includes shallow to very deep, loamy, red and lateritic and red yellow soils. Being part of Indo-Burma Region, a biodiversity hotspot with wide variety of ecosystems and unique geological history and being part of the area where the land mass of the southern hemisphere meets with that of the northern hemisphere, it formulates the transition zone of lowland–highland with the highest diversity of biomes and ecological communities, extremely rich in species diversity, particularly the floristic diversity. It constitutes a geographical 'gateway' for much of the India's flora and fauna, consequent to which, the region is one of the richest in biological values. Around 8,550 floral species have been recorded, of which 2,500 are from Manipur, 2,250 from Nagaland, 2,200 from Mizoram, and 1,600 from Tripura. Horticulturally, *Mizoram* has a rich and unique bamboo genetic diversity (covering 12,544 sq km of the state), contributing to 14 percent of the national bamboo production and *Tripura* is the second largest rubber producer in the country.

The region is predominantly inhabited by Mongoloid, Indo-Mongoloid and Austric races. The main tribes, associated with agriculture and forests are *Angami, Rengma, Lushais* and *Riang.* The traditional farming system is *jhum.* In addition, the tribal communities widely practice terrace cultivation. A typical *jhum* system is a mixed cropping system with tuber crops such as aroids, ginger, vegetables, tapioca, and banana, planted or sown throughout the growing season. Also, a variety of temperate, subtropical and tropical vegetables, which include pea, carrot, chilli, onion, melon, spinach, cucumber, brinjal, tomato, Brassicas, yams and aroids are cultivated in home garden as well as in terraced field. In addition, tea, coffee, cardamom, coconut, arecanut, and rubber are the plantation crops grown mainly in Tripura. Other unique systems evolved are *Zabo farming system,* practiced in the Phek district of Nagaland, which is a composite farming system involving combination of forestry, agriculture, livestock, and fisheries; Alder (*Alnus nepalensis* D.Don) system involving planting of the alder trees to enhance the soil fertility for field or horticultural crops; the *swidden farming* system adopted by the Angami tribal community of Nagaland, involving cultivation of 15 to 30 crop species in the same plot of land after *jhum* forest clearing. Commercial cultivation of temperate fruits in gardens, such as plum, peach, and pear are seen in the higher elevations of Mizoram, Manipur, and Nagaland. Other temperate fruits, such as walnut, almond, and cashewnut are grown in Tripura. The plain and valley lands of Tripura and Manipur are suitable for tropical and subtropical fruits, such as banana, pineapple, citrus, coconut, mango, jackfruit, papaya, litchi, guava, etc. However, banana, pineapple, citrus, papaya, plum, peach, apple, etc., are also widely grown in the hills of Mizoram, Manipur and Nagaland. The cultivation of pineapple is concentrated in Manipur. The plantations of tea, coffee, cardamom, coconut, arecanut, and rubber are concentrated in Tripura.

For most of these crops, the region is important source of genetic diversity. In vegetables, the region is part of the primary center of diversity for cucumber (*Cucumis sativus*) with natural occurance of *Cucumis hardwickii.* In addition, significant genetic variability has also been reported for several cucurbits such as *Cucumis hystrix* Chak., *Luffa graveolens, Momordica cochinchinensis* (Lour.) Spreng, *Trichosanthes ovata* Cogn., *T. khasiana* Kundu, etc. (Sirohi *et al.,* 2005) providing a reservoir of useful genes. In solanaceous vegetables, the Indo-Myanmar region is considered as the centre of domestication of brinjal, *Solanum melongena* with rich diversity in form of varieties having excellent quality of soft flesh, less seeds and large fruit size. Additionally, *S. torvum* Sw., *S. indicum* L. and *S. khasianum* C.B. Clarke, possess resistance to shoot and fruit borer, and root diseases respectively (Kalloo *et al.,* 2005). *Solanum khasianum* is an important species of medicinal value (solasodine content) and so is *S. torvum,* extensively used in the Ayurvedic medicine system. Three tomato varieties namely *Manileima, Manikhamnu* and *Manithoibi,* developed using local landraces, were released by the State Variety Release Committee, Manipur, and found suitable for rice-based cropping systems. *Lycopersicon pimpinellifolium* (L.) Mill. naturalized to the region, has also expressed resistance to late blight and tomato leaf curl virus (Seshadri and Srivastava 2002). Tribals grow a vegetable having red tomato-like fruits slightly bitter in taste belonging to the genus *Solanum* and related to the brinjal. In Manipur, another kind of brinjal, having roundish fruit and intermediate in appearance between tomato and brinjal, is grown. Chillies grow well in the warm to hot and humid climate of Manipur, Mizoram, Nagaland and Tripura. They have long history of cultivation, out-crossing has evolved large genetic diversity in the form of a number of local landraces, with great variability for fruit shape, size, colour and bearing, semi-perennial, perennial habit, and pungency. In spices, the region is known for variability in *Piper nigrum* L. (Lam. Ex Link), ginger and in turmeric. Ginger variety Nadia, is suitable for the region. Lakadong variety of turmeric, found in the region has high curcumin content (7.4 per cent). A unique type of ginger having rhizomes with a bluish-black tinge inside, called black ginger is grown by the inhabitants of Mizoram. Wild relatives of large cardamom and cinnamon are available in the forests.

Among tropical fruits, the region harbors a significant amount of variability in mango, with the dwarf and late-maturing polyembryonic type cultivar, Moresh, which bears sweet fruit with high pulp content and starts fruiting within 2 years from planting, and is free from stone weevil. Distribution of *Mangifera sylvatica* Roxb. also extends to the region. Being the home of several *Citrus* species, *Citrus indica, C. macroptera, C. aurantium* L. and *C. reticulata,* the region presents rich genetic diversity (Bhattacharya and Dutta, 1956). The Indian wild orange *C. indica* is found in the Naga Hills, whereas, lemon, *C. lemon,* is known with a large number of traditional cultivars, such as *Hill lemon, Assam lemon, Nepali oblong,* etc. In addition, a number of tropical and subtropical minor fruits belonging to the genera *Garcinia* L., *Artocarpus* Frost., *Phyllanthus* L., *Annona* L., *Averrhoa* L., *Persea* Mill., *Aegle* Dulac, *Passiflora* L., etc., grow wild in the region. Jackfruit grows abundantly in Tripura. Among temperate fruits available locally are *Malus baccata* (L.) Borkh. Loisel used widely as rootstock of apple, *Pyrus pashia* Buch. and Ham. ex. D. Don, a common rootstock of pear. *Pyrus pyrifolia* Nakai var. *cubhamakai* (*P. serotina* Red) is grown semi-commercially in Manipur. Two species of silverberry *Elaegnus latifolia* L. and *Elaeagnus pyriformis* Hook.f., are known to grow in this region and are edible and used for making a refreshing drink (Pandey, 2002). The fruits of *Docynia indica* Decne., and *D. hookeriana* Decne., acidic, greenish with a red-tinge, are eaten fresh and used in pickles, as well as in jelly preparations. The region offers rich floristic diversity, particularly orchids. Some of the prized orchids occurring in this region are Red Vanda, Blue Vanda, Slipper orchids and Jewel orchids. Further, the local communities of the region have generated rich knowledge about the medicinal value of a large number of indigenous plants. Singh and Singh (2008) listed around 100 plants with medicinal value from Manipur alone.

6. Arid Western Region and Semi-arid Kathiawar Peninsula

The Arid Western region is one of the twelve biogeographical provinces of India (Roger and Panwar 1988). It includes the western part of Rajasthan, and parts of the southwestern Haryana and the Kutch peninsula of Gujarat. There are three principal landforms – the predominantly sand-covered Thar, the plains with hills including the central dune-free part of the region, and the semi-arid area surrounding the Aravalli ranges. It represents a typical hot arid climate characterized by hot summers and cool winters with mean annual rainfall of around 300 mm. The soils are generally sandy to sandy-loam in texture (Sehgal *et al.,* 1992). Floristically not very rich, but being the meeting point of the western and eastern flora, besides the cosmopolitan and tropical species, it has predominance of western elements (65.8 per cent) and eastern elements (34.2 per cent) with an overall dominance of the African element (37.1 per cent) compared to the Oriental elements (20.6 per cent) adapted to extreme agroclimatic conditions. Consequently it is an important center of genetic diversity of plant species domesticated in Africa, West and Central Asia.

It is inhabited by Indo-Aryan and Dravidian races. Protohistory reveals that it is a part of the Indus Valley civilization where settled agriculture evolved from nomadic herdsmanship and has passed through diverse phases of evolution. The main tribes associated with agriculture are Bhils, Rabari, Minas, Garasias and Bhishnois. Recognising the scarcity of water, the locals have developed diverse rainwater harvest and storage systems, such as *Tankas,* traditional stepwell, called '*Vav*' or '*Vavadi*' in Gujarat, or '*Baoli*' or '*Bavadi*' in Rajasthan, and *khadin* (dhora). Because of extreme dry conditions, cultivation of trees and grasses, and their intercropping with food and vegetable crops or fruit trees are the most viable models, integrating multipurpose trees. For sustainability of agriculture productivity and environmental balance, practices, like planting of multipurpose *khejri* (*Prosopis cineraria* Druce.) at the rate of 40 to 100 trees per hectare or Marwar teak (*Tecomella undulata* Seem.) developing wood land and mixed cropping of watermelon, *Citrullus lanatus* (Thunb) Matsum. and Nakai with pearl millet are common.

In horticulture crops, the region contains rich genetic diversity for cucurbitaceous vegetables of African/Asian origin and arid fruits, particularly for drought hardiness, sweetness and better shelf life in *Citrullus lanatus, Cucumis melo var. momordica* Duthie and Fuller. For watermelon, it is an important center of diversity with significant variability in fruit characterstics and wild *Citrullus colocynthis* (L.). In *Cucumis sativus*, it presents specific variability for small (tender) fruit size, drought tolerance and yield. Using this genetic variability, a number of varietal products has been developed like Durgapura Kesar and Durgapura Meetha in watermelon; Durgapura Madhu and Akra Rajhans in muskmelon; and Pusa Do Mausmi in *Momordica charantia* (Sirohi *et al.,* 2005). This diversity is further enriched by the wild relatives, *Cucumis prophetrum* L., *Momordica balsamina* Linn. etc. In fruits, variability is known in mandarin (*Citrus reticulata*) with Kinnow mandarin, sweet orange (*C. sinensis* L. Osbeck) with mosumbi and introduced Malta Blood Red. Other important fruits with significant genetic diversity are *Ziziphus mauritiana* Lam., with local cultivars like *gola, seb* and *mundia; Prosopis cineraria* (*khejri*), with wide diversity for vegetative growth, yield and quality attributes such as cluster bearing, larger pod size, sweet taste and precocity; *Punica granatum* L. (pomegranate) with varieties like *Jodhpur Red* with wider spread, and *Jalore seedless* with greater fruit size and yield (Vashishtha *et al.,* 2005); date palm, believed to have been introduced in the 4th century BCE by a solider of Alexander. Besides, rich diversity is recorded in 'arid fruits', like cactus pear (*Opuntia ficus-indica* (L) Mill.), *Capparis deciduas* Edgew., *lasoda* (*Cordia myxa* Roxb.) and *phalsa* (*Grewia subinaequalis* DC.), which are eaten and used for preparations like pickles, etc. This diversity is further enriched by the wild species, like *Capparis deciduas, Ziziphus nummularia* (Burm.f.) Wight and Arn.; (syn. *Z. rotundifolia* Lam.) and *Z. mauritiana* Lam.; (syn. *Z. jujube* Mill.) (Arora and Nayar, 1984).

The region is also known for rich genetic diversity in seed spices, like coriander, cumin and fennel. A number of varieties, like Rcr41, Rcr435, Rcr436, and Rcr20 and Rcr 684 through selection and mutation respectively in coriander; RZ19 through recurrent selection in UC-19 in cumin; RF101 through recurrent selection in a local germplasm in fennel, and RMt1 through pure line selection in Nagpur local, and RMt303 through mutation in RMt1 have been developed in fenugreek (Ravindran *et al.,* 2005). Further, the region is credited for development of dehydrated products from arid fruits and vegetables, like *chhuhara* from date palm, *sangria* from khejri, *anardana* from pomegranate, Chinese date from '*ber*', and dried *methi* from fenugreek for round-the-year use. In addition, pickles and powder are prepared from almost all fruits for use as culinary supplements.

The Kathiawar peninsula lying between the Gulf of Kuch and the Gulf of Khambhat, including most of Gujarat, and some parts of Rajasthan and Madhya Pradesh are considered the southern extension of the Indus Valley Civilization in continuation of western arid region. Agroclimatically, also it is a continuam of hot arid ecoregion and semi-arid ecoregion with desert saline, and deep to medium black soils. The climate of the area is characterized by hot and dry summers and mild winters with an annual rainfall between 600 to 900 mm and growing period ranging from 90 to 150 days. The landscape consists of main land as plain but with low hills, valleys, forests, and mangroves in the coastal area.The common soilscape is gentle to very gentle slopes with loamy to clayey and medium to deep black soils with inclusion of shallow and medium black soils. The soils are slightly alkaline, and calcareous, and show typical swelling and shrinking properties. Thus despite being ecologically different with comparative greater moisture availability, for horticultural diversity, it can be considered an extension of western arid region, grown under semi-arid conditions. Floristically, the natural vegetation is xeric scrub. Gir Hills, occuping the south-central portion of the peninsula are home to the tropical dry broadleaf forests, as part of the Kathiawar-Gir dry deciduous forests, where climate is tropical monsoon type.

The tribes associated with agriculture in this area are Barda, Bhil, Bavacha, Gond, Koli, Paradhi, Vafhri, Kathodi, Siddis, Kolgha, Kotwalia and Padhar. Like western arid region, the most conspicuous horticultural crops of the peninsula are fruits, such as mango, sapodilla, guava, citrus, ber, banana, pomegranate, date palm, coconut and several minor fruits, vegetables, such as onion, aubergine, cluster bean, okra, tomato, cabbage, cauliflower, chilli and garlic and spices like cumin, turmeric and coriander. Around 270 plants including indigenous and introduced species are used in rural medicine. In continuation of western arid region, the area has recorded significant variability in cucurbits and chilli. In muskmelon, varieties such as Gujarat muskmelon 1 and 2 from landraces obtained from Sabarkantha and Nagpur respectively have been developed (Sirohi *et al.,* 2005). In kitchen gardens, cucurbits offer significant genetic diversity, represented by traditional varieties, like *dudhi* or *nare* in bottle gourd, *dodku* or *gulka* in sponge gourd, *kodhu* or *dangar* in ash gourd, *kachra, kakri* or *kadu* in melons with unique taste; among fruits, in mango by varieties like, *Alphonso, Kesar, Rajapuri, Vanraj*. Besides, the region is very rich in diversity of arid-fruits, such as *Cordia myxa*, where a long-fruited local cultivar Paras is well known; Custard apple, *Annona squamosa* L., is known for variability, where Sel-9 has the highest fruit weight (154 g). Rich variability is seen in semi-arid fruits such as *Ziziphus mauritiana* and *Grewia subenaequalis* (Vashishtha *et al.,* 2005). Among spices, turmeric Uganda is developed through clonal selection in local germplasm; Guj Cor1 is a selection from local germplasm and Guj Cor 2 is a reselection from Co2 in coriander; MC43 is a selection from a local landrace, Gujarat cumin 1, while Gujarat cumin 2, is a selection from UC-19 in cumin; PF-35 is a local selection, whereas Gujarat Fennel 1 and 2 are pure line selections from local germplasm in fennel; and Gujarat Methi1 is a pure line selection from J. Fenu 102 in fenugreek, (Ravindran *et al.,* 2005).

7. The Malwa Plateau Region

The Malwa Plateau is another extension of the Indus Valley civilization. It is part of the Central Highlands of India and covers the western part of Madhya Pradesh and parts of southeastern Rajasthan. Geologically it represents a unique landscape that has a volcanic origin and comprises of round base hills with tops interspersed by the plains. Agroecologically, the area belongs to the hot semi-arid ecoregion with medium and deep black soils. Therefore, the climate of the area is characterized by hot and dry summers and mild winters (Sehgal *et al.,* 1992). Rainfall ranges from about 800 mm in the west, to about 1000 mm in the east with growing period between 90 to 150 days. The common landscape is represented by gentle to moderate slops, consisting of black, brown and bhatori (stony) soils. The region is rich and unique in floristic diversity. As it experiences a tropical climate, the natural vegetation is of the savanna type, represented by tropical dry deciduous forests, scattered with teak (*Tectona grandis* L.f.) and sal (*Shorea robusta* C.F.Gaertn.) and mixed forests.

Agriculture has been very ancient in the region, dating back to 1,500 to 1,200 BC. The region was inhabited by the Dasharn, Dasharh, Kuntal, and Charman tribes and is home to the Mali community, the cultivators of horticulture crops. The Malis, Sainis, and Kushwahas are involved in farming, gardening, horticulture and vegetable growing and selling. Rainfed farming is the common agricultural practice. Historically, the region had plenty of ground and river water supporting cultivation of a large number of horticulture crops. In addition, the locals have developed irrigation systems like reservoirs- primary surface tanks and ponds, *in situ* storage following unique systems like *pat* for rainwater harvesting (Bhitada village of Jhabua district) and inundation irrigation to facilitate cultivation. Opium (*Papaver somniferum* L.) is a traditional horticulture crop, for which the region can be credited for cultivation and introduction to other parts of the world. The tribal areas are very rich in genetic diversity. Among the vegetables, the region has recorded significant variability in *Solanum melongena* for fruit shape, size and color. Similarly high variability was recorded in *Capsicum* fruit color, shape, size, and pungency (Kalloo *et al.,* 2005). In cucurbits, significant variability was recorded and used in developing varieties, like Pusa Nasdar in ridge gourd from a local landrace from Neemach (Sirohi *et al.,* 2005). In okra, rich diversity is represented by wild relatives, like *Abelmoschus tuberculatus* Pal and Singh, *A. manihot* ssp *tetraphyllus* (Hornem.) Borss.Waalk. var. *megaspermus* (large seed), *A. crinitus* Wall., and *Abelmoschus ficulneus* (L.) Wight and Arn., and used in breeding program for traits such as higher yield (Nerkar 1991). The region also offers variability in sweet potato. In seed spices, in ajwain (*Trachyspermum amni* L. Sprague), a small type is being cultivated around Indore.

In fruits, the region has variability in mango, found growing wild with variability in temperature insensitivity and fruit size (large fruited Gadhamar from Pithwara). In addition, several traditional varieties, such as *noorjahan, dasheri, bombai, langra* and *sunderja* are known from the region. The native fruits such as bael [*Aegle marmelos* (L.) Correa ex Roxb.] and Almondette (*Buchnania lanzen* Spreng.) offer large diversity for panicle, fruit size and quality of kernel. *Annona squamosa* (in Sihore), *Embalica officinalis* Gaertn. (in Stana and Panna), *Ziziphus mauritiana* (in Murraina, Bhind, Hoshangabad), *Manilkara hexandra* Dubard (in Mandwa), *Citrus lemon* (L.) Burm. F and *Carica papaya* L. for variability in flesh colour and medicinal properties respectively (in Betul and Badwani). *Syzygium cumini* (L.) Skeels presents significant variability. Leaves and flowers of Mahua (*Madhuca indica* J. F. Gmel.) of Jhabua are geographically associated for extraction of the local brew called *mahua*. In addition, tribals have developed valuable information about the medicinal potential of a number of local plant species, including opium used as a medicine from time immemorial, to cure diarrhea and calm colicky babies. Such knowledge can be used to validate and improve the information for better health care of the rural people (Yadav 2006).

8. The Bundelkhand Region

Bundelkhand region lies between the Indo-Gangetic Plain and the Vindhya Ranges. It is an old landmass composed of horizontal rock beds resting on a stable foundation. The landscape is rugged, featuring undulating terrain with low rocky outcrops, narrow valleys and plains. Surface rocks are predominantly granite of the lower Pre Cambrian/Archaen period. The landscape is a gently sloping upland, distinguished by barren hilly terrain with sparse vegetation, although historically it was thickly forested. Prevailing soil types are a mix of black and red; the latter being relatively recently formed, is gravely and shallow in depth, and unable to retain moisture. For this reason, much of the region suffers from acute ecological degradation due to soil erosion and deforestation. Soil erosion is aggravated by the hilly landscape, high winds and the poor holding capacity of the soils, leading to the widespread growth of gullies, thus making it a complex, diverse, and vulnerable agrarian region, which is socioeconomically heterogenous and ethnically unique? Ecologically, it is a semi-arid to hot sub-humid ecoregion. The climate is characterized by hot summers and mild winters (Sehgal *et al.,* 1992). The annual rainfall ranging between 838–1,251 mm, which has been declining further, is confining to around 40 days, and around 20 hrs. Consequently most of it is lost in runoff, limiting the moisture availability and growing period to 90 to 150 days. It was very rich in floristic diversity, but today, it primarily consists of scrub forest *siris* [*Albizia lebbeck* (L.) Benth., *A. procera* (Roxb.) Benth.], *katai* (*Flacourtia indica* (Burm.f.) Merr.), *gunj* (*Abrus precatorius* L.), *bel* (*Aegle marmelos*), *ghout* trees [*Ziziphus xylopyra* (Retz.) Willd], etc. and scrub brush.

The region is part of Indian ethos from ancient times. The major tribes are Biar, Biyar, Saur, Sawur, Sonta, Soner, Kol, Manjhasi, Mawasi, Agaria, Bhaini, Dhanuk, Saharia, and Bedia. Not withstanding the large number of streams, the depression of their channels and height of their banks render them unsuitable for irrigation. Irrigation is being conducted by means of ponds, tanks, and structures such as lakes and surface-reservoirs constructed, taping the water of many streams, exploiting the sloping topography and building embankment. The traditional farming is rainfed. However, the region has evolved third crop season called '*zaid*', taken in the river beds on residual moisture, predominantly cultivating watermelon, muskmelon and vegetables raised along the dried river beds, and traditional gardens, called *baughs* for fruit cultivation. In vegetables, the region is known for variability in solanaceous crops such as *Solanum melongena*, and landraces such as *Bundelkhand Desi* with tolerance to drought are known nationally (Rai *et al.,* 1993). Rich genetic diversity was recorded among the minor arid fruits. For example, in bel (*Aegle marmelos*) traditional varieties such as *Kagzi Etawah* are known, whereas in *chironji* (*Buchanania lanzan*), diversity has been recorded for panicle, fruit size and quality kernels. Diversity has also been recorded in *aonla* (*Emblica officinalis*), ber (*Ziziphus mauritiana*), and *karonda* (*Carissa carandas* Lour.) (Vashishtha *et al.,* 2005).

9. Upper Gangetic Plains

The Gangetic Plains are watered by perennial rivers like Ganga, Yumuna, Ghagra, Gomati, Gandak, Son and several tributaries with a long stretch of sedimental land spread from foot hills of Western Himalayas to Delta in Bengal in east. It is one of the centres of origin of agriculture and genetic diversity of cultivated species. However, as per the differential ecology and development, it can be divided into three regions- the upper, the middle and the lower Gangetic plains. The Upper Gangetic Plains (UGP) is drier and more arid; with extreme climate and homogenous soil and evidences of very early beginning of agriculture, traced to Neolithic period. In fact, the foundation of agriculture in India was laid here by the early farming communities. It falls between the Western Himalayas in the north and the hills and plateau in the south and comprises the major portion of Haryana and Uttar Pradesh

(UP). The dominant landscape is plains with gentle slope. Ninety-five percent of the eco-region is degraded and converted into agriculture with settlements of dense human population. The landscape of Gangetic Plains based on physical structure and distribution of alluvium are often classified into four divisions, *Bhabar, Terai, Bangar* and *Khadar*. Agro-ecologically, it is characterised as hot sub-humid eco-region with alluvium-derived soil and climatically, by hot to warm summers and cool winters. The region receives a mean annual rainfall of around 1000 mm. As the monsoons are southwest originating in the Bay of Bengal, the UGP is comparatively drier than middle and lower Gangetic plains. The soils of the region are generally deep loamy and have developed on alluvium. The region is rich in floristic diversity. The natural vegetation in the region is tropical dry deciduous forests.

The agriculture in the region is very old dating back to the Vedic Age (1500 BC-600 BC), from the time Aryans settled down as full-time farmers and brought large tracts of fertile land of the region under the plough. Mixed cropping is a traditional cropping system including vegetable crops. The region is an important centre of horticultural genetic diversity because it offers diverse agro-ecological niches for adaptation. In vegetable brinjal (*Solanum melongena*), landraces like *Ramnagar Baingan* (long fruit type is suitable for Ganga River belt), *Dudhiya Baingan* (cluster bearing fruit type), *Jethuwa Baingan* (suited for summer plantation), Haryana Brinjal 14, *Kuchabuchia Baingan* (small cluster bearing type), *Jafrabadi* (deep purple, round oval type with less seeds), *Balfahwa Jathuwa Bhanta* (tall, perfuse branching, cluster bearing, better shelf life type, suited to summer cropping) present great variability. Similarly, in chilli, there is significant variability in fruit shape, size and colour (Kalloo *et al.*, 2005). In Cucurbits, the region is an important centre of diversity for *Cucumis melo* and *Citrullus lanatus*. The round gourd *Praecitrullus fistulosus* (Stocks) Pangalo, a monotypic genus is nearly endemic, was domesticated here and so did the ash gourd *Benincasa hispida* (Thunb) Cong. The region has downy mildew and root rot wilt complex resistant types in *P. fistulosus*. The andromonoecious type in *Legeneria siceraria* from Faizabad and Shahjahanpur are important from breeding point of view. In several of these crops, varieties have been developed through selection from local landraces, for example, in muskmelon, Lucknow safeda from Lucknow local; in *Cucumis sativus* Kalyanpur Green from a local cultigen; in smooth gourd, *Luffa cylindrica* Roxb., Kalyanpur Hari Chikni; in ridge gourd, *Luffa acutangula* (L.) Roxb., Punjab Sadabahar; in *Momordica charantia*, Kalyanpur Barahmasi and Pusa Videsh from Hapur local; and in *Cucumis melo var. utilissimus* Duthie and Fuller, Arka Sheetal from Lucknow local (Sirohi *et al.*, 2005). In addition, a number of wild relatives of these crops, such as *Luffa echinata* Roxb., *Cucumis satosus* Cogn., *Momordica balsamina* L., etc. offer further variability for use in the breeding programs. In potato, the Farakhabad region offers great variability; Agra-red is derived from a local type (Shekhawat *et al.*, 2005). In okra, Mishra *et al.* (2000) reported wild species of *Abelmoschus* from *Terai* belt, in addition to *A. cancellatus* (L.f.) J.O. Voigt (Saharanpur), *A. manihot* (L.) Medik. (Behata and Mahoba), *A. manihot* ssp. *tetraphyllus* (Rampur, Nagina and Dhampur) and *A. tuberculatus* (Sharanpur). In onion, Pusa Madhawi was developed from Muzaffernagar local and Kalyanpur Red Round from Kanpur local (Pandey *et al.*, 2005). Variability has been recorded in some important spices, like coriander and fennel, and varieties, like Pant Haritima and Azad Dhania, and Azad Saunf 1, were developed through selection and mass selection from local landraces respectively.

The region represents rich genetic diversity in case of tropical and arid fruits, particularly mango, with varieties, such as *Bombay Green, Dashehari, Fazli, Langra, Safeda Lucknow, Smarbehisht, Chausa*, etc.; Guava, with wide range of variability in Allahabad, Mirzapur, Kanpur, Unnao and Fatehpur, with traditional varieties, such as *Lucknow 49, Allahabad Safed, Red fleshed, Chitidar, Banarasi, Harijha*, etc. have distinctive features (Mitra and Bose 1999). In arid fruits, like *Agle marmelos*, local cultivars, such as, *Darogaji, Ojha, Rampuri, Azamal, Khamaria, Kagzi Gonda, Kagzi Etawah* offer a lot of variability. In *aonla* or Indian gooseberry (*Embelica officinalis*), a lot of variability for fruit characteristics have been recorded in local landraces, such as *Hathijhool, Basanti red, Deshi, Chakaiya,* resulting selection and release of many varieties from Acharya Narendra Dev University of Agriculture and Technology. In *Zizyphus mauritiana* several forms are available in Western UP and Haryana, and varieties, like Gola Gurgaon, Sanaur 1, Safed Rohtak were developed. In *jamun* or Java plum (*Syzygium cumini*), the region has concentration of variability for fruit size, shape, pulp colour, TCS, acidity and earliness. Similarly, variability is recorded in *Grewia subinaequalis,* the phalsa (Vashishtha *et al.,* 2005). In lemon (*Citrus lemon*), variability has resulted in selection of varieties like Baramasia, Pant lemon and Gandhraj (Karihaloo *et al.,* 2005). Additionally, several biological practices for pests and diseases control using natural products like Neem cake/leaves/powder and spread of cow-dung ash to control the attack of powdery mildew disease etc. are known from the region; locals have developed unique by-products from horticultural species, for example, *Petha* made from ash gourd is geographically associated with Agra.

10. Middle Gangetic Plain Region

The Middle Gangetic Plain (MGP), stretches from west-central part of Indo-Gengetic plains towards Northeast, between Himalayan foot hills in the north to Vindhya ranges in south. Administratively, the area includes parts of eastern UP and parts of Bihar lying on either side of Ganga and Saryu (Ghaghara) within the Himalayan and peninsular rampart on the north and south respectively. It is a flat plain without a hill to break the monotony. The region is part of hot and sub-humid eco-region with climate characterised by hot summers and cool winters. The annual rainfall is between 1400 - 1600 mm, extending the growing period from 180 to 210 days (Sehgal *et al.,* 1992). The soil-scape is represented by alluvium soil, which is calcareous and moderately alkaline in nature. It has attained its present form, when the deep trough was filled with fine alluvium brought down from the Himalayas in the north with an average thickness of 1300 – 1400 meters. The alluvial soil cover is divisible into *khadar* and *bhangar* with patches of *usar.* The *terai* region is a pebbly *bhabar* zone. These alluvial flat plains are irrigated by the Ganges and its tributaries, like Gandak, Sone and Kosi. The original natural vegetation of the region contained moist deciduous forests in sub-Himalayan foothills, whereas, plains comprised remnants of tropical moist deciduous and dry deciduous forests, grasslands, and swampy vegetation in flood prone areas.

Agriculture is as old as in UGP, but with predominance of management of wetland/lowland agriculture practices. It has contributed addition of vegetables and fruits in the humen diet, as reflected by archaeological evidences. The interior areas of the region are still inhabited by a number of tribes with representation of Negrito, Proto-Australoid, Mongoloid, Mediterranean and Nordic races. Dravidian races inhabiting the tribal region are the Hos, the Santhals, the Oraons and the Mundas. The evidences gathered from Senuwar and Chirand, suggest that the Neolithic people of the region cultivated vegetables like field pea. The plant found at the Chalcolithic sites (Narhan) of the region, included jackfruit seeds (*Artocarpus integrifolious* L.), introduced from Western Ghats (Saraswat 1994) suggesting early cultivation of horticultural species. Agriculture is mainly rain-fed with vegetable crops grown are Cucurbitaceous gourds, *bhindi*, radish, carrot, beat root, tomato, cauliflower, cabbage, etc. and fruits such as bael, mango, litchi, jackfruit, banana, etc. India is the largest producer of litchi, *makhana*, guava and okra, and exports litchi. The available genetic diversity is in continuation from sub-Himalayan region and the UGP, as in brinjal and chilli (Kalloo *et al.,* 2005), and cucurbitaceous crops, like *Luffa* Mill., *Momordica* L., *Trichosanthus* L., *Lagenaria* Ser. etc. In *Legeneria,* varieties like Rajendra Chamatkar are developed through selection from local landraces; in ridge gourd, *Luffa acutangula* Singh and Bandhari var. *satputia,* with fruits in cluster is a unique variety endemic to the districts of UP and Bihar (district Bhagalpur); in sponge gourd, *L. cylindrica,* varieties like Pusa Chikni, Pusa Supriya, Rajendra Nenua-1, KSG-14 were developed from local collections. The *Trichosanthes dioica,* another endemic and geographically associated species presents important genetic variation, which has led in development of varieties Rajender Parwal and Rajender Parwal 1 through selection from Bhagalpur local. In muskmelon variety Akrajeet was developed from Basti local of UP (Sirohi *et al.,* 2005). In addition, genetic variability is represented by a large number of wild species, such as *Luffa echinata* Roxb., *Momordica cochinchinenesis, Momordica subangulata renigera* (Wall. ex G.Don) and *M. dioica,* and more. In okra, the region presents maximum variability in fruit and plant morphology along with the wild relatives like *Abelmoschus crinitus.* In tuber crops, 1502 accessions of different tuber crops were collected and maintained at Dholi center. Significant variability has been collected in sweet potato resulting in development of a series of varieties under the name Rajendra Sakarkand 5, -35, -43, -47 and Kiran (Edison *et al.,* 2005). In spices, the region is known for genetic variability in turmeric and ginger. In turmeric short duration variety Rajendra Sonia was derived from a local collection with bold and plump rhizome. In seed spices, the fenugreek variety Rajendra Kranti is a pure line selection from a local collection of Raghunathpur, while variety Rajendra Abha is developed from local high yielding *'Champa Methi'.* In garlic, variety Agrifound White (G41) was developed from a collection from Bihar Sharif of Nalanda district. In coriander, Rajendra Swati is a selection from Muzafferpur collection (Ravindran *et al.,* 2005). Betel vine is also cultivated on commercial scale with significant genetic diversity, represented in the local germplasm like, *Calcuttia, Bangia* with longer internodes, *Ghana Ghatte, Kapoori Bihar* and *Hara Patta* with tolerance to fungal pathogens.

The region is known for the rich variability for several tropical fruits, particularly mango, litchi, aonla, bael, jackfruit, banana, etc. The wild forms in mango are known from Bahriach and Gonda districts. The common varieties of mangoes are *Bathua, Bombai, Himsagar, Kishen Bhog, Sukul, Langra, Sundar Pasand, Fazli, Gulabkhas, Mahmood Bahar* and *Zardalu,* which are being used in state breeding program. Litchi, which has limited distribution in North Muzaffurpur, Darbhanga is known for varieties, like *Bedana, Calcuttia, Purbi, Kasba, Desi, Early Bedana, Shahi, Rose-Scented, Late Bedana, Mandaraji, China,* etc. Similarly, in Bael, the region has several local varieties, likes *Mirzapuri, Ojha, Azamati, Khamaria, Sewan large,* etc. with unique characteristics. In aonla varieties like Banarasi were developed from a local landrace, while banana is known for traditional varieties, like *Dwarf Cavendish, Poovan,* and *Rasthali,* and *Citrus* for *Gola Kagazi.* The *Carissa carundas* also expresses variability for fruit and plant type. The indigenous people have also shown ingenuity in developing by-products of vegetables and fruits and several are geographically associated with region, like *'parval'* sweet from pointed gourd.

11. Lower Gangetic Plain or Delta Region

The Lower Gangetic Plain (LGP) is the extension of the MGP to the southeast, with greater humidity. It is a *Riverine Delta* created as a result of the confluence of the two biggest rivers Ganga and Brahmaputra before going down into the Bay of Bengal. Its major portion falls in Bangladesh and West Bengal. Landscape of the region is the vast river system with three major landforms, the Uplands, Old Aluvial/Deltaic Plains and Young Aluvial Plains (Singh *et al.,* 1998). The common soil-scape is represented by level to very gently sloping Haplustalfs/Sapludalfs (Sehgal *et al.,* 1992). Ecologically, it can be classified into: Tropical humid, Tropical moist humid and Tropical sub-humid ecoregion. Overall climate is characterised by the hot summers and mild winters. The region receives high rainfall both during summer and winter seasons, ranging from 1400 – 2000 mm or even more, with growing seasons ranging from 150 to 270 days. Floristicaly, it is very rich with semi-deciduous vegetation, with upper canopy containing the deciduous species, while the second story is dominated by evergreen species.

It is the largest flood plain (19,389 km2) of the world and the most fertile sub region of Gangetic Plains. Predominantly, fertile alluvial plains are man-made, cleared of forests and are intensely cultivated.

It is now agreed that the foundations of the agriculture-based village life, believed to be one of the foundations of Indian civilization, were laid by the Nishadas or Austric-speaking peoples in LGP region, at least 2000 years before the beginning of the Gregorian (Western) calendar. Aryanization extended to LGP from about the second millennium BC. The aborigins were basically different Dravidian tribes, evolved with migration and integration of diverse racial elements: Northern Indian Aryan longheads, Alpine shortheads, Dravido-Munda longheads and Mongolian shortheads.The main professions in the Ancient Vanga's were agriculture and boating. The availability of moisture throughout the year has made horticulture, including floriculture an important component of agricultural cropping and/or production systems. Horticulture is ancient, as inscriptions from 8th century onwards record the plantation of both betel nuts (*guvaka*) and betel leaves and cultivation of coconut (*narikela*). The region grew a number of vegetables, the *Paryayaratnamala* refers to two varieties of garlic, onion with two forms- red and white, *patola/parval* (*Trichosanthes dioica* Roxb.), *olla* (*Arum indicum* Lour.), *sobhanjana* (*Moringa pterygosperma* Gaertn.), *Kecuka* (*kacu*, a variety of *kanda*), *sthalakanda* (*Vena olakacu*), *ervaruka* or *karkatika* (*kahkunda*), beans- white and black (simba), pumpkin, *karavellack* (bitter gourd), *kusmandake* (a gourd) and *vartaku* (brinjal) etc. and spices like cumin with three varieties, black, white and small; cardamom with two and coriander. Whereas, *Saduktikarnamrta*, refers to *javani*, *satapuspika* (aniseed); and *kustumbari* (coriander), *ela* (cardamom) was referred in *Ramcharita* (Niyogi 2008) and evolved a large amount of genetic variability. *Solanum melongena* was domesticated in neighbouring Indo-Mayanmar region, presents significant level of variability. In Okra, it is considered one of the major centres of variability for fruit and plant morphotypes along with the wild relatives, *Abelmoschus manihot* and *A. crinitus*. The variety Pusa Makhmali, parent of Pusa Sawani, the most popular Indian variety was derived from a West Bengal landrace (Dhankhar *et al.,* 2005). In cucurbits, *Trichosanthes anguina* L. is wide spread, *Momordica cochinensis* is very popular, while *M. dioica* is cultivated. In spices, the region is known for genetic variability in ginger and turmeric. Betel vine is a traditional commercial crop of the region.

Some of the supplementary crops grown in the plains are aroids and cucurbits among vegetables and banana, coconut, mango, papaya among fruits. The *Ramacharita* (Nandi 1939), mentions many trees, flowering plants and medicinal plants, including mango, which has large number of traditional varieties, such as *Bombai, Himsagar, Kishen Bhog, Langra, Malda*; banana presenting traditional varieties, such as *Dwarf Cavendish, Poovan,* and *Rasthali*; and Bael like *Ojha, Azamati*; whereas, Jackfruit (*Artocarpus heterophyllus* Lam.) and *Carissa caronda* of the region are known for their quality fruit and plant characteristics.

12. The Chotanagpur Plateau Region

The Chotanagpur region is a plateau in eastern India, covering much part of recently created Jharkhand state, as well as bordering areas of Orissa, West Bengal, Bihar and Chhattisgarh. Ecologically, the region lies between the moist deciduous forests of the Eastern Ghats and Satpura Range and the lower reaches of the Gangetic Plains. It is composed of Precambrian rocks which are more than 540 million years old. Geohistorically, it was part of the Deccan Plateau, which broke free from the southern continent during the Cretaceous period. The Gondwana substrates attest ancient origin to the plateau. The region has undulating topography presenting a highly dissected landscape of small hillocks and mounds. The dominant area is represented by moderate to gentle slopes with numerous streams dissecting the uplands into a peneplain. Based on altitude, it is divided into, the uplands, medium lands and lowlands. The climate of the plateau is humid and sub-humid, characterised by hot summers and cool winters. The average rainfall varies from 1000 mm to 1600 mm. The cultivation period ranges from 150 to 180 days (up to 210 days in some places). The plateau has predominantly red soil that is derived from peculiar rock formations. Soil content of the region mainly consists of components formed from disintegration of rocks and stones. They can be divided into: red soil, micacious soil, sandy soil and laterite soil (not suitable for agriculture). The region is very rich in floristic diversity as it links the species diversity between Satpura Hill Ranges and Eastern Himalayas (Hora 1949). The forest vegetation is made up of, Tropical Moist Deciduous Forests, Northern Tropical Dry Deciduous Forests and Central Indian Subtropical Hill Forests (Champion and Seth 1968), but today common vegetation is represented by dry scrub forests.

Agriculture is very old in the region, as the earliest settlements of Chalcolithic period extended from Ganga basin. The region is known for non-Aryan, Austrics tribes, which can be classified on the basis of their agricultural activities, starting from hunter-gatherers to settled agriculturists, like *Mundas* and the *Oraons.* Later are believed to have first introduced plough cultivation. Rainfed agriculture is the traditional way of farming, under which the tribal farmers have evolved, a topo-sequential land use, cultivating crops as per fertility and moisture availability. During *zaid* season, which begins from March, ends by the second week of May, people grow mostly vegetables, like *kadu, kohra, bhindi,* French beans, etc. Under agro-forestry, the mixed cropping involves vegetables with cereals and pulses. A number of horticulture crops are grown as per the topography and climatic regimes. Litchi cultivation in Ranchi area and jackfruit all over plateau is nationally known. For cultivation of horticultural crops, locals have developed low cost structure called '*doba*' for conservation of water (Dey *et al.,* 2003). The Plateau has recorded significant amount of variability for fruit characteristics, such as shape, size and colour in Solanaceous vegetables like brinjal and cucurbits like, cucumber, ridge gourd, pointed gourd, etc., used in developing many of the present cultivars. Among the spices, significant variability has been recorded in ginger and turmeric. Some indigenous ginger (*Zingiber officinalis* Rosc.) cultivars are *Maran, Kuruppampadi, Ernad, Wynad, Himachal* and *Nadia.* The region is very rich in the diversity of ornamentals, particularly, the commercially important tropical orchids, represented by a large number of species, for example *Dendrobium*–a group of epiphytic orchids comprises of 11 species is distributed throughout the altitudinal gradient of the plateau (Kumar *et al.,* 2011).

The region is known for important traditional varieties of mango *i.e. Bathua, Bombai, Himsagar, Kishen bhog, Gopal bhog, Sukul, Rani pasand, Safed maldah, Chausa, Fazali, Zardalu,* etc. extending from middle and lower Gangetic plain. In litchi, which is cultivated extensively in the region appreciable genetic diversity has been recorded resulting in collection of 51 accessions (Karihaloo *et al.,* 2005). Some commonly known extant varieties are *Shahi, Rose scented, China, Purbi, Early bedana, Late bedana.* Rich genetic diversity has been recorded in Jackfruit (*Artocarpus heterophyllus*) in the Santhal Parganas for tree characters, fruit behaviors, fruit characters and yield (Nath *et al.,* 2001). Similarly, *Carissa caronda* exhibits variability for fruit and plant type. The tribals have significant amount of knowledge about the medicinal properties of a number of plants and alternative sources of food, like young bamboo shoots and a number of root and tuber crops, which are mostly harvested from nearby forests, leading to evaluation of vegetable cultivation under edible bamboo, for example intercropped with potato (*Solanum tuberosum* L.), tomato (*Lycopersicon esculentum* Mill.), pea (*Pisum sativum* L.) and ginger (*Zingiber officinalis*). The locals have been rearing, tiny lac insect, *Laccifer lacca,* whose sticky, resinous secretion deposits on the twigs and young branches of several varieties of soapberry and *Acacia* Mill., and *Ficus* L. trees, producing commercial '*lakh*'.

13. Bastar Region

The region consists part of the eastern plateau, including the areas of Dandakaranya. Presently, it consists of five southern most districts of the newly formed state of Chattisgarh: Bastar, Kanker, Dantewada, Bijapur and Narayanpur. The region is a hot sub-humid eco-region with red and yellow soils. The climate of the region is characterised by hot summers and cool winters. The annual rainfall ranges from 1200 to 1600 mm with growing period between 150 to 180 days in a year. The dominant landscape in the area is represented by moderate to gentle slopes. The soil is reddish, calcareous and neutral to slightly acidic. The region is very rich in floristic diversity. The natural vegetation comprises of tropical semi-evergreen forests, tropical moist deciduous forests, and tropical dry deciduous forests.

Though the region is crippled with backwardness and poor development in agriculture sector, recent archaeological excavation in Dantewada and Bastar caves have found implements, which potentially date back to over 50,000 years, reflecting agriculture being ancient in the region. But because of respect towards nature, primitiveness of life style and influence of Hindu doctrine and agriculture as the main source of income, about 58 per cent of land area are under forest; only about 19 percent land are under cultivation. Most of the Adivasis now practice some sort of agriculture; only those living in the interior areas, such as the Marias of Abujhmar still do shifting cultivation (called as *Marhan* or *Dippa* in district Kanker). There are seven major scheduled tribes in Bastar- *Gonds, Muria, Maria, Dhorla, Bhatra, Halba* and *Dhurva*. But, because of the richness of floristic diversity in forests, every advasis supplement his cultivated food with vegetables, tubers, fruits, seeds and leaves growing in wild. Jain (1964) based on an ethanobotanical field study during 1960 to 1963 found 88 plants eaten by the tribals of this region. They are classified under vegetables, fruits, nuts, drinks and beverages, grains, oilseeds, pickles, sweets and condiments. This publication includes notes on processing of foods and some food taboos. It lists species variability, which is rich in *Dioscorea* L. spp. (7), for *Diospyros* L. (2), *Ficus* (3), *Grewia* L. (4), *Zizyphus* Adans. (3), offering a reservoir of genetic resources. Otherwise, rainfed agriculture is the traditional way of farming, including horticultural crops, like vegetables both leafy and tuberous, fruits, medicinal and aromatic plants and multipurpose species, such as *Dendrocalamus strictus* Nees (paper pulp), *Diospyros melanoxylon* Roxb. (Beedi leaves), *Schleichera oleosa* Merr. (lac growing) etc. However, they employ very primitive methods of agriculture; hence the agricultural outcome is very low. The farming communities practice terrace cultivation, which evolves the diversity of the same crops in different terraces and facilitate natural cross-pollination among different strains generating unique intraspecific diversity. Farming communities, taking benefit of this system, have made selection in naturally occurring interbreeding populations between wild relatives and cultivated species, for example, in case of brinjal involving wild *Solanum* species. The major horticulture crops are cucurbits like, *Cucumis melo, C. sativus, Lagenaria siceraria, Luffa acutangula, L. cylindrica* and root and tuber crops, like Casava (*Manihot esculenta*), Greater Yam (*Dioscoria alata* L.), aerial yam (*D. bulbifera* L.), sweet potato (*Ipomea batata* (L.) Lam.), elephant foot yam [*Amorphophallus campanulatus* (Roxb.) Bl.], arun [*Colacasia esculenta* (L.) Schott], ginger (*Zingiber officinale*), turmeric (*Curcuma longa* L.), wild tubers (*Dioscorea* L. spp.), *pitkanda* or bitter yam (*D. dumetorum* (Kunth) Pax.), *Bharakanda* or Trifoliate yam (*D. pentakila*), *kulihakanda* (*D. hispida* Dennst); other useful tubers, *Tikhur* (*Zingiber roseum* Roscoe), *keaukanda* or Keth (*Costus speciosus* Sm.), *dhova* or Tamnia (*Xanthosoma* Schott Spp.), *vidarikand* or *didari* (*Pueraria tuberosa* DC.), making the region a very important centre of diversity for root and tuber crops, particularly the tropical and tuber crops, like *Dioscorea* spp. (yams). The other important vegetable crops are cucurbitaceous crops. The Native fruits Chironji (*Buchanania lanzen*) and Caronda (*Carrisa caronda*) offer rich diversity. Caronda has variability for fruit characteristics, particularly the flesh, having white and red colour (Vashishtha *et al.*, 2005).

14. Koraput Region

The Koraput region is a part of North-Eastern Ghats consisting of South-Western Odisha and North-Eastern districts of Andhra Pradesh (Vizagapatnam, Vijanagaram, Srikakulam). The whole region can be divided into four district areas separated by natural barriers, (i) Rayagada area consisting of two fertile valleys of the Nagavali and the Vansadhara slops, rising from a height of 1,300' above mean sea level (MSL) near Ambadala and goingdown to 260' MSL at Gunupur, (ii) Koraput area, from Kasipur to Vizagapatnam plains, 3,000' above MSL, (iii) the Nabarangpur Division extending to Bastar in the west, a plateau 2000' above MSL, with extreme North-East area known as the Pannabeda Mutha having the lowest MSL, some 500' and (iv) the Malkangiri Area. Thus mean elevations of region vary from 150 m to 1000 m. The average annual rainfall is 1521 mm, which is drained by five rivers, namely, Vansadhara, Nagavali, Indravati, Kolab and Mackanand. The undulating topography of the region includes hill peaks of different altitudes creating diverse climatic regimes with varied rainfall patterns. The region is a reservoir of floristic diversity consisting of around 2,500 species of flowering plants. The natural vegetation ranges from tropical semi-evergreen to dry deciduous forest with rich medicinal and aromatic plant species diversity represented by more than 1200 species.

The region is the home of ancient Austrics tribal communities of India *i.e. Gonds, Khonds, Santhals, Lahqulas* and *Kinnaras*. In Andhra, they are referred as Dravidian influenced by Aryan and Negroid cultures and called Adivasi, Girijan and Vanya Jati. The topographic heterogeneity of the region has resulted in a wide diversity in the ecosystems, recognising the region as site of one of the Globally Important Agricultural Heritage Systems (GIAHS). The diverse climatic regimes have made the region a repository of pristine of wild relatives of crops, crop genetic diversity and ethnic agricultural practices. Rainfed farming is the prevalent agriculture practice, including *Podu* cultivation (the slash and burn cultivation), joint cultivation (*pottu vyavasayam*) and Terrace cultivation. The major horticultural crops are vegetables, root and tuber crops, fruits, spices and medicinal plants. It is represented by vegetables, dolichos [*Lablab purpureus* (L.) Sweet], brinjal (*Solanum melongena*), chilli (*Capsicum annuum*), ridge gourd (Luffa *acutangula*), sponge gourd (*L. cylindrica*), bitter gourd (*Momordica charantia*), yam (*Dioscorea alata*), sweet potato (*Ipomoea batatas*); fruits, banana (*Musa balbisiana* Colla), jackfruit (*Artocarpus heterophyllus*), mango (*Mangifera indica*), aonla (*Emblica officinalis*), date palm (*Phoenix dactylifera* L.), karonda (*Carissa caronda*), pomello [*Citrus maxima* (Burm.) Merr.], *halwa tendu* persimmon (*Diospyros kaki* Thunb.) tamarind (*Tamarindus indica* L.); spices, *tejpat* (*Cinnamomum tamala* Nees and Eberm), turmeric (*Curcuma domestica* Valet.; syn. *C. longa* L.), pepper (*Piper nigrum*), ginger (*Zingiber officinale* L.), etc. In vegetable crops, the region has recorded high variability in *Solanum melongena* and chillies for fruit colour, shape, size and pungency (Kalloo *et al.,* 2005). Additionally, wild relatives, *Solanum indicum, S. incanum* Ruiz and Pav., *S. surattense* Burm.f., *S. viarum* Dunal and *S. pubescens* Ruiz and Pav., were recorded from the region (Singh and Varaprasad, 2008). In cucurbits, high variability was recorded in the form of a number of wild relatives, such as *kundru* (*Coccinia indica*), *Cucumis hystrix, Luffa acutangula, L. graveolense, Luffa umbellata* (Klein ex Willd.), *Momordica balsamina, M. cochinchinensis, M. dioica, M. tuberosa, Trichosanthes bracteata* Lam., *T. cordata* Roxb., *Trichosanthes multiloba* Miq., and *Trichosanthes himalensis* C. B. Clarke, (Sirohi *et al.,* 2005). Also, in okra the region presents maximum variability for fruit and plant types along with the presence of *Abelmoschus crinitus, A. ficulneus* (L.) Wight and Arn. (Dhankhar *et al.,* 2005). In root and tuber crops, the variability is enriched by the various yams found in forest areas, like *pitharu kanda* (*Dioscorea belophylla* Voigt ex Haines), potato yam (*D. bulbifera*), *pitta khanda* (*D. glabra* Roxb., *D. glabra var. vera*), *D. hispida, D. intermedia* Thwaites, *D. oppositifolia* L., *D. pentaphylla* L., *D. puber* Blume, *D. tomentosa* J.König ex Spreng., *tonga-alu* (*D. wallichii* Hook.f.), *D. wightii* Hook.f.; Indian Kudzu vine [*Pueraria tuberosa* (Roxb. ex Willd.) DC], *Vigna vexillata*

(L.) A.Rich. (edible root tubers) etc. In spices, the region offers rich variability in ginger having traits of adaptability, maturity, fibre content and dry matter recovery. Several varieties have been developed through clonal selection from local collections, like Suprabha and Suruchi from collections of Kunduli, Odisha and Surabhi from Rudrapur local. Similarly, in turmeric, Roma was derived from T.Sunder and Ranga from Rajpuri local through clonal selection; while Suroma was a mutant selection from T. Sunder (Ravindran *et al.,* 2005). In fruits, the region is known for traditional varieties in mango, such as *Baneshan, Langra, Neelum, Suvarnarekha* and wild species like *Mangifera indica, M. sylvatica* and in *jamun* (*Syzygium cuminii*), by its wild relatives, *S. alternifolia* and *S. zeylanicum.* In addition, wide range of fruits are being harvested from wild with appreciable genetic diversity in Jackfruit (*Artocarpus heterophyllus*), *Carissa inermis* Vahl, lime [*Citrus aurantifolia* (Christm.) Sw.], *makar kenda* (*Diospyros embryopteris* Pers.), *kendu* (*D. exsculpta* Buch.-Ham.), *tendu* (*D. melanoxylon* Roxb.), *D. racemosa* Roxb., *gudaluti* (*D. sylvatica* Roxb.), *tendu* (*D. tomentosa* Roxb.), *Elaeagnus latifolia* L., *kharpet* (*Garuga pinnata* Roxb.), *mirichari* (*Grewia tenax* Forsk.), wood apple (*Limonia acidissima* L.), *Emblica naryanaswami/ officinalis,* tamarind (*Tamarindus indica*) and *Ziziphus horrida* Roth. In addition, the region has many more genera of horticulture and ornamental value, *Argyria, Bauhinia* L., *Clematis* L., *Cyanotis* D.Don, *Cycas* L., *Dysophylla* Blume, *Habenaria* Willd., *Hardwickia* Roxb., *Oleo* etc. Like Bastar region, here also the farmers have taken advantage of naturally occurring interbreeding populations between wild relatives and cultivated species in often cross-pollinated crops like brinjal and pigeonpea. Bastar, Koraput, Southern Eastern Ghats and Kaveri region are the parts of eastern coastal region, considered one of the five major centres of distribution of tropical root and tuber crops (Edison *et al.,* 2005)

15. South Eastern Ghats

The South Eastern Ghats region constitutes southern parts of Andhra Pradesh, comprising of areas south of river Godavari and the Nallmalai-Palkonda ranges, Seshachalam hills, Rayalaseema, Tirupati hills, adjacent semi-arid districts, and the districts of Bellary, Raichur and Kolar in Karnataka. The region is a hot semi-arid eco-region with red loamy soil. The climate of the region is characterised by hot and dry summer and very mild winters with annual rainfall of 500 – 1000 mm and growing period extending from 120 to 150 days. The major landscape of the area is moderate to gentle slop. The soil is generally non-calcareous and slightly acidic in nature in some patches. However, in other patches, the soil is calcareous and moderately alkaline in reaction. Vegetation is represented by tropical dry deciduous and tropical thorn forests in the southern Deccan plateau, flanked by the moist deciduous forests along the lower elevations which receive south west monsoon and by thorn scrub in regions which receive north east monsoon. Dry districts fall on the lee-ward side of the southern Western Ghats mountain ranges.

The hilly areas of the Eastern Ghats in Andhra Pradesh are inhabited by large tribal communities, the major being *Bagatas, Chenchus, Jatapus, Khonds* (*Samantas*), *Kondadoras, Konda Kammaras, Konda reddis, Koyas, Lambadis* (*Sugali*), *Nuka doras* (*Muka doras*), *Porjas* (*Gadabas*), *Savaras* and *Valmikis.* Some Negritos tribes *Irulas, Kodars, Paniyans* and *Kurumbas* are also found in patches. Traditional farming is rainfed with horticultural crops being part of mixed cropping system. The main horticulture crops are vegetables, root and tuber crops, fruits, medicinal and aromatic plants and ornamentals. The main vegetables cultivated in the region are okra, onion, amaranths, pumpkin, cucumber, bitter gourd, brinjal etc. The region has recorded high variability in *Solanum melongena* represented by landraces, *Sanna vanga, Saara vanga, Tella mulaka, Tella vanga* and *Solanum* spp. *S. erianthum, S. nigrum, S. surattense* and in chillies for fruit colour, shape, size and plant type; in cucurbits by *Cucurbita pepo* L., *Cucumis melo var.agrestis, C. pubescens,* Cucumber (*Cucumis sativus*) and bitter gourd (*Momordica charantia*) and its wild relatives *M. balsamina and M. tuberosa; Vigna hainiana* populations were widespread in the region; in root, tuber and bulb crops like onion (*Allium cepa* L.) and its landraces, *Chinna erragada, Erra gada, Tella gadda,* elephant-foot Yam [*Amorphophallus paeonifolius* (Dennst.) Nicolson], potato Yam (*Dioscorea bulbifera*) and wild species, *D. pentaphylla* L., *D. intermedia, D. wightii, D. wallichii;* leafy amaranths, represented by *Amaranthus spinosus* L., *A. tenuifolius* Willd., *A. viridis* L., *A. dubius* Mart. Ex Thell.; and others like okra with variability for various desirable features related with fruit. In spices, significant variability was recorded in ginger and turmeric. The region is also known for variability in seed spices like coriander, where several varieties have been developed from local collections through mass selection. Sadhna was developed from Alour collection, Swathi from Nandyal collection and Sindhu from a local collection (Ravindran *et al.,* 2005). Among fruits, mango is the prime fruit crop with variability being represented by traditional varieties, such as *Banganpalli, Totapari, Cherukurasam, Himayuddin,* and *Suvarnarekha.* In addition, many more minor fruits, such as *Aegle marmelos,* palmyra or toddy palm (*Borassus flabellifer* L.), karonda (*Carissa caronda*) and *C. inermis* Vahl, lime (*Citrus aurantifolia*), pomello (*Citrus maxima* (Burm.) Merr.), khonda mavu (*Commiphora caudata* Engl.), wood apple (*Limonia acidissima* L.), *Emblica officinalis, Phoenix pusilla* Gaertn., *P. sylvestris* (L.) Roxb., *P. robusta* (Becc.) Hook. f., clove [*Syzygium aromaticum* (L.) Merr. and Perry.], *S. alternifolium* (Wight) Walp, *S. zeylanicum* L., rose apple [*S. jambos* (L.) Alston], jamun [*S. cuminii* (Linn.) Skeels], grapes (*Vitis vinifera* L.) and it's wild relatives *V. linnaei* Kurz, *V. pallida* Wight and Arn., *Vitis adnata* (Roxb.) Wall., *V. woodrowii* Cooke and *Ziziphus horrida* Roth present significant diversity. Additionally, there are a large number of underutilised edible greens and fruits which serve as life supporting species during periods of scarcity caused by vagaries of monsoon rains. The forest areas of the region are known for sandalwood (*Santalum album* L.) and red Sander tree (*Pterocarpus santalinus* Blanco) and local farmers have evolved profitable cultivation practices, such as intercropping of marigold with chillies for control of pests and nematodes and use of Neem (*Azadirachta indica* A.Juss.) for control of insect pests.

16. Kaveri Region

The region includes the Coromandel plains (Chengai Anna, South Arcot, North Arcot, Ambedkar), the alluvial plains of the Kaveri delta (Thiruvannamalai, Tiruchirapalli, Thanjavur, Pudukottai, Thiruvarur), and the Javadi hills, Shevaroy hills, Kolli malai, Pachamalai hills, Sirumali hills (Vellore, Kanchipuram, Salem, Erode, Namakkal, Karur and Dindigul). The Kaveri Delta has delta head and delta proper, which include Valar doab, Kaveri doab, main delta plains and marshy low and dune belt near Vedaranyam salt swamps. It is a warm eco-region. The climate is characterised by hot dry summers from March to August with intermittent rains from south west monsoon and mild summer from September to February with good rainfall from northeast monsoon from October to December. The annual rains in the dry Javadi hills, Pachamalai hills and Kolli hills vary from 760 to 1020 mm, with growing period between 90 to 150 days, whereas, the annual rainfall in the deltaic region ranges from 1000 to 1140 mm with growing period extending from 120 to 200 days. The landscape of the region consists of hills, alluvial plains and delta region. The constituent soils are alluvial loamy in the delta region, ferruginous loam in the interiors and patches of black cotton soil in the north western districts and of red sterile soil in Tiruchirapalli and Pudukottai. The local vegetation comprises of tropical dry deciduous and tropical thorn forests. The Tropical thorn forest in Tamil Nadu is differentiated into (i) Southern thorn forest (ii) Carnatic umbrella thorn forest (iii) Southern Euphorbia scrubs and (iv) Southern thorn scrubs.

The major tribal communities called "malayali", in hilly areas are *Irular, Kadar, Kond Kapus, Konda reddis, Malai Padaram* and others. The farmers of the region follow rainfed agriculture in the dry Tamil Nadu hilly outcrop areas of Shevaroy, Javadi, Kolli and Pachamalai hills and irrigated farming along the hinterland of Kaveri basin and delta. To ensure the sustainability of the land, mixed cropping has been adopted by the tribals, including horticultural crops. The major horticultural crops are vegetables, tuber crops, fruits, ornamentals and medicinal and aromatic plants. Crops like tapioca (*Manihot esculenta* Crantz) is cultivated in Erode and Salem for the flourishing starch industry, while plantain and coconut are cultivated wherever irrigation facilities are available. The major vegetables of the region are watermelon (*Citrulus lanatus*), muskmelon (*Cucumis melo*), cucmber (*C. sativus*), pumpkin (*Cucurbita pepo, C. maxima* Wall.), ridge gourd (*Luffa acutangula*), spong gourd (*L. aegyptiaca; syn. L. cylindrica*), bitter gourd (*Momordica charantia*), snake gourd (*Trichosanthes anguina*), brinjal (*Solanum melongena*), chilli (*Capsicum annuum, C. frutescens*), okra (*Abelmoschus esculentus*), canavalia [*Canavalia ensiformis* (L.) DC.], moringa (*Moringa oleifera* Lam.), agathi [*Sesbania sesban* (L.) Merr.], amaranths (*Amaranthus spinosus, A. tenuifolius, A. viridis, A. dubius*), *etc.* The rich genetic diversity in vegetables has contributed to development of a number of varieties, for example in *Cucurbita moschata*, Co1 and Co 2 have been developed through selection from local cultivars, in *Lageneria*, Arka Bahar has been developed from Karnataka local and Co1 from Tamil Nadu local; in *Luffa acutangula*, Co1 and Co2 and in *L. cylindrica* Co1, have been developed from Coimbatore locals (Sirohi *et al.,* 2005). The cultigen variability is further enriched by wild relatives, like *Cucumis melo* var. *anguria, Trichosanthes cucumerina var. cucumerina, Solanum nigrum, S. surattense, Momordica balsamina, M. tuberosa and* wild okra *Abelmoschus angulosus* Wall. ex Wight and Arn. The region offers significant genetic variability in tuber crops, such as aroids, including taro (*Colocasia esculenta, Colocasia* spp.), elephant foot yam (*Amorphophallus paeoniifolius*), tannia (*Xanthosoma sagittifolium* K.Koch), giant taro [*Alocasia macrorrhiza* (Linn.) G. Don], potato yam (*Dioscorea bulbifera, D.alata*), sweet potato (*Ipomoea batatas*), Indian Kudzu (*Peuraria tuberosa*), etc. The elephant foot yam collections include two distinct types, of which '*Karuna*' with small discoid mother rhizome bearing oblong tuberous cormels is from Tamil Nadu (Edison *et al.,* 2005). In onion, Co2 and Co3 have been developed through mass selection from Tamil Nadu local, whereas, MDU1 through mass selection from Somlathi local (Pandey *et al.,* 2005). In seed spices, several varieties have been developed in coriander using local germplasm, such as Co1 from Kovilpatti local by selection, Co2 from collection P2 from Gujarat by reselection, CS287 by reselection from Guntur collection, and Co3, a reselection from Acc 695 of IARI (Ravindran *et al.,* 2005).

The major fruits cultivated in the region are mango (*Mangifera indica*), mangosteen (*Garcinia mangostana*), wood apple (*Limonia acidissima*), lime (*Citrus aurantifolia*), lemon (*Citrus limon*), pomello (*Citrus maxima*); clove (*Syzygium aromaticum*), rose apple (*S. jambos*), jamun (*S. cuminii*), bael (*Aegle marmelos*), karonda (*Carissa caronda*), *Emblica officinalis;* guava (*Psidium guajava*), tamarind (*Tamarindus indica*), *Ziziphus horrida*, etc. The genetic diversity in these fruits is enriched by a large number of traditional varieties for example in mango, *Banganpalli, Bangalora, Neelum, Rumani, Mulgoa;* in banana, *Pachable, Karpurvalli, Monthan, Morris, Mysore poovan, Nendran, Pachanadan, Rasthali, Robusta* etc.; in acid lime, PKM1 seedless lime; and in lemon (*C. limon*), *Galgal* and *Euroka.* In pomegrante (*Punica granatum*), variety Ganesh has been derived from local *Alandi* and G173 from *Ganesh* (Karihaloo *et al.,* 2005). Additionally, the region has very rich genetic diversity of *Tamarindus indica.* It is being further enriched by wild relatives, such as, *Syzygium alternfolia, S. zeylanicum* and *Carissa inermis.*

17. Northwestern Deccan Plateau

It is part of hot semi-arid eco-region situated on the leeward side of the Western Ghats, extending from Satpura-Mahadeo hills in the north to the Bellary-Dharwar in the south, including the leeward districts of Maharashtra, northern dry districts of Karnataka and Andhra Pradesh. The climate of the region is characterised by hot summers and mild winters. The mean annual rainfall ranges from 600 to 1000 mm. The moisture availability period for cultivation ranges from 90 to150 days. The region is characterised by shallow and medium black soils. The common landscape of the region is moderate to gentle slopes. The soil is shallow loamy, skeletal and highly calcareous in nature. In some patches, the soil is clayey, calcareous and moderately alkaline. The main vegetation is represented by dry deciduous forests, thorn forests and scrub jungles, however, because of prolonged dry period, the region has a rich diversity of grass flora.

The major tribes associated with agriculture are Adivasi, Girijan and Vanya Jati, Maria Gond, Katkari, Kolam etc. Rainfed farming is traditional agriculture. Horticultural crops are part of intercropping or mixed cropping system. The major horticultural crops are vegetables, tuber crops and fruits. The local tribes have developed several practices for sustainability and increased productivity, for example, Warli tribes of north Maharashtra to obtain multiple crops from the same plot without irrigation, sow pigeonpea, sorghum or cowpea in the paddy seedling beds. On its borders they plant *ambadi* (*Hibiscus cannabinus* L.), *lal ambadi* (*Hibiscus sabdariffa* L.) or Okra. Cucumber and okra are inter-cropped between rows of maize. *Ambadi, Khorasni,* and wild *Vigna mungo* are cultivated on the borders of fields. This part of peninsular India has species diversity for *Vigna* species extending to Western Ghtas. Warli tribe practices mulching, this is being performed by keeping leaves of khair (*Acacia catechu* Brandis) into water channel, primarily to control brown spot diseases of paddy. In parts, farmers use rhizosphere soil beneath banyan tree, for improving the soil fertility and *'amritpani',* special bio-inoculants prepared from cow dung, cow urine, cow ghee, honey etc. for treating seeds and seedlings for soil enrichment and to facilitate better germination. The region is also known for sacred groves and sanctity to plants to facilitate conservation of biodiversity including agro-biodiversity. The main vegetables cultivated are brinjal (*Solanum melongena*), chilli (*Capsicum annuum*), pumpkin (*Cucurbita pepo*), khira (*Cucumis sativus*), *kharbuj* (*C. melo*), amaranth (*Amaranthus caudatus, A. polygamus, A. viridis*), *Cissus repanda* Vahl, *C. Quadrangularis* L., *Corchorus depressus* (L.) Stocks (leaves as pot herb), *adavi gogu* leaves as pot herb (*Hibiscus aculeatus* Roxb.), Deccan hemp (*Hibiscus cannabinus*), elephant foot yam [*Amorphophallus campanulatus* (Roxb.) Bl.], *yam* (*Dioscorea bulbifera, D. pentaphylla, D. intermedia, D. wightii, D. wallichii*). Significant level of genetic diversity has been evolved in vegetables like brinjal and chilli for plant type and fruit characteristics. In several crops, the local cultivars/landraces have been used for development of varietal products, for example in *Cucumis sativus,* Phule Shubangi has been developed through selection from a local landrace; in okra the Red Wonder of Hyderabad has contributed to development of Co1; in onion, N53, Akra Pragati, Akra Kalyan, Akra Niketan, Agrifound Dark Red, Agrifound White and Red were developed through mass selection in collections from Nasik, and Baswant-780, from collection of Pimalgaon, Maharashtra. In garlic, variety Godavari has been developed through mass selection in a collection from Maharashtra (Pandey *et al.,* 2005). Among spices, the region is known for variability in turmeric, where variety such as Krishna has been developed through clonal selection from Tekurpeta collections. The main fruits cultivated in the region are sweet orange [*Citrus sinensis* (L.), Osbeck], *kondai* or plum [*Flacourtia indica* (Burm. f.) Merr.; syn. *Flacourtia ramontchi*], Lovi-lovi (*Flacourtia inermis* Roxb.), wood apple (*Limonia acidissima*), mango (*Mangifera indica*), *jamun* (*Syzygium cumini*), grapes (*Vitis vinifera*). Significant variability is evolved in the form of traditional varieties, such as *Alphonso, Mankurad, Muloga, Pairi, Banganpalli, Totapari* in mango; *Mosambi, Sathagudi, Malta, Blood red* in sweet lime (*Citrus sinensis*); Nagpur in Mandarin (*Citrus reticulata*); for *kagzi lime,* PKM-1, *Vikram, Baramasi, Sai sarbati,* seedless lime etc. in acid lime (*Citrus aurantifolia*) (Karihaloo *et al.,* 2005). Besides, the region has evolved a large amount of undocumented variability in custard apple (*Annona squamosa*) and *Tamarindus indica*. During the festive days, farmers cook "Bajikoora" which includes vegetables, leafy and tender stems of pulses (pigeonpea, chickpea, green peas, amaranthus, chillies) grown in the village as a ritual symbol meant for the conservation of life supporting flora of the village (Satheesh, 2002).

18. Konkan Region

It includes North Western Ghats, locally referred as Sahyadri Hills and Konkan plains. Konkan plains consist of western coastal plains of Maharashtra, Goa and Uttar Kannada districts of Karnataka. It is a narrow strip, west of the Sahyadri from the river Tapti to Kalinadi. The climate of the region is tropical or with hot summers with heavy to moderate rains from the southwest monsoon. The weather is equitable without extreme hot or cold season. The Konkan region gets annually about 2540 mm of rainfall. The western slopes of the mountain experience heavy rains while eastern slope is the rainshadow region. The region has long growing period of more than 270 days in a year. Along the western slopes of the Sahyadri, the soil is reddish and lateritic, while the coastal plains have alluvium soil. The soil is clayey, acidic in nature and is poor in base saturation. Being part of the Western Ghats, it is part of one of the global biodiversity hot spots, and very rich floristic diversity with about 234 endemic species. The forest types include semi-evergreen forests, montane subtropical evergreen forests, moist deciduous forests, dry deciduous forests, scrub forests and grasslands.

Agriculture is ancient in the region; archeological evidences suggest cultivation and probable domestication of several species (Fuller *et al.,* 2004). The traditional agriculture is rainfed and because of prolonged rainy season, the region has intensive *kharif* cultivation presenting significant genetic diversity in horticultural crops. Adivasies follow mixed cropping (even at the varietal level) to assure the minimum yield and avoid crop failures. The major horticultural crops are vegetables- amaranths (*Amaranthus hybridus, A. paniculatus, A. polygamus, A. spinosus*), bhindi (*Abelmoschus esculentus*), chillies (*Capsicum annuum*), kundru (*Coccinia indica*), pumpkin (*Cucurbita pepo*), cucumber (*Cucumis sativus*), kharbuj (*Cucumis melo*), lab lab (*Dolichos bracteatus* Baker.; *D. uniflorus* Lam.), bottle gourd (*Lagenaria siceraria*), ridge gourd (*Luffa acutangula*), bitter gourd (*Momordica charantia*), brinjal (*Solanum melongena*), snake gourd (*Trichosanthes anguina*); tuber crops- elephant-foot yam (*Amorphophallus commutatus* (Schott) Engl., *A. sylvaticus, A. konkanensis* Hett., S.R.Yadav and K.S.Patil), taro (*Colocasia esculenta var. esculenta*); yam (*Dioscorea bulbifera, D. esculenta*), *Flemingia procumbens* Roxb., sweet potato (*Ipomoea batatas*), tapioca (*Manihot esculenta*), potato (*Solanum tuberosum*), Indian Kudzu (*Peuraria tuberosa*); fruits- pineapple (*Annanas comosus* (L.) Merr.), coconut (*Cocos nucifera* L.), mango (*Mangifera indica*), strawberry (*Fragaria nilgerrensis* Schecht.), banana (*Musa acuminate, M. sapientum* L.), jackfruit (*Artocarpus heterophyllus*); spices- arecanut (*Areca catechu*), tejpat (*Cinnamomum verum* B. and Presl.), *Carum strictocarpum,* turmeric (*Curcuma inodora* Blatt., *C. purpurea* Blatter, *C. pseudomontana* J.Graham, *C. reclinata* Roxb., *Kaempferia galanga* L., Nutmeg (*Myristica malabarica* Lamk., *M. dactyloides* Gaertn.), pepper (*Piper argyrophyllum* Miq., *P. galeatum* C.DC., *P. hookeri* Miq., *P. hymenophyllum* (Miq.) Wight, *P. nigrum, P. trichostachyon* DC), *Vanilla vatsalana,* Ginger (*Zingiber macrostachyum* Dalzell, *Z.officinale* Rosc.). In addition, there are ornamentals, medicinal and aromatic plants and a number of less known food plants, such as *Cissus repanda, Cissus quadrangularis, Corchorus depressus* (leaves as pot herb), *Hitchenia caulina* Baker, (rhizome yields starch) *Flemingia macrophylla* Blume ex Miq., (pods edible), *Ougeinia oojeinensis* (Roxb.) Hochr. (flowers edible) etc. The genetic diversity is futher enriched by wild relatives of crop species, such as, *Abelmoschus angulosus, A. ficulneus, A. manihot ssp. manihot, Cucumis ritchei, C. setosus, Momordica dioica, M. tuberosa, Mangifera sylvatica, Cinnamomum goaense* Kosterm. *etc.*

The region is one the major centres of diversity for *Vigna* spp. with 14 species of the 24 *Vigna* species reported from India and it is possible that two cultivated species *Vigna radiata* (L.) R.Wilczek (Green Gram, *mung*) and *Vigna mungo* (L.) Hepper (black gram, *urd*) might have been domesticated in the region (Fuller *et al.,* 2004). The *Vigna* species occur or cultivated in the region are *V. aconitifolia* (Jacq.) Maréchal, *V. angularis* (Willd.) Ohwi and H.Ohashi, *V. dalzelliana* (Kuntze) Verdc., *V. khandalensis* (Santapau) Raghavan and Wadhwa, *V. mungo, V. radiata* var. *setulosa* (Dalzell) Ohwi and H.Ohashi, *V.*

radiata var. *sylvestris, V. sublobata* (Roxb.) Babu and S.K.Sharma, *V. trilobata* var. *trilobata, V. umbellata* (Thunb.) Ohwi and H.Ohashi, *V. trilobata* var. *pilosa, V. unguiculata* (L.) Walp., *ssp. sesquipedalis* (L.) Verdc., *V. vexillata* (L.) A.Rich var. *sepiaria, V. vexillata* var. *stocksii,* offering valuable genetic diversity. The higher yield in *V. radiata* is due to the trait donated by the wild species *V. sublobata* (Jain and Mehra, 1980). Significant variability is recorded in cucurbitaceous crops like *Cucurbita, Cucumis, Luffa graveolens, Momordica cochinchinensis, M. subangulata, Trichosanthes bracteata and T. cuspidata* Lam. and used in breeding programmes (Sirohi *et al.,* 2005). In okra, the region has natural hybrid between *Abelmoschus esculentus* and *A. caillei* (A. Chev) Stevels, which is naturalised with significant variability. Among fruits, the region is well known for high quality mango varieties, such as *Alphonso, Mancurad, Maussarda, Fernandine, Xayer, Colaco.* In jackfruit high value varieties, *Cappco* and *Rassal* are known from the region. The other collections in Jackfruit, include landraces *Koozha, Navarikka, Varikka,* and *Rudrakha koozha,* originating from the region. In spices, the region exhibits variability in *Piper nigrum,* ginger and cinnamon. In cinnamon, variety Konkan tej is a selection from *Ceylon type* with superior quality.

19. Malabar Region

The region is situated in the southern region of the Western Ghats extending from the river Dakshin Kannada in the north to Kanyakumari in the south, including whole of Kerala, south of Karnataka and the Western Ghats mountain districts of Tamil Nadu. It is a hot, humid-perhumid eco-region. The climate of the region is characterized by hot and mild summers and rainy season. It has an average annual precipitation exceeding 2000 mm facilitating long growing season, which may extend more than 270 days. Besides, there is abundant water availability due to the extensive network of rivers, streams, lakes and backwaters. The landscape consists, high peaks and coastal plains dissected by diverse water bodies. The region has red and lateritic soils, while the alluvium derived soils occur in the coastal plains. The soils are deep, clayey, profoundly to moderately acidic in nature and are poor in base saturation. Being the part of Western Ghats, one of the global biodiversity hotspots, the region is very rich in floristic diversity. There are about 1286 endemic species in this part of southern Western Ghats. The natural vegetation comprises of tropical moist wet evergreen forests, tropical semi-evergreen forests, tropical most deciduous forests, tropical dry deciduous, montane and shola forests.

Agriculture is ancient in the region, archeological evidences suggest that along with north Western Ghats, it may be probable centre of domestication in peninsular India and coastal introduction of several crop species (pepper, coffee, cashew, etc.). The main tribes associated with region are Hill Pulaya, Irulan, Kadar, Kanikar, Kuruman, Malai Pandaram, Pannayan, Ulladan. Jenu Kuruba, Koraga Cholanaickans, Kadar, Kurumbas and Kattunaickans. The agriculture is practiced from ancient times and the region is known as the land of spices, which are cultivated since 3000 years and over time has generated significant genetic variability. The region is known through out the world for quality spices, like cardamom, ginger, pepper, cinnamon, nutmeg, turmeric and others. The region is intensively cultivated with horticultural crops like coconut, vegetables, leaf vegetables, spices, fruits, medicinal and aromatic plants, bamboos, ornamentals, etc. However, it is a hot spot of genetic diversity for spices like pepper and tuber crops. In *Cucurbita moschata,* variety Ambili was developed from Kerala local. In okra, the region has natural hybrid between *Abelmoschus esculentus* and *A. caillei,* which is naturalised to the region and wild relative *A. angulosus* Wall. ex Wight and Arn. In addition, wild relatives, such as *Luffa acutangula* Roxb. f. *amara* (Roxb.) W.J.de Wilde and Duyfjes, *L. umbellata* M.Roem.; *Momordica charantia* L. var. *muricata* (Willd.) Chakrav., *M. dioica, M. tuberosa; Trichosnathes anamalayensis, T. cucumerina* var. *cucumerina, T. nervifolia* L., *T. tricuspidata* var. *tomentosa, T. villosula, T. wallichiana,*

Solanum aculeatissimum Jacq., *S. anguivii var. multiflora, S. erianthum* D.Don, *S. macrocarpum* (Maxim.) Koidz., *S. melongena var melongena, S. melongena var. incanum, S. nigrum, S. pubescens, S. torvum, S. viarum* offer additional variability. Besides the region is rich in several minor tuber crops like arrowroot, *Maranta arundinacea* L., Chinese potato (*Coleus parviflora* Benth.), Queensland arrowroot (*Canna edulis* Ker Gawl.), winged bean (*Psophocarpus tetragonolobus* DC.) and yam bean (*Pachyrhizus erosus* (L.) Urb. used as alternative food crops. In these, variety Sree Dhara of Chinese potato was developed through clonal selection, in CP58 and Rajendra Mishrikand-1 of yam bean, resistance to insect pests and diseases were achieved through seedling selection in local collection (Edison *et al.*, 2005). In laterite dry zones, *tapioca* is being cultivated.

Among fruits, region is well known for different cultivars of banana from red skinned *Kappa*, large yellow skinned *Nedraka* to small yellow skinned *Kadali, Rasakadali, Poovan, Matti, Palayamkodan* to green skinned *Padachi* and *Morris*. The region is also known for traditional varieties of mango: *Mundappa, Plour, Pairi*; in jackfruit *Varikka* landrace has quality fruit preferred by people. The other land races are *Koozha* and *Navarikka*. Coorg Mandarin (*Citrus reticulata*) from Karnataka region is a popular cultivar. Malabar Tamarind, *Garcinia gummi-gutta* also offers considerable variability. A drink, called Kokam has been developed from the region, from fruit juice. Further, a number of wild relatives of *Artocarpus, Diospyros, Garcinia, Syzygium, and Vitis* offer significant variability for use.

The region is the main producer of pepper and cardamom. The region produces 96 percent of the pepper of the country. The other important species are cinnamon, clove, turmeric, nutmeg, ginger and vanilla. The local cultivars present maximum variability as in *Piper nigrum* and several cultivars have been developed through clonal selection from local types, example- Panniyur 4 from *Kuthiravally*, Sreekara from *Karimunda*, Panchami from *Aimpiriyan*, Pournami from *Ottaplackal*, PLD 2 from *Kottanadan* and Panniyur 6 from *Karimunda*. In ginger significant variability is found, and varieties like IISR Varada were derived through clonal selection from Kerala local. The Cardamom (*Elettaria cardamomum* Maton) has three distinct types, Malabar, Mysore and Vazhukka from the region, and a number of cultivars have been developed, like Mudigere-1 by clonal selection from *Malabar type*, PV1 from *Walayar*, ICR-11 from Chakkupalam collections, ICR-13 from *Malabar type* and ICR-14 from *Vadagaraparai* type. The region is also rich in genetic diversity for large cardamom. In turmeric (*Curcuma longa*) significant variability exists for yield, quality attributes and dry recovery. There are well known cultivars like *Alleppey*, which is considered high yielding with quality rhizomes. Suguna Sobha is a clonal selection from local germplasm (Ravindran *et al.*, 2005). Nutmeg (*Myristica fragrans* Houtt.) introduced from Indonesia is naturalized to the region, presents variability due to dioecious nature and sexual propagation for characters like fruit size, shape, mace and seed volume (Krishnamoorthy *et al.*, 1996). High yielding Konkan Sugandha was selected from local seedlings. Cinnamon is indigenous to the region with large number of endemic species, as this region is one of the centres of diversity. The leaf oil contains eugenol, while bark oil has cinnamaldehyde. Varieties like Navasree, a seedling selection from Sri Lankan collections, and Nithyasree from the Indian collection have been produced. Besides, a large number of wild relatives of *Amomum, Cinnamomum, Piper, Curcuma, Zingiber, Myristica* and *Vanilla* offer useful variability for use.

Besides spices and coconut, the region is known for production of quality rubber, cashewnut and tapioca. The southern Karnataka part of the region is also characterised with areca gardens intercropped with coffee, banana, vanilla, pepper etc. The horticultural crops, such as coconut, rubber, tea, coffee, cashew nut play important role in the agro-economy of the region. The region produces 91 per cent of India's rubber. Besides rubber, the other plantation crops are banana and plantain.

References

Arora, R.K. and Nayar, E.R., 1984. *Wild Relatives of Crop Plants of India*. National Bureau of Plant Genetic Resources (NBPGR). Kapoor Art Press, New Delhi, India, p. 88.

Arora, R.K., Nayar, E.R. and Pandey, A., 2006. Indian centre of floristic and economic plant diversity: A review. In: *Hundred Years of Plant Genetic Resources Management in India,* (Eds.) Anurudh K. Singh, K. Srinivasan, S. Saxena and B.S. Dhillon. National Bureau of Plant Genetic Resources (ICAR), New Delhi, India, pp. 1–28.

Bhat, A.R., Ahanger, H.U., Sofi, A.A. and Mir, N.A., 1992. Evaluation of some walnut selections for quality parameters in Jammu and Kashmir. In: *Emerging Trends in Temperate Fruit Production in India,* (Eds.) K.L. Chadha, D.K. Uppal, R.P. Awasthi and S.A. Anand. NHB Technical Communication No. 1, National Horticulture Board, Gurgaon, India, pp. 56–61.

Bhattacharya, S.C. and Dutta, S., 1956. *Classification of Citrus Fruits of Assam*. ICAR Monograph No. 20.

Biodiversity Hotspots, 2008. www.biodiversityhotspots.org/xp/Hotspots/himalaya/Pages/biodiversity.aspx–34k.

Bist, H.S. and Sharma, R.L., 1996. Some promising cultivars of plum for Himachal Pradesh. *Hort. J.,* 9: 107–112.

Borgohain, R., Mazumdar, D., Neog, M. and Saikia, A., 2008. *Bhut jolokia*: A commercially potent chilli from Northeast India. *Asian Agri-History,* 12: 225–229.

Champion, H.G. and Seth, S.K., 2005. *A Revised Survey of the Forest Types of India*. Reprint. Natraj Pub., Dehra Dun, India, xxviii+404 pages.

Chatterjee, D., 1939. Studies on the endemic flora of India and Burma. *J. Royal Asiat. Soc. Bengal N.S. (Science),* 5: 19–67.

Chatterjee, S., Saikia, A., Dutta, P., Ghosh, P. and Worah, S., 2006. Review of biodiversity in northeast India. Draft for discussion, Background paper No. 13. WWF–India, Delhi, India.

Chopra, S.K. and Chanana, Y.R., 1993. Improvement of temperate fruits for subtropical climate. In: *Advances in Horticulture, Vol 1: Fruit Crops,* (Eds.) K.L. Chadha and O.P. Pareek. Malhotra Publishing House, New Delhi, India, pp. 445–462.

Conservation International, 2005. Biodiversity Hot Spots, 1919 M Street, NW, Suite 600, Washington, DC 20036. (202)912–1000, fax: (202)912–1030 www.conservation.org (updated 2nd May 2005).

Convention Concerning the Protection of the World Cultural and Natural Heritage, 1972. General Conference of UNESCO on November 16, 1972. www.unesco.org.

Dabas, B.S., Nayar, E.R. and Dwivedi, N.K., 2006. Arid legumes. In: *Plant Genetic Resources: Food Grain Crops,* (Eds.) B.S. Dhillon, S. Saxena, A. Agrawal and R.K. Tyagi. Narosa Publishing House, New Delhi, India, pp. 255–274.

Dey, P., Sarkar, A.K. and Sikka, A.K., 2003. Management of upland for horticulture development in Jharkhand, Chotanagpur. *Horticulture,* 20: 4–5.

Dhankhar, B.S., Mishra, J.P. and Bisht, I.S., 2005. Okra. In: *Plant Genetic Resources: Horticultural Crops,* (Eds.) B.S. Dhillon, R.K. Tyagi, S. Saxena and G.J. Randhawa. Narosa Publishing House, New Delhi, pp. 59–33.

Dhillon, B.S., Rana, J.C. and Singh, A.K., 2005. Collection, conservation and utilization of indigenous horticultural crops germplasm. In: *Crop Improvement and Production Technologies of Horticultural Crops Vol. 1,* (Eds.) K.L. Chadha, B.S. Ahluwalia, K.V. Prasad and S.K. Singh. Malhotra Publishing House, New Delhi, pp. 10–35.

Edison, S., Velayudhan, K.C., Easwari Amma, C.S., Pillai Santha, V., Mandal, B.B., Sheela, M.N., Vimala, B., Unnikrishnan, M. and Hussain, Z., 2005. Tropical root and tuber crops. In: *Plant Genetic Resources: Horticultural Crops,* (Eds.) B.S. Dhillon, R.K. Tyagi, S. Saxena and G.J. Randhawa. Narosa Publishing House, New Delhi, pp. 228–250.

Fuller, D.Q., Korisettar, R., Venkatasubbaiah, P.C. and Jones, M.K., 2004. Early plant domestications in southern India: Some preliminary archaeobotanical results. *Vegetation, History and Archaeobotany,* 13: 115–129.

Gangwar, A.K. and Ramakrishnan, P.S,. 1989. Ethnobiological notes on some tribes of Arunachal Pradesh, North-Eastern India. *Economic Botany,* 44: 94–105.

Gautam, D.R., Undal, J.K., Sharma, J.N. and Sharma, H.K. 1992. Evaluation of cherry germplasm. In: *Nat. Symp. Emerging Trends in Temperate Fruit Production in India,* Abstr. 12, Dr. Y.S. Parmar University of Horticulture and Forestry, Nauni, Solan, Himachal Pradesh, India, p. 8.

Hora, S.L., 1949. Satpura hypothesis of the distribution of the Malayan fauna and flora to Peninsular India. In: *Proceedings of National Institute of Science,* India, 15: 309–314.

Jain, H.K. and Mehra, K.L., 1980. Evolution, adaptation, relationships and uses of the species of *Vigna* cultivated in India. In: *Advances in Legume Sciences,* (Eds.) R.J. Summerfield and A.H. Bunting. Kew Royal Botanic Gardens, pp. 459–468.

Jain, S.K., 1964. Wild plant-foods of the tribals of bastar (Madhya Pradesh). In: *Proc. Nat. Inst. Sci. India,* 30B(2): 56–80.

Kalita, D. and Deb, B., 2006. Medicinal plants used by the Sonowal Kacharis of Brahmaputra Valley, Assam, India. *Journal of Tropical Medicinal Plants,* 4.

Kalloo, G., Srivastava, U., Singh, M. and Kumar, Sanjeet, 2005. Solanaceous vegtables. In: *Plant Genetic Resources: Horticultural Crops,* (Eds.) B.S. Dhillon, R.K. Tyagi, S. Saxena and G.J. Randhawa. Narosa Publishing House, New Delhi, pp. 19–33.

Kalloo, G., 1993. Eggplant (*Solanum melongena* L.). In: *Genetic Improvement of Vegetable Crops,* (Eds.) G. Kalloo and O.B. Bergh. Pergamon Press, Oxford UK. pp. 587–604.

Karihaloo, J.L., Malik, S.K., Rajan, S., Pathak, R.K. and Gangopadhyay, K.K., 2005. Tropical fruits. In: *Plant Genetic Resources: Horticultural Crops,* (Eds.) B.S. Dhillon, R.K. Tyagi, S. Saxena and G.J. Randhawa. Narosa Publishing House, New Delhi, pp. 121–145.

Kaul, G.L., 1990. Our dry fruit basket. *Indian Hort.,* 35: 46–48.

Krishnamoorthy, B., Sasikumar, B., Rema, J., George, J.K. and Peter, K.V., 1997. Genetic resources of tree spices and their conservation in India. *Plant Genetic Resources Newsletter,* 111: 53–58.

Kumar, Pankaj, Rawat, G.S. and Wood, P.H., 2011. Diversity and ecology of Dendrobium (Orchidaceae) in Chotanagpur Plateau, India. *Taiwania,* 56(1): 23–36.

Mishra, J.P., Negi, K.S., Singh, B., Sharma, A.K. and Kumar, N., 2000. Conservation of biodiversity in wild germplasm of okra. In: *Proceedings of the International Conference on Managing Natural Resources for Sustainable Agricultural Productions in the 21st Century, Vol. 4.* New Delhi, India.

Mitra, S.K. and Bose, T.K., 1999. Guava. In: *Tropical Horticulture, Vol. 1,* (Eds.) T.K. Bose, S.K. Mitra, A.A. Farooqi, and M.K. Sadhu. Naya Prokash, Kolkata, West Bengal, India. pp. 297–318.

Mukherjee, S.K., 1985. Systematic and ecogeographical studies on crop gene pools. I. *Mengifera* L. International Board for Plant Genetic Resources, Rome, Italy.

Nandi, Sandhyakara, 1939. *Ramacharita,* (Eds.) R.C. Majumdar, R.G. Basak and N.G. Banerjee. Published by VRS, Motilal Banarsidass, Rajshahi.

Nath, V., Singh, B. and Rai, M., 2001. Horticulture biodiversity in Santhal Parganas. *Indian Journal of Plant genetic Resources,* 14: 92–98.

Nayar, M.P., Singh, A.K. and Nair, K.N., 2009. Agrobiodiversity hotspots in India: Conservation and benefit sharing, Vol 1: 134–138.

Nerkar, Y.S., 1991. The use of wild species in transferring diseases and pest resistance genes in okra. In: *International Crops Network Series.* 5. Report on International workshop on Okra Genetic Resources, IBPGR, Rome, Italy, pp. 110–113.

Niyogi, Pusa, 2008. Expansion of agriculture in ancient Bengal. In: *History of Agriculture in India (up to C.1200 AD) Vol 1 Part 1,* (Eds.) Gopal Lallanji and V.C. Srivastava. PHISPC (Centre of Studies in Civilization) pp. 637–701.

Pandey, G., 2002. Popularizing under exploited fruits for consumptions. *Indian Horticulture,* pp. 18–21.

Pandey, P.K. and Saini, S.K., 2007. Edible plants of tropical forests among tribal communities of Madhya Pradesh. *Indian Journal of Traditional Knowledge,* 61: 185–190.

Pandey, U.B., Kumar, Ashok, Pandey, Ruchira and Venkateshwaran, K., 2005. Bulbous crops: Cultivated alliums. In: *Plant Genetic Resources: Horticultural Crops,* (Eds.) B.S. Dhillon, R.K. Tyagi, S. Saxena and G.J. Randhawa. Narosa Publishing House, New Delhi, pp. 108–120.

Prakash, S., Kumar, A. and Nautiyal, M.C., 1997. Early maturing apples fetch higher returns. *Indian Hort.,* 44: 22–24.

Rai, M., Rana, R.S., Koppar, M.N., Gupta, P.N. and Thomas, T.A., 1993. Collecting diversity in eggplant germplasm from North-central India. *Indian J. Plant Genet. Resources,* 6(1): 53–59.

Ramakrishnan, P.S., 1992. Shifting agriculture and sustainable development: An interdisciplinary study from north-eastern India. UNESCO–MAB Series, Paris, Parthenon Publ., Carnforth, Lancs. U.K. (republished by Oxford University Press, New Delhi 1993), pp. 424.

Ravindran, P.N., Babu, K.N., Peter, K.V., Abraham, J. and Tyagi, R.K., 2005. Spices. In: *Plant Genetic Resources: Horticultural Crops,* (Eds.) B.S. Dhillon, R.K. Tyagi, S. Saxena and G.J. Randhawa. Narosa Publishing House, New Delhi, pp. 190–227.

Rodgers, W.A. and Panwar, H.S., 1988. Planning a wildlife protected area network in India. Vols. 1 and 2. Department of Environment, Forests, and Wildlife/Wildlife Institute of India, Dehra Dun, India.

Sajem, A.L. and Gosai, K., 2006. Traditional use of medicinal plants by the Jaintia tribes in North Cachar Hills district of Assam, northeast India. *Journal of Ethnobiology and Ethnomedicine,* 2: 33.

Saraswat, K.S., 1994. Plant economy of ancient Narhan (ca. 1300 BC – 300–400 AD) pp. 254–346: In: *Excavations at Narhan 1984–89.* Appendix IV, Varanasi, Benaras Hindu University Press.

Satheesh, P.V., 2002. *Crops of Truth*. Farmers's perception of the agrobiodiversity in the Deccan region of South India, Hyderabad

Sehgal, J.L., Mandal, D.K., Mandal, C. and Vadivelu, S., 1992. Agro-ecological Regions of India. NBSS and LUP Technical Bulletin No. 24. 2nd Edition. National Bureau of Soil Survey and Land Use Planning, Indian Council of Agricultural Research, Nagpur, Maharashtra, India, pp. 130.

Seshadri, S and Srivastava, Umesh, 2002. Evaluation of vegetable genetic resources with special reference to value addition. In: *Proceeding of International Conference on Vegetables*. 11–14 November at Bangalore. Prem Nath Agricultural Science Foundation, Bangalore, pp. 41–62.

Sharma, G.R., 1980. History to Prehistory, Archaeology of the Ganga Valley and the Vindhyas. University of Allahabad, Department of Ancient History, Culture, and Archaeology. Allahabad, Uttar Pradesh, India, pp. 103–110.

Sharma, R.L. and Kumar, K., 1994. Temperate fruit crop improvement in India. In: *Progress in Temperate Fruit Breeding,* (Eds.) H. Schmidt and M. Kellerhals. Kluwer Academic Publishers, Netherlands, pp. 149–156.

Sharma, S.D., Kumar, K., Gupta, S., J.C. Rana, B.D. Sharma and D.S. Rathor, 2005. Temperate fruits. In: *Plant Genetic Resources: Horticultural Crops,* (Eds.) B.S. Dhillon, R.K. Tyagi, S. Saxena and G.J. Randhawa. Narosa Publishing House, New Delhi, pp. 146–167.

Shekhawat, G.S., Gopal, J., Pandey, S.K. and Kang, G.S. 2005. Potato. In: *Plant Genetic Resources: Horticultural Crops,* (Eds.) B.S. Dhillon, R.K. Tyagi, S. Saxena and G.J. Randhawa. Narosa Publishing House, New Delhi, pp. 89–107.

Singh, Anurudh K., 2010. Biodiversity in horticulture and future needs. In: *Savonier 4th Indian Horticulture Congress–2010,* held 18–21 Nov.

Singh, Anurudh K. and Varaprasad, K.S. 2008. Criteria for identification and assessment of agro-biodiversity heritage sites: Evolving sustainable agriculture. *Current Science,* 94(9): 1131–1138.

Singh, L.P., Parkash, B. and Singhvi, A.K., 1998. *Evolution of the Lower Gangetic Plain Landforms and Soils in West Bengal, India* Catena, 33 (2). pp. 75–104. ISSN 0341–8162

Singh, M.S. and Singh, N.R., 2008. Plants used as traditional medicine in Manipur. *Asian Agri-History,* 12: 153–156.

Sirohi, P.S., Kumar, G., Munshi, A.D. and Behera, T.K., 2005. Cucurbits. In: *Plant Genetic Resources: Horticultural Crops,* (Eds.) B.S. Dhillon, R.K. Tyagi, S. Saxena and G.J. Randhawa. Narosa Publishing House, New Delhi, pp. 34–58.

Swift, M.J., Vandermeer, J., Ramakrishnan, P.S., Anderson, J.M., Ong, C.K. and Hawkins, B., 1996. Biodiversity and agroecosystem function. In: *Functional Roles of Biodiversity: A Global Perspective,* (Eds.) H.A. Mooney, J.H. Cushman, E. Medina, O.E. Sala and E.D. Schulze. SCOPE Series, John Wiley, Chichester, U.K., pp. 261–298.

Tripathi, S.P., Lal, H. and Khan, I.A. 1992. Performance of almond cultivars under Pithoragarh agroclimatic conditions. In: *Nat. Symp. on Emerging Trends in Temperate Fruit Production in India* Abstr. 21 Dr YS Parmar University of Hirticulture and Forestry, Nauni, Solan, Himachal Pradesh, India, p. 13.

Tripathi, V., 2008. Agriculture in the Gangetic plains during the First Millennium BC. In: *History of Agriculture in India: Up to c. 1200 AD,* (Eds.) Lallanji Gopal and V.C. Srivastava. pp. 348–365.

Vashishtha, B.B., Saroj, P.L., Kumar, Gunjeet and Awasthi, O.P., 2005. Arid fruits. In: *Plant Genetic Resources: Horticultural Crops,* (Eds.) B.S. Dhillon, R.K. Tyagi, S. Saxena and G.J. Randhawa. Narosa Publishing House, New Delhi, pp. pp.168–189.

Vavilov, N.I., 1934. Le proble'me de l'origine des plantes cultive'es. Annales de l'Institut National de la Recherche Agronomique 36: 239–246 [in French].

Vavilov, N.I., 1935. Theoretical basis for plant breeding. In: D. Love, (Transl.) Vol. I. Moscow. *Origin and Geography of Cultivated Plants: The Phytogeographical basis of Plant Breeding*, Cambridge University Press, Cambridge, UK, pp. 316–366.

Yadav, D., 2006. Ethnomedicinal plant used by Bhil tribe of Babod of Madhya Pradesh. *Indian Journal of Traditional Knowledge,* 5: 263–267.

2013, Biodiversity in Horticultural Crops Vol. 4

Editor: Professor K.V. Peter

Published by: DAYA PUBLISHING HOUSE, NEW DELHI

Pages 61–67

Chapter 3

Impact of Climate Change on Biodiversity

Anita Srivastava and Om Kumar

Biodiversity Climate Change (BCC) Division,
Indian Council of Forestry Research and Education, Dehra Dun
E-mail: srivastavaa@icfre.org; kumarom@icfre.org

It is now well established that Earth's climate is warmer approximately by 0.7°C over past 100 years due to greenhouse gases (GHGs). These gases trap heat from the sun and thus lead to climate change.

Climate is the most important determinant of vegetation patterns globally and has significant influence on the biodiversity. Response of climate related warming predicts a preponderance of poleward/upward shift of species, both of fauna and flora. Warming is also supposed to alter community compositeness and can cause extinction of some important keystone species from ecosystem and in this way may effect distribution and abundance of species. India is a rich mega biodiversity country with two hotspots where forests account for about 19.52 per cent (64.20 million ha) of the geographical area. In the changed scenario, it is important to assess the likely impact of projected climate change on forests and develop strategies towards appropriate mitigation, adaptation and biodiversity conservation in the country.

Climate change and global warming are going to be major issues for the 21st century. Climate change is one of the most important global environmental issues since inception of humanity. Forestry is one of the key sectors in mitigating climate change. Tropical countries are experiencing rapid forest degradation and loss leading to decline in biodiversity, soil erosion, ground water level and shortage of forest products. Afforestation and Reforestation (A and R) projects provide multiple environmental and socio economic benefits, apart from carbon sequestration for global environmental benefit.

The earth's climate is changing because the composition of our atmosphere is being altered, primarily as a consequence of anthropogenic activities. The world's population continues to grow at an alarming rate with a six-fold increase during the 20th century. Despite the fact that most of the world's people still live in unacceptable levels of poverty, our collective wealth is growing and with it, there is a corresponding increase in demand for natural resources, energy, food and goods to consume. In the process, vast quantities of gases and effluents are discharged, which change the atmospheric's composition and its capacity to regulate its temperature. The tropical deforestation is one of the most critical problems facing the developing countries today in term of its long-term catastrophic impact on biodiversity,loss in economic opportunities and resultant social problems consequent to global climate change.

India is a mega biodiversity country where forest and tree cover account for about 23.03 per cent (75.70 million ha) of the geographical area. It is important to assess the likely impact of projected climate change on forest, develop and implement appropriate adaptation and mitigation strategies for both biodiversity conservation and livelihood of forest dependent communities. There is considerable geographical variation in the magnitude of changes for both temperature and rainfall. To understand the effect of climate change, collection and collation of precise data on the links between weather phenomena and phenological phases of forest species at autoecological level should be the first priority. Besides change in the temperature, shift in rainfall amount and distribution (number of rainy days, length of dry season, and rainfall regime, *i.e.* time and season of occurrence of rain) also need consideration to predict migration of species and consequently of vegetation type.

Bio diversity – the source of enormous environmental, economic, and cultural value – will be threatened by rapid climate change. The composition and geographic distribution of ecosystems may change as individual species responds to new conditions created by climate change. At the same time, habitats may degrade and fragment in response to other anthropogenic pressures. Species which cannot adapt quickly enough may become extinct – an irreversible loss to biodiversity.

Biodiversity: Threat Due to Climate Change

Species and Ecosystems have Already Started Responding to Global Warming

Species Ranges

Climate change is likely to have a number of impacts on biodiversity, from ecosystem to species level. The main impact is the effect that temperature and precipitation changes have on species ranges and ecosystem boundaries. Any particular ecosystem consists of an assemblage of species, some of which will be near the edge of their ranges and others will not. Those at the edge of their ranges may need to move due to climate change (Leemans and Van Vliet 2004). Root *et al.* (2003) found that out of 1700 species reviewed globally, 81 per cent of observed shifts in range were in the direction expected by climate change. Flooding and sea level rise will also affect species, ranges and ecosystem boundaries as well as threatening wetlands and coastal ecosystems (Dudley 2003).

Change in Phenology

Climate change is also causing shifts in the reproduction cycle and growing seasons of certain species. It may alter the frequency of pests and diseases out break. Root *et al.* (2003) found that out of 1700 species reviewed, 87 per cent of observed shift in phenology were in the direction expected by climate change. Parmesan and Yohe (2003) found, that the timing of flowering had shifted earlier by an average of 5.1 days per decade over the last half-century.

Changes in Species Interactions

The climate change will have impact on many of the more complex interactions (competition, pollination) and changes in competitive ability may also emerge. For example, early leafing trees may get a two-week head start on their competitions in terms of growth and thus occupy an increasing proportion of woodland (Dudley 2001). The impact of climate change on biodiversity will vary between regions. The maximum changes in climate are expected in the far north and south. In some cases, the strategies which are adopted to mitigate or adapt to climate change can severally impact biodiversity. For example, large sea walls built to protect against storm surges and floods.

Forests Adapt Slowly to Changing Conditions

Observations, experiments and models demonstrate that a sustained increase of just 1°C in the global average temperature would affect the functioning and composition of forests. The composition of species in existing forests will change, while new combinations of species, and hence new ecosystems, may be established. Other stresses caused by warming will include more pests, pathogens, and fires. Because higher latitudes are expected to warm more than equatorial ones, boreal forests will be more affected than temperate and tropical forests. Alaska's boreal forests are already expanding northward at the rate of 100 km per degree Centigrade.(Intergovernmental Panel on Climate Change 2001 and 2007) and U.S. Environmental Protection Agency (2007).

Successive Reports of IPCC concluded that even moderate warming and Climate Change will impact forest ecosystems adversely. Ravindranath *et al.* (2006) made an assessment of impact of climate change on forests of Western Ghats. In India over 70 per cent of the forest grids are likely to undergo change in forest-biome type leading to forest dieback and loss of biodiversity in the transient phase. Ravindranath and Sukumar (1998) predicted the impacts of two climate change scenarios on tropical forests in India – one involving greenhouse gas forcing and the other incorporating the effects of sulphate aerosols-. The first scenario, associated with increased temperature and rainfall, could result in increased productivity, migration of forest types to higher elevations and transformation of drier forest types to moisture types. The second scenario, involving a more modest increase in temperature and a decrease in precipitation in central and northern India, could have adverse effects on forests. In a case study of the south Indian state of Kerala, Achanta and Kanetkar (1996) link the precipitation effectiveness index to net primary productivity of teak plantation. Results indicate that under the climate scenarios generated by the ECHAM3 climate model, the soil moisture is likely to decline and, in turn reduce teak productivity from 5.40 m³/ha to 5.07 m³/ha. The study also shows that the productivity of moist deciduous forests could decline from 1.8 m³/ha to 1.5 m³/ha.

Forests Play an Important Role in the Climate System

Forests are a major reservoir of carbon. World Resources Institute reported that forest ecosystem accounts for about 40 per cent of the total carbon stored in terrestrial ecosystems. Large quantities of carbon may be emitted into the atmosphere during transitions from one forest type to another, if mortality release carbon faster than regeneration and growth absorb it. Forests also directly affect climate on the local, regional, and continental scales by influencing ground temperature, evapo-transpiration, surface roughness, albedo (or reflectivity), cloud formation and precipitation.

Deserts, Arid and Semi-Arid Ecosystems may Become more Extreme

With a few exceptions, deserts are projected to become hotter but not significantly wetter. Higher temperatures could threaten organisms which now exist near their heat-tolerant limits.

Rangelands may Experience Altered Growing Seasons

Grasslands support approximately 50 per cent of the world's livestock and are also grazed by wildlife. Shifts in temperatures and precipitation may reshape the boundaries between grasslands, shrublands, forests and other ecosystems. In tropical regions such changes in the evapo-transpiration cycle could strongly affect productivity and the mix of species.

Mountain Regions are Already Under Considerable Stress from Human Activities

The projected declines in mountain glaciers, permafrost, and snow cover will further affect soil stability and hydrological systems (the most major river systems start in the mountains). As species and ecosystems are forced to migrate uphill, those limited to mountain tops may have nowhere to go and become extinct. Observations show that some plant species are moving up in the European Alps by one to four metres per decade and that some mountaintop species have already disappeared. Agriculture, tourism, hydropower, logging and other economic activities will also be affected. The food and fuel resources of indigenous populations in many developing countries may be disrupted (United Nation Framework Convention on Climate Change 2002).

The Cryosphere will Continue to Shrink

Mountain glaciers are declining, almost two thirds of Himalayan glaciers have retreated in the past decade, and Andean glaciers have retreated dramatically or disappeared. This will impact nearby ecosystems and communities as well as seasonal river flows and water supplies, which in turn would have serious implications for hydropower and agriculture. The landscapes of many high mountain ranges and polar regions may change dramatically. Reduced sea-ice could lengthen the navigation season for certain rivers and coastal areas.

Non-Tidal Wetlands will also be Reduced

Open-water and waterlogged areas provide refuge and breeding grounds for many species. They also help to improve water quality and control floods and droughts. Studies from several countries suggest that a warmer climate will contribute to the decline of wetlands through higher evaporation. By altering their hydrological regimes, climate change will influence the biological, biogeochemical and hydrological functions of these ecosystems, as well as their geographical distribution.

Biodiversity: Reducing the Impacts of Climate Change with Appropriate Mitigation and Adaptation Strategies

The resilience of ecosystems can be enhanced and the risk of damage to human and natural ecosystems be reduced through adoption of biodiversity-based adaptive and mitigative strategies. Mitigation may be described as a human intervention to reduce greenhouse gas sources or to enhance carbon sequestration, while adaptation to climate change refers to adjustments in natural or human systems in response to climatic stimuli or their effects, which moderate harm or exploit beneficial opportunities. Certain identified activities which may promote mitigation of or adaptation to climate change are listed as:

☆ Maintaining and restoring native ecosystems

☆ Protecting and enhancing ecosystem services

☆ Managing habitats for endangered species

☆ Creating refugees and buffer zones, and establishing networks of protected areas taking into account projected changes in climate.

Mitigation Practice in Forestry

Cropland management; restoration of organic soils; rice management and grazing land management are some of the areas where mitigation practices can be adopted. Grazing land managements offer vast potential for carbon sequestration. Reduced emissions from deforestation; afforestation; reforestation and forest management are some of the mitigation practices in the forestry sector. Agriculture and forests may also contribute to mitigation in energy sector through production of biomass feedstocks and energy efficiency measures. There are hidden dangers of competition with other land uses, positive or negative environmental impacts, implications for food security etc. Most mitigation practices in agriculture and forests have synergies with sustainable development and interactions with adaptation.

India's climate change program is required to work with people, government and industry to find sustainable solutions to climate change. Some of the strategies which need address towards mitigation of climate change include development of new technologies; carbon offsets; renewable energy such as biodiesel, wind power, and solar power; nuclear power; electric or hybrid automobiles; fuel cells; energy conservation; carbon taxes; improving natural carbon dioxide sinks A and R projects; deliberate production of sulfate aerosols, which produce a cooling effect on the earth; population control; carbon capture and storage; and nanotechnology. Many environmental groups encourage individual action against climate change, often aimed at the consumer, and there has been business action on climate change.

As policymakers from around the world seek ways to help the poorest to adapt to climate change, priority must be given to the role of biodiversity, an element often neglected from current adaptation strategies (Intergovernmental Panel on Climate Change 2007). Designing, funding and implementing these strategies require cooperation and co-ordination at the global level. In this respect, the Convention on Biological Diversity is working closely with the United Nations Framework on Convention on Climate Change (UNFCCC 1994).

Role of Convention on Biological Diversity (CBD)

Biological diversity is slowly beginning to gain recognition as one among the effective responses to the challenge of climate change. Recognizing the importance of this fact in climate change options, the Convention on Biological Diversity (CBD) has also decided to promote biodiversity in forest management. The resilience of ecosystems can be enhanced and the risk of damage to human and natural ecosystems reduced through the adoption of biodiversity-based adaptive and mitigative strategies. Biodiversity, however, can be a major potential tool too, because the impact of biological process can be much higher.

COP 9 decision (IX/5) of CBD on forest biodiversity urged all parties to promote multidisciplinary scientific research to better understand the impacts of climate change, including mitigation and adaption activities, and environmental degradation on ecosystem resilience, conservation and sustainable use of forest biodiversity and impacts on the livelihoods of indigenous and local communities. This is to maximize positive impacts and avoid negative impacts of climate change, including mitigation and adaption activities, on forest biodiversity; in particular those forests most vulnerable to climate change. International Union of Forest Research Organization's (IUFRO)-led initiative of the Collaborative Partnership on Forests on science and technology needs attention especially their work on climate change research.

Convention on Biological Diversity

The CBD was emerged out of the global concern for conserving biodiversity and was signed by 150 countries at the United Nations Conference on Environment and Development in 1992 in Rio de Janeiro (the 'Earth Summit'), and entered into force on 29 December, 1993. It recognized the sovereign rights of nations over their biological diversity and has three main objectives; (i) conservation of biological diversity, (ii) the sustainable use of its components, and (iii) fair and equitable sharing of the benefits arising from the use of genetic resources, including appropriate access to genetic resources and by appropriate transfer of relevant technologies. There are currently 193 parties and the treaty has been ratified by 168 countries.

The Biodiversity Act 2002

As the national response to CBD, The Biological Diversity Act, 2002 was enacted on December 11, 2002 and came into force since April 15, 2003 with the notification of the Biological Diversity Rules 2003. The Act is primarily aimed to regulate, in the territorial jurisdiction of India, access to genetic resources and associated knowledge with the purpose of realizing equitable sharing of benefits arising out of the use of these resources with the local people. The implementation of the Act is coordinated by three functional bodies namely the National Biodiversity Authority (NBA), the State Biodiversity Boards (SBB), and the Biodiversity Management Committee (BMC). The NBA is the national competent authority to discharge all decisions pertaining to ABS to persons who are not Indian citizens, Indian citizens who are non-residents and a body corporate, association or organization not incorporated or registered in India or incorporated or registered in India under any law which has any non-Indian participation in its share capital or management. The prior approval of the NBA is mandatory before applying for IPRs based on biological resources or traditional knowledge obtained from India.

Further, the Authority, while granting the approval shall ensure equitable sharing of benefits between the applicant and local bodies concerned or the benefit claimers, on mutually agreed terms and conditions. It may develop guidelines to impose and fix criteria for benefit sharing fee or royalty or both or impress conditions including sharing of financial benefits arising out of the commercial utilization of such resources; and shall notify the specific details of benefit sharing formula in an official gazette on a case-to-case basis. The National Biodiversity Authority is also empowered to take necessary measures to oppose the grant of IPRs in any other country on any biological resource or associated knowledge derived/obtained from India.

The National Biodiversity Authority is empowered to advise the Central Government on matters relating to conservation and sustainable use of biological diversity and advise the State Governments in the selection of areas to be notified as heritage sites. Share of monetary benefits will be deposited into the National Biodiversity Fund except in cases where biological resources and knowledge are accessed from specific individual or group of individuals or organizations, in which case the monetary benefit may be directly made to the providers on the direction of the Authority in accordance with the terms of any agreement and in such manner as it deems fit.

The Indian citizens or a body corporate, association or organization registered in India, shall obtain any biological resource for commercial utilization, only after giving prior intimation to the concerned State Biodiversity Board. The State Biodiversity Board, on receipt of such intimation, may, in consultation with the local bodies concerned, prohibit any such activity if it is considered detrimental or contrary to the objectives of conservation and sustainable use of biodiversity or allow use on the basis of equitable sharing of benefits. Exception to this has been provided to the local people and communities of the area, including growers and cultivators of biodiversity, and vaids and hakims.

Climate induced disasters can have devastating effects on the economy, causing huge human losses and can significantly set back development efforts of a region or a country. Those living in developing countries and especially those with limited resources tend to be more adversely affected. As a number of the most vulnerable regions are in India, preparedness and adaptation to disasters have emerged as a high priority for the country. Biodiversity resources can reduce impacts of climate change on people and production. Given the importance of climate change -biodiversity links, it is important to conserve biodiversity, preserve habitats to facilitate the long-term adaptation of biodiversity, improve our understanding of climate change – biodiversity linkages, and fully integrate biodiversity considerations into mitigation and adaptation plans.

References

Achanta, A. and Kanetkar, R., 1996. Impact of climate change on forest productivity: A case study of Kerala, India. In: Paper presented at the *Asian and Pacific Workshop on Climate Change Vulnerability and Adaptation Assessment*, Manila, Philippines, 15–19, January.

Dudley, N., 2001. *A Midsummer Night's Nightmare? The Future of U.K. Woodland in the Face of Climate Change*. The Woodland Trust Grantham, United Kingdom.

Dudley, N., 2003. *A Midsummer Night's Nightmare? The Future of U.K. Woodland in the Face of Climate Change*. The Woodland Trust Grantham, United Kingdom.

Intergovernmental Panel on Climate Change, 2001. *Climate Change: The Scientific Basis.* http: //www.grida.no/climate//ipcc_tar/wgl 409.htm

Intergovernmental Panel on Climate Change, 2007. *The Physical Science Basis: Summary for Policymakers.* http: //www.cpcc.ch/SPM2feb.07/pdf.

Leemans, R. and VAN Viiet, A., 2004. *Extreme Weather: Does Nature Keep Up?* WWF/Wageningen University, The Netherlands.

Parmesan, C. and Yohe, G., 2003. A Globally coherent fingprint of climate change impact across natural system. *Nature,* 421: 37–42.

Ravindranath, N.H. and Sukumar, R., 1998. Climate change and tropical forests in India. *Climatic Change*, 39(2–3): 563–581.

Ravindranath, N.H., Joshi, N.V., Sukumar, R. and Saxena, A. 2006. Impact of climate change on forests in India. *Current Science*, 90(3): 354–361.

Root, T.I., Price, T., Hall, K.R., Schneider, S.H., Rosenzweig, C. and Pounds, J.A., 2003. Fingerprints of global warming on wild animal and Plant. *Nature,* 421: 57–60.

U.S. Environmental Protection Agency, 2007. *Climate Change: Basic Information.* http: //epa.gov.climatechange/basicinfo.htm.

United Nation Framework Convention on Climate Change, 2002. *Biological Diversity and Ecosystems.* http: //unfccc.int/essential_background/background_publications_htmlpdf/ climate_change_information_kit/items/305.php

United Nations Framework Convention on Climate Change, 1994. *UNI.* http: //unfccc.int/essential_background/convention/background/items/2536.php

2013, Biodiversity in Horticultural Crops Vol. 4
Editor: Professor K.V. Peter
Published by: DAYA PUBLISHING HOUSE, NEW DELHI

Pages 69–91

Chapter 4

Role of Sacred Groves in the Perspective of Biodiversity Conservation with Special Reference to Kumaon Himalayas

Tariq Husain, Priyanka Agnihotri and Harsh Singh

Biodiversity and Angiosperm Taxonomy,
CSIR-National Botanical Research Institute, Rana Pratap Marg, Lucknow – 226 001
E-mail: hustar_2000@yahoo.co.uk

Conservation of plant diversity assumes greater importance when the world is facing unprecedented loss of biological diversity. As per estimation of IUCN (2010) 8,692 taxa out of evaluated 12, 873 are facing threat with extinction. Over thousands of years, local people have developed a variety of vegetation management practices and conservation of elements of biodiversity on the religious basis, plant and animal worship and observing taboos on harvesting and hunting of plants and animals are characteristics of many indigenous communities in India. Sacred groves are special sites or areas that have one or more attributes which distinguish them as somehow extraordinary, usually in a religious or spiritual sense.

"Sacred groves are the small or large patches of near natural vegetation dedicated to local ancestral spirits or deities and are preserved on the basis of religious and cultural beliefs and often represented the climax vegetation of the region".

In recent years, the significance of sacred groves (patches of natural vegetation dedicated to ancestral spirits or deities and preserved on the basis of religious beliefs) has assumed immense importance from the point of view of anthropological, taxonomical and ecological considerations (Hughes and Chandran, 1998, Tiwari *et al.*, 1999). Such a grove may consist of a multi-species, multi-tier primary forest, a clump of trees belonging to one species, or even a single old tree, depending on the history of the vegetation and local culture of the region or area. These groves are protected by local communities usually through customary taboos and sanctions with traditional, cultural and ecological implications. It is a tract of usually virgin forests of varying sizes, which are communally protected, and which usually have a significant religious connotation for the protection of community. These are traditionally managed private or community forests and they often represent the relic climax vegetation of the region. Named differently in different parts of India as *Law lyngdhoh* in Meghalaya (Upadhyay *et al.,* 2003), *Kovil kadu* in Kanyakumari (Ramanujam and Praveen 2003), *Dev bhumi* in Uttarakhand (Bisth and Ghadiyal, 2007), *Kavu* in Kerala, *Sarna* and Deorai in Madhya Pradesh (Sinha, 1995), *Oran* in Rajasthan, *Jaherthan* and *Garamthan* in West Bengal, *Deovan* in Himachal, *Ummanglai* in Manipur, etc., these groves are mainly found in areas dominated by tribals and managed by local people for various reasons. The existence of such undisturbed pockets is mostly due to certain taboos, strong beliefs, supplemented by mystic folklores (Gadgil and Vartak, 1975). The groves exhibit diversity in species of trees and other various life forms which are dependent for their existence on trees, huge climbers, epiphytes and other shade loving plants (Vartak, 1983). All forms of vegetation in such a grove, including shrubs and climbers are supposed to be under the protection of the reigning deity of that grove, and the removal of even a small twig, is taboo (Vartak and Gadgil 1973). Irrespective of their origin and size, all sacred groves are islands of greenery in the landscape protecting biodiversity and enhancing the environmental quality and represent a functional link between social life and forest management system of a region (Boraiah *et al.,* 2001). Protected areas are thought to be the cornerstones of biodiversity conservation and the safest strongholds for wildlife (Bruner *et al.,* 2001). They realise that preserving the environment is essential for their survival too. The micro-variations of the ambient harsh environment have been responsible for the rich community knowledge systems and a biodiversity necessary for sustainability of human life here. Therefore, such groups, especially in the hills, have evolved several conservation practices. The institution of sacred forests or groves is ancient and was once widespread. The nurturing of sacred forests (*Dev Van*) is one such practice. In this tradition, local communities dedicate patches of forests to their local deities or ancestral spirits.

International Scenario

A sacred grove is a grove of trees of great religious importance to a particular culture and are found everywhere around the world. Sacred groves were, most prominent in Ancient East and prehistoric Europe, but feature in various cultures throughout the world. They were important features of the mythological landscape and cult practice of Celtic, Baltic, Germanic, ancient Greek, Near Eastern, Roman, and Slavic polytheism, and were also used in Japan, Nepal and Sri Lanka and West Africa. Buddhist temples in Japan and China have tree-gardens. Traditional Chinese honour sacred mountains with trees. Buddhist monasteries and temples of Thailand have sacred groves, while Indonesia has monkey-forests. In America, both pre-Columbian people and the settlers maintained sacred groves. Examples of sacred groves include the Greco-Roman *temenos*, the Norse *hörgr*, and the Celtic *nemeton*, which was largely but not exclusively associated with Druidic practice. During the Northern Crusades, there was a common practice of building churches on the sites of sacred groves.

Excavations at Labraunda have revealed a large shrine assumed to be that of Zeus Stratios mentioned by Herodotus as a large sacred grove of plane trees sacred to Carians. In Syria, there was a grove sacred to Adonis at Afqa. In Latin the term for these demarcated places was Templum. Templum was of course the original root of the word temple. To begin with, those sacred enclosures were the sanctuaries in which religious ceremonies took place. They were in fact open air temples. When later on temples were erected as monumental buildings with columns and all, sacred groves and forests did not cease to exist. They inspired the sense of awe, mystery and in dwelling, being close to Gods.

The most famous sacred grove in mainland Greece was the oak grove at Dodona. Outside the walls of Athens, the site of the Academy was a sacred grove of olive trees, still recalled in the phrase "the groves of Academe." In the world of classical Greece, and then of Rome, these special groves and forests were usually enclosed by stone walls. This enclosure was called in Greek *Temenos*, a cut–off place, or a demarcated place. A better translation would be a sacred enclosure. In the sacred groves and forests of ancient Greece, particularly species of trees were dedicated to particular Gods. Oaks were in the domain of *Zeus*, willows of *Hera*, olivers of *Athena*, the laural of *Apollo*, pines of *Pan* and vine of *Dionysus*.

In central Italy, the town of Nemi recalls the Latin *nemus Aricinum*, or "grove of Ariccia", a small town a quarter of the way around the lake. In Antiquity, the area had no town, but the grove was the site of one of the most famous of Roman cults and temples: that of Diana Nemorensis. A sacred grove behind the House of the Vestal Virgins on the edge of the Roman Forum lingered until its last vestiges were burnt in the Great Fire of Rome in 64 CE. Beautiful pinkish flowers of *Myrtus communis* are considered highly sacred among Greeks, Egyptians, Jews and Persians, and used in religious rites and ceremonies.

Sacred groves have survived in the Baltic states longer than in other parts of Europe. The main Baltic Prussian sanctuary, now considered as a sacred grove, was Romowe. The last extermination of sacred groves was carried out in the lands of present-day Lithuania after its Christianization in 1387 and Samogitia in 1413. A sacred grove is known as *alka*(*s*) in Lithuanian.

Sacred groves feature prominently in Scandinavia. The most famous sacred grove of Northern Europe was at the Temple at Uppsala in Old Uppsala, where every tree was considered sacred - described by Adam of Bremen. The practice of blót - the sacrificial ritual in Norse paganism was usually held in *lund*s or sacred groves. According to Adam of Bremen, in Scandinavia, pagan kings sacrificed nine males of each species at the sacred groves every ninth year. The pagan Germanic tribes (ancestors of the people of modern day Germany, England and Scandinavia) also performed tree-worship and had the concept of sacred groves. It is thought that the idea of sacred trees like the Thor's Oak might have led to the concept of the present day Christmas tree.

The concept of sacred groves is present in Nigerian mythology as well. The Osun-Osogbo Sacred Grove, containing dense forests, is located just outside the city of Osogbo, and is regarded as one of the last virgin high forests in Nigeria. It is dedicated to the fertility God in Yoruba mythology, and is dotted with shrines and sculptures. Suzanne Wenger, an Austrian artist, helped revive the grove. The grove was designated as a UNESCO World Heritage Site in 2005.

Sacred groves are also present in Ghana. One of Ghana's most famous sacred groves - the Buoyem Sacred Grove - and numerous other sacred groves are present in the Techiman Municipal District and nearby districts of the Brong Ahafo Region. They provide a refuge for wildlife which has been exterminated in nearby areas, and one grove most notably houses 20,000 fruit bats in underground caves. The capital of the historical Ghana Empire *El-Ghaba*, contained a sacred grove for performing religious rites of the Soninke people. Other sacred groves in Ghana include sacred groves along the coastal savannahs of Ghana. Many sacred groves in Ghana are now under federal protecttion - like the *Anweam Sacred Grove* in the Esukawkaw Forest Reserve. Other well-known sacred groves in Ghana include the *Malshegu Sacred Grove* in Northern Ghana - one of the last remaining closed canopy forests in the savannah regions, and the Jachie sacred grove.

Sacred groves in Japan are typically associated with Shinto shrines, and are located all over Japan. The Cryptomeria tree is venerated in Shinto practice, and considered sacred. Among the sacred groves associated with such *jinja*s or Shinto shrines is the 20-hectare wooded area associated with Atsuta Shrine (*Atsuta-jingū*) at Atsuta-ku, Nagoya. The 1500-hectare forest associated with Kashima Shrine was declared a "protected area" in 1953. Today it is part of the Kashima Wildlife Preservation Area. The woods include over 800 kinds of trees and varied bird life and plant life. Tadasu no Mori is a general term for a wooded area associated with the Kamo Shrine, which is a Shinto sanctuary near the banks of the Kamo River in Northeast Kyoto. The ambit of today's forest encompasses approximately 12.4 hectares, which are preserved as a national historical site. The Kamigamo Shrine and the Shimogamo Shrine, along with other Historic Monuments of Ancient Kyoto (Kyoto, Uji and Otsu Cities), have been designated World Heritage Sites since 1994. The Utaki sacred sites (often with associated burial grounds) on Okinawa are based on Ryukyuan religion, and usually are associated with toun or *kami-asagi* - regions dedicated to the Gods where people are forbidden to go. Sacred groves are often present in such places, as also in Gusukus - fortified areas which contain sacred sites within them. The Seifa-utaki was designated as a UNESCO World Heritage Site designated in 2003. It consists of a triangular cavern formed by gigantic rocks, and contains a sacred grove with rare, indigenous trees like the Kubanoki (a kind of palm) and the *yabunikkei* or *Cinnamomum japonicum* (a form of wild cinnamon). Direct access to the grove is forbidden.

National Scenario

Indian sacred groves are sometimes associated with temples/monasteries/shrines or with burial grounds (which is the case in Shinto and Ryukyuan religion-based sacred groves respectively in Japan). Sacred groves may be loosely used to refer to other natural habitat protected on religious grounds, such as Alpine Meadows.

The conservation of forest patches, groves and trees probably date back to the pre-epic period in Indian history. In India, sacred groves are found in a wide range of ecological situations from estuaries to mountain localities. According to Gadgil and Vartak (1976) the important regions with sacred groves in India are the North Eastern Himalayas (Khasi-Garo hills), Western Ghats, Aravalli Hills of Rajasthan and Sarguja, Chandes and Bastar area in Central India.

The practice of keeping sacred groves was prevalent among different communities depending on their cultural practices. Gadgil and Vartak, (1976) recorded Western Ghats as one of the important regions with sacred groves. The history of sacred groves in India may be traced back to the hunting-gathering societies which attributed sacred value to patches of forests within their territories, similar to the way they treated several other topographic or landscape features like the mountain peaks, rocks, caves, springs, and rivers. The local people set aside a portion of land around the house as abode of God, Goddess or Serpent and this was also a means for preserving medicinal plants. They attribute the origin of sacred groves to more secular causes, for the preservation of rare and valuable plants. They hold that the practice started well before sixth century AD.

It is inferred that the practice of setting aside patches of forests as sacred groves was strengthened with the spread of shifting cultivation involving clearing of forests. The reason for this could be religious and cultural compulsions as well as subsistence and ecological needs. Historical evidence shows that the human habitation in North Kerala started about 5000 years ago. Even though the Brahmin migration to this region began in the early centuries of Christian era, well organized Brahmin settlements began to appear only by the 7-8th century AD. One of the consequences of the Brahmin settlements was the spread of settled agriculture especially in the coastal plains and the midland, leading to deforestation in these regions. Introduction of the commercial plantations in the highland by the British invariably resulted in the destruction of the highland forests and its fauna. Initially highland cultivation was restricted to the areas with less incline such as Wayanad. But later a major part of the highland forests of Kerala has been cleared for cultivation of tea, coffee, cardamom, and other cash crops. In spite of the massive destruction of forest elsewhere, there were patches of forests in Kerala set aside in the form of sacred groves.

The attitude of pre-British village communities towards forests is reflected in the statement of the British traveler, Francis Buchnan, made near Karwar in Uttar Kannada "The forests are the property of the Gods of the villages in which they are situated, and the trees ought not to be cut without having leave from the Gauda or headman of the villages... who is also priest (pujari) to the temple of the village God." Nevertheless the remains of scores of sacred groves, the leftovers of a loafty tradition, continue to remain in many parts of India. The Gurjan tree (*Dipterocarpus indicus*) has its northern limit in the Western Ghats in a couple of sacred groves of Uttara Kannada. Similarly, myristica swamps, a rare and threatened habitat, belonging to Southern Kerala, has its northern limit once again in a sacred grove of Uttara Kannada. A rare tree species, *Myristica magnifica* and *Pinanga dicksoni,* a beautiful slender endemic palm from Western Ghats are characteristic plants of this swamp. A new species of a Fabaceae climber *Kunstleria keralensis* has been reported from one of the "*kavus*" of Kerala. Another "*Kavu*" has five species of the evergreen tree *Hopea* of which three are endemic to south-west India. Four more threatened species of plants *Blepharistemma membranifolia, Buchanania lanceolata, Pterospermum reticulatum* and *Syzygium travancoricum* have been discovered. In Uttarakhand, Yumnotri, Badrinath, Kedarnath (Garhwal), etc are the sacred places more than 3200 m in altitude. Several other sacred groves are known such as Thatyur, Maanthaat, Danda Ka Deorana from Dehradun, Sem Mukhim Naag from Tehri Garhwal, etc. In Kumaun region, Chitai Gul, Jageswar, Dhaula Devi, Anyar Bunga Aeri, Sidh Baba from Almora, Garanath, Binsar, Minar, Chamarkhan, Bari Goljyu, Gauri Udiyar in Bageshwar district, etc

Sacred groves play an important role in the religious and socio-cultural life of the local people. Rituals and ceremonies are often carried out in the sacred groves to propitiate ancestral spirits and deities for enhancing agricultural yields and for well-being of animals and human beings. Often vows and offerings are made at the sacred groves for wish-fulfilment and various festivals are also held at these sacred groves. A large number of distinct local art forms and folk traditions are associated with the deities of sacred groves, and are an important cultural aspect closely associated with sacred traditions. Ritualistic dances and dramatizations based on the local deities which protect the groves are called *Theyyam* in Kerala and *Nagmandalam*, among other names, in Karnataka. Often, elaborate rituals and traditions are associated with sacred groves, as are associated folk tales and folk mythology.

In India, sacred groves are scattered all over the country, and do not enjoy protection via federal legislation. Some NGOs work with local villagers to protect such groves. They were maintained by local communities with hunting and logging strictly prohibited within these patches. However, the introduction of the protected area category community reserves under the Wildlife (Protection) Amendment Act of 2002 has introduced legislation for providing government protection to community held lands, which could include sacred groves.

Sacred Groves in Kumaon Himalaya

Sacred groves of Kumaon Himalaya are mainly covered by *Rhododendron arboretum, Quercus leucotricophora, Quercus semecarpifolia, Cedrus deodara, Pinus roxburghii, Juniperus communis, Cupressus tortulosa* etc. and most of them are Panchayat forests. These are as follows:

Haat Kali Sacred Grove

This grove is tribute to Maa Kalika, situated near Rawal goan in Gangolihat tehsil, about 77 km. from Pithoragarh city at an altitude of 1668 m (N 29° 39.459' E 080° 02.808') (Figure 4.1). It is rich in folk culture, music and religious traditions and had been chosen by Sankaracharya for the installation of Mahakali shaktipith. The monthly average minimum temperature fluctuated from 0.5 °C to 5 °C and maximum from 15°C to 32°C. The grove is mainly surrounded by trees of *Cedrus deodara* (represented as a sacred tree). There are several bells hanging all around the trees and the outer boundary. This grove is inhabited by *Rawal* communities and *Pant* communities are the priests of the temple. Various religious programme are performed by the local people of the Gangolihat area such as 'Jag' (it is a form of jagran in respect of deity or ancestor and this activity is performed for one day), 'Jaagar' (for two days, it falls in the category of spiritual worship, in the form of a folk song in honour of the various Gods and Goddesses, the famous deities Ganganath, Gorila and Bholanath. Around the burning fire, in a circle, members of the village or family invoke the spirit with measured drum beats and suddenly the Dangariya or medium with *Hurka* (a folk instrument). Coupled with this singing, punctuated by the exotic drum-beats, and the thrill sound of the *thali*, 'Chauras' (it is a form of cultural and traditional activity for four nights) etc. is being performed in this grove. The grove has moist temperate forest with different life-forms and covers an area of ca. 1.7 *ha*. The dominating gymnosperm tree *Cedrus deodara* is continuously increasing its area due to protection by Panchayat of *Rawal* community.

Nakuleshwar Sacred Grove

Nakuleshwar sacred grove (Figure 4.2), *ca.* 250-300 years old, is located in Et guan Shilling near about 8 km from Pithoragarh town with an area of 300-350 sq m. It exists at an altitude of 1523 m (N 29° 32.698´, E 080°13.717´) and is covered by another sacred forest of Thal Kedhar. The monthly average temperature fluctuated from 0.8 °C to 18.1 °C (minimum) and 15 °C to 31 °C (maximum) respectively. Average rainfall recorded during the study period was 1400 mm and average humidity noticed was 64 per cent. The climatic conditions are suitable for luxuriant growth of various micro habitat associations. The grove has a temple of Lord Shiva and contains about 38 stone images of Shiva Parvati, Mahishasur Mardini, Surya Dev, Vaman Dev, etc. The architectural design of the temple is on the pattern of Khajuraho temple but most of the status present here are disfigured and broken because of neglect and age (Singh *et al.,* 2011).

Vaishno Devi Sacred Grove

This grove is located in Jakhani near Gangolihat tehsil about 65 km from the main town of Pithorgarh district. It is more than 100 years old and spread over in an area of 30 *ha*. It houses a small temple of Goddess Vaishno (Figure 4.3). There is a long climb of about 3 km in a straight height from the base village Bhuligaon. It is located at an altitude of 1938 m (29°37.801'N, 080° 03.410'E) and inhabited by local communities such as Upreti and Bhandari. The whole grove is densely covered by *Cedrus deodara, Quercus leuchotricophora* and *Myrica esculentum.* (Agnihotri *et al.,* 2010).

Chandika Devi Sacred Grove

This grove is famous for its biodiversity and deity Maa Chandika Devi (Figure 4.4). The whole grove is covered by *Quercus leucotricophora* tree locally known as Banj Forest. It is located in Chandak near about 10 km from the main town of Pithoragarh at an altitude of 1910 m., N-29°37.761 lattitude, E-080° 12.586 longitude. It covers an area of 3 sq.km and about 250 years old. The grove is maintained by local communities such as Chupal, Bhatt and Bisth. The sacrificing of animals is a common indigenous tantrik ritual and is performed here daily (Agnihotri *et al.,* 2010).

Figure 4.1: Haat Kali Sacred Grove, Gangolihat

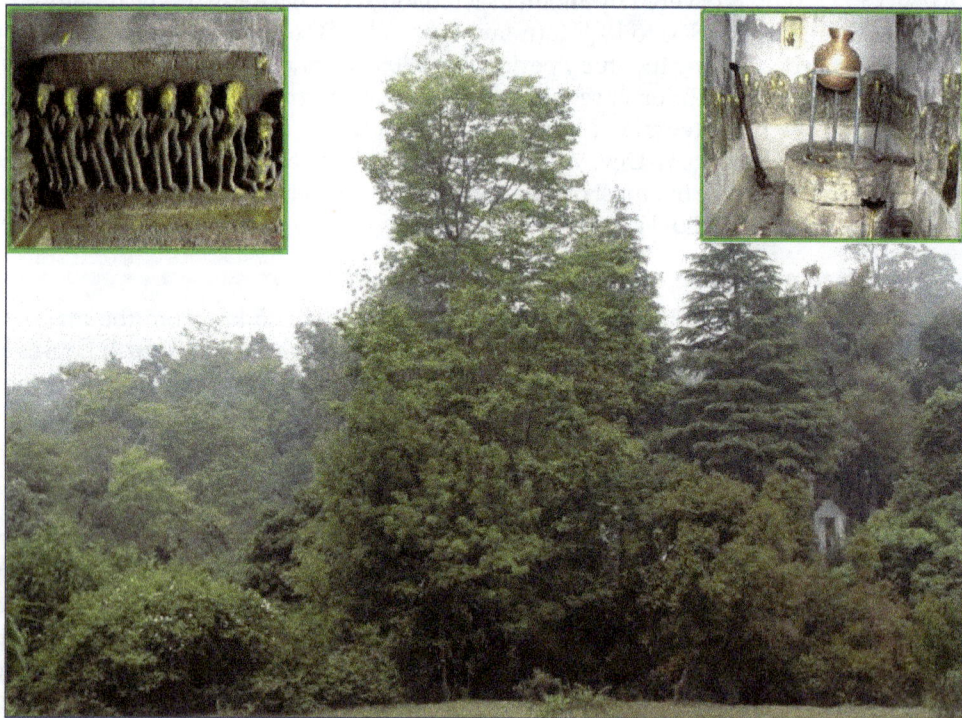

Figure 4.2: Nakuleshwar Sacred Grove, Et gaon, Pithoragarh

Figure 4.3: Vaishno Devi Scared Grove, Jakhani

Figure 4.4: Chandika Devi Sacred Grove, Chandak

Malay Nath Sacred Grove

This grove is situated in Didihat (53 km from the main Pithoragarh) and 2 km by foot in Sherakote village at an altitude of 2000 m. From enquiry with local people, it was known that this grove is 1000 years old, covering an area of about 2.0 *ha*. Seasonal variations are quite similar to Dhwaj sacred grove due to same altitude. Malaynath was the son of Bhaglinga. He was a disciple of Guru Gorakh Nath. Golju was his classmate along with Bhagyalaxmi (who became the wife of mighty king of Chhiplakote). Malaynath ji is the *Isht devta* (God) of Digtar natives (Didihat). His temple is situated on the top of Sherakote village in the grove. The views of Panchachuli and Nanda devi peak are clearly seen from the groves. Sherakote is basically inhabited by *Bora* and *Chuphal* communities. The whole grove is covered by trees of *Rhododendron arboreum* and *Quercus leucotricophora* showing its climax vegetation with humus rich soil.

Dhwaj Sacred Grove

This grove is situated near Totanula, about 15 km from the main Pithoragarh on the route of Didihat. It is a beautiful place with luxuriant growth of lichens to angiosperms; hence showing great phytodiversity. It is about 2 to 3 km from the road by foot at an altitude of 2100 m, below the main temple, is situated the cave temple of Lord Shiva (Baba Khande Nath). Local deities are Jyanti Devi and Maa Kalika. According to local people, this sacred grove is about 600-700 years old and covers an area of *ca*. 1.5 ha. The monthly average minimum and maximum temperature fluctuated throughout the year from 0 °C to 18 °C and 13 °C to 25 °C respectively. In January to February (winter season), for one or 2 days or sometimes a few hours' snow remains in this area and covers whole area of the grove. The broad leaved trees of *Quercus leucotricophora* and *Rhododendron arboreum* are dominant in this grove and support the excellent and luxuriant growth of the non vascular plants. From the hill top where the Jayanti Devi temple is situated, the Panchachuli and Nandadevi peaks can be clearly sighted.

Narayan Swami Ashram Sacred Grove

This sacred grove is tribute to Lord Vishnu, situated on Tawaghat (Dharcula, 92 km from Pithoragarh city) about 16 km from Tawaghat at an altitude of 2715 m (N 29°56.531´; E 080°39.326´). It was established in 1936 by Narayan Swami. It comprises an area of about 200-300 sq m. Winter season is long and temperature reaches upto 0°C to -1 °C and snow falls in January and February while summer season is quite short–April mid to June mid. On the other hand, rainy season starts on mid of June to August last. Moderate climatic conditions make the grove home for many medicinal plants, especially lower plants. It is inhabited by *Bhotia, Boxa* and *Van Rawat* tribal communities. The forest belongs to temperate type of forest dominated by *Cedrus deodara, Quercus semecarpifolia* and *Rhododendron arboreum.*

Chamunda Devi Sacred Grove

This grove contains a small and very famous temple of Goddess Chamunda Devi and is located near Gangolihat tehsil about 75 km from the main Pithoragarh town at an altitude of 1786 m (N-29° 38.997, E-080° 16.221). The whole grove is covered by dense forest of *Cedrus deodara* and harbours a rich floristic diversity. Every half of the year, the local people gather and pray Maa Chamunda by doing *Jagrans, Jags or Jagar* in this grove. There are many local stories and folklores associated with this grove.

Hokra Sacred Grove

This grove is tribute to Maa Hokra Devi, which is a triangular statue, situated 10 km away from Tejam (on the route of Munsiari) at an altitude of 1700 m and comes under Talla Johar area. Hokra is one of the difficult areas to reach. According to B.D. Pande in '*Kumaun Ka Itihas*', this area was inhabited by Hokari tribals. According to the natives of Hokra, earlier this area was inhabitated by judh Jaath but due to ethnic reason (it may be natural disaster) many people of this area died. But, at the present time, there is no such type of tribal group found. Now a days, this area is inhabited by *Mehta* and *Bisht* communities who came from Chitorgarh. Hokra is present in the west of Munsiayri region along with Ramganga river. In the east of Hokra area, Sandhrathi, Quiti, Door, Gogina are present. Bhuyal and Namik Glacier are present in the north and Liti and Sama in the south. The humus rich soil of this grove is suitable for the growth of *Paris polyphylla*. The whole grove is covered and dominated by *Cinnamomum tamala* locally known as "tejpat" with *Zanthoxylum armatum* (Timur), *Fagopyrum esculantum* (Ugal), etc. Summer starts from April to June and temperature fluctuates (20°C to 30°C) throughout the year, while winter is quite long from October to March with temperature -0°C to 5°C. The rainfall is quite lower than other studied sacred groves starting from mid May to August. The famous cultural festival Chaitra Ashtami (in March) is performed and organized by local people of Talla and Malla Johar regions. This cultural festival is based on sacrificing of Goat and buffaloes. In August, another festival, Nanda ashtami was done by whole area of Johar.

Patal Bhuvneshwar Sacred Grove

This grove is the hidden pilgrimage centre, near Gangolihat tehsil in the Pithoragarh district of Uttarakhand. Patal Bhubaneshwar, literally means the sub-terranean shrine of Lord Shiva. The cave temple is 90km from Pithoragarh and 13 km north of Gangolihat at an altitude of 1618 m (N 29° 41.295′, E 080° 05.510′). The way to the main temple is through a tunnel which leads into the cavern through a narrow dark passage of water. Size of the grove is *ca*.1 ha. The holy cave possesses thousands of stone images of deities and ancestors (Lord Shiva along with Garuda, Ganesha, Sheshnag, etc.). According to local beliefs, this sacred grove is more than 2000 years old and it has been described in the Manas Khand of Skand Puran. "*Adi Shankaracharya*" selected this place for his prayer. It is a myth that the cave was formed of the Jatas (hairs) of Lord Shiva. The priest families, the *Bhandaris* are performing religious rites at the grove. But, some communities like *Guro*, *Rawal* also contribute their faith in the grove. The climate of the area is moderate. Summer season is quite short and sometimes temperature reaches upto 32°C. Winter season is long for five months starting from October to February (temperature ranges between 3°C to 0°C). Rainy season is mid May to last August. The nature of soil is acidic due to the presence of *Cedrus deodara* trees. Sometimes the colour of the soil is changing from reddish to black. The whole grove is covered by trees of *Cedrus deodara* along with shrubs such as *Berberis asiatica*, *Pyracantha crenulata*, *Prinsepia utlis* and herbs likes *Geranium ocellatum*, *Cynadon dactylon*, *Galinsago parviflora*, *Arisaema jacquamontii*, etc.

Phytodiversity of Sacred Groves in Kumaon Himalaya

Sacred groves harbour a rich phytodiversity with many rare and endemic flora. The present documented sacred groves of Kumaon Himalaya represent 322 species belonging to 275 genera and 80 families. Trees like *Quercus leucotricophora*, *Rhododendron arboreum*, *Myrsine semiserrata*, *Coriaria neplanesis*, *Acer oblongum*, *Acer caesium*, *Aesculus indica*, *Prunus cerasoides*, *Ficus palmata*, *Ficus roxburghii*, *Quercus incana*, *Q. dilatata*, *Q. semicarpifolia*, *Toona ciliata*, *Viburnum cotonifolium*, *V. cylindricum*, etc. shrubs like *Berberis asiatica*, *B. chitria*, *B. aristata*, *Pyrancantha crenulata*, *Prinsepia utilis*, *Rubus ellipticus*, *Zanthoxylum armatum*, *Woodfordia fructicosa*, *Mahonia nepaulensis*, *Jasminum officinale*, *Hypericum oblongifolium*, *Desmodium heterocarpum*, *Daphne papyracea*, *Callicarpa arborea*, *Boenninghausenia albiflora*, *Buchanania lanzan*, etc., herbs like *Ainsliea aptera*, *Hedychium spicatum*, *Origanum vulgare*, *Bergenia ciliata*, *Viola serpens*, *Viola canescans*, *Polygala persalifoila*, *Goodyera hemsleyna*, *Ophiopogon inermis*, *Urtica dioca*,

Girardinia diversifolia, Duchesnia indica, Fragaria nubicola, Paris polyphylla, Arisaema jacquemontii, Primula denticulata, Centella asiatica, Cynodon dactylon, Corydalis crenuta, Impatiens balsamina, Desmodium elegans, Dicliptera bupleuroides, Eclipta prostrata, Erigeron acer, Geranium ocellatum, Geranium wallichianum, Oxalis latifolia, Oxalis corniculata, Gerbera gossypiana, Valeriana jatamansi, Inula cappa, Lindenbergia grandiflora, etc. Climbers like *Asparagus racemosus, Hedera nepalensis, Rubus niveus, Smilax aspera, Galium rotundifolia, Clematis buchananiana, Clematis montana, Clematis gynandra, Ipomoea* sp., *Ficus sarmentosa, Rosa microphyllus, Stephania elegans, Cissampelos pareira,* etc. making the groves of this area unique and rich in phytodiversity.

Lower plants such as lichens, bryophytes, pteridophytes, etc. also contribute to the phytodiversity of groves. Parmeliod lichens such as *Parmotrema* (*P. austrosinense, P. reticulatum, P. tinctorum, P. nilgherrense, P. ultraleucens, P. hababianum, P. praesorediosum, Parmelinella wallichiana*) are dominant in the grove and represent thinned forest type. Bryophytes like *Fissidens javanicus, F. grandifrons Trachypodopsis serrulata, Pogonatum microstomum, Thuidium assimile, Rhodobryum roseum, Bryosedgwickia aurea, Campylopus goughii, Cratoneuron filicinum, Meteorium buchananii, Erythrodontium julaceum, Hynum cupressiforme, Macromitrium rigbyanum, Vesicularia montagnei, Entodon flavescens, Barbula spp, Entodon luteonitens, Herpetineuron toccoae,* while there are only 7 species of liverworts (*Marchantia paleacea, Plagiochasma appendiculatum, Dumortiera hirsuta, Ptychanthus striatus, Frullania ericoides, Conocephalum conicum* and *Porella hattorii*). Pteridophytes like *Adiantum capillus-veneris, A. edgeworthii, A. venustram, Asplenium dalhousiae, Cheilanthes dalhousiae, Selaginella bryopteris, S. indica, etc.* enhance the phytodiversity of the different groves of Kumaon Himalaya.

Sacred Plant Species

From pre historic time, plants and animals are the parts of our life. Some plant species are grown in sacred places because people thought that ancestors and deities live in these plant species and they protect their life. Plants are the oldest creation of God on earth and the conscious about them is as old as the human civilization. Plant worship and use of plants in worshiping the God is one of the earliest and most religious things in India. Numerous references are available in literature where plants are treated as the abode of God. In the scriptures, these plants are mentioned as *Kalpa vrisksha* and *Chaitya vrisksha*, indicating that worship of the trees is an ancient Indian tradition. These plants are often grown along and within the temples and can be considered as "sacred plants". Various religious ceremonies are based on these trees or plants. In India, there are many festivals, which are based on flora. Basil (*Ocimum species*), Asoka (*Saraca asoca*), Banyan tree (*Ficus bengalensis*), Peepal (*Ficus religiosa*), Kela (*Musa paradisiaca*), Neem (*Azadiaracta indica*), Aam (*Mangifera indica*), Amla (*Emblica officinalis*) and Beal (*Aegle marmelos*) etc., are sacred plant species in India. Many of them like the sacred basil and neem are multi-purpose medicinal plants. These culturally valued species are often ecologically important keystone species, which by their key role in ecosystem functioning contribute to support much biodiversity associated with it.

Uttarakhand, popularly known as *Dev Bhumi* possesses many temples which are devoted to local ancestors and deities such as Golu Dev, Gang Nath, Syamju, Harju, Kot gadi, etc. In Kumaun, sacred groves are known under '*Devta Than*' or simply '*Than*'. People of this state are closely associated with nature and environment. Several studies were carried out in Almora district on the religious or sacred plants (Sharma and Joshi, 2010). Several plant species are present in the religious ground and can be considered as sacred plant species as these plants are protected by deities (in terms of ethnic faith). One such set of species found in the Garhwal region is *Quercus* species. Many ethnic traditional, religious and cultural believes are associated with plant species (folk music, dance, literature and poetry). In spite of this, these plant species play significant role in our daily life. These species are used as fodder, fuel wood and timbers, apart from the fact that they play a key role in nutrient cycling and conservation, as well as in ensuring water balance within the soil. *Cedrus deodara, Osmanthus fragrance, Pinus roxburghii, Ficus* sp. are also found in the sacred areas.

Singh and Pande (2000) mentioned many sacred plant species which are grown on religious ground with special reference to Kumaun Himalaya. These plants are *Olea cuspidata* Wall. ex. G. Don., *Bombax ceiba* L., *Aesculus indica* (Panger), *Bauhinia variegata* L. (Qeuaral), *Celtis australis* L. (Khadig), *Ficus cunia* (Khunia) Buch.-Ham., *Ougeinia oojeinensis* Roxb., *Rhus parviflora* Roxb. (Soutar), *Sapindus mukorossi* (*Ritu*) Gaerth. and *Rhododendron arboreum* (*Burans*). Some important sacred plant species are described as follows:

Deyar or Devdar (*Cedrus deodara* (Roxb. ex D. Don) G. Don)

It is one of the important sacred trees of Kumaun Himalayas and is often seen around the temple at high altitude. Its wood is used in yagya and other religious ceremonies. The bells are hanging on this tree. This tree is seen in Haat Kali, Chamunda Devi, Vaishno Devi, Patal Bhuvnsehwar, Nakuleshwar, Thal ke dhar sacred groves in Pithoragarh. Sometime the dense forests of Devdar are seen in the different temples of Kumaun area.

Bhanj (*Quercus leucotricophora* A. Camus)

It is another sacred tree and helps in retention of water in the soil and helps in preventing soil erosion in the area and often seen in Haat Kali in Gangolihat, Malay Nath in Didihat, Dhwaj near Totanula village and Patal Bhuvneshwar sacred groves. It sometimes covers whole area of the grove and provides shadow to the grove.

Burans (*Rhododendron arboreum* Smith)

It is seen in high altitude of hilly regions. In Narayan Swami Ashram in Dharchula, Malay Nath in Didihat, Haat Kali and Vaishno Devi in Gangolihat, this tree is very common and can be considered as sacred tree. Its flower is offered to God. The red flowers of the tree are the symbol of coming spring season.

Chir (*Pinus roxburghii* Sarg.)

It is also a common sacred tree in Kumaon Himalayas and often seen around the temples such as Haat Kali in Gangolihat. Its wood is often used in *Yagya* and *Bhandara.*

Pangar (*Aesculus indica* (Colebr. ex Cambess.) Hook.)

This tree is also seen in many temples and is seen in Narayan Swami Ashram, Vaishno Devi, Chandika Devi etc. It is found at the altitude of ca. 2000 m.

Kaddig (*Celtis australi* L.)

This tree is often seen near the 'Devta Than' of villagers *i.e.,* Maddyau deity in Gangolihat etc.

Queral (*Bauhinia variegata* L.)

This is another tree commonly found near the temple at low altitude, for example, *Shyamju* in Gangolihat is covered with these trees and can be considered as sacred tree. Its flowers and leaves are used in the temple for distributing *prasad* and *tiika.*

Bedu (*Ficus palmata* Forskl.)

Its leaves are used in religious ceremonies and are seen in the temple of Gangnath in Gangolihat.

Bael (*Aegle marmelos* (L.) Correa)

It is also sacred tree of Rutaceae family, its leaves and fruits are devoted to Lord Shiva. The trifoliate leaf or tripatra of the bael tree is believed to symbolize the three functions of the Lord-the creation, preservation and destruction–as well as his three eyes. This tree is commonly seen in Nakuleshwar sacred grove. It is also used in many diseases.

Timul (*Ficus auriculata* Lour.)

This is a small tree with broad leaves. This tree is considered as sacred tree because its leaves are often used in the rituals and cultural activities. In the old Kumaoni areas, *Pittah* (*Tikka*) are put in the leaves of this tree. Prasad is also given in these leaves. The temples are often seen around the Bhandari Golu Dev in the village Bhuligaun in Gangolihat area, Maddayu in Gangolihat, Haat Kali sacred grove, etc.

Khunia or Khuniyo (*Ficus cunia* Ham. ex Roxb.)

This tree is rarely seen in a few sacred groves of Pithoragarh district.

Ethnobotanical Aspects in Sacred Groves

Vegetation is the most precious gift, nature has provided to us, as it is meeting all kinds of essential requirements of the human beings in the form of food, fodder, fuel, medicine, timber, resins and oil, etc. (Ford, 1994; Jain, 1996).The communities of different sacred groves also possess the knowledge on the surrounding flora. The social and scientific role of ethnobotany is consistently defined and its importance as a tool for conservation strategies at local and regional level is now well recognized by the scientific community throughout the world. Traditional medicine or ethnomedicine, a set of empirical practices, found in a social group are transmitted orally generation to generation to cure the health ailments are strongly related with the religious beliefs and indigenous culture. Traditional medicinal knowledge of indigenous people has fundamental importance in the management of local resources, in the husbandry of regional biodiversity and in providing locally valid models for sustainable life. The ethnic people have emotional and symbiotic relationship with biodiversity of which they are protectors and conservators either on the basis of religious beliefs or social, cultural taboo. Today, when there is enormous loss in the world biodiversity due to overexploitation for economic purposes, urbanization, population growth and changes in climatic conditions, sacred groves are acting as gene sanctuary for several RET and ethnomedicinal plant species.

Various plant species such as *Valeriana jatamansi, Bergenia ciliata, Origanum vulgare, Hedychium spicatum, Berberis asiatica, B. aristata, B. chitria, Mahonia nepalensis, Viola canescans, Asparagus racemosus, Thalictrum foliolosum, Myrica esculentum, Berberis asiatica, Juglans regia, Ficus roxburghii, Ficus palmata, Aegle marmelos, Cedrus deodara, Quercus leucotricophora, Pyracantha crenulata,* are used by local communities for various purposes such as medicine, fodder, fuel, agricultural equipments, edible, household goods, spices, etc. Roots of *Valeriana jatamansi, Berberis asiatica, Asparagus racemosus, Thalictrum foliolosum, Mahonia nepalensis, Rubus ellipticus,* etc., rhizome of *Hedychium spicatum,* bark of *Cinnamomum tamala, Quercus leucotricophora,* etc, flowers of *Rhodendron arboreum,* wood of *Cedrus deodara,* fruits of *Myrica esculentum, Ficus roxburghii, Fagaria nubicola,* etc., whole plant of *Origanum vulgare, Viola canescans, Bidens pilosa,* are used mostly in the treatments of various ailments. *Pyracantha crenulata,*

Quercus leucotricophora, Quercus incana, Cedrus deodara, Pinus roxburghii, etc., are used for making household goods and agricultural equipments. For spices, bark and leaves of *Cinnamomum tamala* and thallus portion of parmeliods are generally used.

Some important ethnomedicinal plants are as follows:

Abutilon indicum (L.) Sweet (Malvaceae) *Ln.* Kanghi

Fresh leaves are cooked and eaten to cure piles. The matured leaves are smeared with castor oil, gently warmed and stuck on abcess for quick healing. The extract of leaves along with ghee is considered a remedy for Diarrhoea. Decoction of leaves is useful for gonorrhoea, inflammation of bladder and enema and vaginal infection. The powdered flowers are eaten in ghee as a remedy in blood vomiting and in cough.

Achillea millefolium L. (Asteraceae) *Ln.* Pharangi

The decoction of whole plant is used in cold, cough, fever and chewable for the relief of toothache.

Ageratum conyzoides L. (Asteraceae) *Ln.* Sadevi

Juice of entire plant is applied on cuts and wounds as haemostatic. Leaves ground with calcium paste are applied on cuts. Root is medicine for antilithic for kidney stone.

Annona squamosa L. (Amaranthaceae) *Ln.* Sharifa

The crushed leaves are applied to the nostrils in hysteria and fits and poultice of leaves with salt is used for ulcers and boils to induce suppuration. Ripe fruits are used orally to cure indigestion.

Asparagus racemosus Willd. (Liliaceae) *Ln.* Satawar, Satmul

Rhizome is chewed and the paste of it applied on the incision of insect bite. Rhizome extract is administered orally as an antiseptic to injuries. Seed powder is dissolved in butter milk and taken before going to bed to facilitate bowel motion. The crushed root is applied on burns.

Bauhinia variegata L. (Caesalpiniaceae) *Ln.* Kachnar

Powdered flowers in combination with "Saunf" are helpful in epilepsy and for regaining lost sexual vigour. Decoction of bark (15-20 ml.) alone or with black pepper is taken as tea for promoting flow of urine, leucorrhoea, easy sheding of placenta after delivery and glandular adenitis. Bark is boiled in equal amount of water. Luke warm decoction is used to wash wounds in foot and mouth disease.

Berberis aristata DC. (Berberidaceae) *Ln.* Kilmora

Tender leaf buds are chewed and kept inside, pressing between the teeth for about 15 min. to relieve dental carries. The extract of the fresh root (after filteration with a clean cloth) is applied externally for the treatment of eye disease.

Capparis zeylanica L. (Capparidaceae) *Ln.* Jakhambel

The root is made into a paste and applied externally to the body in rheumatism. The leaves are applied as a poultice to piles, boils and swelling and also as a counter-irritant.

Celastrus paniculatus Willd. (Celastraceae) *Ln.* Kanghi

Root together with seeds are crushed and made into paste, the paste is applied externally in body ache and joint pain. The crushed root is used for pneumonia and seed oil of this gently warmed and massaged externally in stomach pain. The paste of root is applied on swollen veins.

Celosia argentea L. (Amaranthaceae) *Ln.* Kurdu

The paste of the leaves is applied on scorpion-sting. Seeds are blood purifier and used against diarrhoea and mouth sores. The flowers are useful in blood-dysentry, spitting of blood and menorrhagia.

Centella asiatica (L.) Urban (Apiaceae) *Ln.* Brahmi

Aqueous extract of herb is given ½ teaspoonful twice a day, early in morning and at night after meals for 14 to 28 days to treat stomach-ache, as a blood purifier, tonic, in fever and in leucorrhoea. Leaves are dried in shade, ground and powdered, approximately ½ teaspoonful given thrice a day for 30-90 days to treat mental disorders. Leaf paste is applied on skin ailments.

Cissampelos parierra L. (Menispermaceae) *Ln.* Akanbindi

The aqueous extract of the root is given three times a day for three days in fever. The leaf juice is applied to remove redness of the eyes, twice a day for three days.

Cleome gynandra L. (Cleomaceae) *Ln.* Hulhul

Leaf paste is commonly used externally in headache, rheumatism, neuralgia and to enhance blood circulation. It is applied 2 times daily to cure boils. Seeds are antihelmintic and vapours of boiling seeds are inhaled 3 times a day in whooping cough, congestion and fever.

Cleome viscosa L. (Cleomaceae) *Ln.* Jungali-Hurhur

Leaf is used against boils, ear ache, ulcer, wounds. Seeds are used to control pus formation in boils.

The dried seed powder is mixed with sugar and administered orally twice a day for 7 days to relieve body pain.

Duchesnia indica (Andr.) Focke (Rosaceae) *Ln.* Kiphalia

Ripe fruits are edible and used in stomach problems. Flowers are used in eye disorders and swellings.

Eclipta prostrata (L.) L. (Asteraceae) *Ln.* Bhangra

Leaves along with the seeds of *Foeniculum vulgare* are boiled in coconut oil (*Cocos nucifera*) and the oil extract is applied on the head daily in the morning hours for a week to treat dandruff.

Fumaria indica (Haussk.) Pug. (Fumariaceae) *Ln.* Pithpapra

Whole plant except root is used in medicine for antihelmintic, blood purifier, diuretic, fever, flue, indigestion and liver complications.

Gerbera gossypiana (Royle) Beauv. (Asteraceae) *Ln.* Kapasi

Root is medicine for blood pressure and gastric troubles.

Osbeckia stellata Ker.- Gawl. (Melastomataceae) *Ln.*

Decoction of leaves are used for mouth wash and oral ulcer. Pastes of leaves are used to colour hand and nails.

Prunus cerasioides Don. (Rosaceae) *Ln.* Payan

Ripe fruits are edible and considered as blood purifier. Poultice of pieces of inner bark boiled along with ginger, jayphal and cloves is applied to heal fractured and/or bones.

Rorippa indica (L.) Hiernn. (Brassicaceae) *Ln.* Petu

Seeds are crushed to make powder. One teaspoon in boil water is used against constipation. Seeds are also used in Medicine for Asthma.

Rosa macrophylla Lindl. (Rosaceae) *Ln.* Ghorsepala

The plant is used in agricultural fencing. Flowers are edible.

Raphanus sativus L. (Brassicaceae) *Ln.* Muli

It is generally used in stomach related diseases like constipation, seeds for ringworm and skin erupt. Leaf and root are edible vegetables also.

Rubia manjith Roxb. ex Fleming (Rubiaceae) *Ln.* Majethi

Root is used in dislocation of skin leucod, inflammation, lever complications, menorrhagia and post natal tonic.

Smilax aspera L. (Smilaceae) *Ln.* Kukardara

Paste of root is directly applied on skin to prevent skin diseases and it is also useful for joint pain.

Solanum nigrum L. (Solanaceae) *Ln.* Makoi

Fruit is used in diarrhoea, fever, leaf in dysentery, eye complications, inflammation of scrotum, testicles, kidney, bladder, piles, whole plant in jaundice. Fruits are edible.

Taraxacum officinale Weber (Asteraceae) *Ln.* Kanphul

Root and leaves are dried in shade, powdered and given (2.5 to 5 g). twice a day for 30-45 days in the treatment of migraine, cardiac complaints, jaundice, abdominal complaints, and used as blood purifier. Paste of root and leaves is applied externally on wounds twice a day for a week as an antiseptic.

Thalictrum foliolosum DC. (Ranunculaceae) *Ln.* Mamiri

Root is used in ear ache, eczema, eye disorders, fever, leucod, piles and tooth ache.

Tinospora cordifolia Roem. (Menispermaceae) *Ln.* Giloe

Bark and root are used in medicine for antiseptic, dysentery and fever.

Urena lobata L. (Malvaceae) *Ln.* Boriyal

Paste of leaves is used in cut and wounds, root and bark in hydrophobia.

Valeriana jatamansi Jones. (Valerianaceae) *Ln.* Muskbala

Root is dried in shade, powdered and given approximately ½ teaspoonful twice a day, in morning and at night for 2-3 months in the treatment of hysteria and urinary disorders. The plant is used as a substitute of *Nardostachys jatamansi* by the inhabitants of Tehri. Plant is used in rituals and various religious ceremonies.

Rare and Endangered Taxa of the Sacred Groves

In the last a few decades, many plant species have become threatened due to various anthropogenic pressures and their high demand in Pharmaceutical market. These plants are used in various forms and are utilized by both native and non natives of the area for their benefits. Natural disasters are the main threat in Kumaun Himalayas for the degradation of their biodiversity. According to range of distribution and size of population of plants, they are categories under different names *i.e.*, rare, endangered, vulnerable, low risk and least concerned. Many threatened, endangered and rare species find safe refuge in the sacred groves. The groves also house many species which are not found elsewhere in the region. Several RET plant species such as *Acer oblongum, A. caesium* in Vaishno Devi and Chandika Devi sacred groves (Figures 4.5A,B), *Berberis aristata, Paris polyphylla, Taxus baccata, Valeriana jatamansi* in Narayan Swami Ashram sacred grove (Figure 4.5C) similarly, *Myrica esculantum, Hedychium spicatum, Valeriana jatamansi, Swertia cordata* in Dhwaj sacred grove (Figure 4.5D), *Paris polyphylla* in Hokara sacred grove (1800 m), *Malaxis acuminata, Hedychium spicatum, Bergenia ciliata, Chirita biflora* in Haat Kali sacred grove (Figure 4.5E) and *Mahonia nepaulensis* in Nakuleshwar sacred grove (Figure 4.5F) have been documented from these studied traditionally conserved pockets. Sacred groves are repositories of biological wealth of the country and need attention and care. It is therefore, our collective responsibility to lookout for all opportunities and take all measures to safeguard these islands of biological diversity. According to IUCN status *Taxus baccata, Paris polyphylla, Berberis aristata* and *Valeriana jatamansi* are endangered species while *Bergenia ciliata, Hedychium spicatum, Thalictrum foliolosum* and *Zanthoxylum armatum* are vulnerable species (Arya and Agarwal, 2006, Anonymous, 2003). *Osmunda regalis, Myristica esculentum, Cinnamomum tamala, Malaxis acuminata* and *Adiantum venustum* are threatened plants from Uttarakhand as reported by Joshi *et al.* (1993). Sacred groves are treasure house of rare and endemic flora (Gadgil and Vartak, 1975). Rare lichens such as *Caloplaca himalayana* (Teloschistaceae) were found in Haat Kali sacred grove (on the rock) and Hokra sacred grove (on the bark of *Cinnamomum tamala*) while *Lecanora japonica* was reported from Dhwaj sacred grove (on the bark of *Viburnum cotonifolium*). Most of the genera of Parmeliaceae family (*Parmotrema austrosinense, P. reticulatum, P. tinctorum, P. nilgherrense, P. ultraleucans, P. hababianum, P. praesorediosum, Parmelinella wallichiana, Canomaculina subtictoria, Cetrelia cetrarioides, Canoparmelia caperata, Everniastrum cirrhatum, Flavoparmelia caperata, Hypotrachyna fleilis, Myelochroa xantholepisis, Punctelia rudecta, Ramalina conduplicans, Ramelia reticulata* and *Usnea pseudosinensis*) were found endangered in the area due to their exhaustive exploitation for ethnic and commercial uses as encountered in most of the scared groves. *Parmotrema nilgherrense* was reported from Narayan Swami Ashram and Haat Kali sacred grove which is an endemic species to India.

Sacred Groves and Environment

Conservation

Sacred groves play significant role in biodiversity conservation. These groves are the repositories for varieties of flora and fauna, which are conserved on the basis of religious and cultural beliefs by local people of the area. These are the last refuge of endemic and threatened species. Many keystone species *i.e.*, *Cedrus deodara, Quercus* species, *Rhododendron arboreum,* etc. are found in these sacred groves of Kumaun Himalayas and give shelter to other dependent organisms. Haat Kali scared grove conserves many medicinal plants such as *Malaxis acuminata, Berberis asiatica, Valeriana jatamansi, Viola canescans, Quercus leucotricophora, Rhododendron arboreum, Hedychium spicatum, Urtica dioca, Cedrus deodara,* etc. One of the rare vultures *i.e.*, *Gyps himalayana* was found in Haat Kali and Hokra sacred groves.

Soil and Nutrient Cycle

Sacred groves play crucial role in soil and water conservation. Floristic diversity of the grove binds the soil which prevents soil erosion of the grove. Hokra sacred grove near Tejam receives high rainfall during the months of June-August with a rapid litter decomposition rate which releases high nutrient in the soil of the grove. The soil itself has little nutrients to support a large biomass of the grove. The fine root mat of plants developed the surface layers of nutrients in the soil. Many micro-organisms, invertebrates, fungi, etc. flourish and vast array of species not hitherto indigenous to the groves may also colonise and thrive. The root mat prevents the nutrients from leaching out from the soil. The land surrounding the sacred groves, which is devoid of necessary root mat and litters decomposition, can no longer sustain vegetation (Khiewtam and Ramakrishnan, 1989). This ecological process has affected the development and conservation of sacred groves.

Biogeochemical or Nutrient Cycle

Biogeochemical cycle is one of the important features of sacred groves, as these groves represent small ecosystem of the area. It includes all the basic requirements for functioning the nutrient cycles (Carbon, Oxygen, Nitrogen, etc.). These cycles play significant role in any ecosystem which develops a healthy and balance environment. The groves provide all the essential things for functioning biogeochemical cycles and thus, maintain a rich biodiversity in the grove.

Aquifers

Many sacred groves hold water resource in the form of springs, ponds, lakes, streams or rivers. Not only that, but the vegetative mass of the grove itself retained water by soaking it up like a sponge during wet periods and releasing it slowly in the drought. It is evident that one of the important ecological roles of these groves is to provide a more dependable source of water for the organisms living in and around the sacred groves. The ponds and streams adjoining the groves are often perennial and in some cases, act as the last resorts to many of the animals and birds for their water requirements, especially during dry seasons. Another function may be to reduce the incidence and intensity of forest fire, at least in some climates. In addition, transpiration from the sacred grove vegetation would increase atmospheric humidity and reduce temperature in the immediate vicinity and produce a more favourable microclimate for many organisms (Khiewtam and Ramakrishnan, 1989). In Nakuleshwar sacred grove near Et gaon Shilling village, many aquifers are present due to the presence of *Cedrus deodara, Quercus* sp., *Cinnamomum tamala*, etc., because these trees are able to hold water in the soil. Some aquifers are also seen in other sacred groves *i.e.*, in Patal Bhuvneshwar sacred grove (due to presence of dense forest of *Cedrus deodara*) and Hokra sacred grove (due to presence of *Cinnamomum tamala*). These trees have a great capacity to hold water in the soil and hence increase the availability of water in the area. Thus, the sacred groves play significant role in the environment, which balance the whole ecosystem and conserve bioversity of the region.

Threats to sacred groves in Kumoan Himalaya

Both biotic and abiotic factors are responsible for the loss of biodiversity in the sacred groves; these have been dealt below separately:

Biotic Factors

Commercial Exploitation

Under biotic factors, commercial exploitation of economically important plants including medicinal plants (*Bergenia ciliata, Valarenia jatamansi, Taxus baccata, Berberis aristata, Paris polyphylla, Origanum vulgare, Hedychium spicatum, Malaxis acuminata*, etc.) is the major threat to the biodiversity of the sacred groves. Plants are cut and uprooted completely and good practices for harvesting are not adapted by the local people while they collect these plants from the groves.

Figure 4.5: Threatened Plant Species from the Sacred Groves

A: *Acer oblongum*, B: *Acer caesium*, C: *Valeriana jatamansi;* D: *Hedychium spicatum;*
E: *Bergenia ciliata*, F: *Mahonia nepaulensis*

Figure 4.6: Anthropogenic Pressure in Sacred Groves
A: Grazing and fodder collection; B: Cutting and fuel collection

Human Interference

Human interference is another biotic factor responsible for threat to sacred groves which is the major concern, such as rebuilding of the temples inside the sacred groves and construction of the roads for easy approach to the scared groves.

Developmental Projects

Developmental projects such as building of dams, widening and construction of new roads, buildings, etc. have greatly contributed to the diminishing of the flora and fauna of these sacred groves.

Tourism and Pilgrimage

Tourism is another major concern for the degradation of the sacred groves. The biodiversity of Patal Bhuvneshwar and Haat Kali sacred groves are now degraded due to high rate of tourism.

Removal of Biomass

Removal of biomass by the local people for firewood through lopping of trees and grazing of animals is another major concern to the biodiversity of sacred groves (Figures 4.6A,B).

Invasive Alien Species

Invasive alien species are non-indigenous species which have been introduced intentionally or accidently and compete with native flora for basic resources such as water, light and nutrients. There are several such species, *e.g. Lantana camara, Ipomea carnea, Calotropis procera, Ageratum conyzoides, Parthenium hysterophorus, Eupatorium adenophorum*, etc. which are classified under these categories which degrade the biodiversity of the sacred groves. In Kumaon Himalaya, *Eupatorium adenophorus* and *Ageratum conyzoides* are common invasive alien species which heavily colonize in the sacred groves especially in Patal Bhuvneshwar and Haat Kali scared groves in Gangolihat tehsil..

Abiotic Factors

Temperature

Temperature of the area plays a significant role in the growth of plants. Extremely cold and hot conditions affect adversely the biodiversity of the sacred groves.

Rainfall and Humidity

This is another factor which affects the biodiversity of the groves.

Soil pH

The soil pH of the groves prefers to grow specific flora on the ground. Increase in the acidic value of the soil adversely affects growth of plants. Hence, this is another factor which affects the floral and faunal diversity.

Natural Disasters

Natural disasters such as earth quakes, bursting of clouds, etc. are also responsible for the degradation of biodiversity of the sacred groves.

Conservation Strategies

There is an urgent need to protect these invaluable heritages and treasure houses of plants and animals for the future. So, it is important to conserve our biodiversity, first in small extent after that in large scale. The following conservation measures may be taken for conservation of these important sacred forests and their biological contents.

1. Many rare, endemic and threatened plant species are largely confined to sacred groves only. It is essential that well-preserved sacred groves are immediately brought under protected area net work to ensure the protection of such species and their habitats. Ecological studies on specialized habitat are required and regeneration of the threatened species need to be undertaken for working out suitable strategies to ensure their adequate propagation for continued existence.

2. Management of existing village and community forests may be improved through adequate funding to village durbars and through appropriate management interventions by the government using participatory approach. This will help in meeting the biomass needs of the villagers, which in turn will reduce the anthropogenic presence on sacred groves.

3. A network of community forests representing village reserve/supply may be established and the existing ones be strengthened in areas adjacent to sacred groves. Such forests would act as supply forests and the sacred groves will continue to function as village safety reserve.

4. The degraded sacred groves should be immediately restored or regenerated using appropriate technology. Efforts should be made to utilize the traditional knowledge in regenerating those native species for which the technology may not be available.

5. Nursery techniques and managements must be taken for the native species that are confined to sacred groves and such species should be planted in nearby village reserve forests and in the degraded sacred grove areas.

6. Steps must be taken to raise awareness among the concerned villages regarding importance of sacred grove conservation. There is a need to convert the traditional belief of the tribal people into effective conservation values behind the beliefs need to be explained to the villages.

7. The societies who engage themselves in protection and conservation of biodiversity-rich areas need to be rewarded for their contribution towards welfare of the mankind. The most viable strategy for the protection of sacred groves seems to lie in linking their conservation with economic benefits. Providing rewards and incentives to the people may go a long way in achieving the goal.

References

Agnihotri, P., Sharma, S., Singh, H., Dixit, V. and Husain, T., 2010. Sacred groves from Kumaon Himalaya. *Current Science,* 99(8): 996–997.

Anonymous, 2003. *Conservation Assessment and Management Prioritization (CAMP) Workshop for Medicinal Plants of J&K, Himachal Pradesh and Uttaranchal, Shimla,* 22–25 May, (Eds.) D.K. Ved *et al.* FRLHT, Bangalore, India.

Arya, K.R. and Agarwal, S.C., 2006. Conservation of threatened medicinal and folklore plants through cultivation in Uttaranchal state. *Ethnobotany,* 18: 77–86.

Bisth, S. and Ghildiyal, J.C., 2007. Sacred Grove for biodiversity conservation in Uttarakhand Himalaya. *Current Science,* 92(6): 711–712.

Boraiah, K.T., Bhagwat, S., Kushalappa, C.G. and Vasudeva, R., 2001. In: *Tropical Ecosystems: Structure, Diversity and Human Welfare,* (Eds.) K.N. Ganeshaiah, R. Uma Shanker and K.S. Bawa. Oxford and IBH Co. Pvt. Ltd., New Delhi, pp. 561–564.

Bruner, A.G., Gullison, R.E., Rice, R.E. and da Fronseca, G.A.B., 2001. Effectiveness of parks in protecting tropical biodiversity. *Science,* 291: 125–128.

Ford, R.I., 1994. *Ethnobotany: Historical Diversity and Synthesis in the Nature and Status of Ethnobotany,* 2nd Edn. (Ed.) R.I. Ford, ed. Ann. Arbor Anthropological Papers, Museum of Anthropology, University of Michigan, 67: 33–49.

Gadgil, M. and Vartak, V.D., 1975. Sacred groves of India: A plea for continued conservation. *J. Bombay Natural History Society,* 72: 314–320.

Gadgil, M. and Vartak, V.D., 1976. Sacred groves of Western Ghats of India. *Economic Botany,* 30: 152–160.

Hughes, J.D. and Chandran, M.D.S., 1998. Sacred groves around the earth: An overview. In: *Conserving the Sacred for Biodiversity Management,* (Eds.) P.S. Ramakrishnan, K.G. Saxena and U.M. Chandrashekhara. pp. 69–86.

Jain, S.K. and De, J.N., 1966. Observations on ethnobotany of Purulia district, West Bengal. *Bull. Bot. Surv. India,* 8: 237–251.

Joshi, G.C., Pande, N.K. and Uniyal, M.R., 1993. Inventory of disappearing angiosperms of Kumaun and Garhwal Himalaya: Causes and suggestions. *J. Econ. Tax. Bot.,* 17(2): 421–432.

Ramanujam, M.P. and Praveen, K., 2003. Woody species diversity of four sacred groves in the Pondicherry region of South India, biomedical and life sciences and earth. *Environmental Science,* Springer Netherlands, 12(2).

Sammant, S.S., Dhar, U. and Palni, L.M.S., 1998. Med*icinal Plants of Indian Himalaya: Diversity Distribution Potential Values.* Published by Gyanodaya Prakash Nainital.

Singh, H., Agnihotri, P., Pande, P.C. and Husain, T., 2011. Biodiversity conservation through a traditional beliefs systems in Indian Himalaya: A case study from Nakuleshwar sacred grove. *Environmentalists,* DOI 10.1007/s10669-011-9329-6.

Singh, H., Husain, T. and Agnihotri, P., 2010. Haat Kali sacred grove, central Himalaya, Uttarakhand. *Current Science,* 10: 298.

Singh, S. and Pande, P.C., 2000. Kumaun Himalaya ke dharmik sthalo ke ped paudhey. In: *Kumaun Himalaya Ka Lokvanaspati Vigyan (Ethnoboatny of Kumaun),* pp. 453–458.

Sinha, R.K., 1995. Sustainable utilization and conservation of biodiversity by the tribal societies (Aborigines) of India: A lesson for modern man. *Intl. J. Env. Edn. and Info.,* 14 (2): 195–204.

Tiwari, B.K., Barik, S.K. and Tripathi, R.S., 1999. *Sacred Groves of Meghalaya: Biological and Cultural Diversity.* Regional Centre National Afforestation and Eco-Development Board North-Eastern Hill University, Shillong.

Upadhaya, K., Pandey, H.N., Law, P.S. and Tripathi, R.S., 2003. Tree diversity in sacred groves of the Jaintia hills in Meghalaya, northeast India. *Biodiversity and Conservation,* 12(3): 583–597.

Vartak, V.D., 1983. Observation on rare imperfectly known and endemic plants in the sacred groves of Western Maharashtra. In: *An Assessment of Threatened Plants of India,* (Ed.) S. K. Jain and R.R. Rao. BSI Publ., Howrah, p. 169–178.

Vartak, V.D. and Gadgil, M., 1973. Dev Rahati: An ethnobotanical study of the tracts of forest preserved on grounds of religious beliefs. *Proc. Indian Sci. Cong.,* 60: 341 (Abstract).

2013, Biodiversity in Horticultural Crops Vol. 4
Editor: Professor K.V. Peter
Published by: DAYA PUBLISHING HOUSE, NEW DELHI

Pages 93–119

Chapter 5

Biodiversity in *Anthurium*

Manjunath S. Patil and Anil R. Karale

College of Horticulture,
Mahatma Phule Krishi Vidyapeeth Pune – 411 005, Maharashtra
E-mail: patilmsflori@rediffmail.com

Anthurium is the largest and the most complex genus in the monocotyledonous family Araceae (Croat, 1999) and comprises species with high aesthetic values. Genus *Anthurium* comprises about 1/3rd of the number of species of the family Araceae having 105 genera and more than 3,300 species (Higaki *et al.*, 1994).

The name Anthurium is derived from the Greek- *'anthos'*- the flower, *'aura'*- the tail, referring to the spadix from a taxonomic view point. Like orchids, these flowering stalks, have a long shelf life adding an ethnic beauty to any floral decoration. Besides these, number of anthurium species possess magnificent foliage which are used mostly as pot plants and for cut foliage.

Centers of Biodiversity

Anthurium is a very complex genus due to its large morphological diversity and great phenotypic plasticity. The family Araceae is characterized by both high species diversity and local endemism- *i.e.* endemic species diversity (Croat, 1983, 1988, 1992) with 2/3rd of the species believe to occur in tropical South America (Croat, 1979), which is higher at Ecuador in the Andean and Coastal region than in Amazonian regions. According to Coehio (2004) the genus *Anthurium* is adapted to a number of different tropical wet forest environments, as well as coastal forest with sandy soil and consists of about 1,100 neotropical species, with the greatest concentration of species occurs between 0-1500 m elevations (Croat, 1999). Most of the species originated from Central or South America especially from northern Andes, from Colombia and Venezuela through Ecuador. In Venezuela, about 70 species have been recorded, where they are most frequently found in moist, Andean slope forests at varying elevations. Representation of the genus increases in abundance and species richness in montane and cloud forest belts above 1000 m MSL (Bunting, 1979, 1995 and Croat, 1981).

Figure 5.1: *Anthurium*

Botanical Description

Anthuriums are perennial plants with creeping, climbing, assurgent or arborescent stems. Leaves variable, evergreen, net-veined, with a prominent mid-nerve and lateral nerves and a well defined nerve at or near the margin.

Spathe may either be flat or slightly undulated ending with a prominent tip. Flowers of anthurium are small, insignificant, densely packed on spadix.

The spadix is uniform, cylindrical and subtended in large heart shaped spathe.

The spathe and in many cases spadix are brilliantly coloured ranging from scarlet red, salmon, orange, pink to white. Flowers of anthurium are bisexual and regular with four perianth parts, four stamens and bicarpellary superior ovary. Ovary is 2-celled with 1-2 ovules. Fruit is berry.

Cytology

Chromosome analyses are available for only ~20 per cent of the species in this genus. The majority of counts were recorded by Gaiser (1927) who reported for 43 species and hybrids. Peterson, (1989) opined that some are polyploid with 2n = 60 while, a few species have 2n = 20 to 124 chromosomes. Marchant (1973) extended the number of species counted and his report of ca 124 was the highest chromosome number observed in the anthurium. *Anthurium mexicanum* Engl. (2n=60) a tetraploid member, is the only species where related diploid is not observed. Pfitzer (1957) was the first to indicate the presence of B chromosome in anthuriums. Sheffer and Kamemoto (1976) determined chromosome numbers for 63 species. B chromosomes were found frequently in section *Cardiolonchium* and varied in number from 1, 2, 3. According to them, though 15 was the basic chromosome number of anthurium by previous workers, 15 was considered as a secondary basic number (X_2). Common somatic number was 30 but counts ranged from 2n= 20-90. Most species were part of polyploidy complex based on 30. The basic chromosome nos. of anthurium are n= 15, 16 and 22. The species like *A. andreanum* (2n=30), *A. hookeri* (2n=30) and *A. magnificum* (2n=32) are diploid; *A. scandens* (2n=45) is triploid; *A. digitatum* and *A. wallisii* (4n=60) are tetraploid. The chromosome number of *A. warocqueanum* was 2n= 30 + 3 B chromosomes. These were classified as 2 pairs of large chromosomes, 1 pair of satellite chromosome, 12 pairs of medium to small chromosomes, and 3 B chromosomes. Kaneko and Kamemoto (1979) opined that the transmission of B chromosomes occurs through both pollen and egg.

The *A. affine* karyotype consisted of eight pairs with a centromere in a median position, and seven in the submedian position, one pair of which has satellites on the distal short arm. A majority of such species are 2n = 30, with some 2n = 60, and an isolated count of 2n = 48 in *A. jenmanii* (Sheffer and Kamemoto, 1976). Sheffer and Croat, (1983) worked out the chromosome number in 86 anthurium species from north, central, and South America. Fifty one species had 2n= 24-66, 30 being the most common. The number of somatic chromosomes in *A. longipes* was 2n = 30. Chromosomes with the centromere in the submedian position predominated. *A. bellum* and *A. longipes* belong to section *Urospadix*, which is apparently based on 2n = 30 (Croat and Sheffer, 1983). Thus, *A. bellum*, 2n = 90, may represent a hexaploid. This count disagrees with previous determinations of 2n =56 and n = 28 (Petersen, 1989). Chromosomes of *A. bellum* were apparently smaller compared to the rest of the species analyzed. This decrease in chromosome size may be related to a high ploidy level. The counts 2n = 30 and 2n = 60 for *A. pentaphyllum* var. *pentaphyllum* confirm earlier counts for this species (Sheffer and Kamemoto, 1976; Sheffer and Croat, 1983). The diploid form had one chromosome pair with satellites, while the tetraploid had two pairs of chromosomes with satellites. Paleoneuploidy, polyploidy and B chromosomes had been the basic features of the genus *Anthurium*, but aneuploidy has not been found (Kaneko and Kamemoto, 1979).

Inheritance of Characters

Controlled hybridization indicated that neither white nor red flower colour was dominant and pink was an intermediate heterozygous condition (Kamemoto and Nakasone, 1955).

Maurer (1979) described the techniques of cross-pollination in *A. scherzerianum* and discussed the presence of recessive characters (A= with anthocyanin, a= without anthocyanin, B= whole spathe coloured and b= spotted spathe). When the parents were *Aa/Bb*, the decedents were 9 red (*AB*), 3 red spots on white (*Abb*) and 4 white (*aaB* and *aabb*). The deficit in white plants was provisionally attributed to their lack of vigour.

Iwata *et al.* (1979) opined that the relative concentration of anthocyanins; decide the spathe colour in anthurium species and cultivars. Predominance of cyanidin 3-rhamnosylglucoside results pink to dark red colours whereas, predominance of pelargonidin 3- rhamnosylglucoside results in coral to orange colour. A flavone present in large and variable amount does not have modifying effect on cyanic shades. Sheffer and Kuehnle, (1978) made successful crosses between *A. scherzerianum* and *A. wendlingerii* to produce a hybrid with a grayish orange spathe. Length and coil of the spadix and the length and position of leaf blade were intermediate between the highly contrasting characters of the parental species.

Growth and Developmental Features

Blanc (1977 a,b) and Ray (1987,1988) described two basic types of shoot organization and branching patterns of some Araceae as monopodial and sympodial growths. The monopodial phase corresponds to the juvenile phase for seed propagated *Anthurium andreanum* and *Anthurium scherzerianum* plants (Christensen, 1971). The two species *A. clidemioides* and *A. flexile* in the section *Polyphyllium* have anisophyllous sympodial growth, while other species have triphyllous sympodial growth (Ray, 1988). Young tissue-cultured plants also have first a monopodial growth which corresponds to the juvenile and vegetative phase. After this, the plants have a sympodial phase, with a flower produced for each leaf.

The apparent phyllochron decreases during the monopodial phase from 680 to 250 °C days and increases quite regularly in the beginning of the sympodial phase from 330 °C days for the first leaf to 615 °C days for the seventh leaf. Accumulated day degrees, with a threshold of 14°C; from planting to first flowering were 2143 °C days (219 days) (Dufour and Guenin, 2003).

The Genus *Anthurium* and its Relatives

Anthurium is a beautiful group of tropical aroids. An aroid is a plant that reproduces by producing an inflorescence known to science, as a spathe and a spadix.

Engler (1905) divided the genus *Anthurium* into 18 sections (Table 5.2) with primary characterization based on number of ovules per locule. Further characterizations of sections were based on the leaf shape and texture and inflorescence shape, but characteristic used to separate the majority of the sections was the berry shape.

Table 5.1: Species under Three Groups

Flowering Group	Foliage Group	Fragrant Group
A. andreanum	A. regale	A.angustispadix
A. scherzerianum	A. regnellianum	A.armeniense
A. bakeri	A. robustum	A.brownii
A. brownii	A. amnicola	A.fatoense
A. cinchipense	A. clarinervium	A. fragrantissimum
A. x ferrierense	A.corrugatum	A.hacumense
A.ornatum	A. crystallinum	A.salvadorense
	A. digitatum	A.schlechtendalii
	A. forgettii	A.standleyi
	A. holtonianum	A. ochranthum
	A. leuconerum	A.upalaense
	A. magnificum	
	A. panduratum	
	A. papilionensis	
	A. scandens	
	A. splendidum	
	A. veitchii	
	A. warocqueanum	

Of the known *Anthurium* species not more than 50 are in cultivation and perhaps not more than 10 or 15 known to the trade (Bailey, 1963). Of many reported species, 5-6 species are more popular. They are *Anthurium andreanum* for cut flower; *A. scherzerianum* as potted plant; *A. crystallinum* for foliage purpose, *A. ornatum* for fragrance and *A. grande* as variegated plant, which have great demand in the global floriculture trade.

Horticulturally it is divided into two major sections/groups *viz.*, foliage types and flowering types. Although most anthuriums bear flower, those of the foliage groups have large, handsome velvety leaves, may bear conspicuous or unattractive flowers. The flowering group has remarkably large and ornamental spathe but the foliage is generally not so attractive as those in the foliage section. Cultivars with double flowering spathes are available. *Anthurium* species can be grouped as given in Table 5.1.

Some of the important cultivated species belonging to both flowering and foliage groups are briefly described below.

Flowering Group

Anthurium andreanum–Oil Cloth Flower, Tail Flower, Painter's Palette

A. andreanum is an epiphyte with somewhat creeping growth habit using aerial roots for anchorage. An erect plant with oblong, heart-shaped and long lobed leaves, green in colour and very long as 20-35 cm long, 15-20 cm wide; spathe waxy, heart-shaped, lacquered reddish orange or scarlet, 10-15 cm long and spadix yellow and white. It is suitable for shade net house cultivation and is widely grown for its handsome foliage and coloured spathe. *A. andreanum* var. *rhodochlorum* has bicoloured spadix. Main production areas are tropical and subtropical countries.

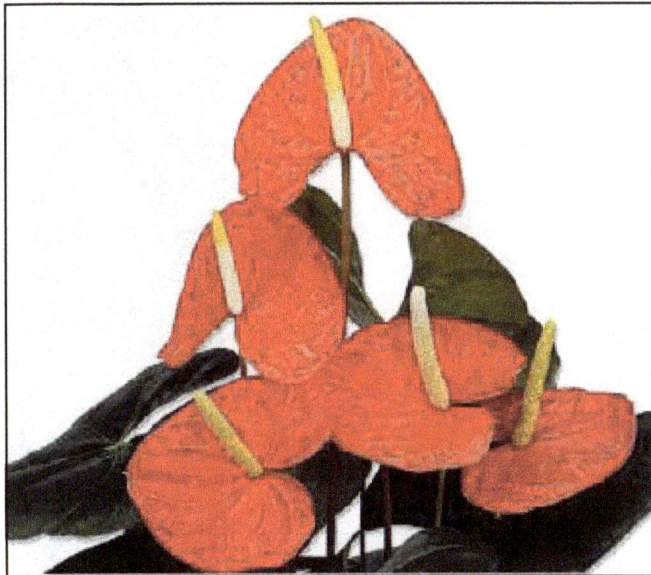

Figure 5.2: *Anthurium andreanum*

Anthurium ornatum

The plants having fragrance like mint but do not possess typical heart shaped spathe which is cup shaped, white and turns purplish rose. Spadix is longer than spathe, leaves ovate, cordate shaped and bright green in colour.

Figure 5.3: *Anthurium ornatum*

Anthurium scherzerianum–Flamingo Flower, Flame Plant

A. scherzerianum is a low-growing herbaceous compact plant. Leaves narrow, 15-20 cm long, 4.6-6.6 cm wide; spathe ovate, brilliant scarlet; spadix spirally twisted, golden-yellow. Flowers chiefly from February to July. This is a popular house plant and needs to keep moist. Main production areas are located in Europe.

Figure 5.4: *Anthurium scherzerianum*

Foliage Group

Anthurium amnicola

In the early 70?s, the species *Anthurium amnicola* (formerly known as *A. lilacinum*) was introduced to the horticultural world. It is a true miniature, not more than six to eight inches in height, with leaves that are narrow and dark green. A small flower that is lavender in colour, which emits a pleasant fragrance and quite frankly, flowers like crazy. The plant is very capable of being in flower year round. Another benefit is that it suckers readily and forms tight clumps. This species paved the way for pot culture anthurium.

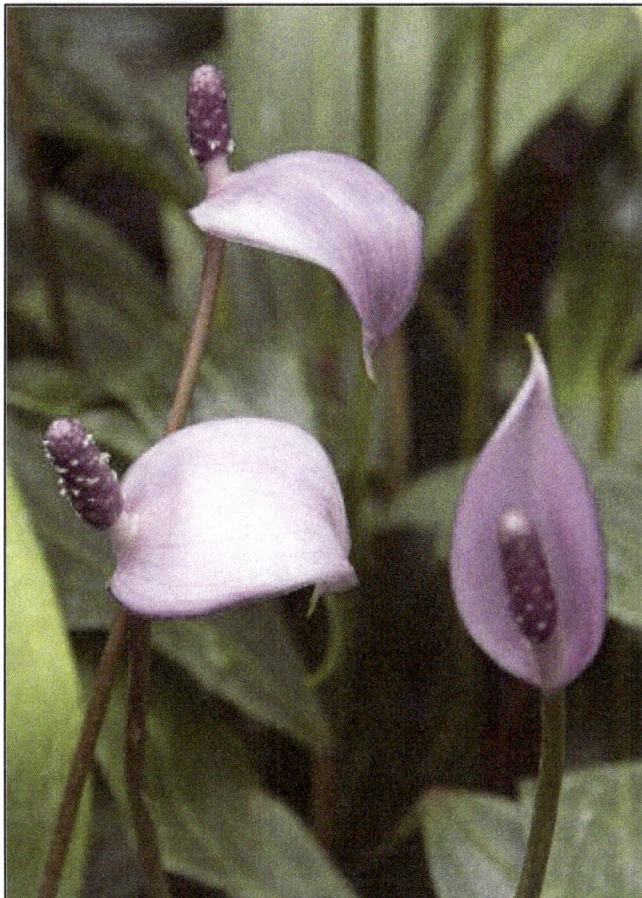

Figure 5.5: *Anthurium amnicola*

A. clarinervium

The species have thick texture to the leaves. That leaf thickness is botanically known as coriaceous, or "leather like" and feels almost like cardboard. The plant spreads and grows laterally producing somewhat velvety (velutinous) leaves with a deep, dark green colour. *Anthurium clarinervium* is not a member of section *Cardiolonchium* which contains the velvet leaved anthurium species. It is found in nature (endemic) in the southern part of Mexico in the state of Chiapas, in well defined areas at an elevation of 800 to 1200 meters a MSL.

Figure 5.6: *Anthurium clarinervium*

A. cordatum

Leaves are broadly sagittate with wide sinus of olive green, spathe green but reddish from inside.

A. crystallinum

Leaves are large, beautiful heart shaped, green with contrasting white veins. Spathe is green and linear.

A. digitatum

The plant produces very attractive palmated foliage and grows well in moderate shady conditions. Conventionally, *A.digitatum* is propagated through seed (if self pollinated) or cuttings. It seldom produces suckers and to obtain a few cuttings, the plant has to attain a minimum age of 5-7 years.

A. forgettii

It is almost similar to *A. crystallinum* but the foliage is oval and larger in size.

A. lindenianum

Leaves are fleshy, sagittate, leathery, glossy green, borne on stiff wiry long petioles.

Figure 5.7: *Anthurium crystallinum*

A. magnificum

The species is exclusively found in (endemic to) Colombia in North Western South America. The petioles of *Anthurium magnificum* are roughly quadrangular (four sided) but may also be 'C' shaped. The cataphylls (which are a modified leaf that surrounds any newly emerging blade) persist semi intact once a new leaf blade opens. New leaf blades are brownish burgundy when first unfurled. Leaves large, heart shaped, velvety, olive green with prominent white veins. The spathe and spadix are green. The spadix yellow-green before anthesis and becomes yellow as it ages near female anthesis. The inflorescence stands erect above the leaves. The peduncle which supports the inflorescence is sharply 9-ridged.

A. podophyllum

Leaves are large, leathery, lobed with finger like segments.

Figure 5.8: *Anthurium digitatum*

Figure 5.10: *Anthurium watermaliense*

Figure 5.11: *Anthurium schlechtendalii*

Figure 5.12: *Anthurium cordifolium*

Figure 5.13: *Anthurium affine*

Figure 5.14: *Anthurium hooker*

Figure 5.15: *Anthurium magnificum* Juvenile

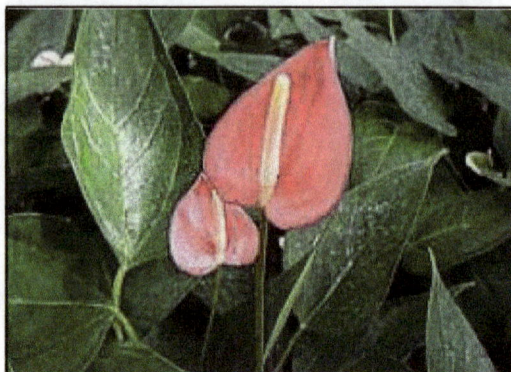

Figure 5.16: *Anthurium antioquiense*　　　**Figure 5.17:** *A. regale*　　**Figure 5.18:** *A. regale*
Spadix Producing Pollen

A. ramoncaracasii

It is closely related to *A. nymphaeifolium* with its spathulate leaf sinus; collective vein originating from the uppermost basal nerve. Robust, terrestrial herb, stem is self elevating or semi-supported near the base. Internodes are 2-3 cm long. It belongs to section *Calomystrium* Schott emend. Engl., with its distinctively cordate, coriaceous leaves (Croat and Sheffer, 1983; Croat and Lambert, 1986), occasionally found amidst humid, montane or cloud-forest of the Venezuelan Andes at 1700-2600 m.

A. regale

Originally identified in 1888, a specimen of *Anthurium regale* is often difficult to locate and difficult to grow. It is not epiphytic, but instead is a terrestrial anthurium growing on the ground. With its coriaceous (leathery) leaves, it can grow stunningly large which may exceed 90 cm, sometimes larger! The species is found along the edge of the cool (not cold) Andes Mountains at relatively low elevations. The *A. regale* is found in a very narrow climatic range in one particular major valley exclusively in Peru *i.e.* Rio Huallaga valley. The average temperature of the region in Peru where *A. regale* is endemic is 20-28°C.

A. scandens

It is a climbing variety with dark green foliage.

A. veitchii

Leaves are ovate-cordate and bright green, spathe is cup shaped, white turning purplish rose.

Table 5.2: Characteristics of some Foliage Species

	A. amnicola	*A. antioquense*	*A. antrophyoides*
Spathe colour	Lavender	Pale lavender	White
Spathe length (cm)	4.3	8.6	9.2
Spathe width (cm)	1.8	2.1	3.8
Spadix colour	Violet-purple	Violet-purple	White
Spadix length (cm)	1.1	4.6	4.6
Leaf length (cm)	13.0 ± 1.0	25.1 ± 1.7	21.7 ± 0.5
Leaf width (cm)	2.8 ± 0.5	6.0 ± 0.4	13.5 ± 0.6
Leaf shape	Elliptic-lanceolate	Elliptic	Trullate

Fragrant Group

Species noted for strong perfume like scent have been mentioned in Table 5.1.

Endemic Species

Lingan and Croat, (2005) reported following species of anthurium as endemic from the Peruvian Andes.

A. chinchipense

Is a member of section *Belolonchium* and is distinguished by its petiole with acute margins, broadly ovate leaves, as well as by the short peduncle and generally deciduous spathe.

A. hamiltonii

Is a member of section *Belolonchium* and is distinguished by its cordate, long-petiolate leaves, cataphylls persist as dark brown mass of fibers and broadly ovate spathe that is erect and held close to the spadix as well as by the greenish to yellow cylindric spadix. This species is the closest to *Anthurium macleanii.*

A. magdae

It is endemic to Peru and close to *A. lutescens*

A. mariae

This species belongs to section *Calomystrium* and characterized by narrow, lanceolate, weakly cordate leaves; cataphylls persisting entire and slender green spadix subtended by subcoriaceous green spathe.

A. piurensis

Member of section *Belolonchium* and is characterized by ovate leaves with straight margins, cataphylls persisting as brown fibers, spathe hooding the spadix and a stipitate, stubby green spadix.

Patterns of Diversity of Flowering Behaviour in the Genus

The structure which is commonly called the anthurium flower is an inflorescence, composed of a peduncle, a coloured bract called the spathe, and a spike of small, perfect flowers, the spadix (Higaki *et al.,* 1994). The sequence of development of flowers on the spadix is of taxonomic significance and varies from species to species correlating in part with the shape of the spadix. The inflorescences are monoecious, with the perfect (male and female) flowers arranged (and opening) in characteristic spirals along the central axis. The flowers are embedded within the spadix, rhomboidal in shape, with four triangular tepals, which enclose the elliptical stigma (Figure 5.19).

In majority of species of *Anthurium,* the inflorescences are protogynous, with gynoecium maturing first from the basal portion of spadix to the top in an acropetal succession (Croat, 1980) *i.e.* maturation of flowers initiate from the basal portion. The female phase for various species varies from 3-12 days. The separation period or interphase between male and female phases in *A. andreanum* species ranges 4-7 days. Pollen production is positively corelated to anther size, flower size and plant height (Bindu and Mercy, 1996). Pollen-shedding habits varied greatly, dependent on the variety. Most varieties shed pollen after their stigmas were no longer receptive. However, some varieties occasionally shed pollen soon after the spathe unfurled. The genus exhibits an unusual pattern of staminal emergence. In most of the species, the lateral pair of stamens emerges first usually one at a time followed by the

**Table 5.3: Section-wise Classification of *Anthurium* (Engler, 1905)
and their Chromosome No. (Sheffer and Kamemoto, 1976)**

Sl.No.	Section	Species	Chromosome No. (2n)
I	Tetraspermimum	A. scandens	24,48,84
		A. trinerve	24,30
II	Gymnopodium	A. gymnopus	30
III	Porphyrochitonium	A. scherzerianum	30
IV	Pachyneurium	A. acaule	
		A. boucheanum	
		A. brownie	
		A. cordatum	
		A. crassinervium	
		A. ellipticum	30
		A. glaziovii	
		A. grandiflorum	30
		A. hacumense	30
		A. hookeri	30,60
		A. joseanum	30
		A. maximum	
		A. recusatum	
		A. seleri	30
		A. tetragonum	
V	Polyphyllum	A. mexicanum	60
VI	Leptanthurium	A. actangulum	30
		A. gracile	30
		A. scolopendrinum	20,40
VII	Oxycarpium	A. pittieri	30
VIII	Xialophyllum	A. pulchellum	
		A. subhastatum	30
		A. triangulum	30
		A. tuerckheimii	
IX	Polyneurium	A. wallisii	30 + 2B
X	Urospadix	A. acutum	
		A. allenii	30
		A. aureum	30,31
		A. bellum	
		A. chiriquense	30
		A. comtum	
		A. gladiifolium	30

Contd...

Table 5.3–*Contd...*

Sl.No.	Section	Species	Chromosome No. (2n)
		A. harrisii	
		A. imperial	
		A. littorale	28
		A. lucidum	
		A. olfersianum	
		A. microphyllum	
		A. sellowianum	
		A. tiranae	28, 29 + 1B
		A. turrialbense	30
XI	*Episeiostenium*	*A. bakeri*	30
		A. consobrinum	
		A. dominicense	
		A. guildingii	
		A. wendlingerii	30
XII	*Digitinervium*	*A. rhodostachyum*	28,29,30,31
XIII	*Cardiolonchium*	*A. clarinervium*	30
		A. crystallinum	30 + 1 B
		A. forgetii	30
		A. grande	30
		A. leuconeurum	
		A. magnificum	60
		A. regale	30 + 1B
		A. splendidum	30 + 2B
		A. velutinum	30
		A. venosum	30
		A. walujewii	30 + 2B
		A. warocqueanum	30 + 3B
		A. wullschlaegelii	30
XIV	*Chamaeerepium*	*A. radicans*	
XV	*Calomystrium*	*A. hoffmanii*	30
		A. lindenianum	30
		A. montanum	30
		A. nymphaeifolium	30
		A. pichinchae	30
		A. ranchoanum	30
		A. roraimense	30
		A. veitchii	30

Contd...

Table 5.3–*Contd...*

Sl.No.	Section	Species	Chromosome No. (2n)
XVI	Belolonchium	A. andreanum	30
		A. concinnatum	30
		A. denudatum	30
		A. flavor-viride	30
		A. gustavii	30
		A. micromystrium	30
		A. patulum	
		A. proderum	30
		A. supianum	90
XVII	Semaeophyllum	A. holtonium	30
		A. signatum	
		A. subsignatum	30
XVIII	Schozoplacium	A. aemulum	30,60
		A. digitatum	30
		A. pedato-radiatum	
		A. pentaphyllum	60
		A. undatum	
		A. variabile	
	Undetermined	A. baileyi	60
		A. ramonense	30
		A. watermaliense	30

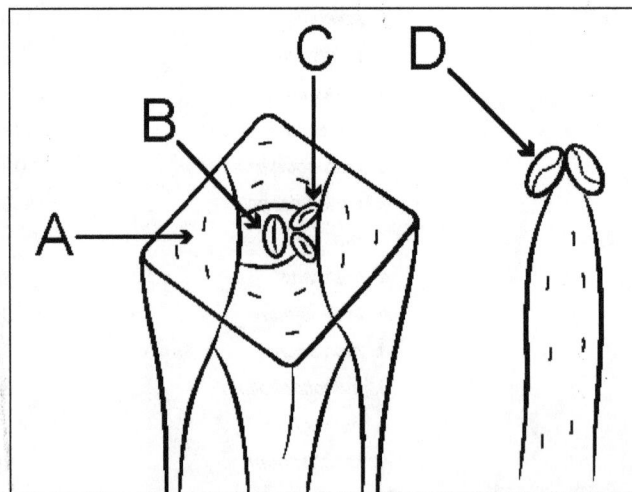

Figure 5.19: Representation of an *Anthurium* Flower (Adopted from Croat 1980)
A: Tepal; B: Pistil (with stigma); C: Anthers; D: Anthers after emergence

anterior then the posterior stamens of the alternate pair. There are also differences in the degree of exertion, the disposition with respect to stigma, degree of retraction and changes in pollen colour. Some species have stamens which are retracted completely after opening; others have stamens which scarcely emerge but instead force the pollen out in long ribbons. Important differences also exist in flower aroma with both fly and bee pollination syndrome exhibited. In *A. bakeri* Hook. stamens may become inverted and may set fruit in greenhouse without the aid of pollinators. *A.bakeri* and *A. scandens* (Aublet) Engler, produce inflorescence almost continuously under greenhouse condition and most varieties shed their pollen in environment controlled greenhouse from December to April. Floral biological aspects of some *Anthurium* species are given in Table 5.4.

Mode and Method of Pollination

The pollination ecology of the neotropical genera bearing bisexual flowers is poorly known and is mostly based on observations and a few studies. Like all members of the araceae they are entomophilous (Grayum, 1990) and olfactory signals are considered critical for the attraction of insect pollinators. A variety of pollinators have been recorded for anthurium, including *Euglossines* (Dressler, 1968) and flies (Madison, 1979 and Grayum, 1990). Most of the species with sweet aromas are probably bee-pollinated. In some cases, the male bees *Euglossa cyanura* collect not only pollen or nector, but also apparently scent compounds in the well known manner of male euglossine bees with the family Orchidaceae (Willams and Dodson, 1972). *Cyclanthura* flower weevils present on the inflorescences in small numbers during the pistillate and staminate phase of anthesis have been identified as pollinators of multiple species of anthurium (Gibernau, 2003) in Costa Rica and hummingbird in *A. sanguineum* (Kraemer and Schmitt, 1999).

The transparent minigrip PE (Polyethylene) bags are proved a good envelop to encase the anthurium flower in order to prevent undesired pollination. Compared to other pollination methods, the pen brush yielded the highest fertility and seed-bearing on the spadix (Chen *et al.,* 1999) which if pollinated, produce berries containing 1 to 2 seeds.

Fragrant Anthurium

Floral scent is an important variable in attracting the insect for pollination. Many *Anthurium* species have a faint yet distinguishable aroma but in some species the aroma may be very strong. The strong scents are either perfume like and sweet or yeasty and rather foul. Species noted for a strong perfume like scent include *A. fragrantissimum* Croat, *A. armeniense* Croat, *A. angustispadix* Croat and Baker and *A. hacumense* Engler. Species with notably strong yeasty or foul scents include *A. brownie* Mast. *A. salvadorense* Croat and *A. schlechtensalii* Kunth. Most of the species with sweet aroma are probably bee-pollinated while, foul scents and dark spathes and spadices suggest a fly-pollination syndrome. In some cases, bees have been seen visiting flowering inflorescences in great numbers (Croat, 1978).

In addition to qualitative differences in scent compounds, temporal aspects are also important. The species (*A. fragrantissimum* Croat) is strongly and sweetly aromatic only during midday and another species (*A. armeniense* Croat) only in the early morning. Scent production in these two species is apparently consistent in quantitative and qualitative terms throughout both the pistillate and, later the staminate phase of flowering. This is true of most of the species for which a perfume like scent has been detected. Other species appear to be aromatic only after the stamens have emerged. This is especially true of the species which have yeasty scents (Croat, 1980).

Table 5.4: Floral Biological Aspects of some Species of *Anthurium* (Croat, 1980)

Species	Spadix			Stigma droplets			Stamens		Pollen		Scent Type	
	Shape	Length (cm)	Colour	Abundance	Duration (Days)	Dry Days Before Anthers	Sequence	Spacing	Fresh Colour	Fading to	Type	Timing
A. binervia	Moderately tapered	7-12.2	Yellow green	Minute	3-5	5	Scattered	Dis-appearing	Pale yellow	White	–	–
A. brownie	Moderately tapered	12-15.5	Dark purple	Large	4-5	1-2	Base-rapid	Open circle	Yellow	White	Spoiled fruit	Pistillate
A. burgeri	Long tapered	9-12.2	Purple brown	Minute	10-14	1-2	Middle-rapid	4 sides of pistil	Orange	White	–	–
A. canasus	Long tapered	7-15	Purple violet	Small	4-8	1-3	Base-slow	Close circle	Pale yellow	White	–	–
A. concolour	Scarcely tapered	5.7-16	Purple violet	Large and runny	5-8	1-2	Scattered	Loose cluster	Purple	White	–	–
A. fragrantissimum	Moderately Tapered	10.3-12	Pale green	Small	3	1-2	Base	Open circle	Golden yellow	White	Perfume	Pistillate
A. hacumense	Slightly Tapered	10-20	Purple violet	Minute	10-14	1-2	Middle to ends	Cluster	lavender	White	–	–
A. kunthii	Long tapered	7-21	Pale green	Minute	1-2	1-2	Base-slow	Tight circle	White	White	Perfume	Pistillate
A. lancetilense	Long tapered	7.5-16	Purple violet	Small	5-10	3-4	Base-rapid	Open circle	Bright yellow	White	–	–
A. lentii	Long tapered	12-25	Violet purple	Minute	1-2	1-2	Base-rapid	Cluster covering	White	White	–	–
A. luteynii	Moderately tapered	7-17	Violet purple	Large and runny	21-28	3-4	Sporadic	4 sides of pistil	Yellow	White	–	–
A. pittieri	Moderately tapered	11-14.5	Green	Small	3-4	10-20	Scattered-slow	4 sides of pistil	Pale yellow	White	–	–
A. rotundi-stigmum	Moderately tapered	10-17	Violet purple	Small	3-7	1-2	Base or scattered	4 eccentric dots	Yellow	White	–	–

Contd...

Table 5.4—Contd...

Species	Spadix			Stigma droplets			Stamens		Pollen		Scent type	
	Shape	Length (cm)	Colour	Abundance	Duration (Days)	Dry Days Before Anthers	Sequence	Spacing	Fresh Colour	Fading to	Type	Timing
A. salvadorense	Slightly tapered	2.5-6.5	Pale green	Small	1-3	1-3	Base-rapid	Close circle	Pale yellow	White	Sweet fruity	Staminate
A. schiechtendalii	Moderately tapered	14-18.5	Green purple	Minute	2-3	1-2	Base-rapid	Close beside	Pale orange	Cream	Yeast	Staminate
A. seibertii	Moderately tapered	10-20	Violet purple	Large	5-10	1-3	Base-rapid	4 clusters	White	White	–	–
A. standleyi	Moderately tapered	10-30	Green purple	Small	3-5	1-3	Base	Close circle	Orange	Cream	Evergreen	Staminate
A. subsignatum	Moderately tapered	5.5-26	Yellow green	Small	5-6	1-2	Base-rapid	Pollen in ribbons	White	White	–	–

Cultivar/Varietal Status

Most anthurium cultivars are asexually propagated clones derived from hybrids of different species involving mainly *A. andreanum* and A. *scherzerianum*. Today hundreds of varieties are avaiable in different colours. Anthurium cultivars have been identified mainly through morphological measurements and description of leaves and flowers. Isozyme banding patterns have also been used as fingerprints to aid the visual identification of anthurium cultivars (Kobayashi, *et al.,* 1987). A successful floral industry is dependent upon a portfolio of flowers that will please the final consumer. Breeding efforts are expanding to include important traits desired by end-users.

Table 5.5: Ideotype of Anthurium

Particulars	Trait/Characteristics
Plant traits	Erect growth Compact habit Produce suckers profusely Stem long about five times the length of spathe, firm and slender
Spathe	Heart shape Bright colour Showy with plenty of blisters Symmetrical overlapping of basal lobes
Spadix	Symmetrical Reclining Shorter in length than length of spathe Orientation at an angle 30º
Flowering quality Colour Size	Fragrant spadix Red shade (bright red, dark red) most preferred followed by pink. 4.5" x 4", 6" x 5"
Vase life	Minimum 10-15 days
Biotic stress	Resistance to diseases like bacterial blight, anthracnose

Collections made by the growers from diverse sources and frequent cross breeding have obliterated the varietal status of anthurium greatly. Based on colour characteristics, cultivars can be grouped as below.

Red

Ozaki, Kaumana, Tropical, Cancan, Altiplano, Tanaka, Micky Mouse

Orange

Nitta, Sun Burst, Sunglow

White

Manoa Mist, Lambada, Angel, Acropolis, Eternity

Pink

Marian Seefurth, Sonata, Rosa

Green

Midori, Pistache

Tulip-types, miniatures with spathe cupped and tulip-type spadix straight and erect and not reclining are also available and treated as novelties.

Cultivars with bi-coloured spathe are extremely variable in size and shape and contain some development of chlorophyll in the spathe. Bicoloured important cultivars are Avenue, Chameleon, Cardinal and Trinidad.

Exclusive

Cascade–white, green spadix

Cheers–light pink, green spadix

Choco–Dark brown-green spadix

Jumbo–Obake-cream

Safari–Brown red white veins

Tequila–Cream green red veins

The double flowering anthurium types produce small and single spadix and enlarge spathe on the same stem. Red, pink and orange doubles are reported.

Genetic Variability and Varietal Development

Genetic variability of *Anthurium crenatum* was measured in eight populations. Polymorphism was found in 55.5 per cent of the loci surveyed with a mean heterozygosis of 0.0148 and 0.234 at the population and species levels respectively (Acosta Mercado *et al.,* 2002).

Under Bangalore conditions, Patil (2009) recommended Sunglow, Simba, Meringue White, Acropolis and Fla Orange cultivars for commercial cultivation.

In Andaman islands, the cultivars Mirage, Agnihotri and Deep Pink were found suitable for sucker production, whereas Honey, Mauritius and Wrinkled Orange for flower production (Shiva and Nair, 2008).

According to Kamemoto and Kuehnle, the varieties 'Honey' and 'Wrinkled Orange' yielded maximum number of flowers (8.5 and 7.6 respectively) per plant per year. The flower stalk length was 42.4 cm in Honey (Singh *et al.,* 2002).

A.andreanum cv. Lady Jane was the most prolific but produced small sized spathes. *A.scherzerianum,* cv.'Amazonica' produced more blooms per plant, maximum peduncle length, spathe length, spathe width and spadix length which are ideally suited for pot plant (Jawaharlal *et al.,* 1998).

'Avenue, Chameleon, and Clove Red' cultivars are suitable for cut flower and pot plant production (Kamemoto and Kuehnle, 1996).

Utilization of Anthuriums

Utilization is the ultimate aim of collection and conservation of plant genetic resources. The collection and conservation of its genetic resources have become paramount importance to provide genetic stocks to breeders and amateurs. The role of nurserymen, hobbyists and Agri-horticultural societies in their domestication and conservation is laudable in the absence of dedicated genebanks to conserve them. The anthurium species may be used in the following ways.

Crop Improvement

Breeding

New cultivars of this outbreeding crop are developed through sexual hybridization, progeny evaluation and selection. Breeding has helped in producing different flower colours and shapes.

Two cultivars, 'Uniwai' (an exceptionally high yielding white) and 'Marian Seefurth' with a rose opal spathe were evolved by clonal selection (Kamemoto and Nakasone, 1963). Kamemoto and Sheffer (1978) made successful crosses between *A. scherzerianum* and *A. wendlingerii* to produce a hybrid with a grayish-orange spathe.

There are good opportunities for interspecific hybridization within the genus. Molecular breeding approaches are used concurrently in a program for varietal development of anthurium as cut flowers.

Breeding for Colour

In recent years, utility of anthurium as pot plants and for cut foliage is being explored for valid reasons, especially for their handsome foliage and coloured sapthe.

Anthurium 'Princess Aiko' (Imperial) is a new, sweet scented, multipurpose cultivar originated from a cross between white *A. antioquiense* and 'Tatsuta Pink Obake' (Kuehnle *et al.,* 2004). It is a high yielding variety with pink, long lasting flowers.

'Ramona' a hybrid from K.P.Holland has a crested spadix, each flower is slightly different.

Anthurium 'Regina' has a unique large purple flower similar in shape (lateral tulip) to 'Princess Aiko'. The parents are 'Marian Seefurth' x *A. formosum* (Kuehnle *et al.,* 2004).

Anthurium 'Centennial' has a unique tulip-type spathe having very good vase life (21 days) and resistant to bacterial blight. It is an offspring of a complex cross between the White x Lavender hybrid 570-77 ([Uniwai x *A. lindenianum*] x *A. amnicola*) and the light pink A494 (*A. andreanum* x *A. antioquiense*). It has an upright, green yellow spadix with a dark green stipe and flower stem. It is notably desirable for flowering potted plants (Kuehnle *et al.,* 2007).

Breeding for Fragrance

Growers express high interest in breeding for floral fragrance notably absent in cultivars. The characteristics of the floral fragrance compounds are sabinene, β-pinene, limonene, 1-8-cineole, phenethyl alcohol and α-pinene (Kuanprasert *et al.,* 1998).

Breeding for Disease Resistance

Anthurium crop is affected by a variety of fungal and bacterial diseases. The wet and humid conditions prevailing in the anthurium growing areas of our country are highly favorable to both bacterial and fungal pathogens.

Fungal diseases namely leaf blight and root rot (*Phytophthora nicotiana* var. *parasitica* and *P.citrophthora*), root rot (*Pythium splendens; P.vexans; P.spinosum, Fusarium* and *Rhizoctonia*), anthracnose and bacterial diseases namely wilt (*Pseudomonas solanacearum*) and bacterial blight caused by the pathogen *Xanthomonas axonopodis* pv. *dieffenbachiae* (previously, *Xanthomonas campestris* pv. *dieffenbachiae* (Nishijima, 1994) are quite devastating.

Resistance to anthracnose (*Colletotrichum gloeosporioides*) has been identified among accessions (Aragaki *et al.,* 1968) and used successfully in breeding. Genetic resistance to burrowing nematode and bacterial blight are not currently available. Anais *et al.* (2000) developed the tools for breeding for resistance to bacterial blight in anthurium with the resistant clone 'A-970', and a reliable method to discriminate the resistant from the susceptible clones.

Using classical hybridization and selection, in breeding program, Kuehnle *et al.* (1995) attempted to transfer the apparent systemic resistance from *A. antioquiense* Engler to the cultivated *A. andreanum* Hort. F$_1$ hybrids had a high degree of resistance to the bacteria. However, due to the small flower of *A. antioquiense* the F$_1$ hybrids have to be backcrossed to the cultivated varieties to obtain resistant plants with horticulturally desirable characteristics (Kamemoto *et al.,* 1992). Using other tolerant germplasm, they released two cultivars with improved resistance, 'Kalpana' and 'Tropic Ice' (Kamemoto and Kuehnle, 1996). After several years, both cultivars eventually succumbed to blight.

Use of Resistant Varieties

In Guadeloupe, a clone apparently resistant to bacterial blight was identified in 1994 growing under a heavily infested shade house. The suspected resistance of this clone encoded A-971 was confirmed by inoculation in the bacteriology laboratory, and shown to be of high level. However, this clone does not have the desirable characteristic to fit the Caribbean local or export cut flower market; nevertheless, it can be used in the breeding program for resistance to *Xanthomonas.*

Need for Genetic Transformation

Development of a new anthurium cultivar usually takes 8 to 10 years, due to the long life cycle of the plant (3 years from seed to seed), slow seed formation (6 months) and perennial nature of the crop. In response to these breeding challenges, a program was initiated to develop genetic transformation method to introduce disease-resistance genes into anthurium (Kuehnle, 1989).

Genetic engineering of anthurium has become feasible after it was shown that, some cultivars can be infected by *Agrobacterium tumefaciens*, a well established vector for plant genetic transformations. Such a method could allow introgression of other desirable traits, such as novel flower colour or nematode resistance.

Kuehnle and Sugii (1991) reported tumor formation and nopaline production in *Anthurium andreanum* Hort. when cocultivated with *A. tumefaciens* strains A281 and C58 in an induction medium containing acetosyringone. Molecular analyses indicated presence of nopaline synthase (NOS) gene of T-DNA in the plant tissue. Etiolated internodes of *Anthurium andreanum* Lind. ex Andre 'Rudolph' (formerly UH265) were suitable for infection by wild-type *Agrobacterium tumefaciens*. Recently, cocultivation of root cuttings with *A. tumefaciens* resulted in plants transgenic for *neo* and *att* with a transformation efficiency of 1.3 per cent (Chen *et al.,* 1997). Two transgenic lines of anthurium 'Paradise Pink', engineered to produce the cecropin like Shiva-1 lytic peptide, were able to significantly resist anthurium blight when compared to a standard resistant cultivar 'Kalpana' (Kuehnle *et al.,* 2004b).

Plant regeneration from tissue cultures is necessary for transformation work. Tissue culture protocols are available and regeneration of *A. andreanum* and *A. scherzerianum* has been obtained via a callus stage from cultured embryos and explants of leaf lamina, petiole, spadix, spathe, roots, and etiolated shoots (reviewed by Geier, 1990; Matsumoto and Kuehnle, 1997; Chen *et al.,* 1997). The impact of classical breeding efforts and transgenic germplasm may take many years to be realized (Kuehnle and Sugii, 1991).

A transformation method to introduce antibacterial genes into this commercial monocot is desirable to complement current hybridization breeding for disease resistance.

Conclusion

Anthurium is the largest and most complex genus in the monocotyledonous family Araceae which is characterized by both high species diversity and local endemism. Horticulturally it is divided into three major groups as foliage, flowering and fragrant types. Genus *Anthurium* consists of 1,100 neotropical species of which only 5-6 have great demand in global floricultural trade. Identification of cultivars is done mainly through morphological description of leaves and flowers. Chromosome analysis is available for only ~20 per cent of species and the basic chromosome nos. are n=15, 16 and 22. Paleoneuploidy, polyploidy and B chromosomes had been the basic features, but aneuploidy has not been found. Monopodial and sympodial growths give shoot organization and branching patterns. Differences exist in leaf shape, colour, flower aroma, male and female phases and pattern of staminal emergence. Development of flowers is of taxonomic significance and varies from species to species. Pollination ecology is poorly known. Most anthurium cultivars are asexually propagated clones and derived from hybrids and frequent cross breeding has obliterated the varietal status greatly. Good opportunity exists for inter-specific hybridization within the genus but resistance to burrowing nematode and bacterial blight is not currently available. Wonderful results can be achieved only through systematic collection, conservation and utilization of biodiversity.

References

Acosta-Mercado, D., Bird-Pico, F.J. and Kolterman, D.A., 2002. Genetic variability of *Anthurium creanatum*. *Caribbean Journal of Science*, 38(1–2): 118–124.

Anais, G., Darrasse, A. and Prior, P.H., 2000. Breeding anthuriums (*A. andreanum* Lind.) for resistance to bacterial blight caused by *Xanthomonas axonopodis* pv. *dieffenbachiae*. *Acta Horticulturae*, 508: 135–140.

Aragaki, M., Kamemoto, H. and Maeda, K.M., 1968. *Anthracnose Resistance in Anthurium*. Technical Program Report, 169: 10.

Bailey, L.H., 1963. *The Standard Cyclopedia of Horticulture*. The Macmillan Co., New York.

Bindu, M.R. and Mercy, S.T., 1996. Pollen studies in *Anthurium*. *Journal of Tropical Agriculture*, 34: 96–98.

Blanc, P., 1977a. Contribution a Petude des Araces. I. Remarques sur la croissance monopodiale. *Rev. Gen. Bot.*, 84: 115–126.

Blanc, P., 1977b. Contribution a Petude des Araces. I. Remarques sur la croissance Sympodiale cher l' Anthurium scandens Engl. *Rev. Gen. Bot.*, 84: 319–331.

Bunting, G.S., 1979. Sinopsis de las Araceae de Venezuela. *Rivista Fac Agron*, 10(1–4): 139–290.

Bunting, G.S., 1995. Araceae: In: *Flora of the Venezuelan Guayana*, (Eds.) P.E. Berry *et al*. 2: 600–679.

Chen, F.C., Kuehnle, A.R. and Sugii, N., 1997. Anthurium roots for micropropagation and *Agrobacterium tumefaciens*-mediated gene transfer. *Plant Cell Tissue Organ Culture*, 49: 71–74.

Chen, J.B., Chen, W.H. and Wu, G.D., 1999. In: *Report of the Taiwan Sugar Research Institute*, 164: 41–57.

Christensen, V.O., 1971. Morphological studies on the growth and flower formation of *Anthurium scherzerianum* Schott and *Anthurium andreanum* Lind. *Planteval*, 75(6): 793–798.

Coehio, M.A.N., 2004. Taxonomia e biogeografia de Anthurium (Araceae) secao Urospadix subsecao Plavescentiviri dia. Tese de doctorado, Departamento de Botanica, UFRGS, Brazil.

Croat, T.B., 1978. *The Flora of Barro Colorado Island.* Stanford University Press, California.

Croat, T.B., 1979. The distribution of the Araceae. *Tropical Botany,* pp. 291–308.

Croat, T.B., 1980. Flowering behavior of the neotropical genus *Anthurium* (Araceae). *American Journal of Botany,* 67 (6): 888–904.

Croat, T.B., 1981. Studies in Araceae III: New species of *Anthurium* from Central America. *Sebyana,* 5 (3–4): 315–341.

Croat, T.B., 1983. A revision of the genus *Anthurium* (Araceae) of Mexico and Central America. Part I Mexico and Middle America. *Annals of Missouri Botanical Garden,* 70: 211–420.

Croat, T.B., 1988. Ecology and life forms of Araceae. *Aroideana,* 11: 4

Croat, T.B., 1992. Species diversity of Araceae in Colombia: A preliminary survey. *Annals of the Missouri Botanical Garden,* 79: 17–28.

Croat, T.B., 1999. Araceae. Catalogue of the vascular plants of Ecuador. *Monograph systematic botany from Missouri Botanical Garden,* pp. 227–246.

Croat, T.B. and Lambert, N., 1986. The Araceae of Venezuela. *Aroideana,* 9 (1–4): 3–213.

Croat, T.B. and Sheffer, R.D., 1983. The sectional groupings of *Anthurium* (Araceae). *Aroideana,* 6: 85–127.

Dressler, R.L., 1968. Observations on orchids and euglossine bees in Panama and Costa Rica. *International Journal of Tropical Biology,* 15: 143–183.

Dufour, L. and Guerin.V., 2003. Growth, developmental features and flower production of *Anthurium andreanum* Lind. in tropical conditions. *Scientia Horticulturae,* 98: 25–35.

Engler, A., 1905. pathoideae. In *Das Pflamzeneich,* (Ed.) A. Engler, 21: 1–330.

Gaiser, L.O., 1927. Chromosome numbers and species characters in *Anthurium.* In: *Proceedings of Trans. Royal Society of Canada,* 21: 1–137

Geier, T., 1990. Anthurium. In: *Handbook of Plant Cell and Tissue Culture, Vol. 5: Ornamental Species,* (Eds.) P.V. Ammirato, D.A. Evans, W.R. Sharp and Y.P.S. Bajaj. McGraw-Hill, New York, pp. 228–252.

Gibernau, M., 2003. Pollinators and visitors of aroid inflorescences. *Aroideana,* 26: 66–83.

Grayum, M.H., 1990. Evolution and phylogeny of the Araceae. *Annals of Missouri Botanical Garden,* 77: 628–697.

Higaki, T., Lichty, J.S. and Moniz, D., 1994. *Anthurium* culture in Hawaii. Research and extension series Hawaiian institute of tropical Agriculture and human resources. Honolulu, Hawaii, 152: 22.

Iwata, R.Y., Tang, C.S., and Kamemoto, H., 1979. Anthocyanins of *Anthurium andreanum* Lind. *Journal of the American Society for Horticultural Science,* 104(4): 464–466.

Jawaharlal, M., Soorinatha Sundaram, K., Balakrishmurthy, G. and Thamburaj, S., 1998. Performance of anthurium cultivars at Yercaud. Paper presented at the *National Seminar on Anthurium Production,* 2–3 June, Chettahalli, Coorg.

Kaneko and Kamemoto, 1979. Karyotype and B chromosomes of *Anthurium warocqueanum. The Journal of Heredity,* 70: 271–272

Kamemoto, H. and Nakasone, H.Y., 1963. Evaluation and improvement of anthurium clones. *Technical Bulletin of Hawaiian Agriculture Experimental Station,* 58: 28.

Kamemoto, H. and Kuehnle, A.R., 1996. *Breeding Anthuriums in Hawaii.* University of Hawaii Press, Honolulu, Hawaii, p. 168.

Kamemoto, H., Kuehnle, A.R., Kunisaki, J., Aragaki, M., Higaki, T. and Imamura, J., 1992. Breeding for bacterial blight resistance in *Anthurium.* In: *Proceedings of 3rd Anthurium Blight Conference,* (Ed.) A.Alvarez. Hawaiian Institute of Tropical Agriculture and Human Resources, 5: 07: 90. University of Hawaii, p. 45–48.

Kobayashi, R.S., Brewbaker., J.L. and Kamemoto, H., 1987. Identification of *Anthurium andreanum* cultivars by gel electrophoresis. *Journal of American Society of Horticultural Science,* 112(1): 164–167.

Kraemer, M. and Schmitt, U., 1999. Possible pollination by hummingbirds in *Anthurium sanguineum* Engl. (Araceae). *Plant Systematics and Evolution,* 217: 333–335.

Kuanprasert, N., Kuehnle, A.R. and Tang, C.S., 1998. Floral fragrance compounds of some *Anthurium* (Araceae) species and hybrids. *Phytochemistry,* 49: 521–528.

Kuehnle, A.R., 1989. Genetic engineering on anthurium. In: *Proceedings of 2nd Anthurium Blight Conference,* (Eds.) J.A. Fernandez and W.T. Nishijima. Hawaiian Institute of Tropical Agriculture and Human Resources, 03: 10: 8. University of Hawaii, pp. 37–39.

Kuehnle, A.R. and Sugii, N., 1991. Induction of tumors in *Anthurium andreanum* by *Agrobacterium tumefaciens. HortScience,* 26: 1325–1328.

Kuehnle, A.R., Amore, T.D., Kamemoto, H., Kunisaki, J.T., Lichty, J.S. and Uchida, J.Y., 2004. *'Princess Aiko' (Imperial) and 'Regina'. Two novelty anthuriums.* College of Tropical Agriculture and Human Resource, University of Hawaii at Manoa, Publication NPH–A–7. Anthurium cultivar release: 1– 2 p.

Kuehnle, A.R., Amore, T.D., Kamemoto, H., Kunisaki, J.T., Lichty, J.S.and Uchida, J.Y., 2007. *'Centennial' Anthuriums.* College of Tropical Agriculture and Human Resource, University of Hawaii at Manoa, Publication NPH–A–11. Anthurium Cultivar Release, p. 1–2.

Kuehnle, A.R. Chen, F.C. and Sugii, N., 1995. Novel approaches for genetic resistance to bacterial pathogens in flower crops. *HortScience,* 30 (3): 456–461.

Kuehnle, A.R., Fugii, T., Mudalige, R. and Alvarez, A., 2004b. Gene and genome mélange in breeding of *Anthurium* and *Dendrobium* orchid. *Acta Horticulturae,* 651: 114–121.

Madison, M., 1979. Protection of developing seeds in Neotropical Araceae. *Aroideana,* 2: 52–61.

Marchant, C.J., 1973. *Chromosome Variation in Araceae: V. Acoreae to Lasieae.* Kew Bulletin, 28: 199–210.

Matsumoto, T. and Kuehnle, A.R., 1997. Micropropagation of *Anthurium.* In *Biotechnology in Agriculture and Forestry, Vol 6,* (Ed.) Y.P.S. Bajaj, Springer-verlag, Berlin, pp: 14–29.

Nishijima, W.T., 1994. Diseases. In: *Anthurium Culture in Hawaii.* HITHAR Research Extension Series (Eds.) T. Higaki, J.S. Lichty and D. Moniz. College of Tropical Agriculture and Human Resources, University of Hawaii, Honolulu, 152: 13–18.

Patil, M.S., 2009. Breeding and development of molecular markers linked to bacterial blight resistance in anthurium. *Ph.D. Thesis,* University of Agricultural Sciences, Bangalore, India.

Petersen, G., 1989. Cytology and systematics of Araceae. *Nord. J. Bot.,* 9: 119–158.

Pfitzer, P., 1957. Chromosomenzahlen von Araceen. *Chromosoma,* 8: 436–446.

Ray,T.S., 1987. Diversity of shoot organization in the Araceae. *American Journal of Botany,* 74 9): 1373–1387.

Ray,T.S., 1988. Survey of shoot organization in the Araceae. *American Journal of Botany,* 75(1): 56–84.

Sheffer, R.D. and Kamemoto, H., 1976. Chromosome numbers in the genus *Anthurium. American Journal of Botany,* 63: 74–81.

Sheffer, R.D.and Croat, T.B., 1983. Chromosome numbers in the genus *Anthurium* (Araceae) II. *American Journal of Botany,* 70: 858–871.

Shiva, K.N. and Nair, S.A., 2008. Performance of anthurium cultivars in Andamans. *Indian Journal of Horticulture,* 65(2): 180–183.

Singh, D.R., Sujatha, A. Nair and Medhi, R.P., 2002. Evaluation of anthuriums under shade net condition at Andaman. *Indian Journal of Horticulture,* 65(2): 23–27.

Williams, N. and Dodson, C.H., 1972. Attraction of male euglossine bees to orchid fragrances and its importance in long distance flow. *Evolution,* 26: 84–95.

2013, Biodiversity in Horticultural Crops Vol. 4

Editor: **Professor K.V. Peter**

Published by: **DAYA PUBLISHING HOUSE, NEW DELHI**

Pages 121–131

Chapter 6

Capsicums and Paprikas

K. Madhavi Reddy

Indian Institute of Horticultural Research,
Hessaraghatta Lake Post, Bangalore – 560 089
E-mail: kmr14@iihr.ernet.in

Capsicum species belong to family Solanaceae. The night shade family, also includes tobacco, potato, tomato, brinjal and petunia. The genus *Capsicum* consists of at least 25 wild species, of which five species, *viz., annuum, baccatum, chinense, frutescens* and *pubescens* are domesticated. Different varieties of chilli belonging to these common species are: *Capsicum annuum* Bell peppers, paprika, New Mexican, jalapenos, Cubanelle, Charleston hot cayenne, Fresno Chile, pimento, Bulgarian carrot chili, Peter pepper, ornamental piquin, tepin, Thai hot and chiltepin; *Capsicum frutescens* Cayenne and tabasco peppers; *Capsicum chinense* Habaneros, Fatalii and Scotch bonnets; *Capsicum pubescens* Rocoto peppers; and *Capsicum baccatum* Aji peppers.

Capsicum annuum pod types are usually classified based on fruit characteristics. A major division between the different *C. annuum* pod types is to classify the fruits as pungent or non-pungent. The pod types *viz.,* bell, pimiento, Cuban and squash have both pungent and non-pungent cultivars, such as Yellow Wax and Cherry. The pungent *C. annuum* varieties include Cayenne, New Mexican, Jalapeno, Serrano, Ancho, Pasilla, Mirasol, de Arbol and Piquin. However, there are also cultivars of Jalapenos and New Mexican which are non-pungent. *Capsicum* species are used fresh or dried, whole or ground, and alone or in combination with other flavoring agents. Paprika is derived from *Capsicum annuum* L. and is used as natural food colorant. Spanish paprika is called pimento and is generally used for coloring purposes. Chilli pepper from cultivars of *Capsicum annuum* L. and *Capsicum frutescens* L. are employed as a flavoring in many foods, such as curry powder and Tabasco sauce. Red or hot peppers from *Capsicum annuum* L. and *Capsicum frutescens* L. are the most pungent peppers and are used extensively in Mexican, Italian and Indian foods.

Chilli crop is introduced to India 500 years back by Portuguese traders and due to the long history of cultivation, out-crossing nature and popularity of the crop, large genetic diversity including local landraces have evolved. Today, numerous landraces of chilli differing in shape, size, color and heat level are found in India as farmers selected as per their needs. Therefore, India is considered as secondary centre of diversity for chilli. In hot chilli, great range of variability for several morphological attributes occur throughout India, particularly in South Peninsular Region, North Eastern Region, in foot hills of Himalayas and Gangetic plains. However, genetic resources of hot pepper landraces in India have not been well documented.

Among the domesticated species, *C. annuum, C. frutescens* and *C. chinense* are the most commonly cultivated species and most of the chillies grown in India belong to these species. In *C. annuum* species, chilli and bell pepper (Shimla Mirch/Bangalore Mirch) are the two broad categories, widely distributed in all South Asian countries. Other types, *viz.,* pickle or stuff type chilli in Eastern Uttar Pradesh, squash type in Dharwad district of Karnataka and tomato chilli in Warangal district of Andhra Pradesh are also popular. In North-eastern India, a few names mentioned include 'Naga Jolokia', 'Bhut Jolokia', 'Bih Jolokia' and the Assamese word 'Jolokia' means the *Capsicum* pepper. Mathur *et al.* (2000) reported the 'Naga Jolokia' to have a very high heat level, *i.e.,* 8,55,000 Scoville heat units (SHUs) and is compared to the hottest chilli pepper on record is the *C. chinense* cultivar Red Savina with a heat level of 5,77,000 SHUs (Guinness Book of World Records, 2006).

Botany

Capsicum is an annual and in suitable climatic conditions, it will be a perennial small shrub, living for a decade or more. *Capsicum* probably got evolved from an ancestral form in the Bolivia/Peru area (Heiser, 1976). Chilli fruits are berries botanically; and chilli types usually are classified by fruit characteristics, *i.e.* pungency, color, shape, flavor, size, and their use (Smith *et al.,* 1987; Bosland 1992). Despite their vast trait differences most chilli cultivars commercially cultivated in the world belong to the species, *C. annuum. Capsicum* species are cold sensitive and generally grow the best in well-drained, sandy or silt-loam soil. Normally direct seeding or transplanting is practiced for its commercial cultivation. Flowering usually occurs three months after planting. Hot and dry weather are desirable for fruit ripening. Fruit is generally handpicked as it ripens, and then allowed to dry in the sun, although artificial drying is often employed. The fruit may be ground intact or after the removal of seeds, placenta parts, and stalks, increasing the fruit color and lowering the pungency.

The typical *Capsicum* flower is pentamerous, hermaphroditic and hypogynous. The corolla is rotate in most species with 5-7 petals, which are 10-20 mm long. The diameter of a *C. annuum* flower is 10-15 mm across. The flower colour is dependent on the species, but most *Capsicum* species have whitish flowers. Flowers are usually solitary at the axils of the branches for *C. annuum*, however, some accessions have clusters of flowers at the nodes. The cluster type is associated with the fasciculated gene, which causes multiple flowers/fruits to form at a node. *C. annuum* starts flowering with a single flower at the first branching node; there can be exceptions where two flowers can be found at some nodes. Then a flower forms at each additional node, at geometric progression. Gradually, more than 100 flowers develop on one plant depending on the cultivar. The rate of fruit set is negatively correlated to the number of fruits set at the early stages of the plant, as the rate of flower production decreases. Fruits from early flowers are usually larger and have greater red colour and pungency content at maturity. Chilli is considered to be a day neutral and warm season crop. Fruits do not set when mean temperatures are below 16°C or above 32°C, as it reduces pollen viability. However, night temperatures are more crucial and the flowers drop when night temperatures are above 28-30°C. Fruit normally reaches the mature green stage 35-50 days after the flower is pollinated.

Pepper flowers are complete, that is they have a calyx, corolla, and male and female sex organs. Most species of pepper are self-compatible. Self-incompatibility has been reported in *C. cardenasii* and in some accessions of *C. pubescens* (Yacub and Smith, 1971). Mating among siblings is required to produce viable seeds within these accessions. Peppers exhibit no inbreeding depression. All species are protogynous and can cross- pollinate. The stigma is positioned slightly below level with the anthers or exerted beyond, in which case the chances for cross-pollination are greater. Studies have shown that cross-pollination can range from 2-90 per cent (Pickersgill, 1997). Therefore, pepper breeders and seed producers must use caution to prevent uncontrolled cross-pollination (Bosland, 1993).

Chilli fruits contain numerous chemicals of importance for nutritional value, taste, colour and aroma; and the two most important groups of chemicals found in the fruits are the carotenoids and the capsaicinoids. The carotenoids contribute to fruit colour and nutritional value; and capsaicinoids are the alkaloids which give characteristic pungency. The diverse and brilliant colours *i.e.* green, yellow, orange and red colours of hot/bell pepper fruits originate from the carotenoid pigments present in the thylakoid membranes of the chromoplasts produced in the fruit during ripening stage. The red colour in pepper comes from the carotenoides capsanthin and capsorubin, while the yellow-orange colour is from ß-carotene and violaxanthin. Capsanthin, the major carotenoid in ripe fruits, contributes up to 60 per cent of the total carotenoids.

Varieties of *Capsicums* across the World

There are more than 400 varieties of *Capsicums* available all over the world. They differ in pungency, size, shape and colors. Peppers are commonly divided into two groups, pungent and non-pungent, also called hot and sweet. Sweet peppers include the bell pepper, paprika, pimiento and the sweet yellow wax peppers (Bosland *et al.,* 1996). Some of the main types grown across the world are given below.

Aji

Aji is South American term used for chilli. It is one of the hottest flavored chile peppers. Generally, aji is bright yellow, red, orange, or purple in color. It is also called as 'Yellow Chile' or 'Yellow Peruvian chile'.

Anaheim

Anaheim peppers are blunt nosed, long and narrow with green or reddish color. Anaheim is available in Mexico. The pungency of Anaheim depends on where they are grown; however, they are moderately hot.

Ancho

The fruit is heart-shaped, pointed, thin walled, and the stem attachment on the fruit is indented. The immature fruit color is dark green and turns red at maturity.

Bell Peppers

Bell peppers are sweet or mild in pungency and originated about 5000 B.C. from South America. The bell group may be the most economically important and has the large number of cultivars (Figure 6.1). In bell or sweet peppers the diversity observed is not as large as in hot peppers. However, variability in bell peppers is found based on plant growth habit, fruit shape, fruit colour both at intermediate and mature stages, fruit size based on fruit length and width, flower position, fruit weight, fruit shape at pedicel attachment, fruit shape at blossom end, number of lobes or locules, fruit pericarp thickness, fruit firmness, fruit surface, pungency level and biotic and abiotic stress susceptibility. The fruit colour at intermediate stage varies from cream, light green, green, dark green, purple and chocolate brown; whereas fruit colour at mature stage varies from red, yellow, orange and orange red. Normally fruits will be blocky square or rectangular, conical or elongated in fruit shape with flat or sunken, round and pointed tip. Fruit size varies from big (250g) to small (50g); fruits will be 3-10cm long and 3-8 cm wide; and fruit shape at pedicel attachment will be round or sunken. Bell peppers of square shape with a flat bottom are preferred. 'California Wonder' is one of the oldest cultivars and is typical of the pod type. Elongated bell pepper type like La Muyo has the most distinguishing characteristics like non-flat blossom end and two-three lobed fruits, instead of the four-lobed fruit types. There are also some cultivars which are pungent, *e.g.* Maxibell.

Bolita

Bolita are dark red, oval shaped fresh chilli fruit with very hot pungency. When dried, the bolita are called Cascabel.

Cascabel

Also called 'Bola Chile' and cascabel chillies are dried form of bolita chillies. Cascabel is a Spanish word which means "round bells" or "rattle" found in Mexico. Cascabel fruit is round and brown or red in color with mild to medium in pungency.

Cayenne

Cayenne is long and slender in shape with dark green color which matures to bright red and it is extremely pungent to taste and used widely in India and other Asian countries. The red Cayennes are used to make chilli powder. There are many varieties of cayenne including Hot Portugal, Large Red Thick, Long Red Slim, Ring of Fire, Super Cayenne, Charleston Hot *etc.* and the Charleston Hot cayenne is considered to be extremely pungent and has attractive fruit. Majority of commercially grown chilli varieties/hybrids in India belongs to this group. The mature red fruit is 15 to 30 cm long by 1.5 to 3 cm wide, and is characteristically wrinkled. It may be straight, crescent or irregular in shape and are highly pungent. Often used as a dried, ground powder, it is also used fresh in salads, hot sauces and culinary dishes.

Cherry

This type has small round, or lightly flattened, immature green fruit that turns red at maturity. They can be hot or sweet.

Chiltepin

It is believed to be the oldest and original form of chilli species available now. Chiltepin has tiny round shaped fruits and are extremely hot in pungency. It is also known as "Chiltecpin" or "Tepin."

Piquin

Sometimes called the bird pepper, it is a small, wild, red pepper that grows throughout Guatemala, Mexico, Southern Arizona, Texas and Northeast India. The fruits are small, less than 1.5 cm wide and less than 3 cm long. In Guatemala, the Indians swallow the tiny fruits as pills to cure stomach ache. Other names are "Chilepiquin", "Chiletepin", "Chile tepin" and "Chiltecpin." Shape can be used to differentiate the eight round tepin from the oval piquin. This pepper, like de Arbol, is reported to have distinct flavors which add to dishes in which it is used. The green fruits are pickled, while the red fruits are dried and ground.

de Arbol

The name is derived from the resemblance of the pepper plant to a tree, although it only grows 72 to 180 cm high. The fruits are 6 to 9 cm long and 0.6 to 1.2 cm wide, and translucent when dried. The calyx or stem end of the fruit is narrow and tapered. Other names for this pepper are 'Pico de Pajaro' and 'Cola de Rata'. The pepper is usually used as a dry powder in red sauces, and gives them a distinctive flavor. In New Mexico decorative wreaths are made using this type of pepper.

Habanero

Habanero chillies in Spanish mean 'the one from Havana' and is commonly used in Mexico and Yucatan and considered the hottest chilli. The fruits of Habanero are yellow, red or white in color. The chilli pods of Red Savina Habanero chilli are considered as the hottest in the world which may even cause tongue blistering if eaten.

Jalapeno

Named after the city Jalapa of Mexico, Jalapenos are extremely popular in Mexico and USA. Jalapenos are cylindrical, oval shaped with very thick flesh with green color and turns to red when ripe. The Mexican cuisine includes "chipotle" which is smoked and dried Jalapenos. Fruits are thick walled, conical shaped, dark green when immature turning red at maturity, highly pungent with corky skin. Different types of Jalapeños are grown in different areas of Mexico. "Peludo" are long and thick "Espinalteco" are pointed, and "Morita" are short and oblong. They are also canned, pickled, made into salsas and eaten fresh. When Red Jalapeños are dried by smoking over hardwood, they are called 'chipotle'.

Jamaican Hot

The name of the chilli suggests that Jamaica is the origin of Jamaican Hot chillies. The chilli is green when immature and it matures to yellow color with extremely hot in pungency.

Mirasol

Means 'looking to the sun' having fruits with erect orientation and are extremely beautiful to look with its rich smooth shinning red skin, oval shaped with high pungency. When dried, it becomes red chilli known as 'guajillo' or 'puya'. Some of the new cultivars of mirasol have fruits that are pendent, or hanging down. Pods are 10 to 12 cm long, 1.5 to 3 cm wide and slightly curved. It is a popular pepper in Mexico, dried and commonly used as a condiment and in soups. 'Guajillo' and 'Cascabel' are types of mirasol pepper. Cascabel means rattle *i.e.* when the dried pod is shaken, loose seed inside rattles.

Mulato

The "Mulato" is brown at maturity. In California, Mulato is also called Pasilla. However, in New Mexico, Pasilla is a long, slender, dried chile pepper pod. The reason for the ambiguity is that Pasilla means raisin in Spanish, so any dried, wrinkled pepper could be a pasilla. "Poblano" is used by U.S. produce managers for all green fruits of the ancho type.

New Mexican

This is the most common pepper grown in New Mexico, also called 'Hatch' chile, because of the famous chile growing valley in southern New Mexico. Fruit length ranges from 12 to 36 cm, and are green when immature. Although most cultivars turn red at maturity, mature fruit color can vary from yellow to orange to brown. Three popular land races of New Mexican types are 'Chimayo', 'Dixon' and 'Velarde', and heat levels vary depending on the specific variety grown. When harvested green, New Mexican chile is eaten fresh, canned or frozen; and if harvested at the red stage, it is usually dried and ground into chile powder (paprika if sweet). This powder is used in many sauces. Both green and red-fruits are used in 7 salsas. This is the type to make chile rellenos, a kind of stuffed pepper.

Paprika

Paprika belongs to *Capsicum annuum* originated from Mexico and used widely in America, California, Spain and Hungary (Figure 6.2). The paprika is usually dried and ground to give seasoning powder having pungent sweet smelling characteristic. The bright red round or elongated shaped fruit of paprika varies in pungency from mild to hot and used extensively in Hungarian cuisine. Paprika is defined in the United States as sweet red pepper powder. This well-known, mild red powder can be made from any variety of *C. annuum* that is non-pungent and has a brilliant red color. Low pungent or non-pungent, New Mexican chiles, *e.g.* 'NuMex R Naky' and 'NuMex Conquistador', are grown for paprika powder in the western United States. European paprika is made from paprika fruit, ground after the seeds have been removed. Spain and Morocco grow a round paprika, about the size of an apricot. Longer, more conical and pointed varieties are also grown. Paprika means pepper in Hungarian, therefore Hungarian paprika can be both pungent and non-pungent.

Pasilla

Pasilla chilli is long thin, with green color which turns dark brown when matures and when dried has raisin-texture. The ripe pasilla is called as "Chilaca" or "Chiles Negro" and is available in Mexico. The pungency of pasilla is from mild to medium, fruits are cylindrical with an undulating shape. Pod length is 19 to 36 cm and 3 to 6 cm wide. Fruit color is dark green, turning brown at maturity. It is usually dehydrated before use in moles and salsas. When eaten fresh, the pasilla and the New Mexican are called "Chilaca". In the United States, produce managers of grocery stores have recently been calling ancho type peppers "Pasilla", causing some confusion.

Pimento

The Spanish paprika is called as pimento chilli. They are also called cherry pepper as the fruit is large heart-shaped red pepper with sweet, aromatic, and succulent fruit. The pimentos are extremely mild in pungency and are stuffed in green olives. Sometimes spelled pimento, this sweet type is characterized by a heart-shaped, thick-walled fruit, which is green when immature and turns red at maturity. The fruits are 9 cm long and 6 cm wide at the shoulders. It is used in such processed foods as pimento cheese and stuffed olives, but can also be eaten fresh. Allspice, *Pimenta dioca,* usually known as pimento or Jamaican pepper outside the United States, is not related to *Capsicum*.

Poblano

The poblano chillies mean "pepper from Pueblo" since the poblano originated from Pueblo in Mexico. Poblano chillies are heart-shaped, green in color when unripe and turn red or brown when ripen. The pungency ranges from mild to moderate. The dried poblanos are called with two different names Mulato and Ancho. The Mulatos are sweeter than Ancho.

Scot's Bonnet or Scotch Bonnet

Closely related to Jamaican hot and habanero, these chillies are equally hot in pungency. Scotch bonnet or Scot's bonnet are irregular in shape with yellow, orange, or red color. It is used frequently in

Figure 6.1: Biodiversity in Bell Peppers

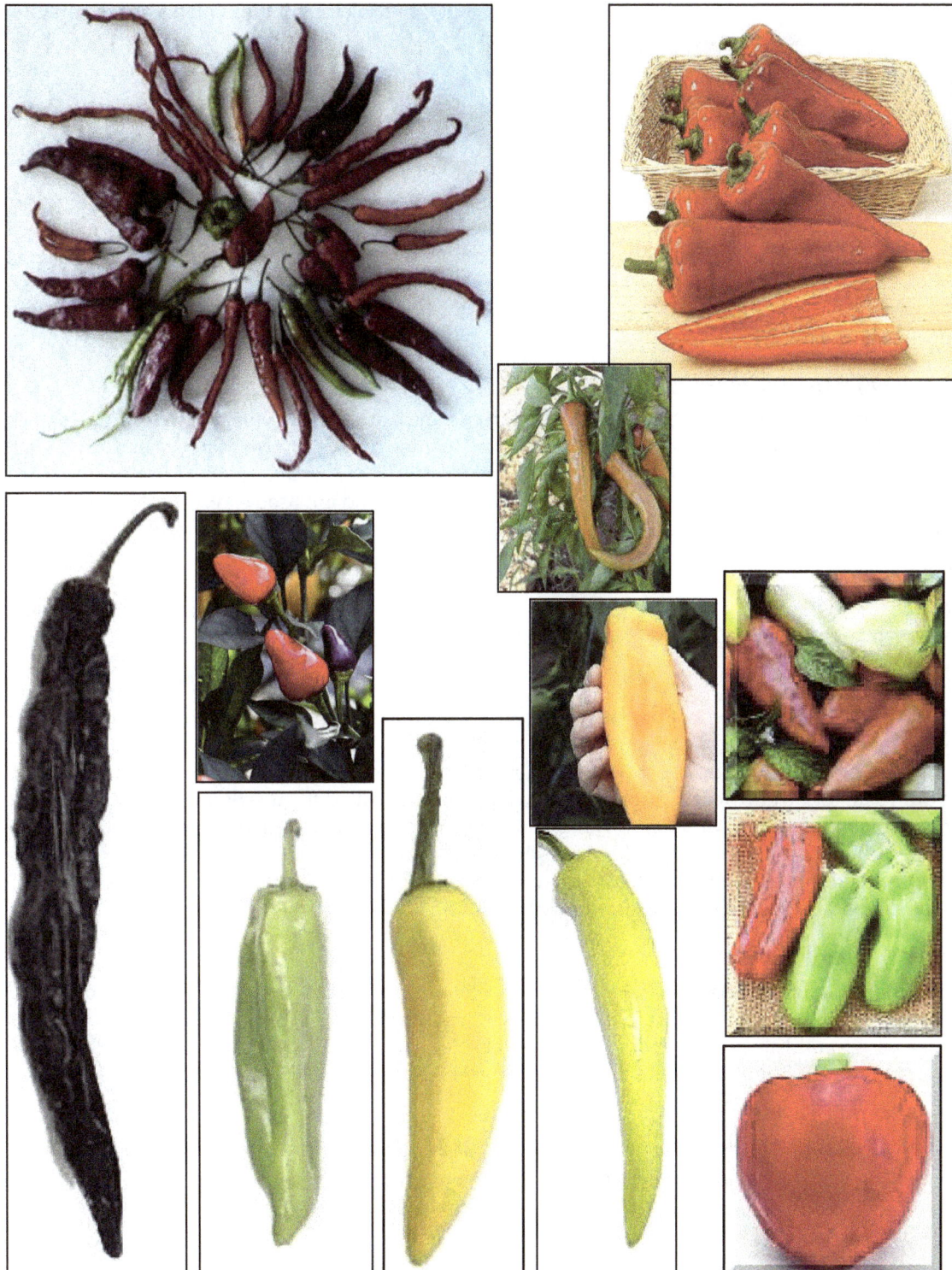

Figure 6.2: Biodiversity in Paprikas

Jamaican cuisine. Scotch bonnet chillies are so hot that when taken will cause blisters to the tongue, dizziness, and severe heartburn.

Serrano

Serrano chillies mean highland or mountain, are small, round in shape with slight pointed at the end. Serrano chillies originated from the foothills of Puebla in Mexico. They are smooth dark green when unripe and turn scarlet red, brown, orange, and then yellow as they ripen with high pungency. The serrano, which literally means, "highland" or "mountain" has fruits 6 cm long, 1.5 cm wide, cylindrical, with medium thin walls and no corkiness. The immature fruit color ranges from light to dark green. When mature, the fruit may be red, brown, orange or yellow. The pungency level of this type is usually higher than Jalapeño. This pepper is used fresh to make "Salsa Verde" and is the pepper of choice in "Pico de Gallo," a hot southwestern relish. Pico de gallo means the "beak of the rooster" as the relish bites you.

Squash/Tomato/Cheese

The fruits are nonpungent, generally flat, 3 to 6 cm long by 6 to 12 cm wide, with a medium to thick walled flesh. Immature fruit may be green or cream, turning red, yellow or orange at maturity. They differ from bells and pimento by generally being a flat fruit. They are usually pickled, but are also eaten fresh in salads. In Guatemala, they are called "Chamborate" and are used extensively in November and December to enhance the typical foods, frambre and Christmas tamales.

Tabasco

Tabasco is the most common pepper type of *C. frutescens*. Fruits are 3 to 6 cm long by 0.75 cm wide, yellow or yellow-green to red and highly pungent.

Yellow Wax

Wax fruits are yellow when immature, with a waxy appearance and turn orange, orange-red or red at maturity. There are two main forms within this group, a long-fruited type and a short-fruited type. The long fruited types are known as Hungarian Wax or Banana Peppers. The fruit length is about 10cm and width 4cm. Common cultivar in the long fruited form is Sweet Banana and short fruited types are Yellow Wax. Yellow wax fruits vary from 6 to 18 cm long and are about 6 cm wide and may be pungent or non-pungent. The fruits are usually conical, or rounded conical. Wax fruits are often pickled but can be used fresh in salads. In Mexico, they are called "Caribe", Guerito", and "Guero", because Guero means light skinned or blond.

Indian Chilli Varieties

Chilli was introduced to India by the great Portuguese explorer Vasco-da-Gama. The spice chilli is blended well in Indian cooking. Chilli became extremely popular in India. The climate in India was favorable to cultivate chillies and soon many varieties of chilli were available in India. Places like Andhra Pradesh, Tamil Nadu, Gujarat and Kashmir became famous for different varieties in chilli.

Naga Jolokia

The Mexican chilli Red Savina Habanero was considered as the hottest chilli with 557,000 Scoville units. The hottest chilli on earth is found in India. The Naga Jolokia or also called "Tezpur Chilli" has been proved by the scientists as the hottest chilli in the world with 8,55,000 Scoville units.

Bird Eye or Dhani Chillies

Birds eye is also called as African Devil Chile. Birds eye chilli is tiny, green and when matured the color changes to bright red. The pungency varies from the place and environmental condition it receives. It is widely available in African countries like Uganda, Zimbabwe, and Malawi and also in other countries including India, China, Mexico, and Papua New Guinea.

Byadagi Dabbi or Kaddi Chillies

Byadagi chillies are grown in Goa and Dharwar in Karnataka. These chillies are also called as *kaddi* (meaning stick-like) chillies and when dried, the skin is wrinkled red in color with aromatic mild pungency. The harvesting seasons of these chillies are from January to May. It is a famous variety of chilli mainly grown in the Indian state of Karnataka. It is named after the town of Byadagi which is located in the Haveri district of Karnataka. The business involving Byadagi chillies has the second largest turnover among all chilli varieties of India. An oil oleoresin extracted from these chillies is used in the preparation of nail polish and lipsticks. Byadagi chilli is also known for its deep red colour and is less spicy and is used in many food preparations of South India. Byadagi Chilli plants begin flowering 40 days after transplantation although the majority of flowers bloom 60 to 80 days after transplanting. The chilli pods are harvested from January to May. The quality of chilli varieties is measured in terms of the extractable red colour pigment; this color is measured in ASTA colour units. Byadagi chilli has an ASTA colour value of 150-200. The higher the ASTA colour unit, the better the quality of chilli and therefore higher the price. The Byadgi chilli has negligible capsaicin content making it less pungent than other chilli varieties with a characteristic flavour. Byadagi chilli oleoresins have excellent export potential. The Union Minister of State for Commerce directed the officials of the Spices Board to take necessary steps to get the "Geographical Indication" (GI) tag for Byadgi Chilli. The Minister also announced that the Spices Board was ready to set up a Spices Park in the region comprising the Dharwad, Haveri and Gadag districts of Karnataka.

Ellachipur Sannam Chillies

Ellachipur Sannam chillies are found in Amravati district, Maharashtra. The Ellachipur chillies are dark red in color when in dry state with very hot pungency. The best time to grow is September to December.

Guntur Sannam Chillies

Guntur Sanman Chillies are cultivated in Guntur, Warangal and Khamman district of Andhra Pradesh. The Guntur chillies are long with thick red skin with very high pungency.

Hindupur Chillies

Hindupur chillies are found in Andhra Pradesh in India. These chillies are extremely pungent and red in color. The harvesting season for these chillies is from December to March.

Jwala Chillies

The most popular variety of chilli in India, Jwala, is long, slender with when unripe has green color and turns to red in ripe stage. Jwala means "Volcano" in Hindi and is extremely pungent. Jwala is found in Kheda and Mehsana in Gujarat.

Kandhari Chillies

The Kandhari chillies are grown in Kerala and Tamil Nadu. They are small and ivory white in color. They are extremely pungent.

Kashmiri Mirch

One of much in demand chillies in India is Kashmiri Mirch. Though there are a lot of wrong claims, the true Kashmir Mirch is grown in Himachal Pradesh and Jammu and Kashmir. It has smooth shinning skin and fleshy with dark red color. They are mild in pungency and have a fruit like flavor.

Mathania Chillies

Are local to Rajasthan with excellent colour intensity and moderate pungency. Fruits are long and thick.

Mundu Chillies or Gundu Molzuka

Mundu chillies are found in Tamil Nadu and Chittoor and Anantapur districts in Andhra Pradesh. Fruits are roundish to conical with round tip, moderately pungent with yellowish red color. Mundu is popularly called as 'Gundu Molzuka' meaning 'fat chilli' in Tamil Nadu.

Nalcheti Chillies

Nalcheti chillies are grown in Nagpur, Maharashtra. The Nalcheti chillies are long, red in color when dried with high pungency.

Tomato Chilli or Warangal Chappatta

Tomato chillies are found in Warangal, Khammam, and Godavari district of Andhra Pradesh. These chllies are short, dark red in color when dried with moderate pungency.

Sankeshwar Chilli

Fruits are very thin (<5cm width), very long (20-25 cm length), curved and wrinkled, mainly cultivated in Maharashtra and Belgaum districts of Karnataka.

References

Bosland, P.W., 1992. Chillies: A diverse crop. *HortTechnology,* 2: 6–10.

Bosland, P.W., 1993. An effective plant field-cage to increase the production of genetically pure chilli (*Capsicum* spp.) seed. *HortScience,* 28: 1053.

Bosland, P.W., Bailey, A.L. and Iglesias-Olivas, J., 1996. *Capsicum Pepper: Varieties and Classification.* Cooperating Extension Service Circular 530, College of Agriculture and Home Economics, NMSU and USDA, Las Cruces NM, pp 1–13.

Guinness Book of World Records, 2006. *Hottest Spice.* www.guinnessworldrecords.com. Accessed 13 Sept. 2006.

Heiser, C.B., 1976. Peppers Capsicum (Solanaceae). In: *The Evolution of Crops Plants,* (Ed.) N.W. Simmonds. Longman Press, London, p. 265–268.

Mathur, R., Dangi, R.S., Das, S.C. and Malhotra, R.C., 2000. The hottest chilli variety in India. *Curr. Sci. India,* 79: 287–288.

Pickersgill, B., 1997. Genetic resources and breeding of *Capsicum* sp. *Euphytica,* 96: 129–133.

Smith, P.G., Villalon, B. and Villa, P.L., 1987. Horticultural classification of pepper grown in the United States. *HortScience,* 22: 11–13.

Yaqub, C.M. and Smith, P., 1971. Nature and inheritance of self–incompatibility in *Capsicum pubescens* and *Capsicum cardenasii. Hilgardia,* 40: 459–470.

2013, **Biodiversity in Horticultural Crops Vol. 4**

Editor: **Professor K.V. Peter**

Published by: **DAYA PUBLISHING HOUSE, NEW DELHI**

Pages 133–144

Chapter 7

Chrysanthemums

P. Narayanaswamy

Dean, College of Horticulture,
University of Horticultural Science, Bagalkot – 587 102 Karnataka
E-mail: deancoh.hiriyur@gmail.com

An ancient Chinese proverb says that "if you would like to be happy for a lifetime, grow chrysanthemums". The chrysanthemum, first described in the 15th century B.C.E in China, was introduced to Japan by the 8th century C.E (www.mountainroseherbs.com). The Japanese were so enthralled by the flower that they made it as the official seal of the Emperor. It still ranks as one of the most important herbs in traditional Japanese medicine and it is thought to have the power of life. There is a "Festival of Happiness" in Japan that celebrates the flower. In the US, it is considered the "Queen of the fall flowers" and is the largest commercially produced flower in the US due to its ease of cultivation, capability to bloom on schedule and its wide diversity of form and color.

In fact, no flower offers a great diversity of forms, vigour of plants, ease of cultivation, and a more extended period of flowering than chrysanthemums. From a production standpoint, for sheer spontaneous display, for the wealth of seasonable colors, for daily cut-flower materials, and for the garden's final burst of splendour, there is really no plant comparable to the chrysanthemum.

Chrysanthemum is the second largest cut-flower after rose among the ornamental plants traded in the global flower market. It is cultivated both as a cut-flower and as a potted plant (pot mums). The basic chromosome number of chrysanthemum cultivars is reported 9 by several workers in different parts of the world *e.g.*, in Europe (Dowrick, 1952), in Japan (Miyazaki *et al.*, 1982) and in India (Nazeer and Khoshoo, 1982). Study of cultivars grown in India has displayed wide range of ploidy levels with 2n= 36, 45, 51 and 75. The commonly grown chrysanthemums have hexaploid complex with average number of 54 chromosomes. It is propagated vegetatively as it has a strong sporophytic self-incompatibility system as shown by almost all the members of Asteraceae family. The wild chrysanthemum is a sprawling, leafy plant with clusters of daisy like flowers at its crown.

In India, chrysanthemum is grown in about 4000 ha and the major growing states are Tamil Nadu, Karnataka, Maharashtra, Rajasthan, Madhya Pradesh and Bihar. Its cultivation is popular around the cities like Delhi, Kolkata, Lucknow, Kanpur, Chennai, Allahabad and Bangalore. In Karnataka, chrysanthemum occupies an area of about 2830 ha with an estimated production of 14556 tonnes (Ramamurthy and Ramakrishnappa, 2000). The utility and popularity of chrysanthemum have increased immensely with the introduction of the techniques for year round blooming based on scientific research in the field of photoperiodism, genetics, biotechnology and molecular biology.

Origin and History

The name "Chrysanthemum", derived from the Greek words *Chrysos* (gold) and *anthos* (flower), was given to the plant by Linnaeus in 1753 AD (Cuming, 1939). The name Chrysanthemum became the genus name and together with many other genera comprise the Asteraceae or Daisy family. The American Horticulturist, Bailey listed *Chrysanthemum morifolium* as the valid name for the garden chrysanthemum varieties we grow today (Bailey, 1947).

The cultivated chrysanthemum is originally native to China (Hemsley, 1889). Its history can be traced back to the Confucian era (550 B.C.). Confucius wrote: "The chrysanthemum has its yellow glory" in his book Li-Ki which treats largely of ancient ceremonies and institutions (Cuming, 1939).

The Chinese as early as fourth century A.D grew ball-shaped progenitors of present- day chrysanthemums in a variety of colours. The chrysanthemum flower appears frequently in Chinese literature, poetry and songs where it is referred to as chu or chu-hwa. Through the ages, it was one of the most popular flowers of China and it remains a popular flower even today.

According to Flower Mirror, a book dealing with flowers written about two hundred years ago in the Ching Dynasty, there were one hundred fifty large exhibitions, incurved, spidery, and quill-petalled type flowers in one regional collection. Some of the cultivar names included "Emperor's Garment," "Golden Dish", "Embroidered Hibiscus", Star of the Sky", "Jade Ball", "Agate Rings", "Full Moon", *etc*. These names are quite different to the western chrysanthemum growers.

The chrysanthemum arrived in Japan from Korea in the year 386 A.D. Both of these nations made distinguished contributions to chrysanthemum breeding and flower improvement. In 797 A.D., the chrysanthemum was recognised as a valuable plant and was recognised as the national emblem of Japan. The flower became the symbol of the sacred ruling dynasty (Woolman, 1953).

In 1689 A.D., Breynius, a native of Holland, wrote a book about the chrysanthemum and is believed to be the first European to refer to this flower. One hundred years later in 1789, Mr. Louis Pierre Blanchard brought a large flowered variety from China to France and then, in 1795, to England. This cultivar (a cultural variety) from China was believed the first variety ever to be cultivated in Europe. After 1820, more varieties were introduced to Europe from China, and by 1826 forty-eight varieties were known (Woolman, 1953).

The first chrysanthemum show in Europe was held in Vienna in 1831. In the year 1843, Robert Fortune sent a Pompon type (the Chusan Daisy) to England from China, which is one of the important parents of chrysanthemum cultivars in Europe today. Later in 1861, he brought seven selected varieties from China and from these plants the incurved and reflexing exhibition type flowers were produced. From 1850 until 1950, English cultivators have played a great part in the continued improvement of the chrysanthemum flower- both in size of the large types and also unusual petal form. In 1930, Mc Gregor sent from China a plant, similar to *Chrysanthemum indicum,* from which were produced the cultivars Sutton's Charm and Cascade (Kyle, 1952).

The Chrysanthemum was introduced to America in 1798 by John Stevens of Hoboken, New Jersey. Dr. H.P. Walcott of Cambridge, Massachusetts in 1879 was the first American reported to raise seedlings of chrysanthemum. The development of chrysanthemum in America was mainly directed towards the production of good commercial varieties (Dowrick, 1953). Among the American breeders, Elmer Smith was the man most responsible for commercial success of large flowered chrysanthemums in the United States and he introduced more than 445 varieties through concerted breeding efforts since 1887.

Much of the credit for the development of modern hardy chrysanthemum called Korean hybrids is attributed to Alex Cumming. Varieties derived from crosses between *C. coreanum* with the garden chrysanthemum *C. morifolium* were produced in 1928. The introduction of germplasm of this species resulted in better hardy garden chrysanthemums, and points the way to future improvement through interspecific crosses (Viehmeyer, 1955).

Species Diversity in Chrysanthemums

A remarkable progress has been made in the production of hardier type of garden flowers by using a number of species for hybridizing. Considerable variation might be expected from these hybrids (Ellen Fan, 1965). These chrysanthemum species include the following types (Figure 6.1):

Chrysanthemum indicum (Mother Chrysanthemum)

From this little blossom, the wide range of garden Chrysanthemum and enormous exhibition types of the florists were largely developed. Foliage and stems have strong odour.

C. morifolium (Florists Chrysanthemum)

A cultigen of Chinese origin and cultivated in China, Japan and Korea for more than three thousand years, these are highly developed horticultural plants and have the greatest number of horticultural varieties (cultivars). These plants are perennial herbs, 60 to 120 cm or more high. Flower

Figure 6.1: Species Diversity in Chrysanthemums

heads have developed into several groups of flower types (Figure 6.1). Its growth habit is regulated by planting date and pinching and photoperiod control. The leaves are petioled or with winged-petiolar base, toothed and deeply lobed, ovate to obovate (inversely ovate).

C. coreanum (Korean Daisy)

Single flowers 6 cm across, open pure white, changing to pink, with occasional carmine-pink tones when mature.

C. nipponicum (Nippon Oxeyed Daisy)

Shrub like, a large part of the plant will overwinter and produce new lateral stems in spring. Leaves crowned on upper part of stem, thick and stiffened, flower head single on the terminal, blooms very late.

C. maximum (Shasta Daisy, Glory of the Wayside)

As a pollen parent, it has been used in the direction of a more compact and earlier flowering type. Rays white, disk yellow, leaves sessile, toothed, oblong lanceolate or oblanceolate (inversely lanceolate).

C. coccineum (Color Daisy, Painted Daisy)

It has proved to be of value as a pollen parent. Some extremely desirable color shades have developed, such as red, pink, lilac and white, with yellow disk, also a greater depth of richness in the crimson shades has been achieved with fern-like vivid-green foliage.

Figure 7.2: Flower Color Diversity in *Chrysanthemum morifolium*

C. arcticum (Arctic Daisy)

A species of low, mound-like growth, heads solitary, 3 to 6 cm across, white tinged rose or lilac, numerous and attractive flowers. It often survives a winter temperature of -38 °C.

C. uliginosum (Giant Daisy)

An erect-growing type can attain a height of 180 cm in the rich moist soil. Stems much branched and finely pubescent above, leaves long-lanceolate and very sharp pointed.

Wild Species of Chrysanthemum Grown in the Indo-Tibetian Border

1. *C. stilliszkai*
2. *C. rkhtsria*
3. *C. atkinsoni*
4. *C. leucanthemum*

Introduced Species or Exotic Species of Chrysanthemum

1. *C. caronanium* (Garland chrysanthemum)
2. *C. carinatum* (Tricolour chrysanthemum)
3. *C. rubellum* (for hardiness)
4. *C. sagetum* (Corn marigold (or) pot plant)
5. *C. boreale* (Evolution of florif, chrysanthemum)
6. *C. cinerarifolium* (Used as insecticide)
7. *C. coccineum* (Perennial, seed propagated)
8. *C. manifolium* (Florist chrysanthemum)

Chrysanthemum morifolium is also an important source of essential oil and sesquiterpenoid alcohol (Hu and Cheu, 1997 and Yang *et al.*, 1997). Certain species like *C. cinereriifolium* and *C. coccineum* are also cultivated as source of pyrethrum, an important insecticide (Chittenden, 1956 and Carter, 1980). Apart from this, seeds of *Dendranthema tanacetum* (*C. corymbosum*) are an important source of acetylenic acid (Tsevegsuren *et al.*, 1998).

Flower Morphology of Chrysanthemum

Chrysanthemum flowers have wide diversity of forms, sizes, shapes, and colors, particularly in *Chrysanthemum morifolium* (Figure 7.2). In single chrysanthemums, there are disc florets in the center, the colored petals are called ray florets. Both ray florets and disc florets have stigmas which connect with ovaries deep in the receptacles at the base of the bloom. Only the disc florets, however, have stamens which support the pollen-carrying anthers. In double chrysanthemums the disc florets are in the minority, and in fully double flowers there are no disc florets. Increase in size of the disc florets produces anemone-type flowers. The small disc florets give rise either to broad, flat florets or to elongation without splitting, which results in quelled or tasselled forms.

Diversity for Flower Morphology in Chrysanthemum

1. Single or Daisy type (Da).
2. Semi double (S-D-) - The ray florets that are arranged in more than 5 rows, but the disc is clearly evident as a daisy-like eye.

3. Button type or baby type (B) - The diameter of the flower is less than 3 cm.

4. Pompon (P) - It is of flat, globular, hairy, or reverse types; from small size to large.

5. Anemone (A) -with center cushion, regular or irregular; single or double.

6. Decorative (D) -This is an aster like flower of the most common garden class.

7. Cactus (Ca) - The florets are longer and narrower than Decorative form.

8. Incurved (I) -Bloom is globular in form. The ray florets are broad and incurved, regularly and smoothly or irregularly overlapping.

9. Reflexed (R) -Bloom globular in form, but the ray florets are broad, short, and reflexed.

10. Spoon (S) -Semi double and double spoon; have disc florets that are rather flat, and ray florets that are regularly arranged and spoon- shaped.

11. Quill (Q) -It has thin, thick, or feathery types, from single to double, small to large or in a cluster of miniatures.

12. Thread (Th) -The disc may or may not show, the ray florets are of an equal length, tubular and delicate; thin or thick.

13. Tube (T) -It has slender and tubular petals; thin or thick; small to large.

14. Spidery (Sp) -It has relatively long, slender and tubular petals of irregular length, sinuous or hooked at the tip, single or double.

15. Exhibition (E) - The blooms often are larger than 18 cm in diameter

Varietal Wealth of Chrysanthemums in India

The success of any breeding programme depends mainly on the extent of genetic variability available in the population. In India, several varieties of chrysanthemum are grown for various purposes. The cultivars grown in northern India are mainly introductions from Australia, Europe, France, Japan or America in addition to those originating within the country. In South India on the other hand, only a few yellow or white small coloured cultivars are grown for use as loose flowers which are probably of Indian origin. Introduction of exhibition types seems to have started in East India, particularly in Kolkatta and Sikkim during British period (Kher, 1977). Names of some well known cultivars grown in India have been given below along with the name of country wherefrom they were introduced into this country or where it originated:

Sl.No.	Country from which Introduced/Originated	Name of the Cultivar
1.	Australia	J.S. Lioyd, Louisa Pockette, Willium Turner and T.W. Pockette
2.	France	Gloria Deo, S.L. Andre Raffaud and Sancho
3.	Japan	Ajina Purple, Kenroku Kangiku, Kiku Biori, Taiho Tozan, Tokyo, Shin Mei Getsu, Senkyo Emaki and Otome Zakura
4.	New Zealand	Gusman Red, Lcicles, Jane Sharpe, Orange Fair Lady and Nancy Ferneaux
5.	United Kingdom	Alfred Wilson, Alfred Simpson, Balcombe Perfection, Beatrice May, May Shoesmith, Maurice White, Leviathan, Pink cloud, Princess Anne and Woolman Centruy
6.	United States	Cassa Grande, Mountaineer, Nob Hill, Snow Ball, Potmac and Peacock

Chrysanthemum Varieties Introduced to India

Varietal Classification of Chrysanthemums

Chrysanthemum varieties are broadly categorised into two types depending on the purpose for which they are used. A cultivar suitable for pot culture may not be fit for growing as cut flower. Similarly, a cultivar may be suitable for cut-flower purpose but not for garland making.

Garland Purpose

Baggi, Basanti, Shanti, Indira, Rakhi, Red Gold, Birbal Sahani, Vasantika, Sharad Mala, Meera and Jaya.

Cut Sprays

Apsara, Birbal Sahani, Jayanti, Jubilee, Kundan, Purnima, Kundou, Jaya, Shard Singer, CO-1, CO-2, Nanako, Megami, Riot, Arctic and Chralia.

Popular Chrysanthemum Varieties in India

Although several varieties are developed in chrysanthemums, the following varieties are popular among farming community in India:

White Charm, KS 16, Basanti Local, Jubilee, Punjab Gold, Red Gold, Gul-e-Shair, Pin Gin, White Prolific, Flirt, Sharad Bahar, Vasantika, Kirti, Maghi, Arka Ravi, Arka Pink Star, IIHR 6, IIHR 13, IIHR Sel-4, IIHR Sel-5, Mother Teresa, Indira, Nilima, Yellow Gold, Ratlam selection, Sonali Tara, Punjab Anuradha, Nagpur Red, Haldighati, Cardinal, Puja, Jaya, Suneet, Gauri, Sonar Bangala, Baggi, Jayanti, Vijaya, Sunil, Ajina Purple, Snowball, Potomac, M-24, Pandhari Rewadi, Agnishikha, Batik, Harvest Home, Gypsy Queen, Navneet Yellow, Gamit, Nanako, Gauri, Rosa, Shabnam, Taruni, Indiana, Kusum, Little Darling, Mini Jessie, Pournima, IIHR-Hybrid-11.

Studies on Varietal Diversity in Chrysanthemums

A study in chrysanthemum denotes significant differences for all the characters. Substantial variability was noticed for weight of flowers per plant followed by number of branches per plant. The higher values of genotypic and phenotypic coefficient of variation were observed for number of flowers per plant while, the lowest values were observed for duration of flowering and diameter of flowers (Chaugule, 1985).

Genetic studies in chrysanthemum by Ponnuswami *et al.* (1985) revealed that lower genotypic variances ranged between 12.7 and 26234.6 compared to phenotypic variances which in turn ranged between 26.0 and 52870.8. Among the various characters studied, number of flowers per plant and earliness to flower exhibited a high degree of genotypic and phenotypic variation.

Brijendra Singh and Dadlani (1989) found that the cv. Sharadmala was the earliest to flower among 14 cultivars tested.

A significant variation for the different characters studied in a trial with thirteen chrysanthemum cultivars indicated a higher variation for plant height in cultivar Pandhari Rewadi, while, the flower yield per hectare was the highest in IIHR Sel-4 (Katwate *et al.,* 1992).

Hemalatha *et al.* (1992) recorded high phenotypic and genotypic coefficients of variation for disc diameter followed by number of branches per plant and number of flowers per plant in chrysanthemum.

The cv. Basanthi produced significantly taller plants (82.6cm) than rest of the cultivars. With regard to number of branches, cv. Co-1 registered higher value, whereas, the cv. Red Gold recorded significantly higher number of flowers per plant (48.2), while Megami recorded significantly higher yield per plant (82.5g) among the cultivars studied. (Kanamadi and Patil, 1993).

Gondhali *et al.* (1997) compared the performance of chrysanthemum cultivars for their flowering behaviour and quality of spray as cut flowers and found that the spray quality of cv. Indira was the best with higher number of flowers per spray.

Open pollinated seedlings of Punjab Gold and Gul-e-Shair cultivars of chrysanthemum numbering 100 seedlings were screened to select better genotype for pot culture. Among the seedlings, thirteen variants of Punjab Gold and four variants of Gul-e-Shair exhibited outstanding performance as potted plants (Arora and Anuradha, 1999).

Misra (1999) studied performance of small flowered cultivars of chrysanthemum for calcareous belt of Bihar and concluded that cv. Suneel was the most outstanding variety with regard to vegetative as well as floral characters. Palai *et at.* (1999) evaluated thirty-three accessions of spray chrysanthemum for their growth and flowering behaviour. ACC-13 showed higher value for the characters like plant height, length of sprays and flower weight. Higher number of flowers per spray and number of flowers per plant were noticed in ACC-22. The yield per square meter was higher in ACC-4.

Gondhali *et al.* (1998) conducted a study in chrysanthemum and reported significant variation in growth and yield characters. The cvs. Indira, Shymal and IIHR Se1-5 registered higher values for yield.

In the variability study, fifty-seven genotypes of chrysanthemum evaluated and recorded higher phenotypic coefficients of variation (PCV) than those of genotypic coefficients of variation (GCV) for all the characters studied. However, higher GCV and PCV estimates were found for number of flowers per plant followed by number of branches per plant and disc diameter. High heritability with high genetic advance was observed for number of branches per plant, disc diameter, number of petals per flower and flower yield (Sirohi and Behera, 2000).

Molecular Markers in the Assessment of Genetic Diversity in Chrysanthemums

Genetic variability is the main driving force used by man to meet not only his food needs but also to produce better cultivars. In the early days, naturally occurring mutations were the main source of new alleles for obtaining more adapted materials. This process led to rapid fixation of traits in elite gene pools and it became more difficult to select for subtle differences in naturally occurring populations. Mendel's work set up the principles and a new scenario for hybridization programs, through which plant breeders could generate genetic variability. However, due to linkages there was need to look for ways to follow up and select for desirable traits in segregating populations. Since 1920's, many researchers used morphological and protein markers with some degree of success mainly in Maize and Drosophila.

Traditionally the diversity in germplasm of crop species was estimated by morphological characters but these are a few in number and highly influenced by environmental conditions. The prime advantage of the use of morphological markers is that they are simple, fast and inexpensive. Then biochemical markers came into existence, but these markers account for a very small fraction of genetic variability, and some are likely to be influenced by environment. In the 1980's, the advent of DNA markers made it possible to develop genetic maps in any crop and to apply them to get a better understanding of genetic variation in many gene pools. The DNA based markers like AFLP and SNPs are becoming more popular since they provide excellent tools to study the genetic diversity. The AFLP technique has recently been employed for genetic diversity studies since it can identify thousands of loci. Also reproducibility is high in AFLP compared to any other molecular marker techniques. The capacity of AFLP analysis to detect thousands of independent genetic loci with minimal cost and time requirements makes it a more efficient marker technology (Maughan *et al.,* 1996).

Studies on Diversity of Chrysanthemums Using Molecular Markers

The genetic variation in chrysanthemum using RAPDs was studied by Wolff and Peters-van (1993). The variations between cultivars were identified by using only two different primers. A family of cultivars derived from one original cultivar by vegetative propagation, had identical fragment patterns.

In order to study genetic variability at the DNA level in chrysanthemum, *Pst*I and *Hind*III genomic libraries were constructed. Probes from both libraries were tested for the presence of RFLPs. Of the probes from the *Pst*I library, 91 per cent appeared to hybridize the low copy genes, while only 31 per cent of these from the *Hind*III library appeared to do so. Genetic analysis was simplified by using locus specific polymerase chain primers to obtain simple polymorphic patterns in a number of cases. The RFLP probes and primers developed were used in the marker-assisted selection (Wolff *et al.,* 1994).

Scott *et al.* (1996) recommended the use of DNA amplification fingerprinting (DAF) to study genetic relationships between closely related chrysanthemum cultivars using arbitrary octamer primers. The phenotypic patterns were established by Unweighed Pair Group Cluster Analysis (UPGCA) using arithmetic means, principal coordinate analysis (PCA) and the average distance between series. DNAs from all cultivars belonging to a series were bulked together to generate profiles containing unique amplified products for each series.

Scott *et al.* (1996) reported that the cultivars especially mutants and sports are difficult to differentiate genetically even by DAF. But it was easily identified by using arbitrary signatures from amplification profiles (ASAP). This involved amplification of genomic DNA with 3-stranded octamer arbitrary primers, all of which produced monomorphic profiles. Products from each of these DNA fingerprints were reamplified using DNA fingerprints and subsequently re-amplified with that of 4 mini decamer primers. The number of ASAP polymorphisms detected have helped to estimate the mutation rate in the mutant characters. It was concluded that the ASAP technique allowed clear genetic identification of somatic mutants and radiation induced sports that were genetically highly homozygous.

Genetic variation in chrysanthemum was examined by Sehrawat *et al.* (2003) using random amplified polymorphic DNA within the thirteen commercial cultivars representing standard, spray and no pinch- no stake. Genetic variation was studied using 60 decamer primers. Of these, 31 primers amplified genomic DNA. The genetic variation was high enough to divide them into two major groups. These groupings were in consistent with their morphological differences and geographical distribution. The first group consisted of Snow Ball, Ajina Purple and Sonar Bangala cultivars, while the second group accounted for Nagpur Red, Haldighati, Cardinal, Puja, Jaya, Suneet, Vasantika, Gauri, Flirt and Baggi.

The similarity among the cultivars was very high for morphological traits showing low genetic diversity and fingerprinting by RAPD using 40 primers was powerful to distinguish the chrysanthemum cultivars (Chatterjee, 2005).

Eleven radiomutants obtained from two chrysanthemum cultivars Ajay and Thai Chen Queen could form into three distinct groups based on RAPD technique which can form a useful tool to supplement the distinctness, uniformity and stability analysis for plant variety protection (Kumar, 2006).

The UPGCA cluster analysis of 11 wild species and 12 cultivars indicated that the wild species had an evolutional tendency from low ploidy to high ploidy and also had a complicated genetic relationship among the cultivars. Based on the evolutional relationships, the flat petal type of *Chrysanthemum* might be the basic type. The germplasm of *C. nankingense* has the closest genetic relationships with all tested cultivars, and the three of *C. chanetii, C. japonicum,* or *C. japonicum* var. *wakasaense* have secondly close relatives with tested cultivars, *C. indicum* var. *aromaticum* has genetic relationship far from other tested varieties. The genetic relationships among wild species and cultivars of *Chrysanthemum* were quite complicated and the ISSR marker technology might be employed to better reveal the genetic relationships among *Chrysanthemum* species at the molecular level (Liu Rui, 2009).

RAPD study of chrysanthemum and its 13 clones showed 0.432 to 0.95 genetic similarity. The cultivars and the somaclones were divided into five clusters (Barakat, 2010).

In a recent study on assessment of genetic diversity in 42 selected chrysanthemum genotypes using morphological and molecular markers, it was revealed that the genotypes were not clearly grouped into separate clusters by morphological dendrogram and PCA possibly due to less number of morphological traits considered for the study as compared to RAPD. The RAPD markers detected only 11 per cent diversity as compared to AFLPs which ranged from 17-114 per cent suggesting a higher genetic variation within the chrysanthemum genotypes. The highest dissimilarity per cent was observed between the accessions 'Fitonia' and 'Usha Kiran' which were collected from distinct sources (Mukund, 2010).

Basically the chrysanthemums are propagated through vegetative means. Hence, the genetic diversity is very low. The estimation of genetic diversity through only morphological traits is very difficult and is very low. The recently developed molecular markers are very effective in detection of genetic diversity especially the AFLPs. The combination of morphological and molecular markers can be a better means of assessing genetic diversity in chrysanthemums.

References

Arora, J.S. and Anuradha, M., 1999. Screening of open-pollinated seedlings of chrysanthemum for pot culture. *J. Ornamental Hort.* (New Series), 2(2): 120–123.

Bailey, L.H., 1947, *Manual of Cultivated Plants, 3rd edn.* The Macmillan Co., N.Y., pp. 985–986.

Barakat, M.N., Fattah, R.S.A., Badr, M. and El-Torky, M.G., 2010. *In vitro* mutagenesis and identification of new variants via RAPD markers for improving *Chrysanthemum morifolium. African Journal of Agricultural Research,* 5(8): 748–757

Brijendra Singh and Dadlani, N.K., 1989. Chrysanthemum varietal wealth. *Indian Hort.,* 36(3): 30–31.

Carter, G.D., 1980. Chrysanthemum. In: *Introduction to Floriculture,* (Ed.) R.A. Larson. *Academic Press.*

Chatterjee, J., Mandal, A.K., Ranade, S.A. and Datta, S.K., 2005, Estimation of genetic diversity of four chrysanthemum mini cultivars using RAPD. *Pakistan Journal of Biological Sciences,* 8(4) : 546–549.

Chaugule, B.B., 1985. Studies on genetic variability in chrysanthemum (*Chrysanthemum morifolium*). *M.Sc. Thesis,* M.P.K.V., Rahuri.

Chittenden, F.J., 1956. *Dictionary of Gardening.* The Royal Horticultural Society, Oxford University Press.

Cumming, A., 1939. *Hardy Chrysanthemums,* Whittlesey House Garden Series, pp. 1–162.

Dowrick, G. J., 1953. The chromosomes of chrysanthemum. III. Mitosis in *C. atratum*. *Heredity*, 7: 219–226.

Dowrick, G.J., 1952, The chromosomes of chrysanthemum. I. *Heredity*, 6: 365–376.

Ellen Fan, B.S., 1965. Further studies on chrysanthemum breeding for the Southwest. *M.Sc. Thesis*, Graduate Faculty of Texas Technological College.

Gondhali, B.V., Yadav, E.D and Dhemre, J.K., 1997. Evaluation of chrysanthemum for cut flowers. *Orissa J. Hort.*, 25(2): 10–13.

Gondhali, B.V., Yadav, E.D. and Dhemre, J.K., 1998. Evaluation of chrysanthemum cultivars for growth and yield. *South Indian Hort.*, 46(3–4): 164–166.

Hemalatha, B., Patil, A.A. and Nalawadi, V.G., 1992. Variability studies in chrysanthemum. *Prog. Hort.*, 24(1–2): 55–59.

Hemsley, W.B., 1889. *The History of Chrysanthemum*. Garden Chronicle, pp. 652–654.

Hu, L.H. and Cheu, Z.L., 1997. Chrysanthemum. In: *Commercial Flowers*, Vol. 1, (Ed.) Bose *et al*. Naya Prakash, Kolkata, pp. 465.

Kanamadi, V.C. and Patil, A.A., 1993, Performance of chrysanthemum varieties in the transitional tract of Karnataka. *South Indian Hort.*, 41(1): 58–60.

Katwate, S.M., Patil, S.S.D., Patil, M.T and Bhujbal, B.G., 1992. Performance of newly evolved cultivars of chrysanthemum. *J. Maharashtra Agri. Univ.*, 17(1): 152–153.

Kher, M.A., 1977. Some notable chrysanthemums introduced from Japan. *Prog. Hortic.*, 9(2): 5–12.

Kumar, S., Prasad, K.V. and Choudhary, M.L., 2006. Detection of genetic variability among chrysanthemum radiomutants using RAPD markers. *Current Science*, 90(8): 1108–1110.

Kyle, F., 1952. *Chrysanthemums*. Ward, Lock and Co., Ltd., London and Melbourne, pp. 1–188.

Liu Rui and Yang JiShuang, 2009. Genetic relationship among 11 wild species and 12 cultivated species of chrysanthemum revealed by ISSR analysis. *Genomics and Applied Biology*, 28(5): 874–882.

Maughan, P.J., Sanghai Maroof, M.A., Buss, G.R. and Huestis, G.M., 1996, Amplified fragment length polymorphism (AFLP) in soybean species diversity, inheritance and near isogenic line analysis. *Theor. Appl. Genet.*, 93 : 392–401.

Misra, H.P., 1999, Evaluation of small flowered varieties of chrysanthemum for calcareous belt of North Bihar. *Indian J. Hort.*, 56(2): 184–188.

Miyazaki, S., Tashiro, Y., Kanazawa, K. and Oshima, T., 1982, On the flower characteristics and chromosome numbers of Higo chrysanthemum. *Bull. Fac. Agril. Saga. Univ.*, 52: 1–11.

Mukund, S., 2010. Characterization of Chrysanthemum (*Dendranthema grandiflora* Tzvelev) germplasm through morphological and molecular markers. *Ph.D. Thesis*, UAS, Bangalore.

Nazeer, M.A. and Khoshoo, T.N., 1982. Cytological evolution of garden chrysanthemum. *Curr. Sci.*, 51(12): 583–585.

Palai, S.K., Mohapatra, A., Patnaik, A.K and Das, P., 1999. Evaluation of spray chrysanthemum for commercial floriculture under Bhubaneshwar conditions. *Orissa J. Hort.*, 27(1): 34–36.

Ponnuswami, V., Chezhiyan, N., Md. Abdul Khader, J.B.M and Thamburaj, S., 1985. Genetic variability in Chrysanthemum. *South Indian Hort.*, 33(3): 211–213.

Ramamurthy, P.B. and Ramakrishnappa, 2000. Floriculture development in Karnataka: An overview. In: *Commercial Floriculture*, (Eds.) H.P. Singh and N.K. Dadlani. Ministry of Agriculture, Govt. of India, p. 47.

Scott, M.C., Caetano, A.G. and Trigiano, R.W., 1996. Chrysanthemum. In: *Commercial Flowers*, Vol. 1, (Ed.) Bose *et al*. Naya Prakash, Kolkata, pp. 578.

Sehrawat, S.K., Kumar, R. and Dahiya, D.S., 2003. DNA fingerprinting of chrysanthemum cultivars using RAPDs. *Acta Hort.*, 624: 479–485.

Sirohi, P.S. and Behera, T.K., 2000. Genetic variability in chrysanthemum. *J. Ornamental. Hort.*, (New Series), 3(1): 34–36.

Tsevegsuren, N., Christie, W.W. and Losel, D., 1998. Chrysanthemum. In: *Commercial Flowers*, Vol. 1, (Ed.) Bose *et al*. Naya Prakash, Kolkata, pp. 466.

Viehmeyer, G., 1955. "*Chrysanthemum Improvement.*" University of Nebraska Agri. Expt. Sta. Bull., 428: 1–20.

Wolff, K. and Peters-van, R.J., 1993. Chrysanthemum. In: *Commercial Flowers*, Vol. 1, (Ed.) Bose *et al*. Naya Prakash, Kolkata, pp. 578.

Wolff, K., Peters–van, R.J. and Hofstra, H., 1994. Chrysanthemum. In: *Commercial Flowers*, Vol. 1, (Ed.) Bose *et al*. Naya Prakash, Kolkata, pp. 578.

Woolman, J., 1953. *Chrysanthemums for Garden and Exhibition*. W.H. and L. Collingridge, Ltd., London, pp 1–109.

www.mountainroseherbs.com

Yang, M.F., Liu, X.D. and Pan, X.F., 1997. Chrysanthemum. In: *Commercial Flowers*, Vol. 1, (Ed.) Bose *et al*. Naya Prakash, Kolkata, pp. 466.

2013, Biodiversity in Horticultural Crops Vol. 4
Editor: Professor K.V. Peter
Published by: DAYA PUBLISHING HOUSE, NEW DELHI

Pages 145–165

Chapter 8

Custard Apple

S.S. Hiwale

Central Horticulture Experiment Station,
Godhra-Baroda Highway, Vejalpur – 389 340, Dist. Panchmahals, Gujarat
E-mail: sshiwale@yahoo.com

Custard apple is a small group of edible fruits of genus *Annona* and family Annonaceae and collectively known as annonaceous fruits. Genus *Annona* has 120 species, 6 of them having pomological significance. Annona fruits are formed by fusion of pistil and receptacle into a large fleshy aggregate fruit. Annonaceous fruits have morphological affinity for each other but each type is unique in its taste, flavor, pulp color and texture.

The Annonaceous fruits originated in tropical America and are widely distributed in tropics and sub tropics. Among annonaceous fruits, custard apple is the most favorite in India. Its plants come up unattended in parts of Andhra Pradesh, Assam, Bihar, Karnataka, Maharashtra, Madhya Pradesh, Orissa, Rajasthan and Tamil Nadu as a shrub or hedge plant. Of late, custard apple has gained commercial significance and exclusive orchards are emerging in Maharashtra, Gujarat, Madhya Pradesh and Rajasthan (Table 8.1).

Table 8.1: Area and Production

State	Area (ha.)	Per cent of Total	Production (mt)	Per cent of Total
Maharashtra	9424	64.45	65968	64.49
Gujarat	1426	9.75	9223	9.01
Madhya Pradesh	3590	24.55	25050	24.48
Rajasthan	180	1.23	2050	2.00
Total	14620	—	102291	—

Uses

☆ As dessert fruit.

☆ In ice creams and other milk products.

☆ As Jam and Jelly.

☆ In Ayurvedic and Unani systems of medicine like seeds as abortifacient and roots as strong purgative.

☆ Seed oil (30 per cent) in soap and paint industry.

☆ Seed Cake (4 per cent N) as cattle feed and as manure.

☆ As insecticide with Neem oil.

Edible Annonas and their Fruit Characters

Other annonnas are cultivated on a limited scale. Bullock's heart is more commonly found in South India than in North India. It is usually associated with gardens and compounds and not commercial orchards. Cherimoya is mostly restricted to Assam and hills of South India. Atemoya and sour sop are cultivated in some gardens as miscellaneous fruits. Atemoya, cherimoya and ilama also provided excellent opportunities for large-scale exploitation in India (Table 8.2).

Custard Apple, Sweet Soup, Sugar Apple (*Annona squamosa*)

Indian name Sitaphal, Sharifa-Plant woody, semi deciduous, fruits 250–300g; globular; green skin; sweet (20 per cent sugar); non acidic; pulp creamy white; distinct segment, 60-80 seeds/fruit.

Bullocks Heart and West Indian Custard Apple (*Annona reticulata*)

Indian name Ramphal- Plant semi deciduous reaching 6-7m height; fruit large (350–400g); Heart shaped; yellowish red; Smooth rind with hexagonal markings, Pulp pale, gritty, flavored, 12.5 per cent sugar and a few seeds (30–40).

Cherimoya and Cherimoyar (*Annona cherimoya*)

Local name HanumanPhal -Semi deciduous, tree reaching to a height of 8m.Fruits weight 250–300g, pale green when ripe, sub-globose, Pine like aroma 18 per cent sugar, segment fused, 10-15 seeds/fruit.

Annona atemoya (*A. squamosa* X *A. cherimoya*)

Local name Lakshaman Phal -Semideciduous, large spreading tree, 5m height, Fruit weight 500g globular green; white smooth pulp is very juicy with excellent sugar acid blend, large segments, 10–15 seeds/fruit.

Sour Soup and Prickly Custard Apple (*Annona muricata*)

Ever green tree, 6-8 m in height, fruits 1.5-3 kg heart shaped; dark green; fleshy pines, pulp is white; fibrous, juicy with mango like flavor; 11-14 per cent sugar.

Ilama, White Annona (*Annona diversifolia*)

Not popular in India. Slender tree; fruit resembles custard apple/cherimoya pulp quality good and highly acceptable.

Table 8.2: Composition of Fruits (g/100g edible portion)

Constituents	Sweet Soup	Cherimoya	Atemoya	Sour Soup
Moisture	75.97	68.71	78.7	80.1
Protein	1.89	1.54	1.4	0.69
Fat	0.57	0.13	0.6	0.39
Carbohydrate	20.82	28.95	15.8	18.23
Fiber	1.41	–	2.5	0.95
Ash	0.75	0.67	0.5	0.58
Energy (kj)	360	460	310	247
Calcium	17.0	9.0	17.0	9.0
Magnesium	22.0	–	32.0	22.0
Phosphorus	54.0	24	–	29.0
Potassium	142	–	250	320.0
Sodium	2.0	–	4.0	22.0
Iron	0.3	0.25	0.3	0.82
Ascorbic acid	35.9	12.20	43.0	16.4
Thiamine	0.10	0.11	0.05	0.07
Riboflavin	0.06	0.11	0.08	0.12
Niacin	0.89	1.0	0.8	1.52

Collection, Introduction and Evaluation of Custard Apple Germplasm

The results revealed significant differences amongst eight cultivars of Custard apple in respect of most of the vegetative as well as physico-chemical characters of fruits under rainfed conditions of Panchmahals. (Table 8.3). As regards to vegetative growth parameters, stem diameter and plant spread were significantly influenced however plant height was found to be non significant but was maximum (4.16m) in Island Gem. However, stem diameter and plant spread were found to be significantly influenced. Maximum stem diameter (103.33mm) was recorded in Island Gem. Plant spread (N-S and

Table 8.3: Evaluation of Custard Apple Germplasm

Variety	Pl Ht m	Stem Dia. mm	NS m	EW m	Fruit Wt. g	Pulp Wt. g	Skin Wt. g	Pulp Skin Ratio	Seed Wt. g	TSS °Brix	Yield/ Plant Kg
Balanagar	2.78	65.1	3.12	3.25	176.25	61.0	75.7	0.81	15.0	20.2	9.74
Washington 98797	3.84	102.5	5.60	5.30	128.00	47.7	59.5	0.80	14.0	18.0	1.68
Seedless Atemoya	3.39	73.4	4.0	4.05	173.50	85.7	58.5	1.46	13.2	20.0	6.29
Pink Mammoth	3.46	96.1	3.9	4.28	152.00	62.5	53.5	1.16	14.0	16.5	25.17
Island Gem	4.16	103.3	5.27	4.66	195.50	99.2	68.5	1.44	6.50	23.0	2.64
Atemoya x Balanagar	3.21	88.3	4.05	3.75	188.75	77.2	66.2	1.16	11.5	20.0	26.84
Local Sitaphal	2.87	67.8	3.52	3.47	132.75	55.7	43.2	1.28	15.7	30.0	20.70
CD 5 per cent	NS	26.8	1.08	0.90	47.13	22.5	18.7	—	4.55	1.44	15.89

Figure 8.1: Fruit Variability in Custard Apple

E-W) was maximum in Washington (5.60m and 5.30 m). Observations on fruit set per tree revealed that it was maximum in Atemoya x Balanagar (263.67).Fruit and pulp weight were the highest in Island gem (195.50 and 99.25g,respectively). Maximum, skin weight was noted in Balanagar (75.75g) and seed weight in Local Sitaphal (15.75g). T.S.S. was maximum in Local Sitaphal (30.00 °Brix) and the least in Pink Mammoth (16.50 °Brix). Yield per plant was the highest in Atemoya x Balanagar (26.84 kg/plant).

Selections Developed by Farmer from Solapur

It is seen from the Table 8.4 that the selection is superior in respect of yield, quality and income per ha. Selection Annona-2 is having a very few seeds. The average weight of fruit is 700-800-g under drip irrigation and pulp percent is 70-75.

Table 8.4: Comparison of Selections Developed by Farmer (Kaspate, Solapur)

Varieties	Average Productivity (tones/ha.)	Income (Rs./ha) (in lakhs)
Balanagar	10-12	1.25
NMK-1	13-15	7-8
Annona-2	16-19	6-8

Leaf and Leaf Area Determination

Studies on leaf area determination by non-destructive method carried out in seven cultivars of Custard apple revealed that correlation coefficient (cv) of the actual leaf area with leaf length, maximum breadth and product of leaf were highly significant for all the parameters. However, maximum values were obtained when actual leaf area was correlated with the product of leaf length and breadth. The values were 0.92, 0.91, 0.98, 0.86, 0.99, 0.99 and 0.77 for different cultivars, respectively, based on which regression equation and factor values were calculated. A linear relationship was established by Y= a + b x for regression method and Y= K x for factor method. The regression equation fitted and is given in Table 8.5 for calculation of leaf area in Custard apple by non-destructive method (Tables 8.5–8.7).

Table 8.5: Leaf Length, Breadth and Actual Leaf Area in Annona Germplasm

Treatments	Length (cm)	Breadth (cm)	L x B (cm²)	Actual Leaf Area (cm²)
Balanagar	7.33	3.23	23.90	29.15
Washington	11.39	5.64	64.67	55.67
Seedless Atemoya	15.51	8.27	129.11	106.58
Pink Mammoth	14.56	6.07	88.89	69.82
Island Gem	16.37	10.05	166.25	137.87
Atemoya x Balanagar	18.18	9.87	182.06	139.27
Local Sitaphal	9.14	3.86	35.42	32.01

Table 8.6: Leaf Length, Breadth and Actual Leaf Area in Annona Germplasm

Treatments	Regression Equation
Balanagar	L x B = (4.29x1.04 x)
Washington	L x B = (10.09x0.705 x)
Seedless Atemoya	L x B = (-2661.54 x 21.44 x)
Pink Mammoth	L x B = (14.91x0.62 x)
Island Gem	L x B = (9.23 x 0.77 x)
Atemoya x Balanagar	L x B = (-3.49x 0.78 x)
Local Sitaphal	L x B = (5.52 x 0.78 x)

Table 8.7: Correlation of Leaf Area in Custard Apple Germplasm

Treatments	Length	Breadth	L x B
Balanagar	0.62	0.77	0.93
Washington	0.72	0.78	0.91
Seedless Atemoya	0.81	0.81	0.98
Pink Mammoth	0.65	0.74	0.86
Island Gem	0.83	0.92	0.99
Atemoya x Balanagar	0.45	0.89	0.96
Local Sitaphal	0.48	0.65	0.77

Climate and Soil

Most annonaceous fruits are acclimatized to tropical climate. Although custard apple withstands heat and drought conditions, high atmospheric humidity is necessary during flowering to improve fruit set. But continuous rains during fruit set are not desirable. An annual rainfall of 60-80 cm is optimum. It cannot stand frost or a long cold period. The trees remain dormant from December to February and shed leaves. When the summer temperature rises above 39.4°C the tree sheds its flowers resulting in low fruit set.

Bullock's heart grows well in humid regions of south India and cannot withstand severe summer. It tolerates frost to some extent. Cherimoya prefers subtropical climate, but it can flourish on higher elevations (2,000m) in tropics, while climatic requirements for atemoya are quite similar to those of custard apple. Sour-sop in contrast, is a fruit of the humid tropics.

Custard apple thrives naturally in rocky terrain with shallow, gravely, well-drained soils. However, they may grow well in arable, red, sandy shallow soil slightly acidic in reaction. Heavy soils are not suitable, especially in waterlogged areas. In Andhra Pradesh, annonas come up chalka-red sandy or gravely soils. They can grow well even on calcareous soils containing lime as high as 50 per cent.

Custard-apple seedlings are found growing wild in India. Since custard apple is a cross-pollinated crop, wide variation in forms and sizes of fruit as well as color of the pulp are available. The natural variability available within the species is often exploited to identify superior genotypes, which are usually named after the place of collection or selection and fruit color. Depending on external fruit color, custard apple is distinguished into green, red and yellow. But green ones are by far more common and popular than the other types. Balanagar, Barbados seedling, British Guinea, Kakarlapahad, Local Sitaphal, Mahaboobnagar, Saharanpur Local and Washington are some of the varieties with green skin. Most of these varieties are not easily identifiable. Some of the traits which distinguish them are fruit shape and size, form of areoles and number of seeds/fruit. But in fruits of a given tree, these attributes vary considerably as pollination and the environment largely influences them.

Ideotypes of Custard Apple

- ☆ Prolific bearing
- ☆ Low seed content
- ☆ Better keeping quality
- ☆ Large fruit
- ☆ Sweet (TSS more than 25 °Brix)
- ☆ Pleasant aroma
- ☆ Resistance to drought
- ☆ Resistance to salinity

Varieties

The varietal or genetic differences get masked confusing the varietal identification. Moreover, variety-specific pulp qualities are not clearly explained. However, some varieties can be recognized by the plant habit and foliage attributes. Two natural hybrids (mostly between custard apple and cherimoya), Israeli Selection and Israeli hybrid have been introduced. Fruits of both are less seeded. A hybrid Arka Sahan has slow ripening (6-7 days), better shelf life (2-3 days), less number of seeds (10/

100g fruit weight) and high Brix (31°). On an average, its fruit weighs 210g each. One 6-years old plant yields 17 kg fruits (Table 8.8).

☆ Exotic – Washington PI 98797, Washington PI 107005,

☆ British Guinea, Barbados seedling.

☆ Exotic hybrids- Island Gem, Bullocks Heart, Pink Mammoth

☆ Selections- Balanagar, Mammoth, Red Sitaphal, Yellow Sitaphal

☆ Indian Hybrid- Arka Sahan

Arka Sahan

A promising hybrid 'Arka Sahan' (*A. atemoya* x *A. squamosa*) has been developed at IIHR, Bangalore. The hybrid yields very sweet, fragrant, low seeded fruits having longer shelf life.

Table 8.8: Salient Characters

Fruit weight (g)	210.7
Rind surface	Smooth
Rind thickness (cm)	0.5
Mesocarp colour	White
Areoles per fruit	54
Edible pulp (per cent)	48.6
Number of seeds per 100g fruit	8.9
TSS (°Brix)	30.8
Acidity (per cent)	0.6
Yield per tree (kg)	16.8
Shelf life (days)	4

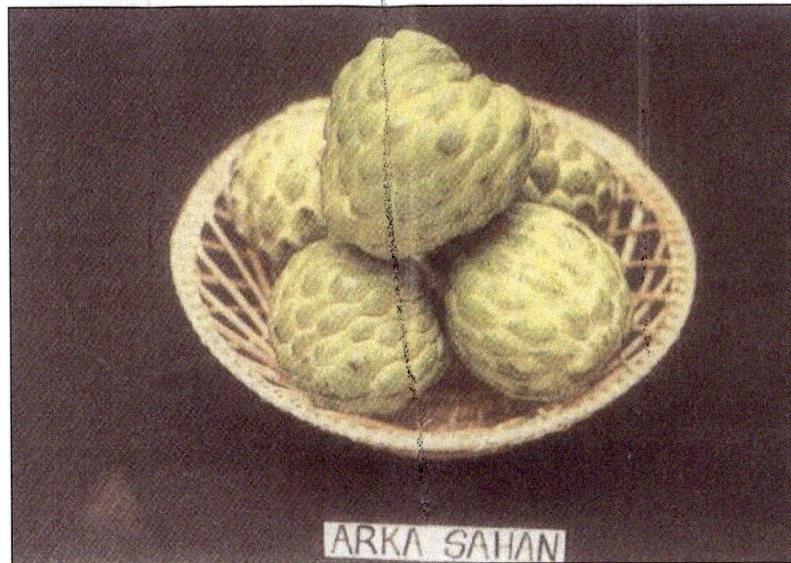

Figure 8.2: Arka Sahan

Propagation and Rootstocks

Seeds traditionally propagate most of the annonas. The seed's viability lasts for 3-4 years. However, fresh seeds germinate better when soaked for 24 hrs and result in the highest germination of 71 per cent (Ratan *et al.,* 1993). Hard seed coat can be softened either by soaking the seeds in water for 2-3 days of keeping them under running water for 50-70 hrs. Treating seeds with GA3 at 5000ppm increased germination (Smet *et al.,* 1999). Seeds are sown 2cm deep either in nursery beds or in pots under partial shade. Regular watering is necessary to maintain good soil moisture. Seeds are slow to germinate and take 3 weeks. Nevertheless it may extend to as long as 8-10 weeks. When seedlings are 10-12 cm tall, they are transferred to pots or plastic containing sand and peat or equal parts of garden soil, sand and decomposed farmyard manure. The 30cm tall seedlings become ready for transplanting.

Seed originated plants are not true-to-type, lack precocity and vigor, whereas grafting or budding helps largely to overcome these drawbacks. *A squamosa, A. reticulata, A. cherimola and A. atemoya* are grafted or budded on their own species and each other. *A. muricata* can be grafted on *A. reticulata* and

Figure 8.3: Grafts of Annonas

Figure 8.4: Island Gem

Figure 8.5: Bullock Heart

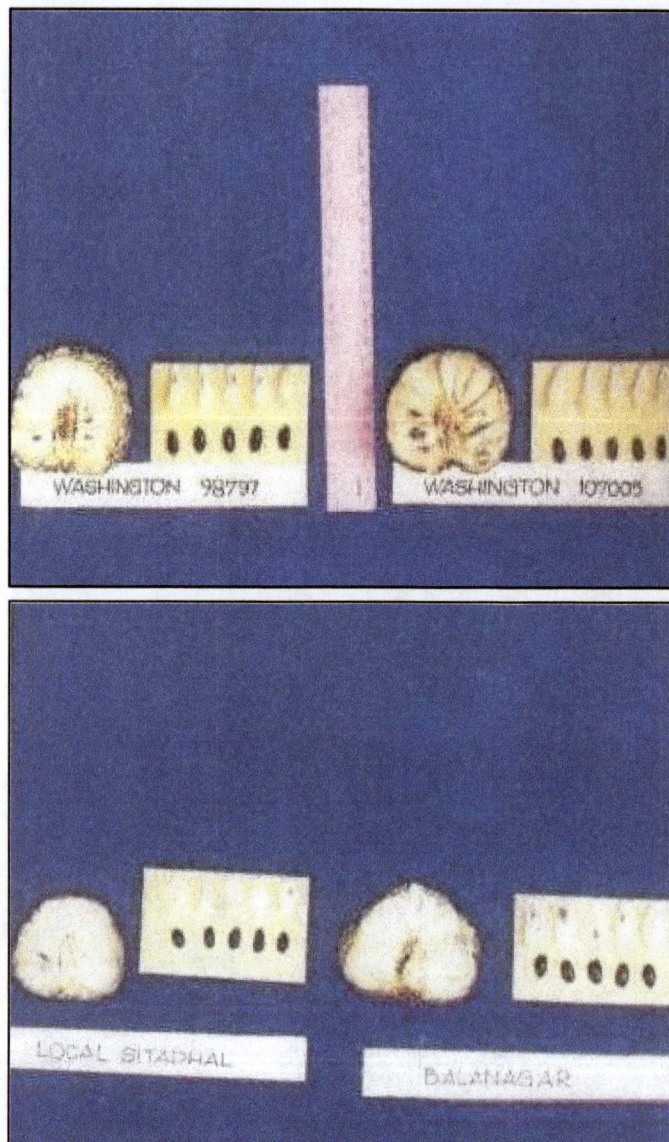

Figure 8.6: Improved Anonnas

A.glabra. However, *A. reticulata*, which promotes vigor and shows good graft congeniality, is commonly employed as a rootstock for most of the annona (Khan and Rao, 1953).

Generally, 18 months old or 30cm tall plants having pencil thickness are selected for grafting. Scion of well-matured wood from which the leaves have dropped at the end of the dormant phase is used to graft either by veneer or cleft technique. Shield or T-budding carried out in spring is also equally effective. Patch and chip bedding are other methods. Large buds, about 4cm length are collected from 1-year-old wood after the leaf drop gives good success. Since propagation by cutting and air-layers give poor results, they are not widely practiced. Commercial production of plants through tissue culture is not yet successful. However, multiple shoot production from leaf explants of seedlings and root initiation from shoots are successful

Propagation studies were undertaken to find out the best method *viz.* bud take percentage, soft wood grafting success, rooting of cutting and air layering in annona (*Annona squamosa* L.) Cv. Balanagar. It was recorded that soft wood grafting was the most successful method for commercial propagation of Annona. Maximum graft take of 58.37 percent was observed in CV. Balanagar followed by budding (33.54 per cent). Percent success in propagation was the least in air layering (8.68). Soft wood grafted plants showed maximum increase in the growth parameters compared to budded plants six months after grafting (Hiwale *et al.,* 2009). Gholap *et al.* (2000) also reported maximum success in softwood grafting fallowed by budding (Table 8.9).

Table 8.9: Effect of Method of Propagation on Per cent Success in Custard Apple Cv. Balanagar

Method of Grafting	Per cent Success
Softwood grafting	51.74
Patch budding	33.54
Cutting	13.62
Airlayering	8.68
SEM	14.09
CD 5 per cent	58.57

Table 8.10: Effect of Time of Grafting/Budding on Per cent Success in Custard Apple cv. Balanagar

Month	Per cent Success	
	Softwood Grafting	Budding
January	28.45	17.4
February	41.72	33.07
March	59.37	41.20
April	30.27	27.17
May	12.03	9.17
June	9.57	7.02
SEM	0.988	7.90
C.D.5 per cent	2.96	23.83

In another experiment on comparative performance of time (January to June) and method (Soft wood grafting and budding) of propagation revealed that soft wood grafting in the month of March gave the maximum success of 59.37 per cent, compared to 41.20 per cent in budding in the same month, when leaf fall starts. Least success was recorded in the month of June (9.57 and 7.0 per cent)in both the methods (Table 8.10). Growth of the plant one month after softwood grafting/budding revealed maximum growth in softwood grafted plants (Table 8.11). Thus for vegetative propagation of Annona by softwood grafting method in the month of March was found the best (Hiwale *et al.,* 2009).

Table 8.11: Effect of Method of Propagation on Growth of Custard Apple cv. Balanagar

Treatment	Pl. Height cm	Stock Dia. mm	Scion Dia. mm	Leaf Area/Plant cm²
Softwood Grafting	356.17	4.91	4.80	311.18
Budding	246.17	4.13	3.58	197.22
CD 5 per cent	NS	NS	NS	NS

Flowering, Fruit Set and Fruit Retentation

Low marketable yield is a major problem in various annonaceous types. The reasons for low fruit set are dichogamy, lack of pollinizers and production of flowers without attractive color, poor pollen germination, environmental and tree condition. Close planting may increase the pollination. Venkatraman (1979) and Thakor and Singh (1965) reported that stigma remains receptive from one day before anthesis to about 2-3 hrs after anthesis but it decreased abruptly so by the time of dehiscence, stigma turned almost unreceptive and observed protogynous dichogamy. Similar trend was reported by Limaye (1966) and Rajput (1985). Ahmed (1935) reported that the best time for artificial pollination was from 5 am to 8 am and also reported that drier and hotter period adversely affected the fruit set. Similar result was reported by Karale (1989). Rajput (1985) reported that there are no insect pollinizers for Annona because the flowers are unattractive. Pollen germination was poor ranging from 11.5 per cent to 20.00 per cent. Work done at CHES, Vejalpur on germplasm evaluation of eight varieties of Annona revealed that maximum No of flowers per plant were produced in Washington (4133.60 no.) however fruit set and retention were the least (0.70 per cent and 0.19 per cent respectively). In cultivar Balanagar though the lowest no. of flowers were produced (350.68) fruit set and retention were found to be highest, 26.29 per cent and 14.66 per cent respectively (Table 8.12), (Hiwale, 2002).

Table 8.12: Flowering Fruit Set and Fruit Retentation in Custard Apple

Variety	No. of Flowers/Shoot	No. of Flowers/Plant	No of Fruit Set/Plant	Per cent Set/Plant	No. of Fruit Retained/Plant	Per cent Retention/Plant
Balanagar	42.29	350.68	91.03	26.29	49.33	14.66
Washington 98797	167.61	4133.60	24.00	0.70	6.33	0.19
Seedless Atemoya	163.13	3140.00	61.00	1.88	9.33	0.30
Pink Mammoth	168.76	2597.36	493.66	25.64	61.66	2.74
Island Gem	145.70	2719.61	33.00	1.21	3.66	0.16
Atemoya x Balanagar	59.21	791.71	404.66	52.26	76.66	9.69
Red Sitaphal	220.38	2095.50	84.00	3.97	13.33	0.78
Local Sitaphal	297.48	2589.64	266.66	10.64	47.66	1.94
CD 5 per cent	59.61	2400.15	69.12	15.31	17.89	1.89

The time of flowering and fruit development in different varieties vary considerably. Custard apple flowers from March to July and sets fruit in June- July and takes 4 months for fruit development where as Cherimoya flowers from May to June and July to September.It takes 6 months for fruit development. On the other hand, Bullocks heart flowers from August to October and takes 8 months for fruit development.

Cultural Practices

Planting

Pits of 60cm x 60cm x 60cm size are dug and left open to sun for a week. They are filled with topsoil mixed with 25-30kg of well-decomposed farmyard manure. Custard apple should be transplanted 5m x 5m apart (400 plants/ha.). However, setting plants closer may be preferred. Plant spaced at 4m x4m (625 plants/ha.) not only accommodates over 50 per cent additional plants/unit area but promotes better fruit set by improving pollination, a problem in annonnas. Due to large canopy, atemoya and bullock's heart require a plant-to-plant spacing of 6m x 6m and cherimoya and sour-sop 8m x 8m spacing. Planting should be carried out preferably in spring so that plants establish roots in summer, start growing as the weather warms up and put up vigorous growth during rains. However, if adequate irrigation facilities are lacking, monsoon is the optimum time for planting. To keep the graft-joint well above the ground is a must. As soon as a young tree is planted, it should be irrigated till it establishes.

Training and Pruning

Annona's require little pruning. It is essential to develop a good crown and better yields over a long period of time. Without pruning, the plants become bushy and their bearing efficiency comes down. Hence, timely removal of misplaced limbs is necessary to build a strong framework. Selective and mild pruning of deadwood and very old branches should be carried out to avoid congestion and encourage well-spaced branching. Severe pruning is detrimental for plant growth. Yellowing of leaves starts as the harvesting season of fruits ends. The leaves begin drop with onset of winter and fresh growth occurs in spring. Flowering occurs singly or rarely in small clusters mostly on current season's growth and occasionally on old wood. Training to a single stem is the only option when rootstock is employed.

Table 8.13: Nutritional Requirement in Custard Apple

Source (g/plant)	Plant Age (Years)		
	1-2	*3-5*	*Above 5*
Nitrogen	75	150	250
Phosphorus	50	100	125
Potash	25	100	125
Organic manure (kg)	25	25-30	50

Manures and Fertilizers

Manures and fertilizers application to custard apple is not common but its plants respond very well to fertilization, increasing vigor, yield and fruit quality. Fertilizer application checks decline and extent longevity of trees. To specify the precise dose common to all soils is difficult, but general recommendations are given. The area below the crown of trees should be cleaned of weeds, loosening of soil in the basin should be carried out and then fertilizers applied in the basin under the tree 30cm away from the trunk. Subsequently it is desirable to irrigate the trees and incorporate the fertilizers. Fertilizer application should coincide with rapid vegetative growth and fruit development. As fruits are born on new as well as old wood, application of slightly higher dose of N is not harmful (Hayes, 1953). Biofertilizers, VAM application can increase the productivity of the plant and soil sustainability.

Application of castor cake and bone meal or super phosphate in the ratio of 2:1was found beneficial (Rao, 1974) (Table 8.13).

Aftercare

After planting, the young plants must be watered and supported by stacks to keep them erect. To start with a 60cm x 60cm basin around the plant is adequate. Regular watering during dry periods, occasional hand digging of the basins to check weeds, to keep the soil loose, attending plant-protection measures, removing of sprouts on stock and building up of a good framework are necessary cultural operations. The basins around the plant should be enlarged, as the plants grow bigger. They should be made little larger than the spread of the plant.

In young orchards, a lot of land remains vacant between the rows for 4-5 years. Hence short duration vegetables – tomato, onion, chilly, okra, brinjal, radish or cowpea, green gram, horse gram, or any green manure crop can be intercropped. These crops should not be raised too near the tree, lest they compete with them for nutrients.

Irrigation

Most of the annonas produce a moderate crop even in the absence of irrigation. Irrigation to plants during flowering and fruit development is essential. Fruit set, yield/plant and quality are superior in irrigated plants with more edible pulp/segment. Plants receiving regular water grow luxuriantly with each bearing. Pruning, fertilization and irrigation are quite essential to get maximum yield. In regions having limiting water, pitcher, trickle or drip irrigation systems help in judicious use of water. Fruits are raised in rain-fed areas in low rainfall areas. land shaping to divert rainwater near the plantation may be taken up. Contour terraces, contour bunds and micro-catchments also help in efficient water use. Ploughing and mulching of the plantation during rainy season help better conservation of moisture. Kulkarni (1993) reported that eight irrigations at 15 days interval starting from 15th April was the best for custard apple in semi arid areas of Rahuri for increasing fruit set and yield of custard apple

Crop Regulation

Some times there is heavy fruit set in custard apple leading to smaller sized fruits or mummified fruits and for obtaining out of season crop, fruiting in custard apple can be regulated by use of manual defoliation or chemical defoliation. Both the treatments produced higher number of fruits as well as resulted in increased fruit weight (Table 8.14).

Table 8.14: Effect of Early Induction of New Growth by Defoliation on Yield of Annona

Variety	Number of Fruits			Weight of Fruit (kg)/trees		
	Chemical Defoliation	Manual Defoliation	Control	Chemical Defoliation	Manual Defoliation	Control
Balanagar	98	99	56	16.17	13.47	7.76
British Guinea	78	125	59	13.55	21.02	9.85
Local Sitaphal	99	85	60	14.32	9.0	8.93
Red Sitaphal	94	81	54	13.15	12.35	7.97
Washington 107005	96	69	38	14.10	7.95	4.70
Mammoth	109	109	61	14.75	14.55	8.80

Figure 8.7

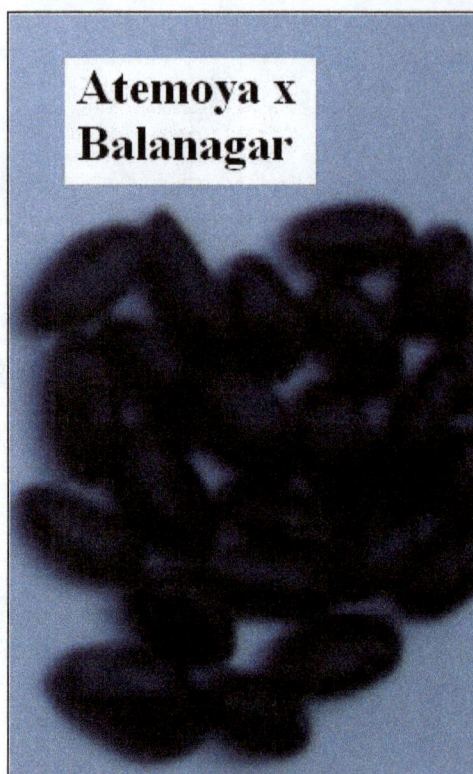

Figure 8.8

Cropping System

To utilize the inter spaces in young orchards and to maximize productivity per unit area various intercrops viz okra, maize, Pigeon pea, green gram, sesamum and moth bean were grown during *kharif* season under rainfed conditions. The vegetative growth parameters were not influenced by the intercropping. Higher productivity per unit area was obtained in Custard apple intercropped with okra with a B. C. ratio of 1:1.54 (Table 8.16). Apart from this, the cropping system also resulted in reducing the run off loss of water as well as reduced soil erosion, there by increasing the sustainability of the soil.

Table 8.15: Mean Morphological Parameters of Custard Apple in Agri-Horti System

Name of Intercrops	Plant Height (m)	Stock Diameter (mm)	Plant Spread (m)	
			NS	EW
Okra	2.89	66.35	2.39	2.33
Maize	2.77	67.81	2.26	2.35
Pigeon pea	2.71	66.36	2.20	2.23
Green gram	2.69	64.30	2.15	2.19
Sesamum	2.75	65.04	2.35	2.40
Moth bean	2.73	66.88	2.12	2.29

Table 8.16: Mean Yield, Net Return and B.C. Ratio in Agri-Horti Production System

Name of Intercrops	Main Crop (q/ha)	Inter Crop (q/ha)	Pure Crop (q/ha)	Net Return (Rs.)	B: C Ratio
Custard apple	26.07	—	—		
Custard apple+ Okra	23.98	22.26	31.67	13964	1.54
Custard apple + Maize	22.85	9.05	15.22	9101	1.23
Custard apple+ Green gram	23.69	1.84	3.06	7276	0.97
Custard apple+ Sesamum	23.55	2.68	3.79	11134	1.53
Custard apple+ Pigeon pea	24.51	1.37	2.01	7186	1.02
Custard apple+ Moth bean	23.19	2.14	2.99	8245	1.12

Rainfall, Run Off and Soil Loss in Different Cropping System

In Custard apple based cropping system, maximum runoff (31.82 per cent) was recorded in cultivated uncropped soil. it was minimum in no tillage (10.54 per cent). Staggered contour trench planting was the best for reducing soil loss effectively (0.93 t/ha). Run-off and soil loss were higher in Custard apple due to total shading of leaves and new growth took 20-30 days to cover the soil and due to higher slope of the land and its soil type. Adopting staggered contour trench planting method resulted in reduced runoff of water and soil erosion losses were also reduced by almost seven times compared to cultivated uncropped, thus resulting in saving the valuable soil and water resources thereby maintaining the sustainability of the soil.

Table 8.17: Rainfall, Run Off and Soil Loss in Different Cropping System

Treatment	Run Off (mm)			Mean	Per cent of Run Off			Mean	Soil Loss t/ha.			Mean
	2002	2003	2004		2002	2003	2004		2002	2003	2004	
Cultivated uncropped	115	147	104.7	122.2	32.63	29.28	33.55	31.82	5.3	7.6	6.9	6.60
Cultivated Cropped	83.94	98.38	66.24	82.85	23.87	19.59	21.28	21.28	1.6	2.15	4.08	2.61
Staggered trench planting	64.94	76.10	61.51	61.51	18.55	15.16	19.74	17.81	0.6	0.98	1.23	0.93
No tillage	46.44	27.46	40.27	38.05	13.26	5.47	12.90	10.54	0.8	1.24	2.12	1.38

Root Distribution Pattern of Custard Apple

Root distribution varies according to the type of soil and method of propagation. Studies were there fore initiated to find out root distribution patter of seedling type six-year-old custard apple cv. Balanagar. Root distribution pattern under semi arid rainfed conditions revealed that the root system is shallow in nature as below 60 cm soil depth not much root activity was recorded and the spread was with in the plant canopy. Maximum root activity on fresh weight basis was observed in 0-30 cm radial distance from tree trunk (70 per cent), which reduced as the radial distance increased. Depth wise the root distribution was maximum at 0-30 cm depth (56 per cent) and subsequently reduced as the depth increased (Table 8.18). The root spread was higher depth wise (up to 120cm), where as, it was just 85 cm radially.

Table 8.18: Root Distribution Pattern of Custard Apple (Fresh wt. basis)

Depth/Radial	0-30	Per cent	30-60	Per cent	60-90	Per cent	Per cent Total
0-30	1.00	40	0.30	12	0.10	4	56
30-60	0.40	16	0.20	8	—	—	24
60-90	0.25	10	0.15	6	—	—	16
90-120	0.10	4	—	—	—	—	4
Total	1.75	70	0.65	26	0.10	4	

Main Root wt. (kg)	Lateral Root wt. Kg	Total Root wt. Kg	Above Ground Biomass Kg	Ratio of Shoot/Root	Rooting Depth cm	Horizontal Spread cm
0.90	1.60	2.5	12.00	4.8	120	85.25

Photosynthetic Efficiency, Conversion Coefficient, Stomatal Conductance and Transpiration Rate in Different Cropping Systems

Agro forestry systems are more productive than the sole cropping systems because of greater light interception, better moisture utilization. Maximum incidental light and its interception were recorded in custard apple intercropped with Sesamum at both the crop growth stages. Maximum photosynthetic activity was recorded in custard apple with maize cropping system. However, transpiration rate, stomatal conductance as well as carbon intake were the least in the same combination. Dry matter production was the highest in custard apple with maize combination. Rate of photosynthesis is higher than rate of evapo-transpiration therefore there is net accumulation in the leaves, resulting in higher biomass production.

Table 8.19: Growth Phase

System	PAR Micromoles/m²/s	EVAP Millimole/m²/s	PN Micromole/m²/s	GS Millimoles m²/s	Carbon Intake (ppm)
C.Apple	301.6	2.46	10.9	268.0	244.0
C. Apple +Okra	593.0	6.66	23.7	1283.4	522.5
C. Apple + *Cajanus cajan*	499.6	6.17	17.4	987	557.6
C. Apple +Maize	528.8	5.29	26.5	620.2	512.8
C. Apple+ Cowpea	528.4	6.35	16.0	993.6	594.0
C. Apple + Sesamum	544.6	6.89	19.5	904.8	520.6

Table 8.20: Fruit Set Phase

	PAR Micromoles/m²/s	EVAP Millimol/m²/s	PN Micromole/m²/s	GS Millimoles m²/s	Carbon Intake (ppm)
C.Apple	1011.6	6.3	27.05	506.0	300.0
C. Apple +Okra	1644.6	14.22	70.35	2009.0	655.4
C. Apple + *Cajanus cajan*	1953.4	15.22	79.05	2176.0	802.8
C. Apple +Maize	1837.0	14.69	82.65	1484.8	547.4
C. Apple+ Cowpea	2620.8	18.25	75.65	1987.2	569.8
C. Apple + Sesamum	2625.8	15.5	62.45	1873.4	555.6

Natural Resources Management

Energy conservation efficiency is the ratio of output of calories captured by vegetation to the input (Solar radiation), in a unit area over a certain period of time. All the intercropping systems resulted in higher production in terms of produce thus making better utilization of natural resources like solar energy, available moisture, higher carbon intake more stomatal conductance, leading to higher photosynthesis and there by biomass accumulation.

Harvesting and Postharvest Management

Custard apple starts bearing fruits at the age of 4 years. Production declines by about 15th year depending upon the maintenance. Custard apple produces single crop in a year during August-October in south India and September-November in north India. On maturing, fruits turn light green. The inter-areolar space widens, the fruit turns creamy-white. Custard apple is harvested manually when they are fully mature but still firm. About 4-5 pickings are required. Occurrence of deformed fruit is due to unfertilized areoles failing to grow. Fruit yield varies widely from tree-to-tree. Normally a 7-year-old tree produces 100-150 fruits, the total yield being 7 tones/ha.

Bullock's heart and cherimoya are ready for harvesting during December-January, yielding 50-75 fruits/plant. Since both have thick stalks, it is necessary to harvest them with the stalks using secateurs. Atemoyas are somewhat early (September-October) and are shy-bearers. Higher yield (50-60 fruits) may be obtained with hand pollination. Sour-sop, the largest annona produces about 25 fruits/tree during June-August in south India.

Annonas are climacteric fruits. Custard apple takes about 3 days to ripen, while others 4-7 days. Prior to ripening the pulp of matured custard apple is not separated into segments on flakes. It is only during conversion of starch to sugar differentiation occurs. In ripe fruits of bullock's heart and cherimoya, the pulp is more or less homogeneous mass of closely cohering carpels, which cannot be separated easily.

Ripe fruits of custard apple and sour-sop are very fragile and with the slightest pressure, the fruits easily get disintegrated into segments. Hence extra care is necessary while handling. Ripe custard apple can be stored about 2 days, but other annonas can be stored for 3-4 days. This may be partly due to the characteristic feature of the rind. In bullock's heart and cherimoya, the carpels are not associated with the external areolar division on the rind and the surface appears to be contiguous or almost fused unlike that in custard apple. Thus the fruits do not split easily along the deep furrows between the inter areolar spaces, the weakest portions of the rind. Custard apple, atemoya, bullock's heart and cherimoya are normally used as fresh fruits. Ripe fruits are popular among the poor. The unripe fruits of custard apple are eaten in Andhra Pradesh after backing the roasting. The row fruits of sour-sop are commonly used to prepare soup or vegetable.

The pulp of custard apple mixed with milk is made into a delightful drink or ice cream. Development of a repulsive off flower on heating beyond 65° C and presence of gritty cells are major constraints in processing custard apple. But its juice is a potential ingredient to prepare squash, syrup, nectar and a fermented alcoholic beverages. Jam, jelly, conserves and tarts can also be prepared from the pulp of custard apple. It is also possible to can the pulp. The sweetish sour flesh of sour-sop is fibrous, juicy with pleasant aroma and is amenable for preparation of ice cream, Sherbet, syrup.

Physiological Disorders

Stone Fruits or Mummified Fruits

In certain neglected plants or those under severe moisture or nutritional stress, the dormancy sets well advance and some fruits turn brown, become hard and without further growth remains on trees for months. Such fruits are termed as " stone fruits" or " mummified fruits ". About 20-25 per cent fruits are found to be mummified in local sitaphal.

Control

☆ Clean cultivation

☆ Application of balanced nutrition

☆ Thinning of fruits to reduce the load, particularly fruits which are set late.

Fruit Cracking

Sudden fluctuation in water supply to the plants may cause cracking of fruits. Irrigating plants after prolonged dry spell may be one of the reasons. An equitable irrigation schedule will help in solving the problem.

Insect Pests and Diseases

Mealy Bug (*Planococus pacificus*)

It is seen affecting the fruit and young branches there by influencing the quality and the market price.

Control

Spraying of 0.05 per cent Phosphomidon or Dichlorvos (Shukla and Tondon, 1984).

Leaf Spot (*Alternaria* spp.)

The affected leaves drop down causing considerable loss in production. British Gunia, Island Gem and Red Sitaphal were moderately susceptible (Annon., 1987).

Anthracnose (*Colletotricum singulata*)

It has been attacking sitaphal in Udaipur area (Annon., 1981).

Control

Fortnightly sprays of Benlate 0.05 per cent or bicor 0.1 per cent can control both the diseases

Tree Decline

Caused by water stagnation. Trees shrivel and drying of old branches takes place and the plants die suddenly.

Control

☆ Stagnation of water should be avoided.

☆ Saline soils with high clay content should be avoided.

☆ Drainage channel should be laid out to carry out excess water.

☆ Drenching with copper fungicide 3 g/liter

References

Ahmad, M.S., 1935. Pollination and fertilization in *Annona squamosa* under dry climatic condition in Egypt. *Hort. Abstract,* 4: 76.

Anonymous, 1981. *Annual Report.* All India Coordinated Project on Arid Zone Fruits, pp. 21.

Anonymous, 1987. All India Coordinated Project on Arid Zone fruits. *Technical Document* No. 20.

Gholap, S.V., DOD, V.N., Bharad, S.G. and Wankar, A.M., 2000. *Crop Research (Hissar),* 20: 158–159.

Hayes, W.B., 1953. *Fruit Growing in India,* 3rd Edn. Kitabishtan, Allahabad, India.

Hiwale, S.S., 2002. *Annual Report.* Central Institute for Arid Hort., pp. 12–13.

Hiwale, S.S., More, T.A. and Bagle, B.G., 2009. Vegetative propagation in Annona (*Annona Squamosa* L.). In: *National Conference on Production of Quality Seeds and Planting Material: Health Management in Horticultural Crops,* New Delhi from 11 March to 13 March.

Karale, A.R., Keskar, B.G., Dhawale, B.C. and Kale, P.N., 1989. The influence of tree condition at the time of hand pollination as a factor on fruit set in *Annona* species. *Maharashtra J. Hort.,* 2: 114–116.

Khan, K.F. and Rao, I.K.S., 1952. *Indian J. Hort.,* 10: 140–144.

Kulkarni, S.S., 1993. Effect of cultural and chemical treatments on fruit set, yield and quality of custard apple. *M.Sc. (Agri) Thesis,* MPKV, Rahuri, MS.

Limaye, S.P., 1966. *Hand Pollination Studies in Atemoya.* Agril. College Mag. Parbahni, 2 (1) 41–42.

Rajput, C.B.S., 1985. Custard apple. In: *Fruits: Tropical and Subtropical.* Naya Prokash, Kolkata pp. 479–488.

Rathore, D.S., 1976. Preliminary studies on the stooling of custard apple (*Annona squamosa* L.). *Indian J. Hort.,* 33: 244–245.

Rao, S.N., 1974. Annona the legendary fruit. *Indian Hort.,* 19(3): 19–34.

Ratan, P.B., Reddy, S.E. and Reddy, Y.N., 1993. *South Indian Hort.,* 41: 171–173.

Samaddar, H.N. and Yadav, P.S., 1970. A note on vegetative propagation of custard apple. *S. Indian Hort.,* 18 (1 and 2): 47–49.

Satyanarayana Swamy, G., 1962. Vegetative propagation in custard apple and Atemoya. *Fruit Nursery Practices in India,* ICAR, pp. 134–139.

Smet, S.de., Damme,P.van, Sheldman, X. and Romero, 1999. *Acta Horticulture,* 497: 269–288.

Shukla, R.P. and Tondon, P.L., 1984. *J. Entomon.,* 9: 181–183.

Thakor, D.R. and Singh, R.N., 1965. Studies on floral biology of Annonas. *Indian J. Hort.,* 22(3–4): 238–53.

Venkataratnam, L., 1955. *Annual Report of Fruit Research Scheme.*

Venkataratnam, L., 1979. Floral morphology and blossom biology on some Annona. *Indian J. Agric. Science,* 29(4): 68–76.

2013, Biodiversity in Horticultural Crops Vol. 4
Editor: Professor K.V. Peter
Published by: DAYA PUBLISHING HOUSE, NEW DELHI

Pages 167–188

Chapter 9

Jujube

Sunil Kumar Sharma

IARI Regional Station,
Agricultural College Estate, Shivajinagar, Pune – 411 005, M.S.
E-mail: sunilkshaerma1959@yahoo.co.in

Jujubes are species of the genus *Ziziphus* Tourn. which belongs to family Rhamnaceae. There are two types of jujubes, *Z. mauritiana* Lam. (Indian jujube or ber), and *Z. jujuba* Mill. (Chinese or common jujube). These two species of genus *Ziziphus* have been cultivated over vast areas in Asia and Africa. Multipurpose value of jujube species including food, honey production, forage and environmental protection was reported from various parts of the world (Sharma, 2009, von Maydell, 1986). The International Centre for Underutilised Crops (ICUC) highlighted *Z. mauritiana* as a priority species for enhanced research attention in the 1990s. India included *Z. mauritiana* in its national programme on underutilised crops (Pareek, 2001). Haryana State Council for Science and Technology has sanctioned a project on conservation of biodiversity of 'ber' germplasm to Haryana Agricultural University. The project included an extensive survey to gather the endangered germplasm developed at various research stations and create a repository created at the university (TNN Jul 1, 2007). Other national programmes also recognized the importance of jujube species in the fruit tree based farming systems (Sharma, 2003 and 2004, Tagiev, 1992). Chinese jujube has been introduced and grown in California and Florida. Ber is also cultivated in Florida (USA) on a small scale. In general, jujubes are popular as fresh fruits and other processed products.

Origin and Distribution

Indian Jujube

The Indian jujube is cultivated for almost 4000 years in India. The precise natural distribution of ber is uncertain due to extensive cultivation, but is cultivated in Southern Asia, Africa and Europe (a likely introduction). It is likely that the original wild species was spread from India through Myanmar.

It would have been moved along with migrations of people. It is thought to have been in Africa for only a few centuries and it is likely that introduction to China and Indonesia is also fairly recent. It was introduced into Guam about 1850 but is not often planted there except as an ornamental, and to Australia (Queensland and the Northern Territory) late in the nineteenth century (Grice, 1998). It reached America during the nineteenth century. *Z. Mauritiana* was introduced to the Negev desert of Israel (Nerd *et al.,* 1990). It was known in Southern Europe for more than 2000 years. The tree is naturalized and forms thickets in uncultivated areas in Barbados, Jamaica and Puerto Rico. Main Asian regions where jujube is cultivated are Lebanon, Iran, Pakistan, Northern India, Bangladesh, Nepal (called Bayar), the Korean peninsula, Philippines, Malayan region, and Southern and Central China. It is also cultivated in Persia, Arabia, North Africa and Asia Minor to the Mediterranean France, Spain and in south eastern Europe.

Chinese Jujube

Z. jujube was likely domesticated in the Yellow River area, for the Yellow River – Huaihe River plain is the main area of cultivation for this species today with much of the production coming from Henan, Shanxi and Shandong provinces. Vavilov (1951) considered the primary centre of origin to be wider than China and the cultivation must have spread quickly through Central Asia. The wild species has a very wide area of distribution from China to Pakistan. Chinese jujube was thought to have reached SW Asia by 2000-3000 BP (de Candolle, 1886). It was taken from the Levantine Coast to Europe in the time of Augustus. It is presumed to spread along the North African coast in the seventh century AD. Chinese jujube has, however, developed a secondary centre of diversity in West Asia and is naturalised in many areas along the Black Sea Coast (Tutin, 1968). Chinese *jujuba* was first from Europe to USA in 1837 by Robert Chisholm and planted in Beaufort, South Carolina and introduced to California and neighbouring states from southern France (Rixford, 1917). Grafted Chinese varieties were introduced in 1906. It is cultivated in southern and western parts of the United States. In the twentieth century superior Chinese materials were introduced to Japan, and also to North Africa by French scientists. *Z. jujuba* (*Z. mauritiana*) was brought to the non-French West Indies from India and Indonesia during the colonial period (Barbeau, 1994).

Taxonomy

Species of the genus Ziziphus were classified based on characteristics of the inflorescences (Hooker, 1875 and Brandis, 1906). Liu and Cheng (1995) suggested that the species are grouped into two Sections, one further divided into two Series:

1. *Ziziphus* - Occurring in temperate zones. Plants glabrous and with deciduous fruiting branchlets.

2. *Perdurans* - Occurring in subtropical and tropical zones. Plants pilose and without deciduous fruiting branchlets.

    2.1 *Series Cymosiflorae* - Occurring widely in subtropical and tropical zones. Flowers in axillary cymes. Ovary and fruit glabrous with thick, hard endocarp.

    2.2 *Series Thyrsiflorae* - Occurring in South and Southeast Asia. Flowers in terminal or axillary thyrses. Ovary and young fruit pilose with very thin endocarp.

The Indian jujube belongs to Section Perdurans Series Cymosiflorae and the Chinese jujube belongs to Section Ziziphus.

Major Cultivated Species

Indian jujube (*Ziziphus mauritiana* Lam.) (synonyms: *e.g.*, *Z. jujuba* (L.) Lam. *Z. jujuba* (L.) Gaertn. (including var. *stenocarpa* Kuntze and forma, *aequilatrifolia* Engl.), *Z. tomentosa* Poir., *Z. rotundata* D.C., *Z. aucheri* Boiss., *Z. insularis* Smith, *Z. sororia* Roem. and Schult., and *Z. orthocantha* D.C.)

The species has a wide range of morphological variations ranging from shrubs to small or medium sized trees which might be erect, semi-erect or spreading. Plant height can vary from 3-4 to 10-16 m. Trees are semideciduous and much branched. The bark has deep longitudinal furrows and is greyish brown or reddish in colour. Plant is usually spinous. Branchlets are densely white pubescent, and tend to be zig-zag. Fruiting branches are not deciduous. Leaf laminae are elliptic to ovate. The apex is rounded, obtuse or subacute to emarginated, the base is rounded and mostly symmetrical. Margins are minutely serrulate. There are three marked nerves almost to the apex. Lower surface is whitish due to persistent dense hairs. Leaves are petiolate and stipules are mostly spiny. Flowers have sepals which are dorsally tomentose and have a 2-celled ovary. Two styles are one mm long and connate for half their length. Flowers are borne in cymes or small axillary clusters. Fruit - greenish, yellow or sometimes reddish - is a glabrous globose or oval edible drupe varying greatly in size. The pulp is acidic and sweet. The species is distributed throughout the warm subtropics and tropics of South Asia. It has spread south-eastwards through Malaysia and eastwards through Indo-China and Southern China. It is widespread in Africa and Southern Arabia while in Africa it has naturalised and so-called 'wild' types are to be found. It adapts to warm to hot tropical climates with low to relatively high rainfall, tolerating poor soils.

Several varietal names are given to wild morphotypes. One variety, *Z. mauritiana* var. *orthocantha* (D.C.) A. Cher. is found south of the western Sahara and in Mauritania.

Chinese jujube (*Ziziphus jujuba* Mill.) (Synonyms, *e.g.*, *Z. sativa* Gaertn., *Z. vulgaris* Lam., *Z. flexuosa* Wall., *Z. nitida* Roxb., *Z. sinensis* Lam., *Z. zizyphus* (L.) Karst., *Z. mairei* Dode, *Z. officinarum* Med., *Z. chinensis* D.C., *Z. chinensis* Watt.)

Plants are shrubs or small trees up to 8-10 m high with stiff branches. Tree forms have a small canopy extending 3.5-4.5 m. Trunks may be short or long depending on genotype. Branches are armed with paired spikes. Older parts of trees can lose their spines. Branchlets are flexuous, green and glabrous when young. Fruiting branchlets are deciduous. Leaves are oblong, obtuse, glandular and have 3-nerves. Stipules form the spines. Flowers are a few in a small axillary cluster or cyme. Styles are two, connate for half their length. Fruit is an ovoid-oblong edible drupe 1.5-2.3 cm long, dark reddish brown to black. Pulp sour to sweet. Chinese jujube spreads westwards to the Mediterranean, throughout the Near East and SW Asia and spread eastwards to Korea and Japan. Like *Z. mauritiana,* this species also naturalised in many Asian countries. It is mostly cultivated in China, India, Central Asia and Southwest Asia.

Z. jujuba var. *spinosa* Hu ex H. F. Chow is typified by possession of small sour fruits and is usually a spiny shrub or small tree (Wang and Sun, 1986). *Z. jujuba* var. *inermis* has unarmed branches and styles not connate (Brandis, 1874). Chinese jujube is adapted to subtropical and warm temperate areas. It prefers a relatively dry climate during the growing season but cool during the dormancy. It can tolerate lower temperatures than Indian jujube and can survive up to -10°C.

Minor Cultivated Species

The following two species of *Ziziphus* are also cultivated on a small scale:

Z. spina-christi (L.) Desf. (Synonyms: *Z. africana* Mill., *Z. amphibia* A. Chev., *Z. nabeca* (Forsk.) Lam., *Z. inermis* A. Chev., *Z. sphaerocarpa* Tul., *Z. spina-christi* Willd.)

Plant is a shrub, often with intertwined branches, or small tree up to 10-15 m tall. Bark is deeply furrowed and scaly, white-brown to pale grey. Branchlets are densely pubescent. Fruiting branchlets are not deciduous. Tends to produce a very deep tap root. Mostly spinous with paired spines. Leaves are ovate-lanceolate, margin crenulated, rounded at base and nearly symmetrical, obtuse. They are minutely pubescent beneath and become glabrescent at maturity. Flowers in short axillary cymes. Peduncles 1-3 mm at flowering and 3-6 mm at fruiting. Sepals woolly dorsally. Fruits subglobose to globose fleshy glabrous drupes, yellow, reddish or redbrown, usually about 2 cm long and 1 cm wide. Flesh astringent to taste. The species is evergreen where water is adequate, but deciduous in the dry season when conditions are drier. It is the best to consider a typical variety, var. *spina-christi*. It is a species of the Middle East through Arabia and West Africa to N. E. Africa, Ethiopia and Eastern Africa. It is wild in the Middle East, especially Iran, Saudi Arabia and also farther west in Turkey. Its edible fruits are gathered for food. It was introduced to Africa by Arab traders along the Mediterranean coast and also via the Horn of Africa (von Maydell, 1986; von Sengbusch and Dippolo, 1980). It is known to be a minor cultivated plant in India and Pakistan and more importantly in Egypt, Syria, the Mahgreb, Saharan oases and Zanzibar. This species is of interest because it has probably hybridised with *Z. mauritiana* according to information from Pakistan, Dahomey and Nigeria. Also it can survive with half the annual rainfall needed by *Z. jujuba* and could be a source of drought resistance.

Z. lotus (L.) Lam. (Synonyms - *Z. nummularia* Aubrev., *Z. saharae* Blatt. and Trab., *Z. lotus* (L.) Desf. subsp. *saharae* Maire, *Z. sylvestris* Mill., *Z. parviflora* Del.)

This species is a spiny shrub growing up to 1.5 m tall and resembling *Z. jujuba*. However, fruiting branchlets are not deciduous and twigs are grey. Internodes on branchlets are less than 1 cm long. Leaves are suborbicular or broadly elliptic to ovate, shallowly glandular-crenate, pubescent beneath and less so above. Flowers are solitary or 2(-3) together. Fruits are subglobose fleshy drupes about 1 cm diameter and deep yellow. It occurs from Asia Minor, south to Arabia, Egypt and along the North African coast and it reaches Cyprus and Greece in Europe. It is also cultivated in S. Portugal and Spain, parts of Italy and Sicily and in Provence, France. In relation to improvement of Indian jujube, *Z. lotus* could be used for earliness of fruit maturity as well as drought tolerance. *Z. lotus* and wild *Z. nummularia* of India are distinct species.

Wild Species

Asia

There are numerous wild species of *Ziziphus* in Asia and they tend to cluster in two regions: China and the Indian subcontinent. There are 14 species in China. *Z. jujube* which belongs to Section *Ziziphus* of the genus is to be found mostly in the central and lower parts of the Huanghe River valley although the primary centre of diversity in China for *Ziziphus* species is South Yunnan and the southeast of Guangxi Province. Pareek (2001) noted the distribution of wild and naturalised *Z. mauritiana* throughout the greater part of India from the lowlands to 1500 m in the Himalayas and also in Sri Lanka. It is associated with dry areas where tree types can be found or bushy types in grasslands. Early writings on the botany of India recorded the species as wild in the Siwalik forests, east of the Ganges and in forests of Central India (Brandis, 1906). *Z. jujube* (*Z. rugosa*) can be found naturalised in the central and eastern sub-Himalayan region, the Central Provinces and western side of the Peninsula. Other species cluster in the north west desert region of India and only a few in the Himalayas. In the drier parts of N. W. India, *Z. nummularia* (Burm. F.) Wight and Arn. (syn. *Z. rotundifolia* Lam.) is another useful rootstock. The species is a thorny shrub producing red, edible fruits. Some wild species with wide distributions, such as *Z. oenoplia* Mill., have become weedy in places such as India, Sri Lanka, Myanmar or Malaysia. According to the literature, a few additional wild species cluster are found in Malaysia (Ridley, 1922) and in Indonesia (Martin *et al.,* 1987). There are three other species used as rootstocks. (i) *Z. xylopyra* Willd. (syn. *Z. rotundifolia* Roth., *Z. cuneata* Wall.) is an erect, small tree frequently unarmed and producing a woody, inedible fruit. It is a species of South India and Sri Lanka. (ii) *Z. rugosa* Lam. is a straggly bush tending to have solitary spines and edible fruit, found in the Central Hills and eastern parts of India. (iii) *Z. oenoplia* Mill. is a scrambling shrub with spines and small black fruits often used for tanning and found in S. India, Sri Lanka and Myanmar. Wild species are sources of diversity for jujube improvement. Pareek (2001) noted the attributes of several wild species of ber (Table 9.1).

Table 9.1: Exploitable Attributes of Wild Species in Ber

Wild Relatives	Exploitable Attributes
Z. nummularia and *Z. lotus*	i) Drought tolerance
	ii) Dwarf tree stature and extensive root system
	iii) Early fruit maturity
Z. jujube	i) Resistance to low temperature damage
	ii) Excellent dehydration quality of fruits
	iii) High vitamin C and P contents in fruits
Z. mistol	i) Resistance to low temperature damage
Z. mauritiana var. *Rotundifolia*	i) Vigorous tree frame
	ii) Wood of marginal timber value

Source: Pareek, 2001.

Africa

Some wild species are also found in Africa. *Z. abyssinica* Hochst. ex A. Rich. is found in scattered tree grassland 400-2200 m from Senegal to Ethiopia and south to Zimbabwe and Mozambique. *Z. mucronata* Willd. grows in open woodland from 0-200 m above sea level from Senegal to Arabia and south to S. Africa and Madagascar. While *Z. spina-christi* is distributed from 0-1300 m and is indigenous in semi-desert wadis from 600-1000 m above sea level in the Horn of Africa and North Africa. *Ziziphus* was probably introduced and through seed propagation in East Africa where it has often reverted from introduced cultivars to wild types.

The *Ziziphus* Genepool

The two major cultivated species are widely distributed in congenial climates in Asia and they cover vast regions. Most of these distributions comprise wild or naturalised materials superimposed on the patterns of distribution in the areas where the species are cultivated. The wild populations are heterozygous and extremely variable from which farmers have selected the best trees in terms of production and propagated them vegetatively, but are now doing so through grafting. A study of wild populations of Chinese jujube has identified promising types to use as germplasm (Liu and Wang, 1991). As a result of long time of local selection many cultivars arose, some of which have become widely recognised. Very little is known about the patterns of genetic variation in the wild populations of any of the other cultivated species of jujube.

Chromosome Numbers

Indian jujube is usually a polyploid with counts of n = 12, 20, 24, 30, 36, or 48. Khoshoo and Singh (1963) looked at a range of cultivars and found n = 24 in most of them, but in two, it was n = 48 and in one it was n = 30. In some wild material, Nehra *et al.* (1983) found n = 48; and also in naturalised 'wild' material it was the same count. Not too many wild species have been counted but from the few that have (particularly *Z. lotus, Z. nummularia and Z. oenoplia*), n = 10, 12, or 36. *Z. lotus* appears to be diploid, *Z. oenoplia* tetraploid and *Z. nummularia* shows a polyploid series. The possibility exists that the genus is tribasic with x = 10, 12, or 13 (Darlington and Wylie, 1955). It is now generally thought that Indian jujube shows a range of polyploids: diploid, triploid, tetraploid, pentaploid and octoploid (Mehetre and Dahat, 2000). Chinese jujube also represents a polyploid series, chromosome counts tending to represent 2n = 45, 60, 90. Chinese jujube exhibited high diversity in chromosome karyotypes, shape, size and surface sculpture of pollen, leaf length and flower diameter, shape, colour, weight of fruit, growth period and soluble solids and ascorbic acid of fruits. The level of ploidy appears to be important for some cultivars of Indian jujube. Powdery mildew resistant genotypes were diploids and a seedless form was octoploid. However, other diploid cultivars were susceptible to powdery mildew.

Biodiversity in *Ziziphus* spp.

Genetic diversity of *Ziziphus* is high in India and about 20 species are found between 8.5-32.5°N and 69-84°E. Economically important species are *Z. nummularia, Z. oenoplia, Z. rugosa, Z. sativa, Z. vulgaris* and *Z. xylopyrus*. The ability of *Ziziphus* species and different varieties/types within *mauritiana* to cross freely has allowed the build up of rich gene pool which depicts heterozygosity in their adaptability to soil and climate; morphological, physiological and phenological traits; chromosome number; tolerance/resistance to biotic and abiotic stresses and genomic DNA. Promising germplasm with distinctive traits such as desired adaptation (Gola, Umran, Thar Bhubhraj, Thar Sevika), diverse quality traits (Banarasi Karaka, Illaichi), high and stable yield (Seb, Ponda) and tolerance to biotic and abiotic stress (Tikadi, Katha, Bawal-Sel-1, Sanaur-2) have been identified which have comparative advantages to improve the productivity of this important arid fruit crop (Awasthi and More, 2009).

Variability in Jujube Cultivars

A wide variation due to cross pollination is exhibited by the jujubes in vegetative, leaf, floral, fruit and quality traits. Variation in morphological and physicochemical characters of cultivated ber types are described below.

Variations in Vegetative Characters

The most appropriate vegetative characters for classification are leaf area and branching habit, while the most dependable fruit characters are apex type, stalk and stylar flesh cavities and shape (Bal, 1992). *Ziziphus mauritiana* genotypes Umran, Illaichi, Desi-1 and Desi-3 plants have a spreading habit, whereas, Kathapal and Desi-2 plants exhibit a semi-spreading habit. The cultivars of Gola group (Gola Gurgaon No. 3, Bhadurgarhia Gola, Dankan Gola, and Kakrola Gola) produce erect plants. Leaf margins of some cultivated and wild forms of ber variously referred to as *Z. mauritiana, Z. rotundifolia* or *Z. nummularia* are serrated except in Desi-3 and Jharber (Gupta *et al.,* 2003). The minimum stomatal density, internodal length and plant height were recorded in Gola budded onto *Z. nummularia* and was the most dwarfing combination. Whereas, scion combination of Ponda budded onto rootstock

Z. mauritiana ecotype- Assam-Gauhati was the most vigorous. The rootstock *Z. mauritiana* ecotype-291 was the most compatible and moderate in plant growth, stomatal density and internodal length thereby giving moderate vigour to trees (Verma *et al.,* 2001). Paired spines were observed in many cultivars of *Z. mauritiana* like Umran, Kathaphal, Gola Gurgaon No. 3, Bhadurgarhia Gola, Dankan Gola and Kakrola Gola (Gupta *et al,.* 2003). The maximum tree height was recorded in Desi Alwar while tree spread was in Sanori No.5 (Saran, 2005). Maximum height of a 15 year old tree of *Z. mauritiana* was recorded in LR-13 (6.59 m); girth was maximum (3.32 m) in LR 11 (Kundi *et al.,* 1989).

Reproductive Variability

Flowering

The peak period of flowering and fruit set in *Z. mauritiana* cultivars Banarsi Karaka, Ponda, Illaichi, Gola and Tikdi were September-October. Tikadi had the shortest duration of flowering (47 days) and fruit set (36 days) but the highest number of fruits/branch (239), fruit set (28 per cent), number of fruits reaching maturity/branch (48) and fruit retention (20 per cent) (Sharma *et al.,* 1990). Whereas, Saran (2005) found the longest period of bloom in Katha Rajasthan (78 days) ranged from 6 August to 23 October and the minimum period of bloom in the Safeda Rohtak (23 days). Flowering occurred from 57 to 75 days, depending on cultivar (Dhaliwal and Bal, 1998). Only one flowering season was observed in Hyderabad from May to July and its total duration varied 68-94 days. Cultivars Gola, Mundia and Akola flowered early; Umran and Seb in midseason; Banarsi and Kaki were late flowering (Babu and Kumar, 1988). The number of flowers per branch is quite high. Umran had the highest number of hermaphrodite flowers (22.2 per cent) followed by Gola Gurgaon with 20.1 per cent (Darbara and Jindal, 1982). Cultivars of *Z. mauritiana* differed in the time of flowering (on set and duration and peak) in Maharashtra, India (Desai *et al.,* 1986). The pollen grains were sub-prolate to prolate spheroidal. Pollen germination was more than 50 per cent in most of the ber cultivars in 25 per cent sucrose solution whereas Illaichi cultivar was sterile. Scanning Electron Microscopic (SEM) and light microscopic studies were carried out on pollen samples of three *Ziziphus* species and six *Z. mauritiana* varieties. Pollen grains differed in size, shape and exine characteristics. Size and shape of pollen within cultivars were quite uniform. Differences in exine pattern, size and P/E ratios could be used for identification of *Ziziphus* genotypes (Diwakar *et al.,* 1996). Size of pollen grains of *Z. mauritiana* Illaichi, Umran and four wild forms was different (Nehra *et al.,* 1984). Pollen diameter in seven cultivars of ber (*Ziziphus mauritiana*) ranged from 20.05 mì in Darakhi 1 to 32.04 mì in Seedless. Similarly, considerable differences were also observed for pollen diameter among the different ploidy levels (Pradeep and Jambhale, 2000). *Z. nummularia* and *Z. mauritiana* were distinct genotypes (Diwakar *et al.,*1992). Studies in India have indicated that in *Z. mauritiana,* some cultivars have cross incompatibility (Teaotia and Chauhan, 1964).

Fruit Set

Umran was compatible as female parent with Sanuar 2, but Sanuar 2 did not set fruit after pollination by Umran (Mehrotra and Gupta, 1985). The requirement for cross pollination, incompatibility and pollen sterility means that fruit set depends on physiological and environmental conditions. The mode and time of anthesis were also cultivar specific. Anther dehiscence started about two hours after anthesis and continued for two to four hours. Peak receptivity of the stigma appeared to be just as the flower opened (Dhaliwal and Bal, 1998). Some cross fertilisation was found necessary for the development of viable seeds and many of the aborted seeds apparently were the result of self fertilisation (Ackerman, 1961). Fruit set in *Z. mauritiana* cultivars took place almost at the same time in Tikadi, Gola, Seb, Umran and the hybrid Umran x Seb cultivars, but Gola fruits matured earlier than those of all other cultivars in the region. Yakobashvili (1973) in a six year study of fruit set, yield and other characters, showed the following were promising varieties: Ta Yan Tszao, Seedling 2, Seedling 1 and 3. Fruit drop studied in seven *Z. mauritiana* cultivars indicated the lowest drop (24.1 per cent) in Ponda and the highest in Illaichi (68.6 per cent) (Vashishtha and Pareek, 1979). Kakrola Gola had considerably higher drop (up to 80.6 per cent) than Kaithali and Umran which did not differ significantly, showing 7.2 and 12.1 per cent drop respectively (Panwar, 1980). Neeraja *et al.* (1995) recorded the highest fruit drop in Seb (87.94 per cent) followed by Umran (83.43 per cent) and Gola (82.14 per cent). Gola was the earliest (108 days) and Umran was the latest (147 days) to ripen under Gurgaon conditions (Singh *et al.*, 1983). Saran *et al.* (2005) studied the bearing behaviour in 35 cultivars of ber at Hisar, India and classified them into three categories based on harvesting time, as early, medium and late. Umran, Katha Bombay, Chhuhara, Illaichi, 2g-3, Kathaphal, Jogia, Ponda, BS-2 and Desi Alwar are late bearing varieties while Gola, Gola Gurgaon No. 2, Gola Gurgaon No.3, Safeda Rohtak, Seo, Katha Rajasthan, Laddu and Akhrota were early bearing varieties and Kaithali, Dandan and Mirchia varieties bear during the mid-season. Kathaphal is late ripening while Gola is early and Kaithali comes in midseason. Seo, Sanaur No.2 and Umran are recommended as early, mid and late cultivars, respectively for commercial cultivation (Chadha *et al.*, 1972) in Northern India.

Fruit and Seed

Considerable variation in fruit length and breadth in different cultivars of *Z. mauritiana* like Banarsi Karaka (5.4 and 3.4 cm), Dandan (5.1 and 3.0 cm), Jogia (4.9 and 3.8 cm) and Umran (4.8 and 3.8 cm) were reported by Ghosh and Mathew (2002). Karaka had the largest fruit (Singh and Singh, 1973). Cultivars Seo Bahadurgarhia, Nari Keli, Desi Alwar and Banarsi Karaka produced fruits weighing > 20 g each (Godara, 1980). Umran produced the heaviest fruits (39.8 g), Gola Gurgaon showed the highest content of pulp (97.2 per cent), total soluble solids in Gurgaon, Haryana (Singh and Jindal, 1980). Pewandi was the best variety for commercial cultivation, having the largest fruit, a high percentage of edible pulp and good skin colour. BS-2 had the highest pulp/stone ratio; and the maximum fruit and stone size was recorded in Ponda (Saran, 2005). Cultivars Dabailing, Daguazao and Linyi Lizao are large fruited, with fruit sizes of 23-26.1 g, Jinsi 3 and Jinsi 4 are small fruited varieties (Chen *et al.*, 2003). The best *Ziziphus jujuba* varieties for uniformity of fruit size are Ta-Pai and Hsueh-pai (Sin'ko, 1974). A lot of variation exists in ber genotypes for pulp/stone ratio. Pulp/stone ratio was maximum in Sanur-6 and minimum in Punjab Chhuhara (Bharad *et al.*, 2002), The pulp/stone ratio was the highest in Umran (Dhingra *et al.*, 1973). Significant genotypic differences were found in the seed characters of *Z. mauritiana* varieties. Saran (2005) observed variation in stone size of 35 cultivars of ber and reported that maximum (3.72 cm) was in Ponda and the minimum (0.53 cm) in Illaichi. Seeds of *Ziziphus mauritiana* trees taken from 5 districts in Yunnan Province, China and Narkum, Myanmar, were tested for their morphological characteristics, germination characteristics, and growth patterns of the seedlings and young trees. A close relationship was found between these characters and their geographical distribution and climatic condition (Wang and Wang, 1994). In Chinese jujube, one group of cultivars gives 85 per cent seed germination (e.g. Yatszao) and the other gives 98 per cent germination, *e.g.* Nikitskii 84, 92 and 94. (Kim and Kim, 1984). The thinnest seed coat wall was observed in Godhan (Singh, *et al.*, 1973). However, the quickness and extent of germination depend upon viability and after-ripening status of the seed, presence of endogenous inhibitors, weathering of the stony endocarp and environmental conditions such as the temperature, moisture, salinity and alkalinity of the growing medium.

Yield

Variation in yields of *Z. mauritiana* cultivars (55.5 to 116.11 kg/tree) was observed by Kumar *et al.* (1986). In the semi-arid subtropical climate of northern India under irrigated conditions, the fruit yield per tree ranges from 80 to 200 kg depending on the varieties and management practices during the prime bearing age of 10 to 20 years (Bakhshi and Singh, 1974). Fruit yield was the highest in Gola (38.4 kg/tree) and the lowest in the local cultivar Sukavani (3.54 kg/tree) at Sardar Krushinagar in Gujarat (Chovatia *et al.*, 1992). Highest yields were obtained from Umran (210 kg/tree) followed by Sanaur No. 2, Dandan and ZG 2 (Gupta, 1977). Thornless gave the highest yield/tree (74.4 kg) followed by Sanaur 5 (71.3 kg) and Sanaur 4 (Singh and Tomar, 1988) at Bhatinda, Punjab, India. Yield in different cultivars of *Z. mauritiana* in Maharashtra (India) indicated maximum fruit, pulp and stone weights in Kadaka when pruned on 25th March (Bharad *et al.*, 2002). Reddy *et al.* (1998) carried out an economic analysis of cost and returns to determine the most profitable cultivars at Dharwad, Karnataka, India. The performance of 11 cultivars showed that Dandan, Sanaur-2 and Chhuhara, with mean fruit yields of 6.78, 6.36 and 6.08 t/ha and were the most promising. Cultivar Pewandi is the best for commercial cultivation in Uttar Pradesh, India (Teaotia *et al.*, 1974). Chinese jujube cultivars yield 50 to 300 kg fruits per year depending on cultivar, location and age of tree (Ciminata, 1996). On the basis of yield, ripening date, frost resistance and storage quality in 7 ber cultivars Ta-Yantszao, Da-bai tszao, Ya-tszao, 93, 58, 107, 52 and 48 are recommended for cultivation (Sin'ko, 1977). In China - Taiwan, fruit yields of 158.6 kg per tree in cultivar Kaolang-1 and 140.8 kg per tree in cultivar Telong have been obtained (Chiv Chu Ying, 1997).

Chemical Variability

A wide range of varietal variability for quality traits was detected in 30 cultivars of ber (*Z. mauritiana*) at Hisar (Bisla and Daulta, 1986). Ber germplasm collection and evaluation at Amer, Bharatpur, Deeg, Tijara and Jhunjhunu of Rajasthan, India indicated significant variation in physicochemical characteristics of 23 genotypes, and ten produced fruits of excellent quality (Lal *et al.*, 2003). Variation in 10 local ber genotypes grown in West Bengal, were noted for soluble solids, total sugars, acidity and ascorbic acid content; ranges of these parameters were 9.53-19.13 per cent, 4.94-12.30 per cent, 0.38-2.60 per cent and 17.25-51.98 mg/100g respectively (Ghosh and Mitra, 2004). The cultivar Umran has 19 per cent TSS and 1.2 per cent acidity. In Kathaphal, the TSS is 23 per cent and acidity was 0.77 per cent while in Gola, the TSS is 17-19 per cent and 0.46 - 0.5 per cent acidity. In Kaithali TSS is 18 per cent and acidity 0.5 per cent (Daulta and Chauhan, 1982). Godara (1980) evaluated 16 cultivars for quality characters. The fruits of Mundia Murhara and Chhuhara had the highest TSS content (22.8 and 22.4 per cent, respectively). HB-1 and LB had excellent flavour and HB- 2, KB1, SB and SB-2 were rated good at Bangalore. Under rainfed conditions of Bawal, Haryana cv. Nazuk had the highest total soluble solids (28.9 per cent) followed by Illaichi (Yamdagni *et al.*, 1985). Kadaka had maximum sugar and minimum acidity content and was considered outstanding (Singh and Singh, 1973). Some cultivars show variation in acidity (0.228-0.78 per cent) and ascorbic acid (80.85-178.04 mg/100 g pulp) (Godara, 1980). Ascorbic acid content of several cultivars ranged 70-165 mg/100g of pulp (Jawanda and Bal, 1978). Bisla *et al.* (1980) observed the highest vitamin-C content (120.15 mg/100g) in cv. Illaichi. The vitamin-C content was the highest in Narikelee, followed by Kaithali 165 and 125 mg/100g fruit pulp, respectively (Gupta, 1977). Godhan was the most nutritive with regard to total soluble solids and ascorbic acid content, whereas Kharki was the sweetest at Hoshangabad, Madhya Pradesh, India. The highest acidity (0.49 per cent) was exhibited by Soni while the lowest acidity (0.25 per cent) was exhibited by Kabra and Amrabati. The highest total sugar content (14.48 per cent) was exhibited by Bekanta and the lowest (8.5 per cent) by Karka (Gupta *et al.*, 2004). Singh and Jindal (1980) found the highest contents of pulp (97.2 per cent), TSS (21.4 per cent) and total sugar (10.7 per cent) in Gola Gurgaon and Kaithli had the highest (113.5 mg/100 g) ascorbic acid content. The TSS content was the highest in Chonchal and Illaichi and the latter cultivar had the highest ascorbic acid content. Munier (1973) studied quality characters. Cultivars grown in China, the United States and Pakistan contained, respectively, 500-600, 300-500 and 45.2-160.8 mg Vitamin C/100 g. Variations in ascorbic acid content of 10 varieties were reported by Ahmad and Malik (1971). Late ripening varieties generally contain more ascorbic acid than early ripening ones. Lomakina (1976) evaluated jujube varieties in south-west Turkmenistan. The best cultivars were Ya-tszao and Ta-Yan-tszao. The soluble protein in fruits and leaves on a fresh weight basis ranged from 9.37 to 26.90 mg and 25.40 to 56.98 mg per gram respectively in 42 cultivars of *Z. mauritiana* (Sudhir *et al.,* 1999). Umran was noted for high sugar and protein contents, followed closely by Kathaphal (Khera and Singh, 1976). In wild samples, the number of amino acids per sample was in the range of 4-8. Of the 17 free amino acids present in wild samples, glutamic acid, amino-butyric acid, threonine, proline and alanine were well represented. In cultivars, the number of amino acids per sample was in the range of 3-10. Of the 22 amino acids represented in these cultivars, arginine, cysteine, cystine, alanine and proline were common. The most widely represented amino acids in cultivars (arginine, cysteine and cystine) were absent in wild taxa. The commonly represented amino acids in wild samples (glutamic acid, amino-butyric acid and threonine) were not common in cultivars.

Genetic Variability

Information on genetic variability, heritability and correlation coefficients derived from data on leaf area, relative water content, stomatal index, stomatal frequency and yield in four year old ber plants of 12 popular ber cultivars, grown at Raichur, were observed. Genotypic coefficient of variation (GCV) and phenotypic coefficient of variation (PCV) were the greatest for stomatal frequency. GCV and PCV values were the lowest for relative water content (RWC). High values of heritability and genetic gain were observed for stomatal index. High estimates of GCV, PCV, heritability and genetic advance were recorded for stone size, pulp stone ratio, fruit weight and yield indicating the effectiveness of improvement through simple selection (Saran, 2005).

Disease Resistance

Forty ber (*Zizyphus mauritiana* [*Ziziphus mauritiana*]) accessions were screened for resistance against black leaf spot disease (caused by *Isariopsis indica* var. *ziziphi*) in Faizabad, Uttar Pradesh, India during 1997-98 to 1999-2000; Tikri during 1997-98 and 1999-2000; Seedless during 1998-99; and ZG-3 during 1999-2000 were found immune. Guli, Seedless and Ber selection-5 during 1997-98; Guli, Darackhi-2, Ber selection-2, 3, 4 and 5 during 1998-99; and Jalandher, Kali, Bagwadi, Banarasi Peondi, Illaichi, Villaiti, Sanour-3, Chhohara, Katha, Seedless, Darakhi-2 and Ber selection-5 during 1999-2000 were resistant. Other accessions were moderately susceptible to susceptible reaction against the disease (Kumar *et al.,* 2003).

Hybridisation

The taxonomic literature on *Ziziphus* refers to reputed hybrids between certain wild species. It is highly likely that a number of the taxa generally accepted will turn out to be stabilised hybrid segregates. Further insight into this could be useful in relation to genetic resources for use in breeding. In practice, improvement programmes have not yet successfully used hybridisation with wild species, although such interspecific hybridisation would be of potential value to expand adaptation of cultivars to wider ecological areas and to introduce resistances to pests and diseases.

Important Jujube Cultivars

The majority of cultivars of both Indian and Chinese jujubes are selections from heterogeneous populations. Superior genotypes were protected and used over millennia by local people and therefore, location specific cultivars are widespread. Those which are used nowadays tend to be propagated clonally and some have become very widespread *e.g.* Umran in Indian jujube; or Sui Men or Li in Chinese jujube. The cultivated ber has more than 300 varieties but only a few are commercially important (Pareek and Nath, 1996). Over 180 named cultivars have been mentioned in the literature (Pareek, 2001). Numerous cultivars were selected during a long period. Important traits of Indian cultivars are given in Table 9.2. A Chinese work published over 300 years ago listed 43 cultivars (Locke, 1948). In China, there are at least 400 cultivars of Chinese jujube (Hayes, 1945) but Qu and Wang, (1993) reported more than 700 cultivars. These can be divided into two groups: the sour type mainly used as rootstocks, medicines or animal fodder, and the cultivated type (Ciminata, 1996).

Table 9.2: Important Traits in Indian Jujube Cultivars

Cultivar	Characters
Umran	Yield, stem girth, fruit size, fruit weight, flesh thickness, flowers/cyme and pulp/stone ratio.
Ponda	Yield, spread, number of flowers per cyme, fruit weight, fruit size and flesh thickness.
Sanori 5	Yield, number of flowers per cyme, fruit weight, size of fruit and flesh thickness.
Laddu	Yield, spread and number of flowers per cyme
Gola	Spread, fruit weight, flesh thickness and stone weight
Chhuhara	Spread, stem girth

Source: Saran, 2005.

There is little information available on the genetic diversity of jujube. The reduction of genetic variability makes the crop vulnerable to diseases and other adverse factors. Morphological variation among thirty five important ber genotypes collected from different parts of India were studied at CCS Haryana Agricultural University, Hisar, India, during 2005-2006. Ten morphological traits *viz.*, powdery mildew, leaf length, leaf breath, leaf area, fruit weight, fruit length, fruit breath, stone weight, stone length and stone breath were analyzed. All the genotypes were classified into VII distinct clusters. Cluster I was the largest with nine genotypes followed by cluster IV (six genotypes). Clusters VI and VII were the most divergent with an inter cluster distance of 5.162, suggesting that the parents for hybridization could be selected especially for powdery mildew resistance (Sanori No. 5, Noki and Mirchia) and smaller stone size (Illaichi and Kishmish) from these diverse clusters to develop useful disease resistant and smaller stone size breeding material for Indian jujube (Saran *et al.,* 2007). Important varieties identified by various workers which can be used by breeders are listed in Table 9.3.

Umran

This cultivar is commercially cultivated on a large scale in Punjab, and Haryana states of India. It was developed from germplasm from Rajasthan at the Fruit Research Station at Bahadurgarh, Punjab. It fetches the highest price. The fruits are large sized, oval in shape and have a roundish apex. They weigh on an average between 30 and 80 g. They are an attractive golden yellow colour which later turns into a chocolate brown at full maturity. The fruit matures in the mid-season (February to March) and ripens during mid-March to mid- April. The fruit is sweet with 19 per cent total soluble solids (TSS) and 0.12 per cent acidity. It has a pleasant flavour and excellent dessert quality. Umran fruits have a good keeping quality and can withstand long transportation. The main reason for its popularity is the long shelf life (15 to 20 days) and excellent organoleptic qualities. The fruit is also known locally as Ketha, Ajmeri and Chamdi.

Table 9.3: Classification of Germplasm of Ber in Different Clusters

Clusters	No. of Cultivars	Name of Germplasm
I	9	Kait Gola Gurgaon-3, Gola Gurgaon-2, Bawal Selection-2, Z.G.-3, Seo, Kathaphal, Sanori No. 3, Sanori No. 1, Safeda Rohtak
II	6	Safeda Selection, Laddu, Umran, Katha Rajasthan, Sua, Dandan
III	3	Noki, Mundia, Murhara, Sanori No. 5
IV	7	Sandhura Narnaul, Thornless, Triloki No. 1, Chhuhara, Banarasi Karka, Narikali, Chanchal
V	4	Kala Gola, Popular Gola, Kaithali, Katha Gurgaon
VI	2	Golar, Mirchia
VII	4	Illaichi, Illaichi Jhajjar, Z.M. elongate, Kishmish

Source: Saran *et al.*, 2007.

Kathapal

This is a late ripening cultivar of ber well known in Gujarat that has small to medium size fruits. At maturity, the fruits remain green on one side while the other side develops a reddish yellow tinge. The average fruit weighs 10 g. They have a high TSS (23 per cent) and an acidity of about 0.77 per cent.

Gola

Gola is an early-maturing cultivar that is grown in Uttar Pradesh, Gujarat, Punjab, Rajasthan, Haryana and Delhi. It starts to bear fruit in the first week of January. The fruits are very attractive, roundish in shape and golden yellow in colour with an average weight of 20 g. The white flesh is very juicy, semi-soft and has a delicious taste. The fruit pulp has a TSS of 17 to 19 per cent and 0.46 to 0.51 per cent acidity. The ratio of pulp to stone is 14.

Kaithali

This is a mid-season cultivar of ber well known in the Punjab that ripens during March. It was developed from the collection from the Kaithali area of Karukshestra in Haryana. The fruit is medium in size with an average weight of 18 g, oval in shape and has a tapering apex. The fruit pulp is quite soft, has a TSS of 18 per cent and 0.5 per cent acidity. Unlike Umran, this fruit does not withstand transportation and has a poor keeping quality.

Tikadi

This ber cultivar is a late maturing type that ripens in Rajasthan during February and March. The fruits are small, weighing on average 10 g. The fruits become edible, with a creamy soft flesh, when the skin turns to a red colour during a short (7 to 10 day) ripening period of the fruits on the trees. The ripe fruit has a high TSS (25 per cent) with a large stone (the pulp to stone ratio of ripe fruit is 6.9).

Jogia

The fruits of this cultivar are also well known in Rajasthan and have a light purple tinge when they are unripe, but are still edible at this time. Because of this, they have an extended harvest period up to early February. The skin surface has coarse ridges and is greenish yellow in colour. The flesh is white, soft, juicy and sweet with a TSS of 19 per cent and a pulp to stone ratio of 14.

Mundia

This is an early Rajasthan, high yielding cultivar that matures during mid-January. The fruits are large, juicy and bell shaped with an average weight of 40 g. They have a yellowish green skin with smooth depressions. The flesh is white and soft with a TSS of 20 per cent and a pulp to stone ratio of 23. Better than varieties Umran or Katha and Chhuhura, although some consumers prefer the more acidic fruits of Sanaur-2, Sanaur-3, Sanaur-4 and Mehrun. Traits identified by various workers are described in Table 9.4.

Table 9.4: Traits Identified in Jujube Cultivars

Traits	Cultivars
Fruit maturity	Early (Gola, Mundia), mid season (Banarsi, Kaithli), late (Umran)
Sweetness (high TSS)	Reshmi, Umran
Pulp texture	Coconut-like (Umran), Juicy (Gola, Aliganj), Melting (Illaichi)
Fruit size	Very large (Ponda), large (Umran), Medium (Mundia, Banarsi, Gola), small (Illaichi)
Fruit shape	Apple like (Seb), cardamomshaped (Illaichi), bell shaped (Mundia), Round (Gola), oblong (Umran)
Fruit colour at	
Maturity	Bright golden (Sanaur), bright yellow (Gola), Greenish yellow with brown blush (Kathaphal)
Acidity	Very low (Umran), low (Gola), Moderate acidic (Sanaur), acidic (Kathaphal)
Shelf life of fruits	Good (Umran, Maharwali), poor (Gola)
Processing uses	Dehydration (Vikas, Raja, Babu, Jeevan, Chinese cultivars, Umran, Bagwari, Chhuhara) Preserve (Umran, Banarsi Karaka, Kaithli) Candy (Illaichi, Umran, Kathaphal, Kaithli) Beverage (Gola, Mundia)
Resistance to fruitfly	Tikadi, Meharun, Illaichi
Resistance to fruit borer	Banarsi Pewandi, Gola Gurgaon, Jhajjar Selection
Tolerance to powdery mildew	Illaichi Jhajjar, Sanaur-5, Safed Rohtak, Kathaphal, Gola, Seb, Meharun
Field resistance to powdery mildew	Dharkhi-1, Dharkhi-2, Guli, Villaiti, Seedless
Tolerance to *Isariopsis*	Safed Rohtak, Sanaur-1, Seo Bahadugarhia, Jhajjar Selection,

Source: Pareek, 2001.

Genetic Erosion

Accelerated selection and wider adoption of clonally-propagated cultivars will lead to a degree of genetic erosion. However, whilst wild and naturalised populations persist in such large geographic areas in the primary centres of diversity, and numerous areas in other parts have become secondary centres, (e.g. tropical Africa for *Z. mauritiana,* Central Asia and South-west Asia and parts of Africa for *Z. jujuba*), there is not a major cause for concern. It is a common practice to top work wild trees with improved cultivars (Singh *et al.,* 1973 and Yadav, 1991), and many rural communities use jujubes for fencing, wind-breaks and other purposes. These activities provide a degree of protection.

Existing Germplasm Collections

Collections of germplasm which are currently maintained are largely geared to maintenance of cultivars and other selections to support the national improvement efforts. Sometimes they represent introductions which have been tested for adaptation for local conditions. More information is required for characterisation and genetic affinities to develop strategic planning to rationalise existing collections so that synonyms of cultivars can be eliminated. The accessions should truly represent specific patterns of genetic variability and be related to improvement needs, including ecological and climatic tolerances. Many of the traits considered in improvement are polygenic and there will always temptation of maintaining a larger number of accessions than actually require. A large number of Institutions in India are holding ber germplasm and are working on their evaluation and improvement (Table 9.5). The germplasm holding of ber at Central Institute for Arid Horticulture (CIAH) consists of a total of 333 accessions, comprising cultivars, indigenous and exotic selections and rootstocks.

Table 9.5: Indian Institutions Holding Ber Germplasm Collections (Number of Accessions)

Institutions	Ziziphus species	Ber Cultivars
Central Institute for Arid Horticulture, Bikaner-334006, Rajasthan, India	7	162
Central Arid Zone Research Institute, Jodhpur-342003, Rajasthan, India	—	68
Indian Institute of Horticultural Research, Bangalore-560089, Karnataka, India	3	32
Central Horticultural Research Station, Godhra-389 001, Gujarat, India.	—	22
Haryana Agricultural University,Hisar-125 004, Haryana, India	—	74
Narendradeo University of Agriculture and Technology, Kumarganj, Faizabad-224229, U.P, India	—	32
Mahatma Phule Agricultural University, Rahuri-413 722, Maharasthra, India	—	87
Gujarat Agricultural University, Sardar Krushinagar-385506, Gujarat, India	—	75
Fruit Research Station, Punjab Agricultural University, Bahadurgarh, Patiala-147001, Punjab, India	—	42
Indian Agricultural Research Institute, New Delhi-110012, India	4	52
Dryland Agriculture Research Station, Haryana Agricultural University, Bawal-123501, Haryana, India	—	36

Source: Pareek, 1988 and Pareek and Sharma, 1993.

Chinese jujube is maintained at a number of sites in Hebei, Shanxi, Henan and Shantung provinces and the Jujube Institute at the Hebei Agricultural University, Baodong, Hebei 071001 and the Chinese Academy of Forestry. Other collections of jujubes are at the Apsheron Experimental Station for Subtropical Crops, Azerbaijan, where in the 1990s the best 40 seedlings of ber were selected from 3762 jujube seedlings, and 25 were included among the elites. The 11 best forms were given varietal names. There are 65 forms of *Ziziphus jujuba* at the Turkmen Experimental Station which include several local, foreign and Soviet bred varieties of jujube. A number of accessions of *Z. jujuba* are held at the Research Institute of Plant Production, Prague, Czechoslovakia. The Fruit-Tree Research Station (FTRS) in Japan has a collection of *Z. jujuba* maintained at the Okitsu Branch (Moriguchi *et al.,* 1994). The National Germplasm System of the USA also holds significant accession numbers of the two major jujubes. Mediterranean European countries have been involved in documenting germplasm of jujubes through a cooperative European Union Project. Turkey maintains a number of accessions of *Z. jujuba* at ARARI, Meneme, Izmir through its national genetic resources programmes. Bangladesh maintains

35 accessions but only two of these cultivars are under cultivation on farms (Saha, 1997). Similar collections are present in Korea, Pakistan and Thailand.

Rootstock Resources

In addition to cultivars, more attention need to be given to wild sources of current or potential use as rootstocks and their maintenance in the collections. A little research on using rootstocks, other than those of cultivars has been done because very limited genetic material of each wild species used has been tested for rootstocks. Table 9.6 shows the situation for ber (Bal *et al.*, 1997). Chinese jujube rootstocks are most frequently wild materials (especially var. *spinosa*) related to the cultivars, but a number of other wild species have been tried in areas of China with more extreme climates (Ming and Sun, 1986).

Table 9.6: Species Used for Rootstocks for Indian Jujube

Compatibility	Species
Most successful	*Z. mauritiana cultivars*
Can be widely used	*Z. mauritiana* var. *rotundifolia* (wild/naturalised)
	Z. abyssinia
Less successful but mostly	*Z. nummularia*
compatible; often cultivar	*Z. xylopyrus*
specific	*Z. spina-christi*
	Z. mucronata
	Z. oenoplia
	Z. jujuba

Source: Pareek, 2001.

Conservation Methodologies

A little attention has been paid to storage of seeds of jujubes for long-term conservation because of the heterozygosity present and the loss of a particular cultivar if seed is propagated. Seeds of jujube species, both cultivated and wild, show an orthodox behaviour when dried and stored at low temperatures. The Millennium Seed Bank of the Royal Botanic Gardens, Kew, UK stores accessions of *Ziziphus*, mostly wild species of arid zones of Africa. The rationalisation of a number of the existing germplasm collections could transfer them into field genebanks. Also more systematic evaluation of existing resources is essential to utilise the variable genepools. Insufficient information on performance accessions is due to the enormity of the task. Thus, proper characterisation and evaluation of germplasm and dissemination of the information to breeders and others are very important.

In situ Conservation

In situ conservation has two approaches (*i*) biosphere reserve; (*ii*) habitat approach. The natural biosphere reserve is a useful solution for species which are endangered and almost on the point of extinction. Habitat approach refers to management of target species in its original habitat, through protected areas, managed forests, natural reserves with multiple uses, preservation plots, wildlife sanctuaries, habitat or national parks and agro-ecosystems through onsite or on farm conservation. At present, much of the conservation is through use of the ranges of cultivars on-farm (Le, 1998). Natural genetic resources programmes are still developing principles, practices and policy issues but those involved with jujube production should also be involved. Many practicable opportunities exist for conservation of *Ziziphus in situ* in natural habitats or in the areas where it grows naturally, such as natural parks, biosphere reserves or gene sanctuaries. This can be achieved by protecting areas from human interference. A gene sanctuary is the best located within the centre of origin of the crop species concerned, preferably covering the micro-centre within the centre of origin. A gene sanctuary conserves the existing genetic diversity present in the population; it also allows for new gene combinations which appear with time. There is a need to establish gene sanctuaries in India for the conservation of *Z. mauritiana* and in China for *Z. jujuba*.

In vitro Conservation

In vitro genebanks can be very useful for clonally propagated material using slow growth in tissue culture and long-term cryopreservation of tissues and/or embryos. No such facilities exist for jujubes yet, however, a technique for *in vitro* storage of plants including *Z. jujuba* was developed in the Nikitsky Botanical Gardens, Ukraine (Mitrofanova *et al.*, 2002).

Current Scenario and Research Needs

The following aspects related to the preservation of biodiversity require immediate attention:

☆ A large number of selections or cultivars are available in both Indian and Chinese jujubes. There is a great deal of confusion in the naming. A modern assessment of the taxonomy of the genus *Ziziphus* is needed in terms of better characterisation using molecular fingerprinting of accessions in all existing germplasm collections including both wild and cultivated taxa. A genepool approach needs to be taken rather than one based solely on morphological types. We may come across those species which may have the required productivity, fruit quality or pest and disease resistance attributes which can be of great use in improvement programmes. Therefore, a long term, well-planned research understanding the broader taxonomy of *Ziziphus* and the patterns of genetic variation in the cultivated species and their wild forms are required.

☆ Very little is known about the diversity in the primary centres of variation of the cultivated species and about any specific patterns of diversity that have emerged in secondary centres. The objective has to be justifiable and meaningful germplasm collections to best serve improvement needs.

☆ The cultivated jujubes represent polyploid series and not enough is known about the distribution of ploidy levels. Polyploid series exist within species and there appears to be a range of self and cross-incompatibilities in *Ziziphus* species.

☆ As jujubes become cultivated more in small or medium sized plantations, there is a need to develop polyclonal cultivars to suit the diverse agroclimatic conditions, and tolerant to pests and diseases. In this respect, knowing more about the genetic relations between *Z. mauritiana* and *Z. jujuba*, and the possibilities of their crossing, would be helpful.

☆ Additionally, adequate networking is required across regions to develop cooperative goals and research. Network linkages can be developed with conservation organisations to maintain specific wild germplasm *in situ* when specific ecotypes or genotypes have been identified and characterised.

Acknowledgement

This manuscript is mainly based on the monograph 'Ber. International Centre for Underutilised Crops' (Azam-Ali *et al.*, 2006) whose copyright is with International Centre for Underutilised Crops, University of Southampton, Southampton, SO17 1BJ, UK [©2006 Southampton Centre for Underutilised Crops], and other sources including authors own experiences and research outputs.

References

Ackerman, W.L., 1961. Flowering, pollination, self sterility and seed development of Chinese jujube. *Proceedings of the American Society of Horticultural Science,* 77: 265–269.

Ahmad, M. and Malik, S.M., 1971. Canning of ber (*Zizyphus jujuba*). *Journal of Agricultural Research,* Pakistan, 9(3): 210–217.

Awasthi, O.P. and More, T.A., 2009. Genetic diversity and status of *ziziphus* in india. *Acta Hort.,* 840: 33–40.

Azam-Ali, S., Bonkoungou, E., Bowe, C., deKock, C., Godara, A. and Williams, J.T., 2006. *Ber.* International Centre for Underutilised Crops, Southampton, UK.

Babu, R.H. and Kumar, P.S., 1986. A note on the seed characters of certain cultivars of ber (*Ziziphus mauritiana*). *Progressive Horticulture,* 18(3–4): 297–298.

Bakhshi, J.C. and Singh, P., 1974. The ber: A good choice for semi-arid and marginal soils. *Indian Horticulture,* 19: 27–30.

Bal, J.S., 1992. Identification of ber (*Ziziphus mauritiana* L.) cultivars through vegetative and fruit characters. *Acta Horticulturae,* 317: 245–253.

Bal, J.S., Singh, M.P. and Sandhu, A.S., 1997. Evaluation of rootstocks for ber (*Ziziphus mauritiana* Lamk.) cv. Umran. *Journal of Research, Punjab Agricultural University,* 34(1): 60–63.

Barbeau, G., 1994. Inventory of tropical fruit trees in Central America and the West Indies. *Fruits,* Paris 49(5–6): 383–389, 469–474.

Bharad, S.G., Mali, D.V., Ingle, V.G. and Sagane, M.A., 2002. Effect of time and severity of pruning on physico-chemical characteristics and yield in ber varieties. *PKV–Research Journal,* 26 (1/2): 31–35.

Bisla, S.S., Chauhan, K.S. and Godara, N.P., 1980. Evaluation of late ripening germplasm of ber (*Ziziphus mauritiana* Lamk.) under semi-arid regions. *Haryana Journal of Horticultural Science,* 9(1/2): 12–16.

Bisla, S.S. and Daulta, B.S., 1986. Correlation and path analysis studies of some quality attributes, disease and yield in ber (*Zizyphus mauritiana* Lamk.). *Haryana Agricultural University Journal of Research,* 16(4) : 348–351.

Brandis, D., 1874. *The Forest Flora of Northwest and Central India.* Herbarium of the Royal Botanic Gardens, Kew.

Brandis, D., 1906. *Indian Trees.* Fifth Impression, 1971. Bishen Singh. Mahendra Pal Singh, Dehra Dun, India, p. 169–172.

Chadha, K.L., Gupta, M.R. and Bajwa, M.S., 1972. Performance of some grafted varieties of ber (*Ziziphus mauritiana* Lamk.) in Punjab. *Indian Journal of Horticulture,* 29(2): 137–150.

Chen-YouZhi, Lin-ZhenHai, Li-MingChun, Wang-ZhiMing and Xie-LinZhong, 2003. The performance of 10 jujube varieties in Zhaoyuan area, Shandong province. *China Fruits,* 6: 26–27.

Chiv Chu Ying, 1997. Effects of various lighting sources on the flowering and yielding date of Indian jujube. *Taichung District Agricultural Improvement Station,* Special Publication, 38: 251–256.

Chovatia, R.S., Patel, D.S., Patel, A.T. and Patel, G.V., 1992. Growth, yield and physico-chemical characters of certain varieties of ber (*Ziziphus mauritiana* Lamk.) under dryland conditions. *Gujarat Agriculture University Research Journal,* 17(2): 56–60.

Ciminata, P., 1996. The Chinese jujube, *Ziziphus jujuba. WANATCA Yearbook,* 20: 34–36.

Darbara Singh and Jindal, P.C., 1982. Studies on flowering and sex ratio in some ber (*Ziziphus mauritiana* Lamk.) cultivars. *Haryana Agricultural University Journal of Research,* 12(2): 292–294.

Darlington, C.D. and Wylie, A.P., 1955. *Chromosome Atlas of Flowering Plants.* Allen and Unwin, London.

Daulta, B.S. and Chauhan, K.S., 1982. Ber: A fruit with rich food value. *Indian Horticulture,* 27(3): 7–9.

De Candolle, A., 1886. *Origin of Cultivated Plants.* Bibliotheque Scientifique Internationale, Paris, 43.

Desai, U.T., Ranawade, D.B. and Wavhal, K.N., 1986. Floral biology of ber. *Journal of the Maharashtra Agricultural Universities,* 11(1): 76–78.

Dhaliwal, J.S. and Bal, J.S., 1998. Floral and pollen studies in ber (*Zizyphus mauritiana* Lamk.). *Journal of Research Punjab Agricultural University,* 35 (1–2): 36–40.

Dhingra, R.P., Singh, J.P. and Chitkara, S.D., 1973. Varietal variations in physico-chemical characters of ber (*Ziziphus mauritiana* Lamk.). *Haryana Journal of Horticultural Science,* 2(3/4): 61–65.

Diwakar-Hegde, Sharma, V.P. and Hegde, D., 1992. A numerical taxonomic study of the ber *Zizyphus* spp. *New Botanist,* 19(1–4): 21–26.

Diwakar-Hegde, Sharma, V.P. and Hedge, D., 1996. Palynological studies in ber (Z. spp.). *New Botanist,* 23(1–4): 145–151.

Ghosh, S.N. and Mathew, B., 2002. Performance of nine ber (*Z. mauritiana* Lamk.) cultivars on top working in the semi-arid region of West Bengal. *Journal of Applied Horticulture,* Lucknow, 4(1): 49–51.

Ghosh, D.K. and Mitra, S., 2004. Postharvest studies on some local genotypes of ber (*Z. mauritiana* Lamk.) grown in West Bengal. *Indian Journal of Horticulture,* 61(3): 211–214.

Godara, N.R., 1980. Studies on floral biology and compatibility behaviour in ber (*Ziziphus mauritiana* Lamk.). *Ph.D. Thesis,* Haryana Agricultural University, Hisar, India.

Godara, N.R., Chauhan, K.S. and Bisla, S.S., 1980. Evaluation of mid-season ripening ber (*Ziziphus mauritiana* Lamk.) germplasm. *Haryana Journal of Horticultural Science,* 9(3/4): 101–105.

Grice, A.C., 1998. Ecology in the management of Indian jujube (*Ziziphus mauritiana*). *Weed Science,* 46: 467–474.

Gupta, M., Mazumder, U.K., Vamsi, M.L.M., Sivakumar, T., and Kandar, C.C., 2004. Anti-steroidogenic activity of the two Indian medicinal plants in mice. *Journal of Ethnopharmacology,* 90: 29.

Gupta, M.R., 1977. Physico-chemical characters of some promising ber cultivars grown at Bahadurgarh (Patiala). *Punjab Horticultural Journal,* 17(3/4): 131–134.

Gupta, R.B., Sharma, Suneel, Sharma, J.R. and Sareen, P.K., 2003. Cytological studies in some varieties of genus *Ziziphus. National Journal of Plant Improvement,* 5(1): 19–21.

Hayes, W.B., 1945. *Fruit Growing in India.* Kitabistan, Allahabad, India.

Hooker, J.D., 1875. *Flora of British India, Part 1.* L. Reeve and Co., London.

Jawanda, J.S. and Bal, J.S., 1980. A comparative study on growth and development of ZG–2 and Kaithli cultivars of ber. *Punjab Horticultural Journal*, 20(1/2): 41–46.

Khera, A.P. and Singh, J.P., 1976. Chemical composition of some ber cultivars (*Ziziphus mauritiana* L.). *Haryana Journal of Horticultural Science*, 5(1/2): 21–24.

Khoshoo, T.N. and Singh, N., 1963. Cytology of northwest Indian trees. *Ziziphus jujuba* and *Z. rotundifolia*. *Silvae Genetica*, 12 Heft.: 141–180.

Kim, W.S. and Kim, Y.S., 1984. Changes in carbohydrates, proteins, RNA and hydrolytic enzyme activity of jujube seeds during germination [in Korean]. *Journal of the Korean Society for Horticultural Sciences*, 25(2): 109–115.

Kumar, P., Singh, H.K. and Saxena, R.P., 2003. Ber germplasm screening and management of black leaf spot disease under Eastern U.P. conditions. *Journal of Applied Horticulture*, 5(1): 43–44.

Kumar, S.S., Babu, R.S. and Reddy, Y.N., 1986. Duration of fruit maturity seasons and yield of certain cultivars of ber (*Ziziphus mauritiana* Lamk.) at Hyderabad. *Journal of Research APAU*, 14(1): 88–89.

Kundi, A.H.K., Wazir, F.K., Abdul, G. and Wazir, Z.D.K., 1989. Physico-chemical characteristics and organoleptic evaluation of different ber (*Ziziphus jujuba* Mill.) cultivars. *Sarhad Journal of Agriculture*, 5(2): 149–155.

Lal, G., Jat, R.G., Dhaka, R.S., Sen, N.L. and Agarwal, V.K., 2003. Evaluation of ber germplasm in semi-arid regions of Rajasthan. *Journal of Ecophysiology*, 6(3/4): 145–148.

Le Thi Thu Hong, 1998. Some aspects of fruit production and genetic conservation in Vietnam. In: *Report of the Second Meeting of MESFIN on Plant Genetic Resources*, INIA, Portugal, p. 287–299.

Liu, M.J. and Cheng, C.Y., 1995. A taxonomic study of the genus *Ziziphus*. *Acta Horticulturae*, 390: 161–165.

Liu, M.J. and Wang, Y.H., 1991. Promising varieties and forms of Chinese wild jujube. *China Fruits*, 4: 23–35.

Locke, L.F., 1948. The Chinese jujube: A promising fruit tree for the southwest. *Bulletin of the Oklahoma Agricultural Experimental Station*, B. 319: 78–81.

Lomakina, M.I., 1976. Chemical composition of the fruit of subtropical and pome fruits in south-west. *Turdy po Prikladnoi Botanike Genetike-i-Selektsii*, 57(1): 150–154.

Malik, S.K., Chaudhury, R., Dhariwal, O.P. and Bhandari, D.C., 2010. *Genetic Resources of Tropical Underutilized Fruits in India.* NBPGR, New Delhi, p.168.

Martin, F.W. Campbell, C.W. and Ruberte, R.M., 1987. *Perennial Edible Fruits of the Tropics: An Inventory.* Agriculture Handbook, USDA, 642: 247.

Mehetre, S.S. and Dahat, D.V., 2000. Cytogenetics of some important fruit crops: A review. *Maharashtra Agricultural Universities*, 25: 139–168.

Mehrotra, N.K. and Gupta, M.R., 1985. Pollination and fruit set studies in ber (*Ziziphus mauritiana* Lamk.). *Journal of Research, Punjab Agricultural University (India)*, 22(4): 671–674.

Ming, W. and Sun, Y., 1986. Fruit trees and vegetables for arid and semi aridareas in northwest China. *Journal of the Arid Environments*, 11: 3–16.

Mitrofanova, I.V., Movcha, O.P., Shishkin, V.A. and Mitrofanov, V.I., 2002. Gene-pool collection in vitro in Nikitsky Botanical Gardens – National Scientific Center. Nikitskogo Botanicheskogo Sada, 85: 30–33.

Moriguchi, T., Teramoto, S. and Sanada, T., 1994. Conservation system of fruit tree genetic resources and recently released cultivars from fruit Tree Research Station in Japan. *Fruit Varieties Journal*, 48(2): 73–80.

Munier, P., 1973. Le jujubier et sa culture [in French]. *Fruits*, 28(5): 377–388.

Neeraja, G., Reddy, S.A. and Babu, R.S.H., 1995. Fruit set, fruit drop and fruiting behaviour in certain ber (*Ziziphus mauritiana* Lamk.) cultivars. *Journal of Research, Andhra Pradesh Agricultural University*, 23(3/4): 17–21.

Nehra, N.S., Chitkara, S.D. and Singh, K., 1984. Studies of morphological characters of some wild forms and cultivated varieties of ber. *Punjab Horticultural Journal*, 24(1/4): 49–59.

Nehra, N.S., Sareen, P.K. and Chitkara, S.D., 1983. Cytological studies in genus *Ziziphus. Cytologia*, 48: 103–107.

Nerd, A., Aronson, J.A. and Mizrahi, Y., 1990. Introduction and domestication of rare and wild fruit and nut trees for desert areas: Advances in new crops. In: *Proceedings of the First National Symposium 'New Crops: Research, Development, Economics'*, Indianapolis, Indiana, USA, 23–26 October 1988. Timber Press, Portland, Oregon, USA, p. 355–363.

Panwar, J.S., 1980. Pattern of fruit drop in ber (*Ziziphus mauritiana* Lamk.) in the arid conditions of Haryana. *Haryana Agricultural University Journal of Research*, 10(1): 57–59.

Pareek, O.P., 2001. *Fruits for the Future 2: Ber*. International Centre for Underutilised Crops, University of Southampton, Southampton, UK.

Pareek, O.P. and Nath, V., 1996. Ber. In: *Coordinated Fruit Research in Indian Arid Zone – A Two Decades Profile (1976–1995)*. National Research Centre for Arid Horticulture, Bikaner, India, p. 9–30.

Pareek, O.P. and Sharma, S., 1993. Genetic resources of under-exploited fruits. In: *Advances in Horticulture, Vol. 1: Fruit Crops, Part 1*, (Eds.) K.L. Chadha and O.P. Pareek. Malhotra Publishing House, New Delhi, India, p. 189–225.

Pradeep, T. and Jambhale, N.D., 2000. Ploidy level variations for stomata, chloroplast number, pollen size and sterility in ber (*Zizyphus mauritiana* Lamk.). *Indian Journal of Genetics and Plant Breeding* 60(4): 519–525.

Qu, Z.Z. and Wang, Y.H., 1993. *Chinese Fruit Trees Record Chinese Jujube*. The Forestry Publishing House of China, Beijing.

Reddy, B.G.M., Patil, D.R., Kulkarni, N.G. and Patil, S.G., 1998. Economic performance of selected ber varieties. *Karnataka Journal of Agricultural Sciences*, 11(2): 538–539.

Ridley, H.N., 1922. *The Flora of the Malay Peninsula, Vol. 1: Polypetalae*. L. Reeve and Co., London, UK.

Rixford, G.P., 1917. *Ziziphus* spp. In: *The Standard Cyclopaedia of Horticulture*, (Ed.) L.H. Bailey. The MacMillan Co. NY., USA.

Saha, N.N., 1997. *Conservation and Utilization of Fruit Plant Genetic Resources: Bangladesh Perspective*, (Eds.) M.G. Hossain, R.K. Aurora and P.N. Mathur. BARC–IPGRI, Dhaka, Bangladesh.

Saran, P.L., 2005. Studies on genetic divergence in ber (*Ziziphus mauritiana* Lamk.) germplasm. *Ph.D. Thesis*, CCS Haryana Agricultural University, Hisar, India.

Saran, P.L., Godara, A.K.R. and Dalal, P., 2007. Biodiversity among Indian jujue (*Ziziphus mauritiana* Lamb.) genotype for powdery mildew and other traits. *Not. Bot. Hort. Agrobot.* Cluj, 35(2): 15–21.

Sharma, S.K., 2003. Optimizing the productivity of natural Sehima pasture by introducing fruit tree and legume components. *Range Mgmt. and Agroforestry*, 24: 38–41.

Sharma, S.K., 2004. Hortipastoral based land use systems for enhancing productivity of degraded lands under rainfed and partially irrigated conditions. *Ugandan Journal of Agricultural Sciences*, 9: 320–325.

Sharma, S.K., 2009. Development of fruit tree based agroforestry systems for degraded lands: A review. *Range Mgmt. and Agroforestry*, 30: 98–103.

Sharma, V.P., Raja, P.V. and Kore, V.N., 1990. Flowering, fruit set and fruit drop in some ber (*Ziziphus mauritiana* Lamk.) varieties. *Annals of Agriculture Research Institute*, 11(1): 14–20. New Delhi, India.

Singh, D. and Jindal, P.C., 1980. Physico-chemical characters of some promising ber cultivars grown at Gurgaon (Haryana). *Haryana Journal of Horticultural Science*, 9(3/4): 114–117.

Singh, J.P. and Singh, I.S., 1973. Some promising varieties of ber. *Indian Horticulture*, 18(2): 3–4, 28.

Singh, K.K., Chadha, K.L. and Gupta, M.R., 1973. *Ber Cultivation in Punjab*. Punjab Agricultural University, Ludhiana.

Singh, R. and Tomar, N.S., 1988. Performance of some jujube (*Ziziphus mauritiana*) cultivars in semi-arid zone of Punjab. *Indian Journal of Agricultural Science*, 58(5): 382–383.

Singh, R.R., Jain, R.K. and Chauhan, K.S., 1983. Flowering and fruiting behaviour of ber (*Ziziphus mauritiana* Lamk.) under Gurgaon conditions. *Haryana Agricultural University Journal of Research*, 13(1): 112–114.

Sin'ko, L.T., 1974. Economic and technological evaluation of *Zizyphus* fruits. *Byulleten'–Gosudarstvennogo-Nikitskogo-Botanicheskogo-Sada*, 2: 28–31.

Sin'ko, L.T., 1977. Agrobiological characteristics of Chinese jujube in the Crimea [in Russian]. *Trudy Gosudarstvennogo Nikitskogo Botanicheskogo Sada*, 73: 98–125.

Sudhir-Kumar, Sharma, V.P. and Kumar, S., 1999. Biochemical study of ber cultivars in relation to protein. *Annals of Agricultural Research*, 20(3): 266–269.

Tagiev, T.M., 1992. Promising areas for cultivating jujube in Azerbaijan [in Russian]. *Sadovodstvo i Vinogradstvo*, 1: 19–20.

Teaotia, S.S. and Chauhan, R.S., 1964. Flowering, pollination, fruit set and fruit drop studies in ber (*Ziziphus mauritiana* Lamk.). II. Pollination, fruit set, fruit development and fruit drop. *Indian Journal of Horticulture*, 2: 40–45.

Teaotia, S.S., Dube, P.S., Awasthi, R.K. and Upadhyay, N.P., 1974. Studies on physico-chemical characteristics of some important ber varieties (*Ziziphus mauritiana* Lamk.). *Progressive Horticulture*, 5(4): 81–88.

Tutin, F.G., 1968. *Ziziphus*. In: *Flora Europea, Vol 2*, (Eds.) T.G. Tutin, V.H. Heywood, N.A. Borges, D.M. Moore, D.H. Valentino, S.M. Walter and D.A. Webb. Cambridge, p. 243.

Vashishtha, B.B. and Pareek, O.P., 1979. Flower morphology, fruit set and fruit drop in some ber (*Ziziphus mauritiana* Lamk.) cultivars. *Annals of the Arid Zone*, 18(3): 165–169.

Vavilov, N.I., 1951. *The Origin, Variation, Immunity and Breeding of Cultivated Plants.* Chronica Botanica, NY, USA.

Verma, M.K., Sharma, V.P., Saxena, S.K. and Jindal, P.C., 2001. Stomatal density as influenced by stionic combinations of ber (*Ziziphus mauritiana* Lamb*). Indian Journal of Horticulture*, 58: 350–353.

von Maydell, H-J., 1986. *Trees and Shrubs of the Sahel: Their Characteristics and Uses.* GT2, Eschborn, Germany.

von Sengbusch, V. and Dippolo, M.F., 1980. *Das Entwicklungs potentia afrikanischer Heilpflanzen – Kamerun, Tschod, Gagun.* Möckmuhl: IFB, Germany.

Wang, M. and Sun, Y.W., 1986. Fruit trees and vegetables for arid and semiarid areas in northern China. *Journal of Arid Environments*, 11(1): 3–16.

Wang-Yun and Wang, Y., 1994. A preliminary study on the geographical provenance of *Zizyphus mauritiana. Forest Research,* 7(3): 334–335.

Yadav, L.S., 1991. Top-working in wild jujube. *Indian Horticulture*, 35(4): 7.

Yakobashvili, V.K., 1973. Some varieties and promising seedlings of jujube. *Subtrop.–kul'tury,* 3: 78–81.

Yamdagni, R., Gupta, A.K. and Ahlawat, V.P., 1985. Performance of different cultivars of ber (*Ziziphus mauritiana* L.) under rainfed conditions. *Annals of the Arid Zone*, 24(2): 175–177.

2013, Biodiversity in Horticultural Crops Vol. 4   *Pages 189–201*
Editor: Professor K.V. Peter
Published by: DAYA PUBLISHING HOUSE, NEW DELHI

Chapter 10

Papaya

Sunil Kumar Sharma

Indian Agricultural Research Institute,
Regional Station, Agricultural College Estate, Shivajinagar, Pune – 411 005, M.S.
E-mail: sunilksharma1959@yahoo.co.in

Papaya (*Carica papaya* L.) is primarily grown in tropics for fresh fruit and the latex, papain, containing cysteine endopeptidases, which have many industrial uses. Papaya fruit is also known as 'paw paw' or 'papaw' in some countries. While in Australia, yellow fleshed cultivars of *C. papaya* are known as 'paw paw', and the red and pink fleshed cultivars are known as 'papaya'. Both these terms refer to the same plant species. However, another plant species, *Asiminia triloba* (family Annonaceae) is also called 'paw paw' in North America. In India, the plant is commonly known as 'papaya'. It is among major fruit crops in India which contributed about 36 per cent to world production in 2008-09 from about one forth of total global area (98 thousand hectare) under cultivation. It is mainly cultivated in Andhra Pradesh, Gujarat, Karnataka, West Bengal and Chhattisgarh. These states covered more than 63 per cent of total area and contributed about 87 per cent of country's total papaya production. These states have average productivity of 53.72 tones/ha as against the national productivity of 37.11 tones/ha (National Horticulture Board, 2010). Although *C. papaya* has limited biodiversity in wild, but being an open pollinated crop, it is rich in biodiversity of cultivated varieties. Other genera of the family *Caricaceae* have rich biodiversity in wild in their places of origin. Since some species in these genera show resistance against viral diseases in *C. papaya,* and are being used in resistance breeding, the existing biodiversity in papaya becomes important.

Origin and Distribution

The center of origin of the family *Caricaceae* is Tropical Central America, except for the genus *Cylicomorpha* which has centre of origin in Africa. *C. parviflora* Urban is believed to be originated in Eastern Africa (Kenia, Malawi, Tanzania) and *C. solmsii* (Urban) Urban in Western Africa (Nigeria, Congo, Cameroun, Central African Republic). Major countries where origin of various species of other five genera in the family are attributed are Argentina, Bolivia, Brazil, Central Chile, Colombia, Ecuador, El Salvador, French Guyana, Guatemala, Mexico, Nicaragua, Panama, Paraguay, Peru, Uruguay and Venezuela (Table 10.1). *C. papaya* is likely originated in Tropical Central America (Badillo 1993). Several species of *Vasconcellea*, which is the most important genus of the family, can be found in wild in a broad ecological range. *V. parviflora* is found in the dry coastal tropical lowlands, *V. weberbaueri* in the humid subtropical forests, and *V. chilensis* in the temperate regions. The centre of diversity of the genus *Vasconcellea* is located in the Andean highlands from Colombia to Peru with a hot spot in Southern Ecuador and Northern Peru. Papaya has interesting story of its distribution. It might have spread to the islands of the pacific at an earlier time as it is indicated by the presence of two Hawaiian names which were not ordinarily given to plants introduced by the Europeans. Its introduction to Hawaiian Islands is not known. Some authors believe that it might have been brought in between 1800 and 1823 by Don Moris while others believe that it came to the islands via Asia and the other South Sea islands before the Europeans reached there. The first papaya introduced to Hawaiian Islands was the large fruited type. Introduction of 'Solo' papaya from Barbados and Jamaica on 7th October 1911 by Gerrritt P. Wilder completely transformed papaya cultivation in Hawaii. Within a decade or so, the 'Solo' papaya completely replaced the large fruited varieties for commercial cultivation (Ram 2005). Spanish and Portuguese sailors took it to Panama and then to the Dominican Republic in the sixteenth century (Singh, 1990), where its cultivation spread to warmer elevations throughout South and Central America, southern Mexico, the West Indies and Bahamas, and then to Bermuda. It was disseminated into Asian tropics during the 1600s through seeds taken to the Malay Peninsula, India and Philippines. The Dutch traveler, Linschoten in 1576 described fruit brought from the Philippines to Malaya and then to India. The introduction of papaya in Asia from the outside world is supported by the various literature where its Burmese name, *thimbawthi*, means fruit brought by sea going vessels (Watt, 1989). Papaya seeds were taken from India to Naples, Italy in 1626 and to China as an Indian plant in 1565. Documents show wide distribution of papaya in the Pacific Islands by the 1800s (Nakasone 1975). Presently, papaya is a popular crop in tropical regions of the Pacific Islands and has become naturalized in many areas.

Table 10.1: Centre of Origin of Various Genera in the Family *Caricaceae*

Sl.No.	Genus	Origin
1.	*Carica*	Tropical Central America
2.	*Cylicomorpha*	Eastern Africa (Kenia, Malawi, Tanzania) and Western Africa (Nigeria, Congo, Cameroun, Central African Republic)
3.	*Horovitzia*	Southern Mexico (Oaxaca)
4.	*Jacaratia*	Colombia, Bolivia, Paraguay, Argentina, Brazil, Ecuador, Peru, Panama, Mexico, El Salvador and Nicaragua
5.	Jarilla	*Mexico/Guatemala*
6.	*Vasconcellea*	Southern Mexico to northern part of South-America, Central Chile, Brazil, Panama, Venezuela, French Guyana
		Argentina, Bolivia, Paraguay, Uruguay, Colombia
		Ecuador and Peru

Biodiversity

Biodiversity is the variation of life forms representing totality of genes, species and ecosystems of a region. It is a representative of species diversity and richness that is often used as a measure of the health of a biological system or genus. Ecosystems function efficiently in rich biodiversity. Some measure of biodiversity is essential for sustaining life on earth. Therefore, collection, conservation, documentation, evaluation and utilization of papaya biodiversity are essentially required for future advancement of papaya cultivation in terms of developing cultivars suitable for growing under different agro-ecological regions. There are the following potential uses of biodiversity occurring in the family *Caricaceae:*

☆ Domestication of new plant species producing tasty and high quality fruits.

☆ Development of new variety or species by exploiting natural hybridization.

☆ Using the existing biodiversity in breeding programmes for papaya improvement for cold adaptability, and for resistance against viral diseases by using genes from highland papayas. *Vasconcellea cauliflora, V. cundinamarcensis* (*pubescens*), *V. quercifolia, V. stipulata, V. goudotiana* and *V. parviflora* have been tried for *Papaya Ringspot Virus* resistance breeding in papaya.

☆ *V. cundinamarcensis* and *V.* x *heilbornii* cv. 'Babaco' have underutilized potential for their use as a source of papain.

The following is the account of biodiversity in the family *Caricaceae*.

Biodiversity Hot Spots

Ecuador is one of the biodiversity hot spots for the genus *Vasconcellea,* for 15 out of 21 species are found there (Badillo, 1983). Southern Ecuador, covering 39,987 km² or 15 per cent of the total land area of the country, comprises the following nine native *Vasconcellea species: V. candicans* (A. Gray) A. DC., *V. cundinamarcensis* (Solms-Laub.) V. Badillo, *V. microcarpa* (Jacq.) A. DC, *V. monoica* (Desf.) A. DC., *V. parviflora* A. DC., *V. stipulata* (*V.* Badillo) V. Badillo, *V. weberbaueri* (Harms.) V. Badillo, *V.* x *heilbornii* (*V.* Badillo) V. Badillo, and *V. palandensis* (V. Badillo *et al.*) V. Badillo. (Van den Eynden *et al.,* 1999). Therefore, Ecuador is generally considered as an important centre for *Vasconcellea* research (Soria, 1999, National Research Council, 1989).

Morphological Diversity

Plant

Papaya is a fast growing tree-like herbaceous dicotyledonous plant that grows up to 10 m in fertile, well drained soil with sufficient moisture. Plant is perennial that may produce fruit for more than 20 years, but economic life is not more than three years. Stem is simple thick spongy with no lateral branches but sometime dividing into several erect stems bearing heads of leaves. It is marked with scars of falling leaves. Most of *Vasconcellea species* plants are semi-lignose shrub to tree-like species with an erect habit, except for *V. horovitziana*, a climber in growth habit. The stem is medullose in the genus *Vasconcellea*, contrary to the hollow stem of the *Carica* genus. In *V. stipulate,* stipules are converted into spines.

Leaves

Leaves are borne in terminal cluster and arranged in alternate whorls. Leaves are 60 cm or more in length, long, round, deeply palmately dissected into 5-7 lobes that are also lobed. They are multicostate with reticulated venation, and stipulated. Petiole is long and hollow. In the genus *Vasconcellea,* leaves are generally entire or lobed. Heart-shaped and compound palm-shaped leaves can also be found in *V. palandensis* and *V. candicans* respectively.

Sexes

Cultivated (*C. papaya*) plants are polygamous, with staminate, pistillate, and hermaphrodite flowers. Wild plants, mostly *Vasconcellea spp.,* are frequently dioecious, except *V. monoica and V. cundinamarcensis* which are monoecious. Higgins (1916) reported variability in tree forms and sexes. Sex forms are based on floral composition that may consist of various admixtures of normal and teratological forms of flowers. The pistillate tree is extremely stable, always producing pistillate flowers. Conversely, the staminate and hermaphrodite trees can be sensitive to fluctuation of environmental factors, and can go through seasonal sex-reversals (Hofmeyr, 1939). Dioecious cultivars are more stable and vigorous, and grow better in the subtropical areas of Florida, Australia and South Africa when compared with hermaphrodite cultivars, which are not adapted to cool winters.

Inflorescence

Since papaya is a polygamous species, many forms of inflorescence are reported. Flowers are typically pentamerous with a small calyx and tubular corolla. Male inflorescences are always multi-flowered and usually long-pedunculated, and paniculate having many-layered cymose inflorescences. The corolla of the flower forms a slender tube two-third of its length and terminates in five free petals. The flower is unisexual, having functional stamens and lacking a functional pistil. They show diplostemonous stamens with filaments free or fused above the corolla mouth. Female inflorescences are single-flowered or pauciflorous with short peduncle. The flowers have large functional pistils but entirely lack stamens. The petals are free from one another. Female flowers show an incompletely 5-locular ovary with 5 entire or branched stigmata. Hermaphrodite flowers are bisexual with pistil and stamens on a relatively short inflorescence. The petals are fused for one-half to three-fourth of their lengths, forming a rigid tube (Giacometti, 1987). Oschae *et al.* (1975) classified flowers and its parts into following five types:

Type I: Typical pistillate

Type II: Pentandria like I above except having 5 stamens attached to the ovary

Type III: Intermediate that is mostly unstable in nature

Type IV: Hermaphrodite or elongate which produces long fruits

Type V: Staminate flowers hanging on the long peduncles

According to above flower types, plants are also categorized into four groups:

Group A: Pistillate or female plants producing Type I flowers

Group B: Hermaphrodite or bisexual may bear flowers of Type II, III, IV or V. Mostly, it has Type II in summer, Type IV in winter, Type III during transition periods, and rarely Type V

Group C: Summer sterile hermaphrodite produces Type IV in winter and pseudo-type IV in summer (an aberrant of group B)

Group D: Staminate or male producing Type V flowers, which are usually born, but occasionally Type IV flowers appear.

Fruits

Fruits vary greatly in shape, size and colour among the various species (Figure 10.1). They are baccate with crested seeds often embedded in gelatine-like sarcotesta. Fruits from pistillate trees are spherical whereas those from hermaphrodite trees are pyriform, oval or elongate.

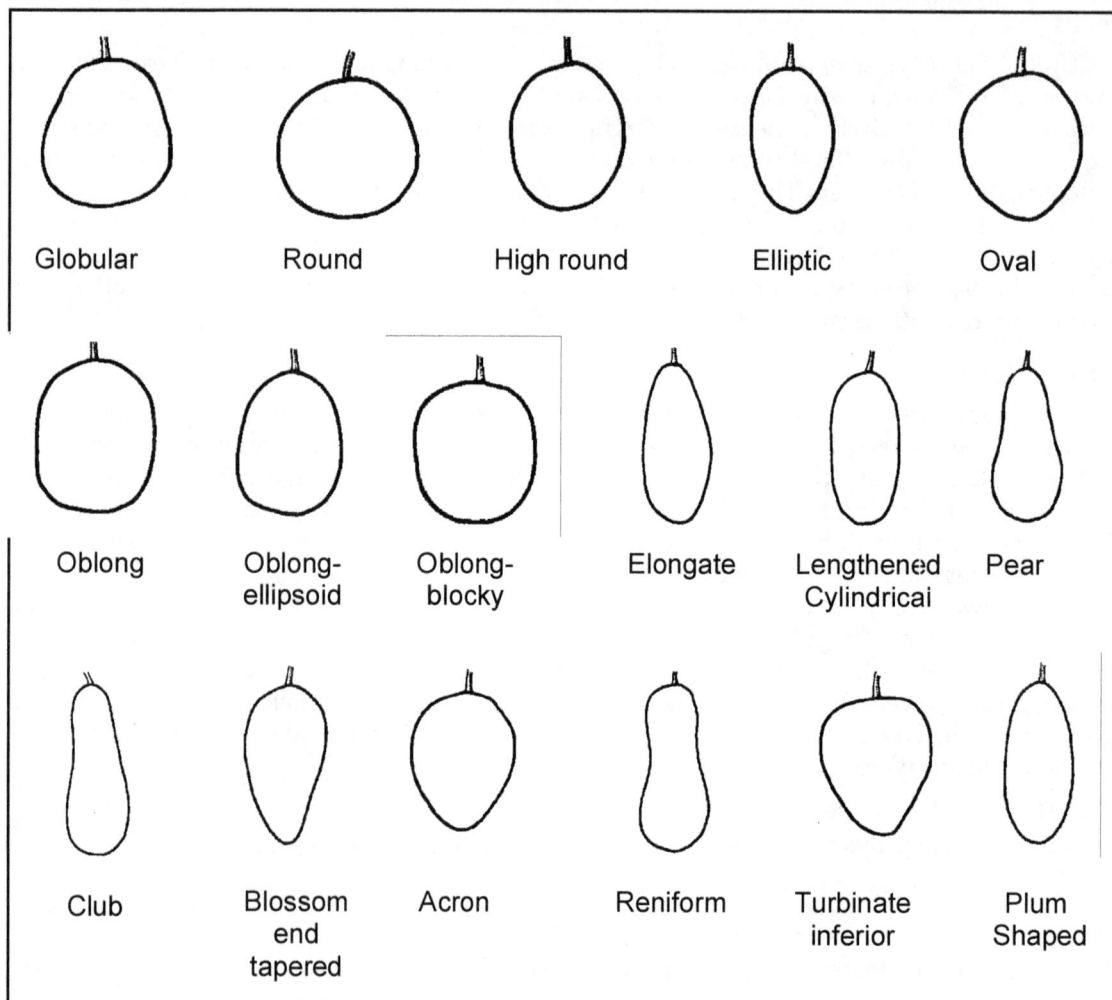

Figure 10.1: Fruit Shapes

Genetic Diversity

The papaya, a diploid plant ($2n = 2x = 18$), belongs to family *Caricaceae*. According to the earlier classification by Badillo, (1967), the family *Caricaceae* had four genera, *Carica* with about 21 species, *Jacartia* with six species, *Cylocomorpha* with two species and *Jarilla* with one species. There was a major change in the classification of genera and species in the family *Caricaceae* in 2000 (Badillo *et al.,* 2000). All species of genus *Carica* (except the species *papaya*) were rehabilitated in other genera of the family. Now, the genus *Carica* is left with only one species, *papaya*. The family has six genera comprising of 35 species. *Vasconcellea* is the largest genus in the family with 21 species. *Genus Jacaratia* has seven species followed by *Jarilla* that has three species. Genus *Cylicomorpha* has two species, while genera *Carica* and *Horovitzia* have one species each (Table 10.2). Till 2000, there was a consensus about 20 species in the genus *Vasconcellea* (Badillo, 1993). With the discovery of another species, *V. palandensis* (V. Badillo *et al.*) V. Badillo, in Southern Ecuador, total number of species in the genus Vasconcellea became 21

(Badillo *et al.*, 2000). Due to easy natural hybridization between species of the genus *Vasconcellea*, proper species identification is difficult, *e. g.*, existence of the hybrid *V. x heilbornii* with its different varieties (Kempler and Kabaluk, 1996, Drew *et al.*, 1997). It is believed that this species is a natural hybrid between *V. pubescens* and *V. stipulata*. *V. microcarpa* (Jacq.) A. DC. has four subspecies, namely, subsp. *Baccata*, subsp. *Microcarpa*, subsp. *Pilifera*, and subsp. *Heterophylla* (Table 10.3).

Table 10.2: Distribution of Genus and Species in the Family Caricaceae

Sl.No.	Genus	Number of Species
1.	Carica	1
2.	Cylicomorpha	2
3.	Horovitzia	1
4.	Jacaratia	7
5.	Jarilla	3
6.	Vasconcellea	21

Table 10.3: Nature of Species in Various Genera of the Family Caricaceae

Sl.No.	Genus	Number of Species
1.	*Carica*	*papaya* L. - Edible fruits
2.	*Cylicomorpha* Wild	*parviflora* Urban *and solmsii* (Urban) Urban
3.	*Horovitzia*	*cnidoscoloides* (Lorrence et Torres) V. Badillo - Wild
4.	*Jacaratia* Wild	*chocoensis* A. Gentry et Forero, *corumbensis* O. Kuntze, *digitate (Poeppig et Endl.) Solmns-Laub.* - Edible fruits, *dolichaula* (J.D. Smith) Woodson, *heptaphylla* (Vellozo) A. DC. -Wild cultivated, *mexicana* A. DC., *spinosa* (Aublet) A. DC. - Edible fruits
5.	Jarilla Wild	*caudata (Brandegee) Standley, chocola* Standley, and *heterophylla (Cerv.) Rusby*
6.	*Vasconcellea* Wild	*candicans* (A. Gray) A. DC. - Edible fruits, *cauliflora* (Jacq.) A.DC. - Edible fruits, *chilensis* (Planch. ex A.DC.) A.DC. - Fodder, *crassipetala* (V. Badillo) V. *Badillo* - Edible fruits, *cundinamarcensis (Solms-Laub.)* V. Badilo, - Wild cultivated, Edible fruits, *glandulosa* A. DC., *goudotiana* Triana et Planch. - Wild cultivated, Edible fruits, *horovitziana* (V. Badillo) V. Badillo, *longiflora* (V. Badillo) V. Badillo, *microcarpa* (Jacq.) A. DC. [4 subsp. subsp. Baccata, subsp. Microcarpa, subsp. Pilifera, subsp. Heterophylla] - Edible fruits, monoica (Desf.) A. DC. - Edible fruits and leaves, *omnilingua* (V. Badillo) V. Badillo, *palandensis* (V. Badillo et al.) V. Badillo - Edible fruits, *parviflora* A. DC. - Edible fruits, *pulchra* (V. Badillo) V. Badillo, *quercifolia* (St.-Hil.) A.DC. - Edible fruits, *sphaerocarpa* (García-Barr. et Hern.) V. Badillo - Edible fruits, *sprucei* (V. Badillo) V. Badillo, stipulata (V. Badillo) V. Badillo, - Edible fruits, *weberbaueri* (Harms) V. Badillo, V. X *heilbornii* (V. Badillo) V. Badillo [Dif. Varieties - var. fructifragrans var. chrysopetala cv 'babacó'] - Wild cultivated and Edible fruits

Variation in Domestication

While *Carica papaya* produces edible fruits, other species in the remaining five genera are considered wild, except for a few species in the genera *Jacaratia* and *Vasconcellea*. *Jacaratia corumbensis* O. Kuntze, *J. digitate* (Poeppig et Endl.) Solmns-Laub. and *J. spinosa* (Aublet) A. DC. produce edible fruits. *J. heptaphylla* (Vellozo) A. DC. is a wild cultivated species. Out of 21 species in the genus *Vasconcellea*, the following seven are wild: *V. glandulosa* A. DC., *V. horovitziana* (V. Badillo) V. Badillo, *V. longiflora* (V. Badillo) V. Badillo, *V. omnilingua* (V. Badillo) V. Badillo, *V. pulchra* (V. Badillo) V. Badillo, *V. sprucei* (V. Badillo) V. Badillo, *V. weberbaueri* (Harms) V. Badillo. Two species *V. cundinamarcensis* (*Solms-Laub.*) V. Badilo and *V. glandulosa* A. DC., *goudotiana* Triana et Planch, and the natural hybrid *Vasconcellea* X *heilbornii* (V. Badillo) V. Badillo are wild cultivated that produce edible fruits. Leaves of *V. chilensis* (Planch. ex A. DC.) A. DC. are used as animal fodder. The species *V. monoica* (Desf.) A. DC. produces edible fruits and leaves. The remaining nine species of genus *Vasconcellea* are also wild but produce edible fruits.

Besides their use as edible fruits, these species mostly come naturally in the wild. However, *V. cundinamarcensis* and *V.* x *heilbornii* cv. 'babaco' are currently cultivated on a commercial scale.

Varietal Biodiversity

Papaya varieties cultivated in India show great biodiversity based on their sex expression (dioecious – male and female flowers on separate plants; or gynodioecious – female and hermaphrodite plants only, no male plant), height (tall, medium or dwarf), objective of cultivation (fresh fruit or papain), flesh colour (yellow, orange, pink or red), etc (Tables 10.4–10.6). Papaya cultivars grown in India are originated from five horticultural forms (Bangalore, Honey Dew, Ranchi, Ceylon and Washington). Considerable homogeneity is noted within the members of one horticultural form, while slight structural differences are observed in the chromosomes of different forms. The sequence of chromosomes in different forms varies when arranged from the longest to the shortest types.

Table 10.4: List of Gynodioecious Papaya Cultivars

Main Gynodioecious Papaya Cultivars in India
Red Lady
Surya
Coorg Honey Dew
CO-7
CO-3
Pusa Delicious
Pusa Majesty

Table 10.5: Classification of Papaya Cultivars Based on Flesh Colour

Yellow	*Yellowish Red/Orange*	*Pink/Red*
CO-2	Washington	Red Lady
Pusa Majesty	CO-1	Pusa Nanha
Pusa Dwarf	Pusa Giant	CO-3
CO-4	Pusa Delicious	CO-7
CO-5	Coorg Honey Dew	Surya
CO-6		

Old Cultivars

These varieties were cultivated in India before systematic breeding work started. Now, their commercial cultivation are rare and confined to isolated areas only. Most of the modern papaya cultivars are derived from them.

Washington

It was the most popular dioecious, table purpose variety of Bombay. Tree is fairly vigorous and tall. Stem is having purple colour on nodes. Petiole is dark purple that grows darker towards lamina. Flower colour is deep yellow. Fruit size is medium to large (1-1.5 kg). Fruit shape is ovate to oblong having distinct purple colour rings at its

Table 10.6: Classification of Papaya Cultivars Based on Main Product

Fruit	Red Lady
	Surya
	CO-7
	CO-3
	Pusa Delicious
	Coorg Honey Dew
Papain	CO-5
	CO-6
Both fruit and papain	CO-2
	Pusa Majesty

top connected with the fruit stalk. Pulp colour is orange, and very sweet, has agreeable flavour of fine consistency and better keeping quality.

Honey Dew (*Madu Bindu*)

Honey Dew was a table variety of India. It is also used for papain production. Plants are of medium height, and bear fruits low on stem. Fruits are elongated. Pulp is extra fine, sweet, and has agreeable flavour. Keeping quality is medium.

Ranchi

It was a high yielding variety from Bihar (now Jharkhand) and popular in south India. The good size fruits are oblong with dark yellow pulp, sweet taste and good flavor. Many popular Pusa cultivars are derived from this cultivar.

Cyelon

This variety is one of the five horticultural forms of papaya in India.

Bangalore

This variety is also one of the five horticultural forms of papaya in India.

Pusa Bihar

IARI Regional Station, Pusa (Bihar) started working on papaya in mid sixties and released a number of good varieties.

Pusa Giant (Pusa 1-145V)

It is a dioecious selection from cultivar Ranchi. The plant size is vigorous, over 2 m tall. Thick trunk is tolerant to strong winds. It starts fruiting at a height of about 90 cm within nine months after transplanting. Fruits are large sized (2 to 3 Kg), oblong with thick orange flesh. Sweetness is low (7 to 8.5° Brix). Fruits are used for canning.

Pusa Delicious (Pusa 1-15)

It is a table purpose, gynodioecious selection from cultivar Ranchi. Plant size is medium and high yielding. It starts fruiting at a height of above 1.2 m. It starts yielding fruits eight months after transplanting. Fruit size is medium (1 – 2 Kg), round to oblong in shape, flesh thickness 4 cm and TSS ranging from 10-13° Brix. Fruits are excellent in taste with good flavour and deep orange colour.

Pusa Majesty (Pusa 22-3)

The gynodioecious, Pusa Majesty, was developed by sibmating of cultivar Ranchi. It is suitable for papain production. Plant is of medium stature (about 2 m). It is claimed to be tolerant to viral diseases and root knot nematodes. Fruiting starts at about 50 cm within 250 days after transplanting. Fruit size is medium to big (1 to 2.0 Kg). Fruit shape is oblong, yellow colour flesh is firm and thick (3.5 cm). Fruit is moderately sweet (9° Brix) with good keeping quality.

Pusa Nanha (Mutant Dwarf)

It is a dioecious mutant of papaya strain Pusa 1-15 with gamma radiation. It is a dwarf (height 1 m) and precocious variety. It is resistant to lodging. Fruiting starts at a height of 40 cm within 240 days of planting. Fruit size is small to medium, oval, and 3.5 cm thick flesh is of blood red to orange colour. Sweetness is low, TSS (6.5 to 8° Brix).

Pusa Dwarf (Pusa 1-45)

This variety was also developed by selection and sibmating of Ranchi. The draught hardy plant is dwarf, bearing fruit at 25-30 cm above the ground. It is suitable for high-density plantings. Fruits are of medium size (0.5 to 2.0 Kg), oval in shape, pulp is yellow-colour, TSS is low (6.5 to 8° Brix).

Tamil Nadu Agricultural University, Coimbatore

Tamil Nadu Agricultural University, Coimbatore started papaya breeding work in late sixties. They released a series of papaya cultivars in the name, CO, for Coimbatore.

CO-1

It is a dioecious, table variety developed by sibmating and selection in cultivar Ranchi. Plant is dwarf, bearing first fruit at 60-70 cm height. Fruit size is medium (1.5 Kg), shape is spherical with flattened base, nipple slightly raised and ridges present at the apex. Firm flesh is orange yellow, juicy and sweet (12° Brix). Fruits have good keeping quality. Fruits are practically free from objectionable papain odour.

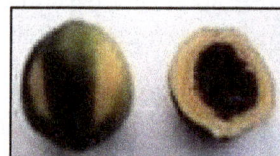

CO-2 (Selection 7)

This dioecious variety was developed by selection and sibmating of 13 introductions. Plants are of medium height. It is a dual-purpose variety for fruits and papain extraction, yielding about 4-6 g dry papain/fruit. Fruit yield ranges 80–90 fruits per tree. Fruit size is medium to large (1.5 to 2.0 Kg). Fruits are oblate, light green when ripe, and have shallow furrows. Flesh is orange, soft to firm, moderately juicy. Pulp is 3.8 cm thick. TSS is 11.4 to 13.5° Brix.

CO-3

It is a table purpose, gynodioecious hybrid of CO-2 and Sunrise Solo. Tree yields 90-120 fruits/year. Fruits are medium sized (0.5 to 1.5 Kg), pyriform in shape. Pulp is red, moderately thick (2.7 cm), sweet (TSS 14.6° Brix) with good keeping quality.

CO-4

It is a dioecious hybrid of CO-1 and Washington. Plant size is dwarf. Tree has purple petiole and stem. Yield is 80-90 fruits per annum (200 t/ha). Fruit colour is yellow with purple tinge. Fruits are of medium size (1.3 to 1.5 kg), flesh thickness is 4 cm., and TSS is 13.2 to 13.5° Brix.

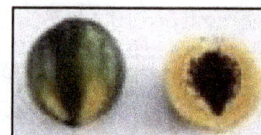

CO-5

This variety is a dioecious selection from Washington. The variety is good for papain production (14 g/fruit) that can be harvested in all seasons. Plants are of medium height. Fruit bearing height is 90 cm. Petiole is pink in colour. It produces 75-80 fruits per tree in two years and 14 g papain per fruit. Fruit size is medium to large (1.5-2.0 kg) with yellow pulp colour. TSS is 13° Brix.

CO-6

It is a dioecious selection from Pusa Giant. Plant height is low. Fruits can be harvested within eight months. It is good for papain (7-8 g/fruit) and fresh fruit (80-100 fruits per 2.5 years) both. Fruit is medium to large in size (1.9 to 2.1 kg) with yellow flesh. TSS content is 13.6°Brix.

CO-7 (CP 81)

It is a gynodioecious hybrid of CP 85 (Pusa Delicious x CO-3) x Coorg Honey Dew. It can be grown in plane as well as up to an altitude of 1000 m above MSL. First bearing height is about 50 cm. It produces about 100 fruits/tree with the yield potential of 340 t/ha in full life. Fruits are oblong small to medium size (1.15 kg) with an attractive and firm red colour and small cavity. The TSS is 16.7 °Brix.

GBPUA&T, Pantnagar

A systematic papaya-breeding programme was started at GBPUA&T, Pantnagar in 1972 to breed suitable varieties for Tarai region of UP and north Indian plains. With the utilization of genetic variability and following ear-to-row method of selection and sibmating, the following superior strains could be developed:

Pant Papaya-1

It is a dwarf selection (125-135 cm) bearing first flower at a height of 45-60 cm. Fruit is oblong of medium size (1-1.5 kg) with good quality. It is susceptible to Anthracnose infection.

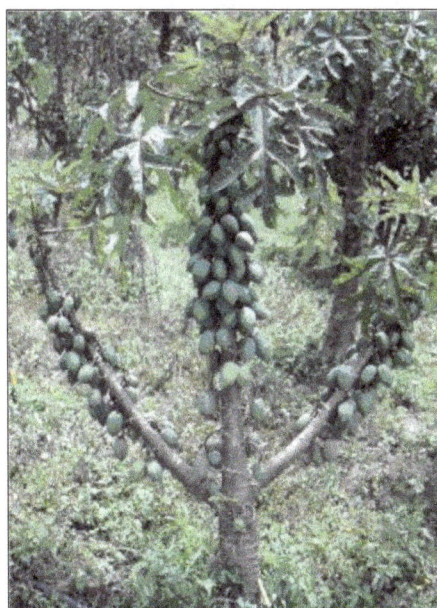

Vasconcellea Plant   Red Lady Plant

Pant Papaya-2

This variety is tolerant to frost as well as water logging. Plants are vigorous and of medium height (180-220 cm), bearing first flower at a height of 90-100 cm from the ground. It produces medium to large size fruits (1-2 Kg) of good quality with the yield potential of 30-35 fruits per plant.

Pant Papaya-3

It is tolerant to frost as well as water logging. Plants are of medium height (225-250 cm), bearing first flower at a height of 115-130 cm from the ground on strong stem. Small to medium size (0.5 to 0.9 kg), fruits are of excellent quality. It has large number of fruits per plant (45 to 50), although of small size.

IIHR Hessarghatta (Bengaluru)

Two prominent papaya cultivars were released from IIHR, Bengaluru. The variety Coorg Honey Dew was released from its Chetalli Regional Research Station while another variety, Surya, was released from the Head Office, Bengaluru.

Coorg Honey Dew

It is a gynodioecious selection from Honey Dew cultivated for table as well as processing purpose. Trees are of medium height, heavy bearer with low fruiting height. Fruit is greenish yellow, long to oblong in shape, slightly nippled, ridges are prominent at the apex. Fruit size is medium to large (1.25 to 2.5 Kg) with big central cavity. Flesh is orange with good flavour, soft, moderately juicy and good in taste with 13.5° Brix TSS. Keeping quality is moderate.

Surya

It is an advance generation gynodioecious hybrid of (Line 9 x Kamia Solo) and (Sunrise Solo x Pink Flesh Sweet). Average plant yield is about 60 Kg. Fruits are medium sized (600-800 g). The pulp is deep pink and firm. The TSS is about 13-14° Brix. Keeping quality is good.

Imported Cultivars

There are many papaya cultivars available in the Indian market imported from abroad. The important ones are:

Red Lady (Taiwan 786)

It is a gynodioecious F1 hybrid of unknown pedigree. It is cultivated for table as well as processing purpose. The fruits are oblong with a tasty sweet pulp. The plant starts bearing fruits from 100 cm above the ground level. The fruit weighs between 1-3 kg and has excellent keeping quality.

Sinta (Philippines)

It is a gynodioecious hybrid, claimed to be moderately tolerant to PRSV. Plant is medium in stature, round to oblong fruit shape with good TSS.

Creating Biodiversity by Transgenic Means

Biodiversity was created unintentionally by developing two transgenic (TG) papaya cultivars, SunUp and Rainbow, resistant to Papaya ringspot virus (PRSV) at Hawaii, USA. They are being cultivated successfully in Hawaii.

SunUp

SunUp is resistant to PRSV under Hawaiian conditions for Hawaiian strain only. This variety was developed by transferring PRSV resistant gene to cv Sunset.

Rainbow

Since Sunset was a red colour cultivar, the resultant TG SunUp was also a red colour cultivar. However, the preferred papaya cultivar in Hawaiian region was yellow colour, Kapoho. Therefore, SunUp was crossed with non-TG Kapoho. The hybrid, Rainbow, is the main TG papaya cultivar in Hawaiian region.

Laie Gold

It is a hybrid developed from crossing selfed (F2) Rainbow with non-transgenic cultivar Kamiya to serve a niche market on Oahu Island.

Red Kamiya

It is red flesh progeny of Laie Gold developed for niche market Oahu Island.

Threat of Erosion in Biodiversity

The highland papaya complex (comprises of genera of the family *Caricaceae* other than *Carica papaya*) is a susceptible fruit species with big potential but under an increasing threat of genetic erosion (Scheldeman *et al.,* 2001). Five species of the genus *Vasconcellea, V. horovitziana, V. omnilingua, V. palandensis, V. pulchra,* and *V. sprucei,* have been placed on the International Union for the Conservation of Nature and Natural Resources (IUCN) Red List of Threatened Species. Most significant threats identified by IUCN are habitat destruction resulting from deforestation and conversion of forests into croplands or grasslands (IUCN, 2003). Personal observations in Ecuador by Van Droogenbroeck *et al.* (2006) suggested that even more species of *Vasconcellea* are endangered.

References

Badillo, V.M., 1967. Esquem de la *Caicaceae. Agonomid Tropical,* 27: 245: 73.

Badillo, V.M., 1971. Monografia de la Familia Caricaceae. *Universidad Central de Venezuela, Maracay, Venezuala.* 221 pp.

Badillo, V.M., 1983. Caricaceae. In: *Flora of Ecuador No. 20,* (Eds.) G. Harling and B. Sparre. Göteborg University, Göteborg, Sweden, 48 pp.

Badillo, V.M., 1993. Caricaceae. Segundo Esquema. *Revista de la Facultad de Agronomía de la Universidad Central,* Alcance 43. Maracay, Venezuela. 111 pp.

Badillo, V.M., Van den Eynden, V. and Van Damme, P., 2000. *Carica palandesis* (Caricaceae): A new species from Ecuador. *Novon* 10: 4–6.

Badillo, V.M., 2000. *Carica* L. vs. *Vasconcella* St.–Hil. (Caricaceae) con la rehabilitacion de este ultimo. *Ernstia* 10(2): 74–79.

Drew, R.A., O'Brien, C.M. and Magdalita, P.M., 1997. Development of *Carica* interspecific hybrids. *Acta Horticulturae,* 461: 285–291.

IUCN–The World Conservation Union, 2003. The IUCN Red List of Threatened Species. *IUCN Species Survival Commission.* IUCN, Gland, Switzerland and Cambridge, UK. Available at website, http://www.iucnredlist.org/

Giocometti, D.C., 1987. Papaya breeding. *Acta Hort,* 196: 53–60.

Kempler, C. and Kabaluk, T., 1996. Babaco (*Carica pentagona* Heilb.): A possible crop for the greenhouse. *Hortscience,* 31(5): 785–788.

Nakasone, H.Y., 1975. Papaya development in Hawaii. *Hort Science,* 10: 198.

National Horticultural Board, 2010. *Indian Hort. Database.* Ministry of Agriculture, Govt. of India, New Delhi, 240 p.

National Research Council, 1989. *Lost Crops of the Incas. Little–Known Plants of the Andes with Promise for Worldwide Cultivation.* National Academy Press. Washington D.C., USA, 415 pp.

Ram Mansa, 2005. *Papaya.* Indian Council of Agricultural Research, New Delhi, 189 p.

Scheldeman, X., Romero, J. and Van Damme, P., 2001. Highland papayas in Southern Ecuador: Need for conservation actions. In: *Proceedings of the International Symposium on Tropical and Subtropical Fruits.* Cairns/Australia, November 26–December 1, 2000. *Acta Horticulturae.*

Singh, I.D. 1990. *Papaya.* Oxford and IBH Publishing Company Prívate Limited, New Delhi, 224 p.

Soria, N. and Viteri, P., 1999. Guia para el cultivo de babaco en el Ecuador. *INIAP, COSUDE. Quito, Ecuador.* 48 pp.

Van den Eynden, V., Cueva, E. and Cabrera, O., 1999. Plantas Silvestres Comestibles del Sur del Ecuador – Wild Edible Plants of Southern Ecuador. *Ediciones Abya–Yala, Quito, Ecuador,* 221 pp.

Van Droogenbroeck, B., Kyndt, T., Romeijn-Peeters, E., Van Thuyne, W., Goetghebeur, P., Romero-Motochi, J.P. and Gheysen, G., 2006. Evidence of natural hybridization and introgression between *Vasconcellea* species (Caricaceae) from Southern Ecuador revealed by chloroplast, mitochondrial and nuclear DNA markers. *Ann. Bot.,* 97(5): 793–805.

Watt, G., 1889. *A Dictionary of the Economic Products of India.* Cosmo Publication, pp. 158–164.

2013, Biodiversity in Horticultural Crops Vol. 4   *Pages 203–220*
Editor: Professor K.V. Peter
Published by: DAYA PUBLISHING HOUSE, NEW DELHI

Chapter 11

Potato

K.M. Indiresh[1] and H.M. Santhosha[2]

[1]Professor and Head,
Department of Vegetable Crops,
Post Graduate Center, UHS Campus, GKVK, Bangalore – 65
E-mail: indiresh_kabali@yahoomail.com
[2]Ph.D Research Scholar
University of Horticultural Sciences, Bagalkot, Karnataka

The potato is a starchy, tuberous crop from the perennial *Solanum tuberosum* of Solanaceae family (also known as the nightshades). The word potato may refer to the plant itself as well as the edible tuber. In the region of the Andes, there are some other closely related cultivated potato species. First introduced outside the Andes region four centuries ago, today potatoes have become an integral part of much of the world's cuisine and are the world's fourth-largest food crop, following rice, wheat, and maize. Long-term storage of potatoes requires specialized care in cold warehouses and such warehouses are among the oldest and largest storage facilities for perishable goods in the world. Among the major potato growing countries of the world, China ranks first in area, followed by the Russian Federation, Ukarine and Poland. India ranks fourth in area in the world. The present area under potato in India is about 1.4 million hectares. India produces a total of about 25-28 million tonnes of potatoes every year and ranks fifth in production also after China, Russian Federation, Poland and Ukarine. From each hectare of land, it produces about 16-19 tonnes of potatoes. In European and American countries the potato productivity is about 30-40 tonnes per hectare. The states of Uttar Pradesh, West Bengal and Bihar account for nearly 3/4 of the area and 4/5 of the potato production in the country. The highest area and the production are in Uttar Pradesh followed by West Bengal and Bihar. The highest productivity of the crop is in West Bengal followed by Gujarat. Potato is one of the principal cash crops and it also contributes to Indian economy in several ways.

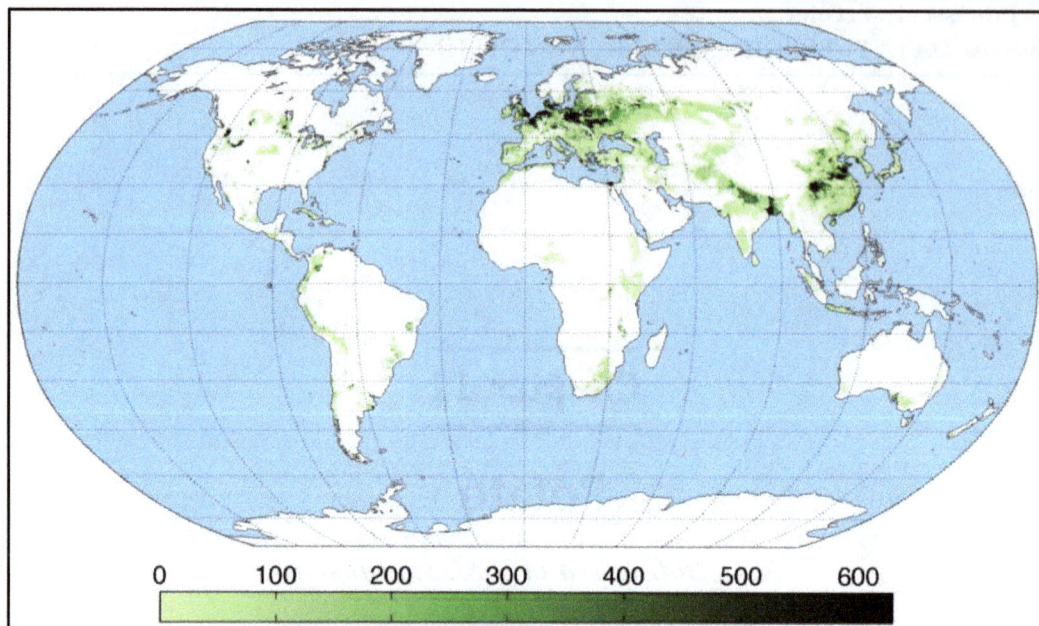

Figure 11.1: Average Regional Potato Output (Kg/ha)

A Fascinating Journey of Potato from Andes to Himalayas

Potato, the tuber crop that changed history, is grown in more than 148 countries and is one of the worlds major non-cereal food crops. Yet till 16th century it was unknown to the people of Europe, Asia and North America. The crop has a very fascinating history stretching to nearly 7000 to 9000 years back-some well documented while other chronicled from the archaeological and historical evidences.

Origin

Potato has its origin in South America, where it grows wild in nature. It is believed that cultivated potato originated from Andes of Peru and Bolivia in South Americas in the Basin of Lake Titicaca on Peru-Bolivian borders, from its wild diploid ancestors many of which may be extinct now. The species are grown in a wide variety of habitats from semi-desert conditions of northern Argentina, southern Bolivia and Mexico to the high rain fall sub-tropical forests of central and south America. Thus potato shows a wide adaptation to an altitude right from the sea level to nearly 5000 masl.

Archaeological Evidence

Spectacular ceramics were excavated dating from the Moche cultures in Northern Peru (c.AD 1-600) and the Chimpu peoples (c.AD 900-1450), as well as Hauri or Pacheco urns from the Nazca valley in Southern Peru (c.AD 650-700). These ceramics, depicting many forms of potatoes, were from coastal areas. Therefore, it is presumed that the potters obtained potatoes during trade or other means from farmers in the highlands where potatoes were actually cultivated. Actual remains of potatoes were also recovered infrequently from tombs, dwellings and rubbish heaps including chuno or tunta, from some archeological sites.

Historical Evidence

The conqueror of Peru, Francisco Pizarro, may well have been the first European to see potatoes in 1533, but there is no actual (historical) record of this event. The first historical record is of 1537, when a band of Spaniards led by Jemenez de Queseda penetrated into the highlands of what is now Columbia. This was followed by accounts of Lopez de Gomara (1552) for potatoes in Southern Peru. Potatoes in Chile received first mention by Sir Francis Drake in 1578.

The native names also indicate potato's ancient and wide spread cultivation. Thus in the Chibcha language of Central Columbia, the names iomza, iomuy, etc. were used, in quechua, the language of the inca empire, the usual name was papa. In Bolivia, the Aymara Indians used the word amka and choque, whilst in chile, the arucanians gave it the name poni. The Spaniards adopted the name papa for the potato, which was used through their South American colonies. In Europe neither batata nor papa for potato ever adopted because the Spaniards first encountered sweet potato and not having name for similar tuber, they used Indian word batata.

Early History

In South America where it originated, potato was main source of food for centuries for the people in the high Andes and Southern Chile. Potatoes were dried by Andean Indians to make chuno, a powder prepared from repeated freeze drying, thawing and trampling by men and women folk to squeeze out water and finally dehydrating under hot sun. It was widely used during food shortage and between successive crops during period of scarcity.

Spread in Europe

Though the Spanish chroniclers wrote detailed accounts of the history of their conquest and the life and customs of the people; no contemporary accounts exist on the first introduction of potato into Europe. Apparently there were two introductions into Europe, firstly into Spain in c.1570 and secondly into England in c.1590 (Figure 11.2). The Spanish introduction rests on market records from the Hospital de la sangre in Seville, whilst the English records are extremely complex; nevertheless, potato is described and figured in Herbal of Gerard, 1597. The early European potatoes came from Andes perhaps from the Northern Columbian part and were adopted to short (12 hours) days of Andes and not to long (16-18 hour) days of Europe. These potatoes further evolved through several centuries of unconscious selection in Europe to adopt it to the long summer days of Northern Europe and it was not until the late 18th and early 19th centuries, potato cultivation started on a large scale to spread into Central and Eastern Europe.

Spread in Asia, Africa, etc.

The potato's global voyage began in the 17th century. While stay at home, Europeans may have had misgivings about the new crop, the sailors, soldiers, missionaries, colonial officials and explorers quickly carried it to their foreign outposts. Thus, Belgian, British, Dutch, French, Portuguese and Spanish sailors carried the potato first to ports in Asia and South Pacific while trading, whaling and fishing and later inland to their homes. Dutch and French missionaries contributed to its further spread to Taiwan and later to China where it was known by names as Earth Bean, ground nut and tuber with many children. The trade route to China passing across Eastern Europe, over the Urals and into the Steppes of Asia turned out to be perfect environment for potatoes.

In the early years of 17th century, most probably Portuguese sailors took the potato to India, however, they might have been carried by the Britishers to the hills of the Northern India and to Sri Lanka where it flourished in the colonial home gardens.

• South America – Center of origin	8. Holland – Taiwan <1650
1. South America – Spain 1570	9. Taiwan – China <1650
2. South America – UK 1590	10. Spain – Philippines <1700
3. UK – India <1610	11. UK – New Zealand 1773
4. Portugal – India <1610	12. Holland – Java 1794
5. India – Sri Lanka <1610	13. Holland – Russia <1800
6. UK – Bermuda 1613	14. UK – South African continent 1830
7. Bermuda – Virginia, USA 1621	

Figure 11.2: Potato's Journey from Centre of its Origin

The potato arrived in Africa relatively late. A few grew in South Africa as early as 1830, but British and German colonists and missionaries did not introduce potatoes into East Africa until about 1880. In North and West Africa, the two world wars were main stimuli for crop introduction. With supply lines from Europe cut, armies and colonial personnel were forced to grow their own food items. While Africa is not a major producer in terms of volume, more African countries grow potatoes today than any other continent.

In Northern America, potato was completely unknown until the early 17th century. In fact, this continent first received potatoes from England via Bermuda in 1621 where it was introduced in 1613. The first potatoes were grown in Virginia. Later in the century, there were more introductions from England and Ireland, but no records of an introduction were made from South America before Goodrich (1863) obtained some varieties in a Panama market.

Spread in India

The earliest reference of potato in India occurs on account of the voyage of Edward Tery (1655), who was chaplain to Sir Thomas Roe, British Ambassador to the court of the Moghul Emperor Jahangir from 1615-1619. Terry in his description of Indian soil and its produce wrote "in the northern most part of this empire, they have a variety of pears and apples; every where good roots of carrot and potatoes are grown".

Early potato introduction to India was *S. tuberosum ssp. Andigena*. There was enormous confusion on the identity and nomenclature of these introductions as these were known by different local names in diverse dialects. As a result, during the initial period of potato research in India, efforts were made directed towards identification of such local desi varieties.

Potato Biodiversity

The potato is believed to have originated in the Andes Mountains of South America (and even today the Andes are a source of great diversity of potatoes whose relatives are not found elsewhere). Spanish explorers took the plant home to Europe around 1570, and it is said that the British explorer Sir Walter Raleigh introduced it to England a few years later.

There is a great deal of genetic diversity in potatoes. Some of the varieties are very strong at resisting diseases. But the potatoes growing in Ireland and Europe in the mid-1840s represented a very limited number of varieties. They lacked resistance to a fungus named *Phytophthora infestans*. The result was a terrible famine in Ireland, where so much agriculture depended on a single crop.

Would more diversity have prevented the Irish Potato Famine? Because the potatoes in Ireland are essentially biological copies of each other, it would have been difficult (but not impossible) to use genetic diversity to quickly replace them with varieties which resisted the blight. The economics of trade between Irish farmers and British landlords had a lot to do with the seriousness of the tragedy. But farmers learned one important lesson about genetic diversity. They stopped relying on just one crop. They began to appreciate the importance of planting a number of food crops, each serving as insurance in case another one failed. This is a lesson that Andean farmers had learned long ago. They encouraged diversity (and still do) by growing many different potato varieties in the same field.

The Andean farmer conserves and maintains the existing genetic diversity of potato cultivars by means of clonal propagation of tubers. However, surveys of traditional farms showed that botanical seed propagation was used for disease elimination, stock rejuvenation and creation of new cultivars. Electrophoretic surveys based on 542 tubers collected from 18 markets sampled in the Cusco area disclosed a total of 229 different cultivars from diploid, triploid and tetraploid forms of *Solanum tuberosum* L. These could be classified by isozyme cluster analysis into four major groups and six minor groups. However, they did not agree with groups based on flesh or skin color. It is therefore concluded that all genotypes belong to a single, large gene pool with considerable gene flow between cultivars of different groups. When the samples were grouped by the three most common tuber skin colors, namely red/pink ('Q'ompis type'), purple ('Yana Imilla' type), and yellowish/brown ('Yuraq Kusi' type), similar allozymes were observed in all three classes. The structure of the isozymic phenotypes within each group indicate that they may have been derived as segregants after outcrossing of diverse parental types. In order to provide further evidence for the origin of new types by hybridization, two segregating diploid progenies were generated by crossing purple by yellow skin types. In the resulting F_1, most of the tuber phenotypes observed in the Andean varieties were reproduced in these crosses. It can be concluded that the Andean potatoes form a large and plastic gene pool amplified and renovated by outcrossing followed in some cases by human selection of desirable phenotypes.

Taxonomy

☆ Botanical name–*Solanum tuberosum*

☆ Family–Solanaceae

☆ Origin–South America

☆ Folk name–Irish potato, batata, papas

☆ Chromosome no. 2n= 4x= 48.

☆ Mode of pollination- self pollination.

The ancestor of potato that is *Solanum tuberosum* is derived from *Solanum stenotomum* (2n=24), but the cultivated *Solanum* is amphidiploids (tetraploids). Commercially cultivated potatoes are 4n in nature, and amphidiploids have been generated by *Solanum stenotomum x Solanum sparcifolium*.

Cytology

Basic chromosome number of potato is 12. Right from diploid to hexaploid species are available. Majority species are diploids (74 per cent), 12 per cent are tetraploids, and 14 per cent are other species. Triploid species available are derived from spontaneous process between diploid and tetraploids, these triploids are sterile. Some pentapliods also present, they are developed by crossing hexaploid with tetraploids.

Botany

In potato plant, 3 distinct parts are present, they are:

Haulm

It contains both foliage and stem. It is also referred to as aerial stem. Stem is a hollow, mostly cross section is triangular in nature. The basal portion of the stem is solid and round and is straight or narrow. These stems arise directly from seed tuber and it produces stolons. In future these stolons will develop into tubers, *i.e.* tip of stolons enlarges to form a big tuber. Main stem is arising from seed tuber and produces stolons and tuber called as main stem. Sometimes from main stem, we find another branch called as secondary stem.

Stolons

Modified underground stem, if plant get sufficient photosynthesis, the tip of stolon produces bulging portion. Starting point of stolon bulging is known as tuber initiation or tuberization. Each plant produces 5- 10 tubers. These tubers have rose end and distal end. Eyes of potato tuber are nothing but the leaf scar with a subtended lateral bud having an undeveloped internodes. Tubers are important as 75-85 per cent of the total dry matter produced by the plant is stored in them

Root

Roots are shallow and spread to 30-45 cm both length and breadth. The tuber formation occurs earlier at low temperatures and gets delayed at high temperatures. Tuber yield is good if plants are grown in short days with low night temperature. Tuber initiates after 45-50 days after planting and at that time the night temperature should be 15-18°C for 15-20 days. If temperature is more than 20°C tuber initiation is affected.

Floral Biology

Its flowers are large, bisexual and arranged on an inflorescence described as a cyme. The floral parts are regular in their arrangement. The flowers are bisexual *i.e.* the stamens (male parts) and the pistil (female parts) are on the same flower; the flower consists of the four basic whorls (calyx, corolla, stamens and pistil) and each whorl is composed of five parts.

The Floral Whorls

The calyx is large in relation to other parts. The parts, sepals, are five in number and are joined by a calyx tube. They are green due to presence of chlorophyll. The corolla consists of five petals which are joined at the base by a short corolla tube. The corolla is brightly purple which attracts insects for pollination. There are five stamens, each arising opposite a corolla segment.

Perianth

The perianth is the outer whorl which is red, orange or yellow in colour; the perianth lobes are arranged in two whorls of the three outer (the calyx) and three inner (corolla) lobes; such a perianth is referred to as perigone, and is characteristic of monocotyledons; the perianth segments, the tepals are fused towards the lower end and is either sepals or petals.

The Stamens

There are 5 stamens, also arranged in two whorls of three each; each stamen consists of a filament and an anther; each anther has two lobes and the filaments are attached at the middle of the anther.

The Pistil

The pistil is the female whorl/gynoecium. It has a single stigma, a single style and a single ovary.

The ovary is superior (*i.e.* it is situated above the other floral whorls on the receptacle). The pistil is situated in the middle of the flower. It consists of an ovary with three lobes, which represent the three fused carpels; a thin, long style; a simple stigma situated at the end of the style.

Mode of pollination is self pollination, anthesis occurs at 5 am to 6 am. Anther dehiscence will be at 5.30 am to 7 am. Stigma receptivity is observed next day afternoon to anthesis. Pollen remains viable from a day of anthesis to next morning. Mature buds are emasculated 12 to 16 hours before they open in the afternoon and pollination in the next morning.

Classification of Potato Species

The genus *Solanum* contains 200 species. According to Dunal (1882), genus *Solanum* is divided into 2 sub- genera, namely, Patchystemonum and Leptostemonum. Patchystemonum has five sections like, Tuberarium, Morella, Dull camera, Micranthes and Lycianthes.

According to Bitter (1912), section tuberarium is divided into 2 sub sections : Basarthum (Estolinifera) and Hyperbasarthum. Hyperbasarthum has cultivated species, is grouped under series Tuberosa of sub-section Potato. Corella (1962) grouped species into 26 series containing 159 species. Hawkes (1990) grouped species into 19 series into cultivated potatoes. This cultivated species of diploids have been noticed in nature. Cultivated diploids (2n= 24) species are:

1. *Solanum ajanhuiri*: frost resistance is seen in South Peru and North Bolivia at high altitudes.
2. *Solanum goniocalyx*: tubers have bright yellow flesh, seen in Central to North Peru at high altitudes.

3. *Solanum phureja*: Absence of tuber dormancy indicates that it has become specially adopted to regions that are free from long periods of drought or frost, but also grown under higher altitudes.

4. *Solanum stenotomum*: Tubers are produced in 5-6 months with definite dormancy period. It is probably an ancestor to all the cultivated potatoes, seen in Central Peru to Central Bolivia at very high altitudes.

5. *Solanum vybini*

6. *Solanum lanciforum*

7. *Solanum chaekoemis*

Cultivated triploids: (2n=36) Species are:

1. *Solanum chaucha*: A triploid form derived from natural crosses between *Solanum tuberosum ssp andigena* and *Solanun stenotoum*. *Solanum chaucha* has corolla lobes which are in general about 3 times as broad as long when spread, seen in central Peru to central Bolivia at high altitudes.

2. *Solanum juzepczukii:* A natural triploid between *Solanum acaule* and *Solanum stenotomum*, seen in Central Peru to South Bolivia at very high altitudes. It is frost resistant.

Mainly there are 4 tetraploid (2n=48) species and they are:

1. *Solanum tuberosum* ssp. *tuberosum*

2. *Solanum tuberosum* ssp. *andigena*

3. *Solanum acule*

4. *Solanum columbinum*

Only 1 pentapliod (2n=60) cultivated species is present; *Solanum curtilobum*–a cross between *Solanum juzepczukii* and *Solanum tuberosum ssp andigena*. *Solanum demissum* is the only hexaploid (2n=72) cultivated species present in nature.

Propagation

The tuber is not only the principal means for potato propagation, but also a major human food source. Potatoes are mainly propagated by vegetative methods (cloning). This is the primary commercial propagation method. Vegetative reproduction ensures a uniform crop, contrary to what would happen with sexual propagation. Sexual propagation of potato is accomplished by planting its true seed, but a high variability exists and that is why it is not commonly used. However, sexual seed is becoming more and more popular; especially in places were disease pressure is very high and maintaining disease free seed is becoming a problem.

Asexual, Vegetative or Clonal Propagation

When potato growers talk about seed, they are talking about the tuber and not the botanical or sexual seed. Potato tubers are actually a modified stem with approximately 70-75 per cent content of water and remaining 25-30 per cent of dry mater. They have nodes or eyes from which the new growth begins. The new stems growing from each eye are called sprouts. Sprouts grow from the tuber after a period of dormancy after they are harvested. This varies largely between cultivars. After this dormancy is broken, sprouts grow and when planted, they give rise to the plant stems and from there all the

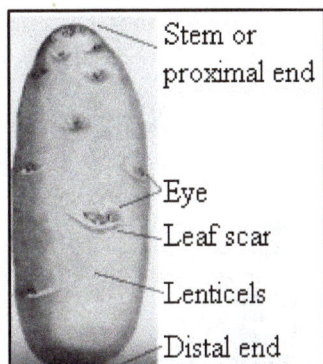

**Vegetative Seed can be Either
a Whole Tuber or a Cut Tuber**

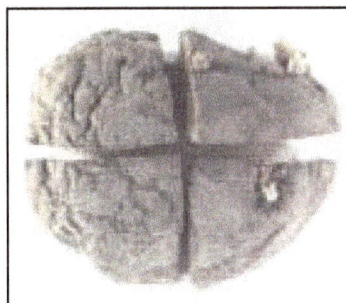

**Potato Tuber Cut to Form
Seed Pieces**

vegetative parts of the plant. Underground, lateral shoots called stolons are formed, from which the new tubers will be formed.

Research shows that a seed piece has the adequate amount of carbohydrate levels for shoot initiation and growth. If the tubers are cut, the usual procedure is to let the cut pieces to suberize or cure, for about 10 days. Suberization in tubers allows them to develop a corky layer around the seed piece that prevents the seed piece decay by entry of several kinds of pathogens.

Old Seed	New Seed
Rapid emergence	Slower emergence
More stems, More tubers, smaller tuber size	Fewer stems
Earlier tuber initiation	Later tuber initiation
Earlier maturity	Later maturity
Earlier senescence	Prolonged vigor of plants
Less potential for high yield	Higher final yield in a long season

Physiological Age of Seed (PAS)

The PAS is not the chronological age of the seed piece; instead, is the influence of the growing environment of the seed. Physiological age of seed is influenced by growing conditions, handling, storage and cutting procedures. Physiological age of the seed will have an impact on how the new crop grows.

It is very important to manage the physiological age of the seed because it has a big impact on how the new crop is going to look like and it will probably, along with many other factors, determine weather the crop will be of a high quantitative and qualitative value.

Other than seed age, there are some pros and cons when talking about vegetative reproduction, some of them include the fact that cloning assures genetic purity and product uniformity. It also favors high yields. Some of the disadvantages are that cloning favors disease spread (*e.g.* viruses, bacteria, fungi), and also a significant amount of storage space, transportation and heavy planting equipment are required.

Propagation by Botanical or Sexual Seed

Most potato cultivars produce fruit but some are pollen sterile or fail to set fruit for other reasons. If fruit is established, they usually are small, up to 1.25 cm in diameter and are green colored, resembling a small tomato.

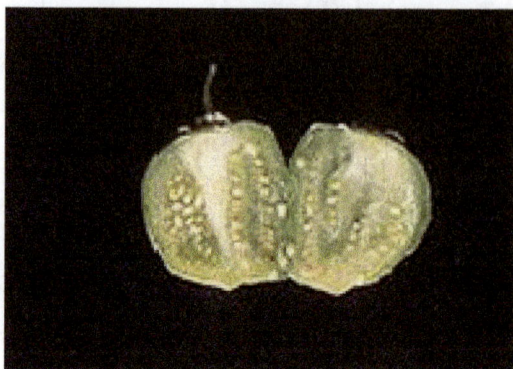

The fruit contains true seeds of potato plants, containing approximately 300 seeds per fruit. When the potato plant reproduces, usually through self-pollination, the chromosomes (along with the genes they carry) are randomly distributed to the seeds. Each seed will develop into a plant with unique characteristics. This is a process very useful in crop improvements in breeding programs, but its genotypic variation is of little value to growers because a new plant could be totally different to the mother plant and no uniformity would be seen in the field. Some of the advantages which true potato seeds have, are that they do not carry any diseases and that seed can be stored in small places contrary to what happens with tubers.

Tissue Culture

Tissue culture permits a very rapid propagation. Under traditional propagation, one tuber yields approximately 8 daughter tubers in one growing season, while with tissue culture, 100,000 identical plantlets can be produced in eight months, that when transferred to the field, could produce 50 tones of potatoes.

Each plant has a root system, leaves and terminal points or growth points. There is an apical meristem found at the apex of a potato stem and it also has lateral growth points. Each of these buds has a meristem which allows it to become a different plant.

The process of this technique is very simple. Disease free plantlets are grown in test tubes on a nutrient media. Each plantlet is cut into 3 to 10 nodal sections after 18-60 days. Each new cutting is planted in a new test tube. This can be repeated until the desired number of plantlets are obtained. Plantlets are then removed from the tubes and grown in sterile soil and let them complete their entire growth cycle. Tubers produced are collected and stored later to be sold to growers. This is also the process followed to obtain certified seeds. This first seed lot would be called nuclear seed and then after harvesting the product of this seed, you get Generation 1 (G1) and so forth.

Meristems have no vascular system, therefore are less prone to viral, fungal and bacterial infections. For this reason, this method is used to maintain disease free seed stock, which can be then stored *in vitro* and be used when needed.

True Potato Seed (TPS)

True Potato Seed (TPS) was first evolved through sexual reproduction by Ramanujam in1957 and subsequently its commercial viability was tested. However, high heterozygosity of seedlings of TPS and uniformity in crop hampered its commercial adoption by farmers.

How to Use TPS

Initially TPS is sown in nursery for raising seedlings which are latter transplanted in properly prepared seed bed but the process becomes labour intensive and very expensive. The TPS was tested and tried at International Potato Centre and Central Potato Research Centre Institute (CPRI) Shimla and recommended ways for eliminating raising seedlings and then transplanting. The method suggested two stage programmes for raising healthy seeds, *i.e.*

- ☆ Production of tuberlets in nursery beds in the first year.
- ☆ Storing them in cold store and planting them thickly as commercial crop in the following year. The method however, needs testing field for its yield and economic viability. A seed rate of 50-150 g/ha is used for sowing.

Advantages with TPS

1. This technique is used for disease free seed production. It is the technique for virus free seed production.

2. Cost of tubers used in conventional method of planting is very high whereas the production of tuberlets in nursery for planting in next year is relatively low.

3. Higher cost required for storage of huge bulk of conventionally used tubers is reduced as tuberlets (being very small in size) require very little space for storage. This also reduces cost in transplantation.

4. The viral infiltration in the seed tuber is also less.

5. The cost of tuber treating chemicals is also reduced because of relatively lesser volume of tuberlets.

6. By this method, the disease free potato seed can be produced and prevention of diseases to new areas can be ensured.

Disadvantages

1. Production of large number of small sized tubers (50 per cent tubers will be small in size).
2. Cost of cultivation will be very high if raised through true potato seed method.
3 Duration also will be more (140-150 days).
4. Tuber yield is not uniform.

Genetics

Genetics of characters in potato are as follows:

Characters	No. of Genes	Type of Gene Action
Skin colour	Digenic	Brown> > red
Position of eyes on tuber	Monogenic	Flat eye incompletely> deep eyes
Flesh colour	monogenic	Deep yellow colour is incompletely > white colour
Stolon length	monogenic	Long stolon axis > short axis
Tuber shape	Monogenic	Long tuber axis> short axis
Tuber diameter	Polygenic	–
Late blight disease	Monogenic	Recessive character

Crop Improvement in Potato

Breeding Uniqueness

☆ Potato is propagated asexually.

☆ In potato, transmission of diseases is via tubers.

☆ Easy maintenance and multiplication of the parent material in the original vegetative state through vegetative propagation.

☆ Complex tetrasomic inheritance due to auto tetraploidy.

☆ Diverse source of germplasm including wild relatives for resistance to biotic and abiotic stresses.

Breeding Goals

☆ Higher tuber yield

☆ Earliness

☆ Photoperiod insensitivity.

☆ Responsiveness to fertilisers

☆ Better keeping qualities (resistance against shrinkage, rottage, accumulation of sugars, especially reducing sugars and reasonable dormancy)

☆ Round, medium sized tubers with shallow eyes and free from greening for general consumption.

☆ Tubers with high vitamin-C and protein content.

☆ High specific gravity tubers (dry matter content) suitable for french fries, chips and dehydrated products.

☆ Low sugar content of tubers for chips and french fries to avoid browning

☆ Tubers resistance/tolerance to biotic stresses like late blight, early blight, charcoal rot, wart, common scab, bacterial wilt, soft rot, viral diseases, nematodes, aphids, potato tuber worm and abiotic stresses like heat, drought, frost, soil salinity.

Varieties Developed by Introduction in Potato

☆ Uptodate- Medium type, tubers are large, oval shape with white flesh, with good keeping and cooking quality, 30-32t/ha. It was introduced from Northern Ireland. It is like a most popular variety.

☆ Atlantic-Medium maturing, tubers white, with medium deep eyes, resistant to wart. yield is 28-30t/ha. It was introduced from Australia.

☆ FTL-1533-: Medium maturing resistant to late blight, tuber dry matter (21.2 per cent),30-35t/ha yield. It is a processing variety. It was introduced from Australia.

☆ Craigs Defiance: It is the oldest variety. It is used as breeding line.

☆ Alley- Medium maturing, tuber white, oval long flattened with flat eyes, resistant to late blight, tolerant to frost, 30t/ha.

Varieties Developed by Clonal Selection in Potato

☆ *Kufri Red*: It was released in 1958. It is medium, round, red colour in cortex, medium deep eyes. It is suitable for North Eastern plains. Kufri Red potato is a clonal selection from Darjeeling Red Round; it was developed from a single disease-free plant.

☆ *Kufri Safed*: is a clonal selection from the potato variety Phulwa. It is medium, round, white, deep and pickled red-purple eyes.

Some Hybrids

☆ *Kufri Chandramukhi* (S-4485 x Kufri Kuber): Plants medium tall, spreading open and vigorous. stems a few, thick, colored at base with well developed wavy wings. Foliage: Grey-green. Leaves open, rachis green. Leaflets ovate-lanceolate, smooth glossy surface with entire margin. Flowers: Light red-purple. Profuse flowering. Anthers lemon-yellow, poorly developed and pollen sterile. Stigma round and slightly notched. Tubers: White, large, oval, Slightly Flattened, smooth skin, fleet eyes and dull white flesh. Early in maturity (80-90 days) and Yield potential is 25 t/ha.

☆ *Kufri Badshah* (K. Jyothi x K. Alankar): Plants tall, erect, medium compact and vigorous. Stems a few, thick, uniformly colored, well developed straight wings. Leaflets ovate, smooth surface with entire margin. Moderate flowering. Anthers orange-yellow, poorly developed, high pollen stainability. Stigma round and notched. Tubers are white, large, oval, smooth skin, fleet eyes and dull white flesh tubers turn pale purple on exposure to light. Medium is 100-110 days and yield potential is 50 t/ha.

☆ *Kufri Sheethman* (Phulwa x Craigs Defiance): A frost-resistant variety; suitable for Punjab, Delhi, Rajasthan, Haryana, Uttar Pradesh and Madhya Pradesh.

☆ *Kufri Sindhuri* (K. Red x K. Kundan)- Late Maturity (110-120 days), yield is 40 t/ha. Moderately resistant to early blight, tolerant to leaf roll. Slow rate of degeneration. Can tolerate temperature and water stress to some extent. Medium dry matter content in nature.

Heterosis Breeding in Potato

Kufri Jyothi (Phulwa × CP 1787). It is observed for many characters like earliness, tuber yield, tuber size, tuber weight and F1s are raised in the field.

Inter Specific Hybridisation

It is mostly used for disease resistance transfer.ex; Kufri Kuber- selection from *S. curtilobum* x *S. tuberosum* and late blight resistant breeding lines- *S.demissum* x *S.tuberosum.*

Mutation Breeding

It is not always possible to make crosses with wild potato species to introduce new genetic variation. Some desired traits are absent or never found in the total potato population. In that case, mutation breeding might bring a solution. With this method, plant material is exposed to a mutagenic treatment like X-rays. This will cause a lot of changes in the DNA resulting in other plant properties. In a number of horticultural crops, this has led to interesting improvements like new flower colours. The tetraploid nature of potato however makes the discovery of mutants quite difficult.

Sources of Resistance in Potato

☆ *S. demissum* (late blight,PLRV).

☆ *S. acaulae* (PVX,PLRV,potato spindle tuber viroid).

☆ S. *chacoense* (PVA,PVY,late blight,tuber moth).

☆ *S. spegazzinii* (fusarium,wart,cyst nematode).

☆ *S. stoloniferrum* (PVA,PVY).

☆ *S. vernei* (high starch).

Molecular Breeding for Crop Improvement in Potato

During the last two decades, new methods involving techniques of molecular biology have increasingly been used for crop improvement programmes. These techniques mainly include production of transgenic crops and the development of DNA- based molecular markers to be used for indirect-marker assisted selection (MAS). Significant progress has been made in both these areas at international level. For instance, last year, as much as more than 150 million acres of land were under transgenic crops worldwide, despite ongoing debate, and despite the protest against the genetically modified crops in general and GM food in particular. Although initially the most important traits that were improved in the transgenic crop included resistance against herbicides, insects and viruses. Efforts are now being made to produce GM food crops with improved nutritional quality and also crops for molecular farming (for industry).

Potato is one of the seven transgenic crops that are already being grown commercially, although the area occupied globally by transgenic potato (insect resistant, virus resistant) in 2002 was less than one million hectares (less than 1 per cent). Transgenic potato with disease resistance and improved starch are also being grown in China, but only marginally. This situation is likely to change in the next few years, since more transgenic potatoes for additional desirable traits are being regularly produced.

For instance, transgenic potatoes with increased biomass were produced in the recent past. In another study, for the first time, a hybrid Bt cry gene (not fusion gene that was used in the past) was used for production of transgenic potatoes that were resistant to both coleopteran and lepidopteran insects. An individual cry gene, due to its specificity, provides resistance against only one of the several groups of the insects, *e.g.* coleopteran, Lepidopteran and dipteral. In India also, potato with improved protein should become available within the next 2-3 years.

Future of transgenic research in potato seems to be bright. For instance, an important recent study conducted at the University of Wiscosin, leading to the production of blight resistant transgenic potato, involved isolation from a wild species of *Solanum,* a gene for resistance against potato late blight (*Phytophthora infestans*). Another important recent study involved in the identification of a number of genes for protease inhibitors (Kunitz-type) that are exposed in potato tubers. These and similar other genes will be used in future for developing potatoes that would be resistant against a number of insects and pathogenic fungi (e.g. *Fusarium moniliforme*).

In the area of MAS also, DNA-based markers have been used for introgression of a variety of major desirable genes, mainly through backcross breeding programmes, and the method will be increasingly used in future. Pyramiding of genes for resistance against rice by scientists at PAU, Ludhiana in collaboration with IRRI (Manila) is one of the several examples of successful use of MAS in plant breeding. There is a need for development of markers in potato, which may be used for selection of difficult traits through MAS in future. While biosafety issues are being raised against transgenic crops and are being addressed world over, no such issues are involved with the technology of marker assisted breeding. However, the difficulties in the finer molecular dissection of quantitative traits, the interactions (Q X Q and Q X E) in which QTLs are generally involved and the cost involved in MAS may not allow its routine use in plant breeding, in the immediate future.

Genetically Modified Varieties

Genetic research produced several genetically modified varieties. 'New Leaf', owned by Monsanto Company, incorporated genes from *Bacillus thuringiensis*, which conferred resistance to Colorado potato beetle; 'New Leaf Plus' and 'New Leaf Y', approved by US regulatory agencies during the 1990s, also included resistance to viruses. McDonald's, Burger King, Frito-Lay, and Proctor and Gamble announced they would not use genetically modified potatoes, and Monsanto published its intent to discontinue the line in March 2001. The starch content of 'Amflora', waxy potato variety from the German chemical company BASF, has been modified to contain only amylopectin, making it inedible but more useful for industrial purposes. In 2010, the European Commission cleared the way for 'Amflora' to be grown in the European Union. Nevertheless, under EU rules, individual countries have the right to decide whether they will allow this potato to be grown on their territory. Commercial planting of 'Amflora' was expected in the Czech Republic and Germany in spring of 2010, and Sweden and the Netherlands in following years. Another GM potato variety developed by BASF is 'Fortuna'. In 2010, a team of Indian scientists announced they had developed a genetically modified potato with 35 to 60 per cent more protein than non-modified potatoes. Protein content was boosted by adding the gene AmA1 from the grain amaranth. They also found 15 to 25 per cent greater crop yields with these potatoes.

The researchers expected that a key market for the GM potato would be the developing world, where more than one billion people are chronically undernourished.

Some of the Examples of Indigenous and Exotic Varieties of Potato

Indigenous Varieties	*Exotic Varieties*
☆ Agra Red	☆ Ally
☆ Coonoor White	☆ Arran Counsal
☆ Coonoor Red	☆ Great scot
☆ Darjeeling Red Round	☆ Italian White Round
☆ Gola type A,B, and C	☆ Majestic
☆ Sathoo	☆ President
☆ Phulwa Purple Splashed	☆ Red Rock

Growth and Cultivation

Potato growth has been divided into five phases. During the first phase, sprouts emerge and root growth begins. During the second, photosynthesis begins as the plant develops leaves and branches. New tubers develop during the third phase, which is often (but not always) associated with flowering. Tuber formation halts when soil temperatures reach 26.7°C; hence potatoes are considered a cool-season crop. Tuber bulking occurs during the fourth phase, when the plant begins investing the majority of its resources in its newly formed tubers. At this stage, several factors are critical to yield: optimal soil moisture and temperature, soil nutrient availability and balance, and resistance to pest attacks. The final phase is maturation: The plant canopy dies back, the tuber skins harden, and their sugars convert to starches.

New tubers may arise at the soil surface. Since exposure to light leads to greening of the skins and the development of solanin, growers are interested in covering such tubers. Commercial growers usually address this problem by piling additional soil around the base of the plant as it grows ("hilling", or "earthing up"). An alternative method used by home gardeners and smaller-scale growers involves covering the growing area with organic mulches such as straw or with plastic sheets.[

Correct potato husbandry can be an arduous task in some circumstances. Good ground preparation, harrowing, ploughing and rolling are always needed, along with a little grace from the weather and a good source of water. Three successive ploughings, with associated harrowing and rolling, are desirable before planting. Eliminating all root-weeds is desirable in potato cultivation. In general, the potatoes themselves are grown from the eyes of another potato and not from seed. Home gardeners often plant a piece of potato with two or three eyes in a hill of mounded soil. Commercial growers plant potatoes as a row crop using seed tubers, young plants or microtubers and may mound the entire row. Seed potato crops are 'rogued' in some countries to eliminate diseased plants or those of a different variety from the seed crop.

Potatoes are sensitive to heavy frosts, which damage them in the ground. Even cold weather makes potatoes more susceptible to bruising and possibly later rotting, which can quickly ruin a large stored crop. At harvest time, gardeners usually dig up potatoes with a long-handled, three-prong "grape" (or graip), *i.e.*, a spading fork, or a potato hook, which is similar to the graip but with tines at 90 degree angle to the handle. In larger plots, the plow is the fastest implement for unearthing potatoes. Commercial harvesting is typically done with large potato harvesters, which scoop up the plant and surrounding earth. This is transported up an apron chain consisting of steel links several feet wide, which separate some of the dirt. The chain deposits into an area where further separation occurs.

Different designs use different systems at this point. The most complex designs use vine choppers and shakers, along with a blower system or "Flying Willard" to separate the potatoes from the plant. The result is then usually run past workers who continue to sort out plant material, stones, and rotten potatoes before the potatoes are continuously delivered to a wagon or truck. Further inspection and separation occur when the potatoes are unloaded from the field vehicles and put into storage.

Immature potatoes may be sold as "New Potatoes" and are particularly valued for taste. These are often harvested by the home gardener or farmer by "grabbling", *i.e.* pulling out the young tubers by hand while leaving the plant in place.

Potatoes are usually cured after harvest to improve skin-set. Skin-set is the process by which the skin of the potato becomes resistant to skinning damage. Potato tubers may be susceptible to skinning at harvest and suffer skinning damage during harvest and handling operations. Curing allows the skin to fully set and any wounds to heal. Wound-healing prevents infection and water-loss from the tubers during storage. Curing is normally done at relatively warm temperatures 50°C to 60°C with high humidity and good gas-exchange if at all possible.

Storage

Storage facilities need to be carefully designed to keep the potatoes alive and slow the natural process of decomposition, which involves the breakdown of starch. It is crucial that the storage area is dark, well ventilated and for long-term storage maintained at temperatures near 4°C. For short-term storage before cooking, temperatures of about 7°C to 10°C are preferred.

On the other hand, temperatures below 4°C convert potatoes' starch into sugar, which alters their taste and cooking qualities and leads to higher acrylamide levels in the cooked product, especially in deep-fried dishes — the discovery of Acrylamides in starchy foods in 2002 has led to many international health concerns as they are believed to be possible carcinogens and their occurrence in cooked foods are currently under study as possible influences in potential health problem.

Under optimum conditions, possible in commercial warehouses, potatoes can be stored for up to ten to twelve months. When stored at homes, the shelf life is usually only for several weeks. If potatoes develop green areas or start to sprout, these areas should be trimmed before using.

Commercial storage of potatoes involves several phases: drying of surface moisture; a wound healing phase at 85 to 95 per cent relative humidity and temperatures below 25°C; a staged cooling phase; a holding phase; and a reconditioning phase, during which the tubers are slowly warmed. Mechanical ventilation is used at various points during the process to prevent condensation and accumulation of carbon dioxide.

When stored in the home, mature potatoes are optimally kept at room temperature, where they last 1 to 2 weeks in paper bag, in a dry, cool, dark, well ventilated location. If mature potatoes are refrigerated, dark spots can occur and conversion of starch into sugar can give rise to unpleasant sweet flavour when cooked. Only new potatoes can be refrigerated, and should be kept so, where they have a shelflife of 1 week. If kept in too warm temperatures, both mature and new potatoes will sprout and shrivel. Exposure to light causes them to turn green.

Uses of Potato

Potatoes are used to brew alcoholic beverages like vodka, potcheen, or akvavit. They are also used as food for domestic animals. Potato starch is used in the food industry as, for example, thickeners and binders of soups and sauces, in the textile industry, as adhesives, and for the manufacturing of papers and boards. Many companies are exploring the possibilities of using waste potatoes to obtain polylactic acid for use in plastic products; other research projects seek ways to use the starch as a base for biodegradable packaging.

Culinary Uses

Potatoes are prepared in many ways: skin-on or peeled, whole or cut up, with seasonings or without. The only requirement involves cooking to swell the starch granules. Most potato dishes are served hot, but some are first cooked, then served cold, notably potato salad and potato chips/crisps. Common dishes are: mashed potatoes, which are first boiled (usually peeled), and then mashed with milk or yogurt and butter; whole baked potatoes; boiled or steamed potatoes; French-fried potatoes or chips; cut into cubes and roasted; scalloped, diced, or sliced and fried (home fries); grated into small thin strips and fried (hash browns); grated and formed into dumplings, Rösti or potato pancakes. Unlike many foods, potatoes can also be easily cooked in a microwave oven and still retain nearly all of their nutritional value, provided they are covered in ventilated plastic wrap to prevent moisture from escaping; this method produces a meal very similar to a steamed potato, while retaining the appearance of a conventionally baked potato. Potato chunks also commonly appear as a stew ingredient.

References

Gerhart, U. Ryffel, 2010. Making the most of GM potatoes. *Nature Biotechnology,* 28: 318.

Gupta, 2003. Molecular breeding for crop improvement. *Trends in Plant Science,* 21: 1003–1012.

Hagman, J., 1990. Micropropagation of potatoes: Comparisons of different methods. *Crop Production Science,* 9 : 94.

Kleinkopf, G.E. and Olsen, N., 2003. Storage management. In: *Potato Production Systems*, (Eds.) J.C. Stark and S.L. Love. University of Idaho Agricultural Communications, p. 363–381.

Pandey, S.K., Kaushik, S.K. and Paul Khurana, S.M., 2003. A fascinating journey of potato from Andes to Himalaya. In: *Symposium on Potato Research Towards National Food and Nutritional Security.* October 2–3, CPRI, Shimla

Quiros, C.F., Ortega, R. and Brush, S.B., 2002. Increase of potato genetic resources in their center of diversity: The role of natural outcrossing and selection by the Andean farmer. *Genetic Resources and Crop Evolution,* 39(2): 107–113.

Srebniak, Rasmussen and Jolanta, 2002. Cytogenetic analysis of an asymmetric potato hybrid. *J. Appl. Genet.,* 43(1): 19–31.

2013, Biodiversity in Horticultural Crops Vol. 4   *Pages 221–228*

Editor: Professor K.V. Peter

Published by: DAYA PUBLISHING HOUSE, NEW DELHI

Chapter 12

Sponge Gourd

D.K. Singh and Mangaldeep Sarkar

Department of Vegetable Science,
College of Agriculture, GBPUA&T, Pantnagar – 263 145, U.S. Nagar, Uttarakhand
E-mail: dks1233@gmail.com

Sponge gourd is a common vegetable belonging to family Cucurbitaceae grown throughout India. It is more commonly cultivated in Europe and America. It is a sub-tropical to tropical plant, which requires warm summer temperature and long frost free growing season. It is an annual climber which produces fruit, containing fibrous vascular system. China, Korea, India, Japan and Central America are the main countries which produce sponge gourd commercially. The fruit is edible when young. The dry inner fibrous portion of the mature fruit is used for bathing purpose, cleaning utensils, in making shoe soles and also as filters in factories. Some industrial oil is also extracted from its seed. Immature fruit, used as vegetable, is good for diabetes. The crop is commonly called as loofah, vegetable sponge, bath sponge or dish cloth gourd. The number of species in this genus, *Luffa* varies from 5 to 7. Only two species, *Luffa cylindrica* and *Luffa acutangula* are domesticated. Loofah vines are very large, lending themselves to training on a stout vertical trellis that encourages the development of straight fruit. The principal food used is immature fruit prepared like summer squash. Young shoots and leaves are used as greens. When grown to maturity, fruit of smooth loofah produces phytosponges. After harvest they are soaked in water to encourage decay of the outer fruit wall and inner pulp, and then washed thoroughly to remove extraneous matter. The remaining fiber is dried in the sun and bleached white. The crop is monoecious in sex form. The fruit is 20-50cm long and almost cylindrical in shape. The yellow coloured flowers open in early morning hours (4-8 am). Fruit contains higher protein and carotene than ridge gourd. Fruits contain a gelatinous compound called luffein. Pusa Chikni, Phula Prajakta, Pusa Supriya and Pusa Sneha are the varieties developed in India. Exogenous application of ethrel (250ppm) enhances the female flower production in sponge gourd. Loofah is closely related and has similar cultural requirements as the cucumber. The blood circulation, the sponge induces on the skin, has been credited as a relief for rheumatic and arthritic sufferers. The versatility of the loofah goes beyond the sponges. The young fruit, when small (around 18 cm) are delicious, used in soup and stew. Older fruits have been reported to develop purgative chemicals. Because loofah has a compact network of close fibers, its resiliency makes it useful for many products like filters, slipper shoes, baskets etc. Small pieces of sponges are good for scraping vegetables like carrots without having to remove the valuable nutrients by peeling them. Loofah is environmentally safe, biodegradable and a renewable resource.

Early History, Origin and Spread of Cultivated Sponge Gourd

The exact place of origin of the sponge gourd is difficult to pinpoint. There is no Sanskrit name for this plant. There is very little variation in either vine or fruit-a trait not generally associated with plants with a long history of cultivation. These suggest that it is a comparatively recent addition to the existing list of vegetable crops. However it is considered that the sponge gourd is indigenous to tropical Asia, probably India (Choudhury, 2003).

The loofah gourd originated from subtropical regions of Asia, most probably from India as primary centre. This was domesticated and known in very early times. Loofah was mentioned in Kautilya's Arthashastra, c. 350-300 B.C. in India. It was also described in early Egyptian and Chinese literature. Rich diversity in vine and fruit morphological characters occurs in North Eastern region including Sikkim, West Bengal, Western, Central and Southern India. *L. hermaphrodita* considered to be originated from *L. graveolans* is another potential species distributed in parts of North-Central India. *L. acutangula* var. *amara* grows in peninsular India, is wild relative of sponge gourd and *L. echinata* grows in natural habitats in Western Himalayas, Central India and Gangetic plains. The wild species *L. graveolens* is the progenitor of the cultivated smooth gourd species *L. cylindrica* and ridge gourd, *L. acutangula*. The wild dioecious species, *L. echinata* and monoecious *L. operculata* were evolved from the wild *L. graveolens*. The centers of diversity are in South and South East Asia. In India, there is great diversity occurring in Loofah gourds. The species *L. echinata* var. *longistylis* (Edur.) Clarke is widely distributed in North Western Himalayas and upper Gangetic plains, extending South to Tamil Nadu and sporadic in Eastern Himalaya and the species *L. umbellata* M. (klein) Roem. is confined to the Eastern coast or Coromandal belt. *L. acutangula* var. *amara* (Roxb.) Clarke is mainly distributed in peninsular tract (Swarup, 20006).

The name 'Loofah' or 'Luff'a' is of Arabic origin because sponge characteristic has been described in Egyptian writings and Chinese name 'Szkua'-dish cloth gourd or towel gourd signifies its mention in early Chinese literature. Sanskrit name 'Koshataki' indicates its early cultivation in India (Bose *et al.,* 2002).

The Genus *Luffa* and its Relatives

The genus *Luffa* Mill. belongs to family Cucurbitaceae, subfamily cucurbitoidae, tribe Benincasae and Subtribe Luffinae C.Jeffr. Of the four known Luffa species, two are cultivated, *L. acutangula* (L.)Roxb., the ridge gourd, and *L. cylindrica* Roem., the smooth gourd. The other two are wild related species, *L. echinata* Roxb. and *L. graveolens* Roxb. All *Luffa* species are monoecious except *L. echinata* which is dioecious. All species of the genus *Luffa* have chromosomes 2n=2x=26 and are cross compatible. Though the *Luffa* species can be crossed among themselves, they have remained isolated in nature. The isolation of species may be genic and not chromosomal as indicated by cytological investigations. The amphidiploid produced between *L. acutangula* and *L. graveolens*, (2n=52) had slight pollen sterility partly due to univalents. The induced amphidiploids showed quadrivalents, trivalents, bivalents and univalents at meiosis. The smooth gourd and ridge gourd are cross compatible. The total species under this genus are *L. acutangula, L. cylindrica, L. echinata, L. operculata* (grown in tropical America), *L.graveolens, L.umbellata, L.pentalldra, L.giganta, L.scabra* and *L.narylandica* (Peter and Pradeepkumar, 2008).

Genetic Diversity Analysis by Biochemical Characterization

Seed protein electrophoresis is being utilized as an additional approach for species characterization. The use of seed protein profile for resolving taxonomic and evolutionary problems has been greatly expanded because of stability, uniformity, addictiveness and biosystematic approaches. This makes it unique and powerful tool in studies. Electrophoresis is basically a process of forced diffusion within an electric field. Protein molecules are moved through a medium *i.e.* gel, paper or cellulose by applying on electrical gradient (Pierce and Brewbaker, 1973).

Ishihara *et al.* (1997) studied the isolation and molecular characterization of four arginine/ glutamate rich polypeptides (5k-6.5k, 12.5k- and 14 k-AGRPs) from the seeds of sponge gourd. It was assumed that the 4 AGRPs might occur in the protein bodies within cells of the seed.

Singh *et al.* (2003) studied molecular characterization of translation inhibitor protein in sponge gourd. A ribosome inactivating protein (RIP), Luffein has been isolated from the seeds of sponge gourd by ammonium sulphate fractionation followed by cation exchange and gel-filtration chromatography. The gel filtration and SDS-PAGE method help in determination of molecular mass of Luffein, which is 28 KDa (approx.)

A thesis was done under the supervision of Dr. D.K. Singh in 2006 on "Morphological and Biochemical Characterization of Germplasm of Sponge gourd" to find out the protein profiling pattern among the germplasm lines through SDS-PAGE.

Ecological Adaptations

Climate

Smooth gourd is a warm season crop. It prefers warm and humid climate. The optimum temperature requirement is 24-27°C.

Soil

Loam and sandy loam soil rich in organic water is the best suited to loofah gourd. The optimum soil pH is 6.0-70. Water logging is harmful to the crop. Smooth gourd can be grown in a wider range of soil than ridge gourd.

Seed Sowing and Planting

Generally smooth gourd is direct seeded. Transplanting of seedlings in small Polybags is also sometimes practiced as it economizes seed quantity, ensures better plant stand and growth, and enables to raise early crop. The rainy season crop is grown on raised beds. In case of bed planting, seeds are sown on both sides of the bed. The width of the bed is maintained at 150-250 cm and 50-60 cm. furrows are on either side for irrigation. About 2-3 seeds are sown in a hill on the side of the bed and the distance between hill is about 100-150 cm. Seeds (about4-6) are also sown in pits.

The dimension of the pit is 45 cm x 45cm and 60 cm deep and the distance between the pits is 100-150 cm and the row distance is about 200-250 cm. The average seed rate is about 5-6 kg/ha. In Southern region the seeds are sown in December –January and also in June – July for a rainy season crop. The crop can be grown almost throughout the year in Maharashtra and Karnataka. However, it grows the best in *kharif* season. Seeds are sown in January – March and June – July in most parts of Northern India. In West Bengal, seed sowing is done in November – January and also in June – July, seeds are sown in April – May in the hills.

Manures and Fertilizers

About 20-25 tonnes/ha of FYM or compost are applied to the soil at the time of land preparation. The fertilizers required are about 100kg. N, Half quantity of N is applied at the time of bed/pit preparation and the other half quantity of N about 30-45 days after seed sowing.

Training

The vines are allowed to trail on the beds. In the villages, the plants grow on the walls of houses or huts. The plants in kitchen garden or in small growing areas are supported on trellis. During the rainy season, it would be useful to have mulching on beds with dried grasses or straw to avoid rotting of fruits which come on contact with the wet soil.

Intercultivation

About two to three weedings and light intercultivation with a hoe during the early stage of vine growth are necessary for a good crop. Delayed weeding should always be avoided to ensure vigorous and healthy growth of vines.

Irrigation

The summer crop is irrigated at 4 to 6 days interval. However excessive watering should be avoided. The rainy season crop requires much less irrigation depending on the rain fall and in some areas there may not be any need to irrigate the crop. Water logging is harmful to the crop.

Harvesting

The immature fruits are harvested after 60-90 days after seed sowing depending on the variety and the season. The fruit attains marketable stage after 5 to 7 days of anthesis. Fruits should not be fibrous at harvesting time.

Yield

Average yield is about 8-12tonnes/ha. The improved cultivars and F_1 hybrids yield higher, about 20-25 tonnes/ha (Swarup, 2006)

Morphology, Floral Biology and Fruit Set

The gourd family has a distinct set of morphological features as easily distinguished from other plant families. Generally the plants have a fairly long tap root with lateral roots, confined to top layer of 60cm. This crop is adapted to grow in river beds to utilize subterranian moisture and also some of them have xerophytic habit. (Bose *et al.*, 2002). Leaves are simple, mostly 3-5 lobed, variously shaped, palmate, cordate or reniform. Tendrils are borne on the axils of leaves, simple or bifid. Inflorescence is racemose. Flowers mostly unisexual, large and showy, mostly monoecious (staminate and pistillate flowers separately in the same plant). Staminate flowers mostly on long pedicels and are borne in racemes having campanulate, showy corolla, calyx forming a perianth tube, calyx lobes alternating with corolla lobes. Filaments free. Pistillate flowers borne singly in short peduncles, 1 to 5 carpels usually three, thick short style terminated, 3 lobed or divided pistillate stigma (pistillate and hermaphrodite flowers similar). Usually staminate and pistillate flowers borne in different axils, or may be on the same node. Hermaphrodite form, which bears only bisexual flowers, like 'Satputia' cultivar of *L.acutangula* is rare. Fruits are essentially (inferior) berry or many seeded pepo. Seeds are black, flat and not pitted like ridge gourd. In smooth gourd, white patches are present on leaf, male flowers are larger, ovary is cylindrical and tomentose and fruits are smooth (Peter, 2008). Fruit setting starts 45-55 days after sowing of seeds.

Varieties

1. *Pusa Chikni:* A selection from Bihar collection early fruiting cultivar, flowering in about 45 days. The fruits are smooth and dark green colour,more or less cylindrical,15to 20 fruits per vine, suitable for both Spring – Summer and rainy seasons;released by IARI,New Delhi.

2. *Pusa Supriya*: A selection from a weal type, released by IARI, New Delhi.

3. *Pusa Sneha*: It is a variety developed by selection, released by IARI, New Delhi.

4. *GFESSMG-108:* It is a selection from a local cultivar, released by MPKV, Rahuri.

5. An extremely long cultivar 'Yizhangquang' introduced from Zhejiang province of China produces many female flowers. Fruits grow up to 1m long at 20 days after flowering and a single plant produces 50kg fruits in a season.

The greatest need in breeding smooth loofah is for standardization of fruit characteristics. In China there is ample diversity in smooth loofah types. Grown commercially in the Yangtze River regions are long-fruited cultivars, *e.g.* "Xian-si-kua" and "Hu-Lu-qing", which produce fruits up to 150 cm in length. In the past 20 years, research relevant to commercial loofah sponge production has been conducted in tropical West Africa and India. Popularity of sponges in the USA has led to recent investigation on production practices for temperate regions of that country.

Spread of Biodiversity: Germplasm Collection and Exchange

The Indian subcontinent is considered the centre of origin for a number of wild and cultivated cucurbitaceous vegetable crops. Chakravorty (1982) reported that out of 110 genera and 640 species in the world, 36 genera and 100 species are found in India. This includes 38 species apparently to be endemic. It seems that Malayan and Chinese elements have played a great role in the formation of cucurbitaceous flora of India. Many cultivated and wild species of Cucurbitaceae date back to pre-historic times are associated with man's culture. The germplasm collections are being maintained at NBPGR, New Delhi and Project Directorate of Vegetable Research, Varanasi (U.P.). A large diversity in Loofah gourd occurs in India even today. It may be worthwhile to collect, maintain and evaluate the existing germplasm available in the country. A rich diversity of smooth gourd occurs in Indo Gangetic plains, Terai region (Foot hills) and North Easter plains.

One field experiment was carried out at G.B. Pant University of Agriculture of Technology, Pantnagar (U.S. Nagar), Uttarakhand, India in summer season, 2008 to detect the variation for different characters in smooth gourd germplasms. The experiment was laid out in randomized block design (RBD) using three replications. The result is given in Table 12.1.

Days to First Male Flower Anthesis

The anthesis of first male flower was the most advanced in PSG-115 (51.67 days) and the most delayed in PSG -93 (67.67 days) among the parents.

Table 12.1: Diversity Among Various Smooth Gourd Genotypes for the Following Characters

Sl.No.	Characters		Genotypes		
1.	Days to anthesis of first male flower	Earliest	PSG-115 (51.67 days)	Most delayed	PSG-93 (67.67 days)
2.	Node number of first male flower	Lowest	PSG-115 (4.33)	Upper most	PSG-82 (10.67)
3.	Days to anthesis of first female flower	Earliest	PSG-07-04 (54.67)	Most delayed	PSG-93 (77.33)
4.	Node number of first female flower	Lowest	PSG-115 (12.00)	Upper most	PSG-82 (26.67)
5.	Average fruit weight (gm)	Maximum	PSG-07-04 (255)	Minimum	PSG-161 (121.33)
6.	Fruit length (cm)	Maximum	PSG-07-04 (21.99)	Minimum	PSG-161 (18.36)
7.	Fruit diameter (cm)	Maximum	PSG-07-04 (4.56)	Minimum	PSG-161 (3.56)
8.	Fruit flesh thickness (cm)	Maximum	PSG-07-04 (0.57)	Minimum	PSG-93 (0.37)
9.	Number of fruits/plant	Maximum	PSG-161 (19.16)	Minimum	PSG-07-04 (12.4)
10.	Days to first fruit harvest	Minimum	PSG-115 (69.33)	Maximum	PSG-199 (83)
11.	Days to last fruit harvest	Maximum	PSG-82 (133.67)	Minimum	PSG-199 (83)
12.	Duration of fruit harvest	Maximum	PSG-115 (57.67)	Minimum	PSG-199 (22.33)
13.	Number of primary branches	Maximum	PSG-199 (8.32)	Minimum	PSG-161 (6.99)
14.	Main vine length (m)	Maximum	PSG-199 (10.91)	Minimum	PSG-115 (8.91)
15.	Fruit yield/plot (kg)	Maximum	PSG-07-04 (15.31)	Minimum	PSG-82 (8.76)
16.	Fruit yield (q/ha)	Maximum	PSG-07-04 (95.86)	Minimum	PSG-82 (52.23)

Node Number of First Male Flower

The first male flower opened at the lowest node in PSG-115 (4.33) and at upper most node in PSG-82 (10.67) among the parents.

Days to First Female Flower Anthesis

The anthesis of first female flower was recorded earliest in PSG-07-04 (54.67 days) and latest in PSG -93 (77.33 days) among the parents.

Node Number of First Female Flower

Node number of first female flower was the lowest in PSG-115 (12.00) and the highest in PSG-82 (26.67) among the parents.

Fruit Length

The fruit was the longest in PSG-07-04 (21.99 cm) while the shortest in PSG -161 (18.36 cm) among the parents.

Fruit Diameter

The maximum fruit diameter in parents was recorded in PSG-07-04 (4.56 cm) and minimum in PSG-161 (3.56 cm).

Flesh Thickness

The flesh was the thickest in PSG-07-04 (0.57 cm) while the thinnest in PSG-93 (0.37 cm) among the parents.

Fruit Weight

Among the parents, the highest fruit weight was recorded in PSG-07-04 (255.00 gm.) while the lowest in PSG-161 (121.33 gm.)

Number of Fruits per Plant

The number of fruits per plant was maximum in PSG-161 (19.16) and minimum in PSG- 07-04 (12.16) among the parents.

Days to Harvest First Fruit

The harvest of first fruit was recorded the earliest in PSG-115 (69.33 days) and the latest in PSG-199 (83.00 days) among the parents.

Days to Harvest Last Fruit

The harvest of last fruit was recorded the latest in PSG-82 (133.67 days) and the earliest in PSG-199 (111.33 days) among the parents.

Duration of Fruit Harvest

The duration of fruit harvest was longer in PSG-115 (57.67 days) and lesser in PSG-199 (22.33 days) among the parents.

Number of Primary Branches

Among the parents, the number of primary branches was observed to be higher in PSG-199 (8.32) and the lowest in PSG -161 (6.99).

Maine Vine Length

The main vine length was recorded was in PSG-199 (10.91 cm) and minimum in PSG-115 (8.91 cm) among the parents.

Fruit Yield per Plant

Among the parents, PSG-07-04 gave the highest fruit yield per plant (15.31 kg) and PSG-82 gave the lowest fruit yield (8.76 kg).

Fruit Yield/ha

Among the parents, PSG-07-04 gave the highest fruit yield per hectare (95.86 q.) and PSG-82 gave the lowest fruit yield (52.23 q/ha).

Conclusion

Currently, most sponge gourds are produced in tropical or semi tropical environments as in Taiwan, Koria, El Salvador, Guatemala and Clombia. Whole sale prices of $0.40 to $0.50 per sponge, coupled with raising demand for Loofah products and a desire for new high value crop, have stimulated interest among some North American growers (Davis, 1996). Cultivars of smooth gourd are confined to a few places in tropical countries. Bitterness is the most serious undesirable character reducing quality and acceptability of fruits. Commercialisation of crop depends very much upon the development of new varieties devoid of bitterness. Gynoecious hybrids have to be developed for commercial purposes.

References

Bose, T.K., Kabir, J., Maity, T.K., Parthasarathy, V.A. and Som, M.G., 2002. *Vegetable Crops Vol 1.* Naya Prokash, Kolkata.

Chakravorty, H.L., 1982. *Fascicles of Flora of India, Fascicle II, Cucurbitaceae.* BSI, Howrah.

Choudhrary, B., 2003. *Vegetables.* National Book Trust, New Delhi.

Davis, M.J., 1996. *Vegetable Production and Marketing.* Texas Agriculture Extension Service, 6 (7): 1.

Ishihara, H., Sasagawa, T., Saka, Nishikawa, M., Kimura, M. and Funatsu, G., 1997. Isolation and molecular characterization of four arginine/glutamate rich polypeptides from the seed of sponge gourd. *Bioscience, Biotechnology and Biochemistry,* 61: 168–170.

Peter, K.V. and Pardeep Kumar, T., 2008. *Genetics and Breeding of Vegetable Crops.* ICAR Publication, New Delhi.

Pierce, L.C. and Brewbaker, J.L., 1973. Application of isozyme analysis in horticultural science. *Hort. Science,* 8: 17–22.

Swarup, V., 2006. *Vegetable Science and Technology in India.* Kalyani Publishers, New Delhi.

Singh, Ranjit C., Alam, Anis and Singh, Vinod, 2003. Purification, characterization and chemical modification studies on a translation inhibitor protein from *Luffa cylindrica. Ind. J. of Biochem. and Biophy.,* 40: 31–39.

2013, Biodiversity in Horticultural Crops Vol. 4   *Pages 229–253*
Editor: Professor K.V. Peter
Published by: DAYA PUBLISHING HOUSE, NEW DELHI

Chapter 13

Tamarind

V. Ponnuswami, M. Prabhu, S.P. Thamaraiselvi and J. Rajangam

Horticultural College and Research Institute, Periyakulam
E-mail: swamyvp200259@gmail.com

Tamarind (*Tamarindus indica* L.) also called as 'Indian date' is a multipurpose tree known for drought tolerance and used primarily for its fruits, which are eaten fresh or processed, used as a seasoning or spice, or the fruits and seeds are processed for non-food uses. Tamarind is the name derived from the Arabic word 'tamr' which means ripe, dry date and the Persian word 'hind' in reference to the river Indus. The original name could have also arisen from the Indian word, 'thamar' meaning fruit.

Origin and Distribution

Many authors have proposed the origin of tamarind being, India (Morton, 1987), the Far East or Africa (Coates-Palgrave, 1988), Ethiopia (Troup 1921) and the drier savannahs of tropical Africa through sub-Sahelian Africa to Senegal (Brandis, 1921; Dalziel, 1937; Irvine, 1961). The movement of tamarind to Asia is thought to have taken place in the first millenium BC and it has been mentioned in the Indian Brahmasamhita Scriptures between 1200-200 BC.

Taxonomy

Tamarind botanically, *Tamarindus indica* Linn. belongs to the Fabaceae, Sub-family Caesalpiniaceae. The chromosome number is 2n=24 (Anitha Karun, 1985).

The vernacular names in different Indian languages are Assamese – Tetali ; Hindi, Punjabi, Urdu – Imli, Amli ; Bengali, Gujarathi –Ambli, Marathi – Chinch; Kannada – Amli, Hulr ; Malayalam, Tamil –Puli ; Oriya – Teetuli and Telugu – Amlika.

Geographical Distribution

Salim *etal.* (1998) documented the geographical distribution of tamarind.

Africa

Tamarind is endemic to tropical Africa, particularly where it continues to grow wild as in Sudan. It is also cultivated in Cameroon, Nigeria and Tanzania.Tamarind occurs widely throughout tropical Africa, where it is frequently planted as a shade tree (Storrs, 1995). It is commonly found in woodlands, and is well adapted to the arid and semi-arid zones (Albrecht, 1993, quoted by Hong *et al.,* 1996).

Asia

The tamarind has also long been naturalized in Indonesia, Malaysia and Philippines and the Pacific Islands. Thailand has the largest plantation of the ASEAN nations, followed by Indonesia, Myanmar and the Philippines. Tamarind is now widely spread throughout semi-arid South and Southeast Asia (Gamble, 1922; Chaturvedi, 1985). It is presently cultivated in home gardens, farmlands, on roadsides, on common lands and on a limited.plantation scale in India and Thailand, where the species is more economically important. It reached India likely through human transportation and cultivation several thousand years prior to the Common Era. It is grown through out India, except sub-Himalayan region and North West Punjab and is the most commonly grown in the drier warmer areas of the South and Central region, where it thrives the best. It grows in the dry and intermediate zones, up to an elevation of about 600 m through natural regeneration or sometimes as a planted tree (Gunasena, 1999). Tamarind is often used as a roadside or avenue tree grown along canals, particularly in the North and South dry zones (Macmillan, 1943).

The Americas

The tamarind is thought to be introduced into tropical America, mainly Mexico, Bermuda, the Bahamas, and the West Indies by either Portuguese or Spanish colonists or perhaps by African slaves or seamen much earlier, in the 16's CE. In the United States, it is a large-scale commercial crop common in the tropical climes notably South Florida, and as a shade and fruit tree, along roadsides and in dooryards and parks. Tamarind is also found growing throughout the Caribbean islands including Jamaica, Cuba, the Greater and Lesser Antilles and the Dominican Republic. Commercial plantations are also found in Brazil and other Latin American countries.

Botany

Tamarind is a long lived tree and its fruiting capacity increases with age. The productive life is 50-70 years and the normal lifespan of the tree is approximately 150 years.

Root

Tamarind produces a deep tap root and an extensive lateral root system, but the tap root may be stunted in badly drained or compacted soils. The tap root is flexuous and lateral roots are produced from the main root at different levels.

Stem

It is beautiful, slow growing and long lived evergreen tree with spreading habit bearing pinnately compound leaves and grows up to a height of 30m with a spread of 15-20m. Tamarind timber consists of hard, dark red heartwood and softer, yellowish sapwood. The bark is brownish-grey, rough and scaly. A dark red gum exudes from the trunk and branches when they are damaged.The crown has irregular vase-shaped outline of dense foliage. The branches droop from a single, central trunk as the tree matures and is often pruned to optimize tree density and ease of fruit harvest.

Leaf

Leaves are evergreen, bright green in colour, elliptical ovular, arrangement is alternate, of the pinnately compound type, with pinnate venation and less than 5cm in length. At night, the leaflets close up. The pinnate leaves with opposite leaflets are giving a billowing effect in the wind. Laminae are glabrous or puberulent, glaucous underneath and darker green above. Venation is reticulate and the midrib of each leaflet is conspicuous above and below. Leaflets are in pairs, each narrowly oblong, rounded at the apex and slightly notched and asymmetric with a tuft of yellow hairs; at the base obliquely obtuse or subtruncate. At the leaf base is a pulvinus and two small stipules 0.5-1.0cm long which are caducous early on; stipules are falcate, acuminate and pubescent. A permanent scar is seen after leaf fall. Leaflets fold after dark due to presence of lupeol synthesized when light and degraded in the dark (Ali *et al.*, 1998).

Flowering and Fruiting Phenology

In tamarind, terminal vegetative shoots which bear flowers in the next flowering season are produced annually. Two types of terminal shoots have been observed by Nagarajan *et al.* (1997), short ones with an erect habit and long ones with a drooping habit. This has been identified as a useful character to evaluate genotypes, since terminal shoot length and foliage production are highly correlated. Shoot growth continues through the rainy season into the dry season. The new leaves appear in March to April. Flowers emerge on new shoots produced in the spring or summer season. In an inflorescence, the flowers only open on alternative days. In general, flowering and fruiting of tamarind take place in the dry season. An extended spell of dry weather may be essential for fruit development, and trees which grow in the humid tropics without this dry spell often do not bear fruit. At higher altitudes, shoots grow mainly in spring and flower throughout the summer.

In India early, mid and late flowering types of tamarind have been identified and those with delayed flowering habit are reported to be high yielders (Usha and Singh, 1994). Mass flowering is common in tamarind during the flowering season. However, in some trees, flowers can be seen at any time during the year and may be due to genotype x environment interactions (Nagarajan *et al.*, 1997).

The period from flowering to pod ripening is 8-10 months. Ripe fruits, however, may remain on the tree until the next flowering period. In most of the tamarind producing countries, the fruits are harvested from February to March/April, but sometimes the harvesting period may extend to June (Coronel, 1992). In India, fruits are harvested from April to May although in Kerala and other parts of South India fruit collection may be over by the end of February.

Pods ripen in the spring at high altitudes. In Thailand the fruiting season is December to February. In the Philippines, the fruiting season is from May to December with a peak in August to October (Coronel, 1992). North of the equator, fruit ripening is late in autumn or winter (December to January). In Zambia, fruiting is in the following cool dry season after the main dry season (July to November). In Ethiopia, fruiting is during the dry season, September to April (FAO, 1988). Mahadevan (1991) observed a noticeable tree-to-tree variation in flowering and fruit ripening in India. Similar observations were also observed in Sri Lanka.

An experiment was conducted to evaluate the effectiveness of foliar sprays of cycocel (1500 ppm), Ethrel (500 ppm), Triacontanol (20 ml tree-1), IBA (150 ppm), Planofix (100 ppm), micronutrient mixture

(0.5 per cent), ZnSO$_4$ (0.5 per cent) + boric acid (0.3 per cent) + FeSO$_4$ (0.5 per cent) and urea (1.5 per cent) on flowering, pod set and fruit retention at Tamil Nadu Agricultural University, Coimbatore. All the treatments with growth regulators and chemicals exhibited significant effects on flowering, pod set and retention. Treatment with foliar feeding urea resulted in maximum number of flowering (75.7 per cent) and cycocel resulted in maximum pod set (32.3 per cent) and retention (54.7 per cent) per unit area (Ilango and Vidyalakshri, 2002).

Flowers

Blooming occurs in summer and flowers are yellowish with reddish streaks. New vegetative flush during May is followed by flowering. Flowers are borne in lax racemes which are a few to several flowered (up to 18), borne at the ends of branches and are shorter than the leaves, the lateral flowers are drooping.

Flowers are irregular 1.5cm long and 2-2.5cm in diameter each with a pedicel and jointed at the apex. Bracts are ovate-oblong, and early caducous, each bract almost as long as the flower bud. There are 2 bracteoles, boat shaped, 8mm long and reddish. The calyx is long with a narrow tube (turbinate) and 4 sepals, unequal, ovate, imbricate, membranous and coloured cream, pale yellow or pink. Corolla of 5 petals, the 2 anterior reduced to bristles hidden at the base of the staminal tube. The 3 upper ones are a little longer than the sepals, 1 posterior and 2 lateral, these 3 obovate to oblong, imbricate, coloured pale yellow, cream, pink or white, streaked with red.

Flowers are bisexual. Stamens are 3 fertile and 4 minute sterile ones. Filaments of fertile stamens are connate and alternate with 6 brittle-like staminodes. Stamens are united below into a sheath open on the upper side and inserted on the anterior part of the mouth of the calyx tube. Anthers are transverse, reddish brown and dehisce longitudinally. Pollen grains are dimorphic, radially symmetrical, tricolporate, oblate spheroidal in shape and sticky (Perveen and Qaiser, 1998). Nagarajan *et al.* (1997) showed pollen dimorphism in tamarind with two distinct sizes of pollen grains, 40-42 ìm and 22-25 ìm.

Pollen and Seed Dispersal

The presence of nectar in the tamarind flower suggests that pollination is carried out by insects. Studies carried out in Sri Lanka revealed that tamarind pollen grains are sticky and the flowers produce nectar. Sticky pollens are not efficient for wind pollination. Pollination is carried out by red ants (*Oecophylla smaragdna* Fab,), flies and bees. Honey bees (*Apis* spp.) are common visitors to tamarind flowers particularly between 08.00-11.00 hrs and 16.00-18.00 hrs. Observations of honey bee activity on flowers and their visitation patterns suggest that they are effective pollinators. Thus, pollination is mostly by honey bees (Prasad, 1963; Nagarajan *et al.,* 1997 and Thimmaraju *et al.,*1977).

Anthesis and Stigmatic Receptivity

Under Indian situations, flower opening begins as early as 5.30 am and continues up to 8.30 am, with peak at 6.30 am. Mature anthers are reddish brown in colour and dehisce by longitudinal splitting. Anther dehiscence occurs between 10am to 11.30am Stigma remains receptive from one day before anthesis to two days after anthesis. The presence of cleistogamy was recorded (Anitha Karun, 1985). Observations carried out in Sri Lanka revealed that anthesis starts at 16.30- 21.00 hrs and flowers completely open by 04.00hrs. Anther dehiscence occurs from 7.00-11.00 hrs. Tamarind stigmas are receptive for nearly 48hrs with peak receptivity on the day of anthesis (Thimmaraju *et al.,* 1977; Nagarajan *et al.,* 1997).

Fruit Set

Tamarind flowers mid-May to mid-July but natural fruit set is low: 3-5 per cent compared to 70-90 per cent in controlled experiments. Pollen limitation and floral abnormalities are some of the reasons for low fruit set under open pollinated condition (Karale *et al.,* 1997).

In spite of profuse flowering, fruit set in tamarind is very low under open pollination. Studies carried out in Sri Lanka revealed that about 10-15 per cent of flowers developed as fruits although Nagarajan *et al.* (1997) observed only 1-2 per cent of flowers developed as fruits. This may be due to the short-lived nature of the flowers (about 48 h) and also due to pollinator limitation, thus many flowers appear not to be pollinated during their short stigmatic receptive period. All unpollinated flowers drop within two days. In contrast, in controlled cross pollination, fruit set was more than 75 per cent whilst controlled self pollination resulted in 2-6 per cent fruit set. (Nagarajan *et al.,*1997). Usha and Singh (1996) also reported that controlled cross pollination results in higher fruit set and retention in tamarind than when open or self pollinated. Fruit set was only 36 per cent with open pollination whereas it increased to 56 per cent with cross-pollination.

Fertilization and Ripening

Under Indian conditions, the time taken for complete development of flower bud from its visible initiation to anthesis varies from 16 to 20 days (Bajpai *et al.,* 1962). Fruit development in tamarind has three distinct stages: growth, maturation and ripening. The indehiscent pods ripen about 8-10 months after flowering and may remain on the tree until the next flowering period (Chaturvedi, 1985; Rama Rao, 1975).

Fruits

The fruits called pods begin to ripen from February to April. The fruits are 8-20cm long, brown, irregularly curved pods, when fully ripe. The pulp dehydrates to a sticky paste enclosed by a few coarse stands of fibre. The pod has an outer epicarp which is light grey or brown and scaly. The pulp is firm, soft, thick and blackish brown. The pulp is traversed by formed seed cavities, which contain the seeds. The outer surface of the pulp has three tough branched fibres from the base to the apex.

Seeds

The pods may contain 1 to 12 large, flat, glossy brown, obovate seeds of 11-12.5 mm (Morton, 1987) embedded in the brown, edible pulp. Seed chambers are lined with a parchment like membrane. The fleshy cotyledons make up most of the seeds volume and weight, and serve as the sole food storage organ. Seeds average 1800 to 2600 numbers per kg.

Biochemical Composition

Fruits

In general, the dried tamarind pulp of commerce contains 8-18 per cent tartaric acid and 25-45 per cent reducing sugars of which 70 per cent is glucose and 30 per cent fructose. Tamarind fruit contains a biologically important source of mineral elements and with a high antioxidant capacity associated with high phenolic content can be considered beneficial to human health. The phenolics include gallic acid equivalent of 626-664 mg per 100g (Parvez *et al.*, 2003). The fruit is a good source of calcium, phosphorus and iron. The pulp constitutes 30-50 per cent of the ripe fruit, the shell and fibre account for 11-30 per cent and seed about 25-40 per cent. Fruit compositions are variable, depending on locality.

Chemical Composition of Tamarind Fruit (Gunasena and Hughes, 2000)

Constituents	Amount (per 100g)	Constituents	Amount (per 100g)
Water	17.8-35.8 g	Phosphorus	34.0-78.0 mg
Protein	2.0-3.0 g	Iron	0.2-0.9 mg
Fat	0.6 g	Thiamine	0.33 mg
Carbohydrates	41.1-61.4 g	Riboflavin	0.1 mg
Fibre	2.9 g	Niacin	1.0 mg
Ash	2.6-3.9 g	Vitamin C	44.0 mg
Calcium	34.0-94.0 mg		

Seeds

The seed comprises the seed coat or testa (20-30 per cent) and the kernel or endosperm (70-75 per cent) (Coronel, 1991; Shankaracharya, 1998). Tamarind seed is the raw material used in the manufacture of tamarind seed kernel powder (TKP), polysaccharide (jellose), adhesive and tannin. The seeds are also used for other purposes and are presently gaining importance as an alternative source of protein, rich in some essential amino acids. Unlike the pulp, the seed is a good source of protein and oil.

Leaves and Flowers

The leaves are used as a vegetable by indigenous people in producing countries. They contain 4.0-5.8 per cent proteins while the flowers contain only 2- 3 per cent. The leaves are also a fair source of vitamin C and β-carotene and the mineral content is high, particularly in potassium, phosphorous, calcium and magnesium. Leaves contain tartaric acid and maleic acid; the latter is found in excess and increases with the age of the leaves.

The leaves are also used as fodder for domestic animals and by wild animals, including elephants. According to Kaitho *et al.* (1988), crude protein content of the fodder tends to vary with the locality and season. Wild animals prefer tamarind leaves to other fodders due to its high crude protein (12-15 per cent) content.

Uses

Tamarind is a nutritious fruit with a variety of uses. The fruit of the tamarind is most commonly consumed as raw or cooked. The unique sweet/sour flavour of the pulp is popular in cooking and flavouring. All the parts of the tree have some commercial uses.

Fruit Products

Tamarind is used for the preparation of various processed products and some examples of value additions are cited below:

Tamarind Beverage

Tamarind fruit pulp is used for the preparation of beverages in different regions. Good quality ready to serve beverage, syrup and concentrate can be prepared with a shelf life of six months at ambient storage (Kotecha and Kadam, 2003 a). The carbonated tamarind beverage has 12.5 per cent juice, 16° Brix and 0.4 per cent acidity (Lakshmi *et al.,* 2005).

Pulp

Tamarind is valued mostly for its fruit especially the pulp, used for a wide variety of domestic and industrial purposes (Kulkarni *et al.,* 1993). Sweet tamarind is often eaten fresh directly from the pod. More commonly, the acidic pulp is used as a favourite ingredient in culinary preparations such as curries, chutneys, sauces, ice cream and sherbet in countries where the tree grows naturally (Dalziel, 1937; Eggeling and Dale, 1951; Little and Wadsworth, 1964). In Sri Lanka, tamarind is widely used in cuisine as an alternative to lime and also in pickles and chutneys (Jayaweera,1981). It is also used in India, to make 'tamarind fish', a sea-food pickle, which is considered a great delicacy. Immature tender pods are used as seasoning for cooked rice, meat and fish. In India, the pulp is eaten raw and sweetened with sugar (Lotschert and Beese, 1994). It is desirable to remove the pulp without using water when the pulp is used in confectionery.

Seed

Tamarind seed is a by-product of the commercial utilization of the fruit, however, it has several uses. Seed contains a polysaccharide called 'jellose' which has been used as a stabiliser in ice cream, mayonnaise and cheese and as an ingredient or agent in a number of pharmaceutical products (Morton, 1987). In India, seed kernels are used in times of food scarcity in Chennai, Andhra Pradesh and Madhya Pradesh either alone or mixed with cereal flours (Shankaracharya, 1998).

Tamarind xyloglucan, commonly known as 'tamarind gum', is the major component of TKP. It forms a stiff gel and is used for thickening, stabilizing and gelling in food. It is commercially available as a food additive for improving the viscosity and texture of processed foods (Sone and Sato, 1994). Seeds give amber coloured oil, free of smell and sweet to taste, which resembles linseed oil. It could be used for making varnishes, paints and burning in oil lamps (Watt, 1893).

Leaves and Flowers

The leaves, flowers and immature pods of tamarind are also edible. The leaves and flowers are used to make curries, salads, stews and soups in many countries, especially in times of scarcity (Benthall, 1933). In India, leaves are made into a dish called 'Chindar'. The seedlings are also eaten as a vegetable. Young leaves of tamarind are used as a seasoning vegetable in some Thai food recipes because of their sourness and specific aroma (Coronel, 1991). Flowers are an important nectar resource for honeybees in South India. The honey is of golden yellow colour and slightly acidic in flavour (NAS, 1979; Sozolnoki, 1985).

Medicinal Uses

The laxative properties of the pulp and the diuretic properties of the leaf sap were confirmed by modern medical science (Bueso, 1980). Tamarind products, leaves, fruits and seeds are extensively used in traditional Indian and African medicine (Jayaweera, 1981; Parrotta, 1990). Several medicinal properties are claimed for preparations containing tamarind pulp, leaves, flowers, bark and roots (Bueso, 1980).

Pulp

Tamarind pulp alone or in combination with lime juice, honey, milk, dates, spices or camphor is used as a digestive and a carminative, even for elephants, and as a remedy for biliousness and bile disorders and febrile conditions. It is said to improve loss of appetite. Tamarind is used in the treatment of a number of ailments, including the alleviation of sunstroke, *Datura* poisoning (Gunasena and Hughes, 2000), and the intoxicating effects of alcohol and 'ganja' (*Cannabis sativa* L.). It is used as a gargle for sore throats, dressing of wounds (Benthall, 1933; Dalziel, 1937; Eggeling and Dale, 1951; Chaturvedi, 1985) and is said to aid the restoration of sensation in cases of paralysis. The fruits are reported to have anti-fungal and anti-bacterial properties (Ray and Majumdar, 1976; Guerin and Reveillere, 1984; Bibitha *et al.*, 2002; Metwali, 2003; John *et al.*, 2004).

Seed

The seed is usually powdered and is often made into a paste for treatment of most external ailments. In Cambodia and India, it was reported that powdered seeds were used to treat boils and dysentery (Rama Rao,1975; Jayaweera, 1981). Seed powder has also been externally applied on eye diseases and ulcers. Boiled, pounded seeds are reported to treat ulcers and bladder stones and powdered seed husks are used to treat diabetes (Rama Rao, 1975). The seed can also be used orally, with or without cumin seed and palm sugar, for treatment of chronic diarrhoea and jaundice.

Leaves

Tamarind leaves are usually ground into powder and used in lotions or infusions. The leaves, mixed with salt and water, are used to treat throat infections, coughs, fever, intestinal worms, urinary troubles and liver ailments. Internally, leaves and pulp act as a cholagogue, laxative and are often used in treating 'congestion' of the liver, habitual constipation and haemorrhoids. Leaf extracts also exhibit anti-oxidant activity in the liver.

Industrial Uses

Tamarind pulp is used as a raw material for the manufacture of several industrial products, such as Tamarind Juice Concentrate (TJC), Tamarind Pulp Powder (TPP), tartaric acid, pectin, tartarates and alcohol (Anon, 1982 a; 1982 b). Tamarind Kernel Powder produced from seeds is another commercial product and is often reported upon in commercial digests (e.g. Mathur and Mathur, 2001).

Construction

The seed is also used as filler for adhesives in the plywood industry and a stabiliser for bricks, as a binder for sawdust briquettes and a thickener for some explosives. Ground, boiled and mixed with gum, the seeds produce a strong wood cement (Benthall, 1933; Rama Rao, 1975). A composite material of tamarind seed gum is suitable for construction applications such as false roofing and room partitioning.

Paper Making

Uses of xyloglucans as alternatives to currently used wet-end additives in paper making were studied by Lima *et al.* (2003). Xyloglucans improved the mechanical properties of paper sheets without affecting the optical ones.

The fruit pulp may be used as a fixative with turmeric (*Curcuma longa*) and annatto (*Bixa orellana*) in dyeing, and it also serves to coagulate rubber latex. The seed testa contains 23 per cent tannin, which when suitably blended is used for tanning leather and imparting colour fast shades to wool.

Wood

Tamarind wood has many uses including making furniture, wheels, mallets, rice pounders, mortars, pestles, ploughs, well construction, tent pegs, canoes, side planks for boats, cart shafts and axles, and naves of wheels, toys, oil presses, sugar presses, printing blocks, tools and tool handles, turnery, etc.(Coates-Palgrave, 1988; Troup, 1909). Tamarind heartwood is considered to be a very durable timber and is used in furniture making as it takes on a good polish (Jayaweera, 1981).

Value Added Products

Tamarind Kernel Powder

The major industrial use of the seeds is in the manufacture of Tamarind Kernel Powder (TKP). It is prepared by decorticating the seed and pulverising the creamy white kernels. In India, TKP is used as a source of carbohydrate for the adhesive or binding agent in paper and textile sizing, and weaving and jute products (Anon, 1976; Shankaracharya, 1998) as well as textile printing (Khoja and Halbe (2001).

Pectins

Polysaccharides obtained from tamarind seed kernels form mucilaginous dispersions with water and possess the characteristic property of forming gels with sugar concentrates, like fruit pectins. However, unlike fruit pectin, tamarind polysaccharide can form gels over a wide pH range, including neutral and basic conditions. Tamarind polysaccharide can be useful as a gel formation agent, and may be substituted for fruit pectins. Tamarind polysaccharide does not contain galacturonic acid and methyluronate and is therefore not regarded as true pectin; it is termed 'jellose' (Rao, 1948).

Tamarind Juice Concentrate

A process for the preparation of tamarind juice concentrate was developed by Central Food Technological Research Institute, Mysore, India. The process involves extraction of tamarind juice and concentrating the juice in a vacuum evaporator to total soluble solids of 70°Brix. The tamarind concentrate is shelf stable at room temperature and can be diluted and used in various food preparations.

Tamarind Pulp Powder

Tamarind pulp cannot be kept indefinitely because of browning in storage due to phenolics and non enzymatic browning (Kotecha and Kadam, 2003 b). Here it is processed into pulp powder. Tamarind Pulp Powder (TPP) is prepared by concentrating, drying and milling the pulp into a powder form. It is one of the convenience food products produced commercially by several manufacturers in India. Starch is the major ingredient in tamarind pulp powder (20-41 per cent). On a small scale, the fruit pulp is made into a refreshing drink after dissolving in water and squeezing by hand. Tamarind pulp is enjoyed as a refreshing drink and beverage in most of countries. Tamarind drinks in polypacks are commercially available in Thailand, Indonesia and many countries of Africa.

Tamarind Pickle

Pulp is used commercially to prepare tamarind pickle. The pickles are commonly used in Asia as an accompaniment to curries or other main meals. Pickles are hot, spicy and have a salty-sour taste, and can be preserved for several months. Preservation is due to presence of salt, increased acidity and spices. The preparations of pickles are simple and can be done at household level.

Antioxidants from Tamarind Seed Coat

Extraction of antioxidant compounds from the seed coat of sweet Thai tamarind was reported by Luengthanaphol *et al.* (2004). The antioxidants were extracted by supercritical fluid extraction and by solvent extraction using ether or ethyl acetate. The mixture obtained by solvent extraction with ethanol showed epicatechin yield of 150mg/100g and the highest antioxidant activity in terms of peroxide value as compared to the supercritical fluid extraction.

Crude tamarind seed coat extracts were stable following heat treatment at 100° C for two hours. Antioxidative activity of the extracts was lower at pH 5.0 than at pH 3.0 or 7.0.

Species and Cultivars

Tamaraindus is a monotypic genus belonging to family Fabaceae. On the basis of fruit size and shape, Bailey (1947) recognized two types of tamarind.

1. East Indian type having long pods with 6-12 seeds,
2. West Indian type having shorter pods containing 1-4 seeds.

Seeds also exhibit a wide range of variation in shape, size, colour and the ornamentation of the seed coat. On the basis of pulp colour, there are two distinct phenotypes of tamarind based on mainly their pulp colour:

1. The yellow or brown pulp type, turning dark brown on storage, it is harvested after full maturity.
2. In red tamarind, all parts of pistil, staminal sheath and filaments have bright red colour (Karale, 1998). The red type is sweeter than brown type because it has lower content of free acids and is generally harvested when fruits are immature and green. It is mostly preferred for making preserves.

Tamarind Pods

Tamarind has been recorded over a century ago as a variable species especially for pulp colour and sweetness (Von Mueller, 1881). Since there is such extensive variation in characters like foliage, flower and pod production and timber quality, there is considerable scope to improve the species. Improvement holds the key for boosting productivity and yield of the orchards and involves development of genotypes possessing desirable characters like fast growth, good tree form, high yield and resistance or tolerance to major pests, diseases and drought (Radhamani *et al.,* 1998). Since the variation in pod length and pod width was found genotypic (Shivanandam and Raju, 1988; Hanamashetti, 1996). Similarly for other trials, the potential for improvement depends on sampling the genetic variability available within and between populations. Hence, knowledge of genetic variation and structure of a species and genetic parameters of important traits are essential to develop effective improvement and conservation strategies.

The apparent high observable variability within and between populations indicates that speedy benefits may be obtained by selecting superior trees within provenances and propagating such stocks as clones. A considerable variation in growth characters such as shoot length, root length, germination percentage, plant height and pinnae per plant were also observed in different populations (Challapilli *et al.,* 1995; Bennet *et al.,* 1997; Divakara, 2002; Shanthi, 2003). Recent studies on isozyme analysis of ten tamarind populations from Coimbatore, India identified 12 loci with 25 alleles (Shanthi, 2003). All populations had more than 50 per cent polymorphic loci (range from 50 -75 per cent). The mean number of alleles per locus was 1.85 (range from 1.6-2.0). Mean heterozygosity observed among 10 populations ranged from 0.085-0.154 whilst mean heterozygosity expected according to the Hardy-Weinberg equilibrium ranged from 0.163-0.258.

The mean total genetic variation of the 10 tamarind populations was 0.322. The total genetic variation within population was 0.291 whilst variation between populations was 0.03. The relative extent of gene differentiation among populations was identified as 0.115 (Shanthi, 2003). Even though this showed a low level of genetic diversity, it was mainly because the populations sampled were from a small geographic region. A high level of within population variation and low level of between population variations also revealed that the species is highly outcrossing, and there is ample individual variation to use in genetic improvement. Based on the assessment of candidate 'plus' trees and their progenies, Shanthi (2003) concluded that pod size, tree height, number of flowers, number of inflorescences and number of tertiary branches could be used as measures for evaluating tamarind trees to select superior trees for further evaluation.

The experiments carried out in Coimbatore revealed that the highest *phenotypic* variation occurred for pod length and the lowest for seed weight whilst the maximum *genotypic* variation was observed for pulp weight. The highest phenotypic and genotypic coefficients of variation were recorded for pulp weight (Chundawat, 1990; Challapilli *et al.,* 1995; Divakara, 2002; Shanthi, 2003). A maximum heritability of 0.5 and the highest genetic advance (percentage) were recorded over a mean of 42.5 for pulp weight. The highest genetic advance was observed for pulp weight. Pod length showed the highest positive phenotypic correlation with pulp weight. Hence, selection based on pulp weight and pod length is useful in tamarind improvement programmes. Further, because of presence of significant positive correlation coefficients between pod length and pod width, they are suggested as selection criteria for identification of superior trees.

Genetic improvement through use of superior clones was described by Kulkarni *et al.* (1993). Tamarind has a relatively long generation time and is primarily outcrossing, thus any conventional breeding approaches would require considerable investment in time and money. There are trees with very high yielding potential exceeding 800 kg/tree/year and such trees could be selected for vegetative propagation by air layering or grafting methods to produce fast growing trees for local (home garden) and commercial (orchard) use. Such trees can be grown with comparative ease and minimum management and should prove to be profitable due to their commercial value for small-scale farmers in African and Asian countries.

If quick growing and high yielding strains are selected for different uses, tamarind would rapidly become a desirable tree for small scale plantations in a short period, particularly since it continues to bear fruit for many years. More traits specific research and combining desirable characters together are needed to develop cultivars. An emphasis should be to select for agroforestry systems and small scale plantations. The desirable ideotypes should be developed to fit into various niches of these systems as well as for more intensive commercial production. This will require more attention to tissue culture since it offers a way of cleaning stocks from diseases. Then planting and distribution systems need to be developed to provide materials to growers and even to exchange internationally.

Selection

The selection of elite trees is an important step. These trees can be selected using the following characteristics: acidity of the pulp, content of tartaric acid and sugar, real value of pulp, pod bearing ability (flowering and fruit maturing), pod size, pulp, fibre and seed weights, and number of seeds. In India, local farmers usually identify mother trees which consistently produce large number of fruits for their seed collection and propagation. Some of the trees selected on this basis are reported to have large pods, 25 cm long and 5 cm wide (Jambulingam and Fernandes, 1986).

The major breakthrough in recent years is the identification of tamarind types with less acidic pulp, commonly referred to as sweet tamarinds. In Thailand and the Philippines, farmers are growing the sweet types on a limited plantation scale and they are also known locally in India. Selections have been made from natural stands growing in these countries. In Thailand, more than 50 sweet tamarind cultivars are grown, while in the Philippines, eight selected cultivars are popular among the farmers. These sweet types have created a resurgence of interest in many countries of the Southeast Asian region and encouraged researchers to undertake studies on tamarind.

The sweet tamarind is attributed to a point mutation. Occasionally isolated branches on a tree may bear sweet fruits while others bear normal sour ones. Bud sports have been propagated vegetatively and form the basis for a range of recent cultivars.

Participatory Improvement

Improvement of multipurpose trees like tamarind should also be participatory, involving farmers in all phases of the programme. As tamarind is presently a smallholder crop, this approach is even more important at this stage, before steps are taken for commercialisation. The local people with their indigenous knowledge could provide information on why they favour a particular selection, and characterisation based on users' perspectives will be an essential starting point for improvement of tamarind. The preferences will also vary from country to country. As such the use of Participatory Rapid Appraisal (PRA) or Diagnosis and Design (D and D) could provide technical information on reproductive biology, propagation techniques, existing conservation, management practices and handling and processing technologies. The approach may be the most important for countries which do not have strong national programmes on genetic improvement of underutilised tree species.

Ideotypes

It is essential to identify different ideotypes for different purposes, localities, environments and cultural practices. There is no universal ideotype suitable for all sites and end users. Ideotype identification should focus on specific services and products that could be marketed, either locally or internationally.

Genetic Resources

Tamarind populations are also not considered endangered and hence there is low priority for its conservation. However, several countries in Africa prioritised tamarind for conservation. In Sudan, prioritisation for genetic conservation has been studied by collecting information through on-market surveys of non-wood products in relation to export and home consumption (FNC/FAO, 1995). These surveys revealed that tamarind products used for home consumption ranked number one among the species studied. Other countries in Africa, (Burkina Faso, Cameroon, Chad, Côte d'Ivoire, Gambia, Guinea-Bissau, Kenya, Mauritania, Nigeria and Senegal) have prioritised it for conservation based on its utilization and value.

The wide variability that exists in tamarind in terms of acidity of the pulp, pod bearing ability and pod size have to be used for genetic improvement in countries of Africa, and South and Southeast Asia. In order to incorporate such traits, germplasm collections have to be available in each region for scientists to work on. The collections should be subjected to continued evaluation to select and release desirable types. In this respect, an African Tamarind Network has been formed to study genetic variation in the African tamarind under the sponsorship of ICRAF, Nairobi, Kenya. From such diverse germplasm, plants with sweeter, juicier pulp and other desirable characteristics can be selected for tree improvement programmes. In recent years less acidic, sweet tamarind types have been selected in the Philippines and Thailand and are becoming popular.

Studies in Karnataka which evaluated the correlation between different attributes of growth and fruits in 17 genotypes, and others in Thailand, are now standard guides for evaluation (Challapilli *et al.,* 1955; Birdar and Hanamashetti, 2001). The length of the fruit is also reported to be positively correlated with fruit weight, pulp, and number of seeds. Fruit thickness was negatively correlated with fibre weight, seed weight and seed number. Since fruit length, weight and thickness are measures of fruit size, the larger the fruit, the heavier the pulp weight (Shivanandam and Raju, 1988).

Thailand

In Thailand, wide genetic diversity in tamarind germplasm of both sweet and sour types are observed in all regions. Germplasm collection was undertaken in 1986 and 1987. A total of 1811 accessions were collected during this period. Several other accessions, such as No: 86-2-23-017, No: 86-13-008 I and No: 87-2-01-035 were selected for improvement for industrial purposes (Feungchan *et al.,* 1996). Some of the names given to the selections relate to the district from where the clone was first selected (*e.g.* Jae Hom) or the name of the grower (e.g. Muen Chong).

Fruits with red and brown coloured pulp have also been found in the accessions from Thailand. At the University of Khon Kaen, cultivar selection is more advanced than in other countries. Several accessions are selected which are potentially high yielding. In Thailand, there are more than 50 sweet tamarind cultivars under cultivation. They include Muen Chong, Sri Tong, Nam Pleung, Jae Hom, Kun Sun, Kru Sen, Nazi Zad and Sri Chompoo. A cultivar commonly grown and popular with Thai farmers is Makham Waan.

In this study, it was also found that excellent accessions had RVP values of over 21 and the number of pods was about 13-15 per kilogram. Most of the best accessions were found in provenances along the Mae Kong River, but the role of the river in all these cases has not been fully understood.

Thailand is also paying attention to the percentage of tartaric acid and sugar, the main determinants for taste and flavour. Accessions were selected for improvement for industrial purposes. One accession (No: 86-2-23-017) has given the highest content of tartaric acid (up to 11.2 per cent) and has potential for commercial extraction of tartaric acid. Another accession (No: 86-13-008) had a sugar content of 39.9 per cent and could be valuable for fructose production. Yet another accession (No: 87-2-01-035) gave the highest concentrated tamarind flesh and suitable for processing into sauces, drinks and confectionery.

Philippines

Variation is observed in fruit forms in the Philippines. Some are oblong and short, while others are long and curved. Sweet and sour types are found in different parts of the country. Several trees bearing sweet fruits are found in Cavite and Laguna. The University of the Philippines at Los Banos long ago selected and identified sweet tamarind varieties, namely Cavite, Batangas, Bulacan, and Laguna. These varieties have long plump fruits with thick sweet pulp. Some of the sweet types have pods 7-10 cm long and 2-3 cm wide, weighing 20-30g, containing thick sweet pulp with 6-7 seeds. Forty six accessions of tamarind germplasm are presently available in the Institute of Plant Breeding, Los Banos, Philippines. The varietal characteristics of tamarind lines introduced from the Philippines were evaluated at the US Department of Agriculture, Subtropical Horticulture Research Station at Miami, Florida (Knight, 1980) and the station developed a cultivar called 'Manila Sweet'.

India

In India, most of the area under tamarind cultivation is planted with unselected inferior cultivars. Selections from tamarind provenances are available. The Bharath Agricultural and Industrial Foundation (BAIF), Pune, India attempted to supply improved planting material to smallholders by selecting superior trees from among existing natural populations. The parameters for selection of superior trees were based entirely on pod characters such as pod length, pod colour and pulp yield per pod and the association of fruit characters has been worked out (Karale *et al.*, 1999), as well as genotype x environment effects of different locations on fruit size grades (Parameswari *et al.*, 2000).

Under this programme, 15 superior tamarind provenances were identified and a multiplication stand was established in the BAIF's Central Research Station in Pune. Among them, selections such as Prathisthan from Maharashtra and Periyakulam (PKM-1) and Urigam from Tamil Nadu are well established and preferred by farmers.

Some of the important varieties released in India are as follows:

PKM-1

It was released from Horticultural College and Research Institute, Periyakulam. It is mainly propagated through grafts. It is an early bearing, high yielding (263kg pods/tree), high pulp recovery (39 per cent) with sweet pulp, less fibrous pods and gives 26tonnes pods/ha at a spacing of 10 x 10m. It is highly suitable for high density orchards (160 plants/acre is recommended instead of 40 plants/acre under conventional planting). The variety has purple pigmentation in its terminal buds and the stems are dark brown in colour. Fruits are borne in clusters of 4-5, slightly curved with brown pulp.

Urigam

It has long and fleshy sweet pulp. It bears only 2-3 pods per bunch and yield is low in comparison to PKM-1.

PKM-1

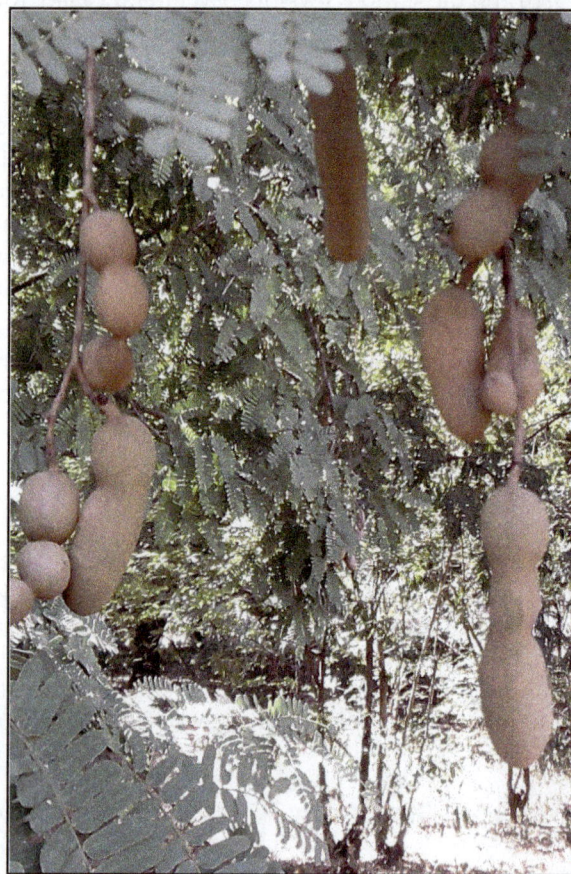

Urigam

Yogeshwari

A high yielding red type was released by Marathwada Agricultural University, Parbhani, Maharashtra.

Prathisthan

It is released by Fruit Research Station, Aurangabad and is a sweet type and has 61 per cent pulp, 12 per cent seed and 27 per cent shell (Anonymous, 1984).Studies were also undertaken to select high yielding cultivars based on their flowering pattern. In seedling populations, early, mid and late flowering tamarind types were identified. Duration of flowering is longer in late flowering trees than in mid and early flowering trees. In mid and late flowering trees natural cross-pollination is greater than with early flowering trees which are mostly self-pollinated under natural conditions. Hence mid and late flowering trees can be selected and are most suitable for selection for improvement (Usha and Singh, 1994). There is also interest in India in using fruit tree genotypes for ameliorating marginal farmland, *e.g.* in semi-arid areas of Gujarat or wastelands as in Tamil Nadu. Pareek (1999) lists selections 'Yogesharai', 'PKM1', 'T3' and 'TIB' as suitable cultivars for arid zones.

Identified Cultivars

In India, Thailand and the Philippines, two major types of tamarind are recognised based on the sweetness of the fruit pulp. These are 'sweet fruit' and 'sour fruit' types. Sometimes the cultivars are differentiated by pulp colour, which may be red or brown. The red coloured pulp is of superior quality due to its sweet taste. Generally unselected tamarind cultivars are grown and these vary in fruit size and degree of sweetness. In the Philippines, eight recognised cultivars are grown by farmers while over 50 are grown in Thailand. Specific sweet and sour cultivars also exist in the east and northern parts of Uganda and Kenya.

Genetic Conservation

Although wild and naturalized tamarind is widely available, some genetic erosion is occurring in most areas due to rapid deforestation resulting from rural development and urbanisation. Population pressure on limited ecosystems and resources is causing rapid and damaging shifts in land use patterns with increasing areas being planted with perennial cash crops. Selective felling of individual trees of tamarind for timber and fuelwood in rural locations can also cause genetic erosion as farmers' selections are lost. Instances of loss of germplasm were reported in different countries. The development of the Mahaweli River Development Project in the dry zone of Sri Lanka removed 200,000 ha of dry zone forest in which tamarind thrived in forest ecosystems. In Southern Malawi, deforestation is reported to be the major cause for the loss of wild and semi wild fruit trees such as *Tamarindus indica, Ficus* spp., *Annona senegalensis* and others. Here, tamarind is one of the most important indigenous fruit trees for consumption and sale in the Mangochi district of Malawi, but the number of trees are decreasing at a high rate and becoming rare: this is now of concern to the local communities. (Maliro and Kwapata, 2002). A strategy for such fruit trees should be promoted to include conservation, enhanced domestication and systematic germplasm maintenance in working collections to back up improvement activities.

Options for genetic conservation include *ex situ* management, one of the easiest and the most economically and socially acceptable method for tamarind conservation (Singh, 1995). In this way, phenotypically superior germplasm is collected from different regions and multiplied through vegetative propagation and maintained in clonal orchards. Seed germplasm banks can be maintained since seeds may be stored under conservation conditions for long periods. However this is a heterozygous material and cannot quickly replicate superior phenotypes. Conservation through utilisation is likely to be the most successful method since access to trees provides a sustainable incentive for local farmers to conserve them. Hence there is need to develop strategies for on-farm conservation (including home gardens).

On-farm conservation should be linked to the establishment of protected areas as practised for conservation of forest genetic resources (Collins *et al.,* 1991). This method of *in situ* conservation has inherent problems for tamarind as individual trees are often grown in villages, roadsides, boundary plantings etc. and not in mixed stands as in natural reserved forests. Protected areas and on-farm conservation can be linked to wood lots or patches of trees when grown in different locations.

Cultivation

Planting

The trees may be grown as an orchard crop in a pure stand, as an agroforestry species in mixed cropping systems including home gardens, or as a hedgerow tree. Tamarind can be established in the field by directly sowing seed (Chaturvedi, 1985). Seeds may be sown directly to establish plantations, hedge rows, or home gardens. On suitable sites, direct seeding of tamarind may indeed be a more economical method of establishment, as it eliminates the cost of growing seedlings in the nursery and is less time-consuming than planting seedlings. Also, germinating seedling has an undisturbed root system and does not suffer from transplant shock. Seeds should be planted as soon after collection as possible to minimize loss of viability. The seed should be planted not deeper than 1.5 cm. For planting on the square, on the triangle, or in lines for orchard establishment, a few seeds are dibbled in 5 cm apart at each planting station. Planting stations should be approximately 4-5 m apart. In Tamil Nadu and Andhra Pradesh, the seeds are sown at random in discrete patches, at 4-5 m. When the seedlings attain 10-20 cm, thinning is done by retaining the one with the best growth.

Transplanting nursery grown plants into the field can have advantages over direct seeding as the plants are already established. But the transplantation shock may result in increased vulnerability to drought, insects, diseases and other problems. Regular care during the first 2-3 years following transplanting, is thus very important. In order to provide space and an ideal medium for the development of a vigorous and deep tap root system, planting should be done in 1 x 1 x 1m pits filled with well rotted organic manure at the time of planting.

A technique for raising large quantities of planting stock for use in roadside plantings has been described by Swaminath (1988). In this method, nursery seedlings are pricked out and transplanted into 12 m x 0.6 m x 0.6 m trenches

Planting stock

dug two months beforehand. The trenches are filled with silt to 10-15 cm from the top. The seedlings are planted 60 cm apart and watered daily by flooding the trenches. After three months, the watering is reduced only once a week. After about seven months, when the height is about three metres and girth 6-8 cm, watering is stopped and the seedlings are dug up with a ball of intact earth. They are then tied with paddy straw and kept in a nursery and watered until they have recovered from the uprooting shock. When a new flush of leaves appears, they are ready for transplanting to their permanent roadside positions.

Overgrown seedlings can be more effectively transplanted as stumped seedlings, with stem and tap root pruned to lengths of about 5 cm and 20-25 cm, respectively (Troup, 1921). When seedlings or grafted plants are transplanted to the field, their size may range from 0.4-2.0 m tall. In some districts of Tamil Nadu, seedlings are encouraged to grow up to 2 m tall by applying manure and removing axillary buds. As these field planted seedlings can be tall and lanky, they should be staked for at least a year, or until they can support themselves. The planting holes should be well prepared and be at least 30-50 cm3 in size.

Field Spacing

Grafted plants of 'PKM-1', 'Pradishtan', and 'Urigam' cultivars are being grown in Tamil Nadu either as isolated trees or in commercially viable grooves in dry tracts. A spacing of 10 x 5 m is recommended for commercial planting, and about 250 trees will be enough to cover a hectare. In Central, Eastern and North-eastern Thailand, the recommended tree to tree spacing in sweet tamarind orchards is 10-12 m (Yaacob and Subhadrabandu, 1995). In the Philippines, the spacing recommended for seedlings is 16-18 m and for vegetatively propagated plants 8-10 m. In Sri Lanka, unproductive scrub forest lands have been successfully planted with tamarind by the Janata Estate Development Board at a spacing of 3 x 3 m in lines (Rodrigo, 1992).

Time of Planting

The best time for field planting is at the beginning of rainy season, as soon as there is sufficient moisture in the soil. This will reduce the need for frequent watering until the plants are firmly established in the soil. In the initial stages of growth, the plants may need some watering, especially during the hot summer months. If irrigation is available, field establishment may be undertaken at any time of the year, even in the dry season. However, it is advisable to provide partial shade to the newly established plants if planting is carried out during the dry periods. Seedling growth in the field is initially fast (about 1.2 m in the first two years) but slows later. In the Philippines, the planting time is from May to June, and in Sri Lanka from April to June, but preferably from October to December to coincide with the onset of the main rainy season (Gunasena, 1997).

Irrigation

Irrigation is not normally practised in tamarind cultivation, but promotes better growth during establishment and the early stages of growth, especially during the dry seasons (Yaacob and Subhadrabhandu, 1995). Where irrigation facilities are available, watering should be done and repeated as the need arises in the early stages of growth. In later years as the deep tap root system develops, the need for watering becomes less. Flowering and fruiting are promoted by irrigation. In dry areas, the use of water harvesting techniques during the rainy season should be considered as it encourages subsequent growth and fruiting (Chundawat, 1990). Mulching during the dry season will also help to reduce water losses from evaporation. Mulches around the trees also help in weed control and water conservation.

Fertilisers and Manures

In experimental plots in India, 25 g urea and 25 g single super phosphate per plant are used, although response of tamarind to fertiliser application is reported as poor (Gupta and Mohan, 1990). In India, inorganic fertilisers are not normally applied to tamarind trees, but 5 kg of farmyard manure is applied to the planting hole at the time of planting. Every year thereafter 5 kg of farmyard manure and 5 kg of neem cake are applied per tree in March and April. The commercial sweet tamarind growers in Thailand use inorganic fertilisers, mostly urea at 100-200 g/tree, which supports high yields. In the Philippines, the general recommendation is to apply 100-200 g/tree of ammonium sulphate, about a month after planting and an equal amount at the end of the rainy season. The amount of fertiliser is gradually increased as the trees grow. When the tree begins to bear fruits, about 500g of a complete fertiliser containing high amounts of nitrogen and potassium is applied per tree twice a year. A full bearing tree should receive 2-3 kg of an NPK complete fertiliser mixture each year, the formula depending upon the nutrient content of the soil (Gunasena and Hughes, 2000).

Ilango and Vijayalakshmi (2002) reported that foliar spray of Cycocel (1500 ppm), Ethrel (500 ppm), Triacontanol (20 ml tree-1), IBA (150 ppm), Planofix (100ppm), micronutrient mixture (0.5 per cent), $ZnSO_4$ (0.5 per cent) + boric acid (0.3 per cent) + $FeSO_4$ (0.5 per cent) and Urea (1.5 per cent) significantly improved flowering, pod set and retention, yield and quality attributes of tamarind compared to non-sprayed trees. Treatment with foliar feeding of urea resulted in maximum number of flowering (76 per cent), with Cycocel resulted in maximum pod set (32 per cent) and retention (55 per cent) per unit area, and with Triacontanol (20 ml tree^{-1}) resulted in maximum pod yield (8.3 kg tree^{-1}).

Pruning and Training

Initial training and pruning of young plants during the first years are essential for development of well-formed trees. Tamarind is a compact tree and produces symmetrical branches. Young trees should be pruned to allow 3-5 well spaced branches to develop into the main scaffold structure of the tree. In the Philippines, young trees are pruned in the early stages of growth to train 2-3 lead branches and to remove the very low branches, thus developing a desirable frame. Bearing trees require very little pruning other than maintenance pruning to remove dead, weak and diseased branches and water sprouts (Salim *et al.,* 1998). In closely planted orchards, regular pruning is needed to rejuvenate fruiting wood and control size of trees.

Intercropping

Tamarind allows intercropping with a variety of annual crops. Vegetables and legumes can be grown during the rainy season in the interspaces in the first three to six years to augment farm income and improve soil fertility. Once fully established, the trees will provide regular income to the grower even when other annual crops fail at times of protracted drought. The intercropping period is usually limited to four years for vegetatively propagated trees. In Thailand's central delta, intensive cropping is practiced in the plantations of the sweet cultivars, which are always vegetatively propagated and transplanted. Intercropping can be extended up to even 8-10 years for the slower maturing seedling propagated plantations. In Central America, Mexico and Brazil where there are more than 4000 ha of well maintained tamarind orchards, intercropping is practised in the early stages. Intercrops, and where they are not used, cover crops, are useful to control competition from weeds and to conserve moisture. Intercrops should be fertilised in their own right, to reduce competition for nutrients and to maximise their yield. There is a wide choice of intercrops which can be grown depending on the soil, climate and local market demand.

Maturity and Harvesting

Trees grown from seed may take more than seven years to start bearing and up to 10 or 12 years before an appreciable crop is produced. Unselected trees in India are late bearers and may take 10-14 years before fruiting. It is reported that in Madagascar, trees will begin to bear in the fourth year and in Mexico in the fifth year. The sweet cultivars planted in Thailand bear in about 3-4 years. These differences in the onset of bearing may be associated with genetic or ecological factors. According to Coronel (1991), plantations in the central delta of Thailand become dwarfed due to high water table, which prevents growth of a deep root system. Such stress conditions lead to early bearing. In all cases, pod yield stabilizes at about 15 years and continues for up to 50 or 60 years

Tamarind seed pods fill at maturity. The pulp becomes brown to reddish brown and the testa becomes brittle and cracks easily. The pulp dehydrates and becomes sticky and the seeds become hard, dark brown and glossy. Mature fruits have brown shells, while immature ones are green. Tapping them with the finger can help to identify the mature fruits. A hollow and loose sound will be produced as the pulp shrinks with maturity and the shell becomes brittle. At this stage, the fruits are mature and ready for harvesting. However, it is not always easy to determine whether the fruits are ready for harvesting, as the testa colour only changes slowly as the pods mature. Individual fruits on the same tree also mature at different times making it necessary for harvesting to be done selectively.

The fully ripened fruits can remain on the tree for six months. Fruits are harvested at two stages depending upon the market. Green pods are harvested for flavouring and ripe pods for processing. The fruits of sweet types are also harvested at two stages, half ripe and fully ripe. At the half ripe stage, the pulp is yellowish green and has the consistency of an apple, particularly in sweet forms. At the fully ripe stage, the pulp shrinks, due to loss of moisture, and changes to reddish brown and becomes sticky. The fruits are green initially but turn brown at maturity. The shells are brittle and easily broken. The pods should be allowed to ripe on the tree until the exocarp becomes dry and hard and separates from the pulp without adherence.

In the Philippines, fruits are harvested at two stages, green for flavouring and ripe for processing. The fruits of sweet types are also harvested at two stages, half-ripe and ripe. At the half ripe stage the pulp is yellowish green and has the consistency of an apple, particularly in sweet forms. At the ripe stage, the pulp shrinks, due to loss of moisture, and changes to reddish brown and becomes sticky.

In most countries, the sour tamarind ripe fruits are harvested by shaking branches and pods are collected on a mat. Sweet tamarind pods fetch a high price both in local and foreign markets and are carefully harvested by hand picking. Sometimes, bamboo ladders are used to pick the fruits. If the whole fruit is marketed, both sweet and sour types should be harvested by clipping to avoid damaging the pods (Coronel, 1991). The use of poles in picking is not desirable as knocking can cause damage to the pods. Generally, the fruits are left to ripen on the tree before harvesting, so that the moisture content is reduced to about 20 per cent. Mexican studies reveal that the fruits begin to dehydrate 203 days after fruit set, losing approximately half of the moisture. When the fruits are left unharvested they may remain hanging on the tree for almost one year after flowering and eventually abscise naturally. Sometimes they remain on the tree until the next flowering period (Chaturvedi, 1985).

Yield

The yield of tamarind varies considerably in different countries, depending on genetic and environmental factors. In India, the average production of tamarind pods per tree is 175 kg and of processed pulp is 70 kg/tree. The best yielding elite trees in Bangalore have pod yields ranging

250-350 kg/tree and pulp yields of 100-175 kg/tree (Kulkarni *et al.*, 1993). Yields up to 170 kg/tree/yr of prepared pulp are reported in India and Sri Lanka (Coronel, 1991). Rao (1995) reported that Periyakulam 1 (PKM1), an improved cultivar in Tamil Nadu, yields about 263 kg/tree, 59 per cent more than unselected cultivars.

References

Ali, M.S., Ahmad, V.U. and Usmanghani, A.K., 1998. Chemotropism and antimicrobial activity of *Tamarindus indica. Fitoterapia*, 69(1): 43–46.

Anonymous., 1982a. *Some Recent Developments.* Central Food Technological Research Institute, Mysore, India.

Anonymous., 1982b. Tamarind juice concentrate plant starts in Mysore. *Indian Food Industry*, 1: 43–44.

Benthall, A.P., 1933. *Trees of Calcutta and its Neighbourhood.* T.Spink and Co. p 513.

Birdar, S.R. and Hanamashetti, S.I., 2001. Correlation studies in tamarind (*Tamarindus indica* L.) genotypes. *Journal of Plantation Crops*, 29(3): 64–65.

Bueso, C.E., 1980. Soursop, Tamarind and Chironka. In: *Tropical and Subtropical Fruits*, (Eds.) S. Nagy and P.E. Shaw. AVI Publishing, Westport, Conn., p. 375.

Chaturvedi, A.N., 1985. Firewood farming on the degraded lands of the Gangetic plain. *U.P. Forest Bulletin* No. 50. Lucknow, India Government of India Press, (1): 286.

Chundawat, D.J., 1990. *Arid Fruit Culture.* Oxford IBH Publishers, New Delhi, p. 158–163.

Coates-Palgrave, K., 1988. *Trees of Southern Africa, 10: Tamarindus indica L.* C.S. Striuk Publishers, Cape Town, p. 278–279.

Collins, N.M., Sayer, J.A. and Whitmore, T.C., 1991. *The Conservation Atlas of Tropical Forests of Asia and the Pacific ICUW.* Macmillan Press, London.

Coronel, R.E., 1991. *Tamarindus indica* L. In: *Plant Resources of South East Asia, Wageningen, Pudoc. No.2. Edible Fruits and Nuts,* (Eds.) E.W.M. Verheij and R.E. Coronel. PROSEA Foundation, Bogor, Indonesia, p. 298–301.

FAO, 1988. *Fruit Bearing Trees.* Technical notes. FAO–SIDA. Forestry Paper, 34: 165–167.

Feungchan, S., Yimsawat, T., Chindaprasert, S. and Kitpowsong, P., 1996. Tamarind (*Tamarindus indica* L.) Plant genetic resources in Thailand. *Thai Journal of Agricultural Science*, Special Issue No. (1): 1–11.

FNC/FAO, 1995. *Forest Products Consumption Survey in Sudan.* Summary Findings. Forests National Corporation. Forestry Development Project.

Gunasena, H.P.M., 1999. Tamarind: An unexploited species in Sri Lanka. Paper presented at the *National Workshop on Fruits for the Future*, Gannoruwa, Sri Lanka. 4 November, Gannoruwa, Peradeniya, Sri Lanka.

Gunasena, H.P.M. and Hughes, A., 2000. *Tamarind.* International Centre of Underutilized Crops. Southampton, UK.

Hong, T.D., Linnington, S. and Ellis R.H., 1996. *Seed Storage Behaviour: A Compendium.* Handbook for Genebanks No.4 International Plant Genetic Resources Institute, Rome.

Ilango, K. and Vijayalakshmi, C., 2002. Effect of growth regulators and chemicals on pod set and retention in Tamarind (*Tamarindus indica* L.). *My Forest*, 38(2): 133–137.

Jambulingam, R., Chellapilli, K.L. and Murugesh, M., 1997. Variation in natural populations of *Tamarindus indica* L. in Tamil Nadu. In: *Proceedings of National Symposium on Tamarindus indica L., Tirupathi (A.P.)*. Organised by Forest Department of Andhra Pradesh, 27–28 June, p. 35–39.

Jayaweera, D.M.A., 1981. *Medicinal Plants (Indigenous and Exotic) Used in Ceylon, Part 3: Flacourtiaceae–Lytharaceae*. A publication of the National Science Council of Sri Lanka, p. 244–246.

John, J., Joy, M. and Abhilash, E.K., 2004. Inhibitory effects of tamarind (*Tamarindus indica* L.) on polypathogenic fungi. *Allelopathy Journal*, 14(1): 43–49.

Kaitho, R.J., Nsahlai, I.V., Williams, B.A., Umunna, N.N., Tamminga, S. and Van Bruchem, J., 1988. Relationships between preference, rumen degradability, gas production and chemical properties of browses. *Agroforestry Systems*, 39: 129–144.

Karale, A.P., Wagh, A.R., Pawar, B.G. and More, T.A., 1999. Association of fruit characters in tamarind. *Journal of the Maharashtra Agricultural Universities*, 24(3): 319–320.

Khoja, A.K. and Halbe, A.V., 2001. Scope for the use of tamarind kernel powder as a thickener in textile printing. *Man Made Textiles in India*, 44(10): 403–407.

Knight, T., Jr. 1980. Origin and world importance of tropical and subtropical fruit crops. In: *Tropical and Subtropical Fruits*, (Eds.) S. Nagy and P.E. Shaw. AVI Publishing, Westport, Conn., p. 1.

Kotecha, P.M. and Kadam, S.S., 2003a. Preparation of ready-to-serve beverage, syrup and concentrate from tamarind. *Journal of Food Science and Technology (Mysore)*, 40(1): 76–79.

Kotecha, P.M. and Kadam, S.S., 2003b. Studies on browning in tamarind pulp during storage. *Journal of Food Science and Technology (Mysore)*, 40(4): 398–399.

Kulkarni, R.S., Gangaprasad, S. and Swami, G.S.K., 1993. Tamarind economically and important minor forest product. *Minor Forest Products News*, 3(3): 6.

Lakshmi, K., Vasanth Kumar, A.K., Jaganmohan Rao, L., and Madhava Naidu, M., 2005. Quality evaluation of flavoured RTS beverage and beverage concentrate from tamarind pulp. *Journal of Food Science Technology*, 42(5) : 411–415.

Lima, D.U., Oliveira, R.C. and Buckeridge, M.S., 2003. Seed storage hemicelluloses as wet-end additives in papermaking. *Carbohydrate Polymers*, 52(4): 367–373.

Little, E.L. and Wadsworth, F.W., 1964. Common trees of Puerto Rico and the virgin islands. *Agriculture. Handbook* 249, Washington DC, US Department of Agriculture, 548 p.

Lotschert, W. and Beese, G., 1994. *Tropical Plants*. Collins Photo Guide. Harper Collins Publishers, p. 223.

Luengthanaphol, S., Mongkholkhajornsilp, D., Douglas, S., Douglas, P.L., Pengsopa, L. and Pongamphai, S., 2004. Extraction of antioxidants from sweet Thai tamarind seed coat: Preliminary experiments. *Journal of Food Engineering*, 63(3): 247–252.

Macmillan, H.F., 1943. *Handbook of Tropical Plants. 362. Medicinal Plants*. Anmol Publications, New Delhi, India.

Mahadevan, N.P., 1991. Phenological observations of some forest trees as an aid to seed collection. *Journal of Tropical Forestry*, 7(3): 243–247.

Maliro, M.F.A. and Kwapata, M.B., 2002. Impact of deforestation on diversity of wild and semi wild edible fruits tree species in southern Malawi. *Doscov. Innov.* (Special edition), p. 98–104.

Mathur, N.K. and Mathur, V., 2001. Industrial polysaccharides-9-tamarind seed polysaccharide and tamarind kernel powder. *Chemical Weekly*, 46(51): 143–150.

Morton, Julia F., 1987. *Fruits of Warm Climates*. Creative Resources Systems, Inc., pp. 115–121.

Nagarajan, B., Nicodemus, A., Mandal, A.K., Verma, R.K., Gireesan, K. and Mahadevan, N.P., 1997. Phenology and controlled pollination studies in Tamarind. *Silvae Genetica* 47: 237–241.

NAS, 1979. *Tropical Legumes: Resources for the Future*. Washington DC, p. 117–121.

Parameswari, K., Srimathi, P., Sasthri, G. and Malarkodi, K., 2000. Influence of locations on size grades of tamarind. *Progressive Horticulture*, 32(2): 131–137.

Pareek, O.P., 1999. Arid zone fruits research in India. *Indian Journal of Agricultural Sciences* (Special issue), 68(8): 508–514.

Parrotta, J.A., 1990. *Tamarindus indica* L Tamarind. SO–ITF–SM–30 June, USDA Forestry Service. Rio Piedras, Puerto Rico, p. 1–5.

Parvez, S.S., Parvez, M.M., Nishihara, E., Gemma, H. and Fujii, Y., 2003. *Tamarindus indica* L. leaf is a source of allelopathic substance. *Plant Growth Regulation*, 40(2): 107–115.

Perveen, A. and Qaiser, M., 1998. Pollen flora of Pakistan–X. Leguminosae (Sub-family Caesalpinioideae). *Tropical Journal of Botany*, 22: 145–150.

Radhamani, A., Nicodemus, A., Nagarajan, B. and Mandal, A.K., 1998. Reproductive biology of tropical trees. In: *Forest Genetics and Tree Breeding*, (Eds.) A.K. Madal and G.I. Gibson. CBS Publishers, New Delhi, India.

Rama Rao, M., 1975. *Flowering Plants of Travancore*. Bishen Singh Mahendra Pal Singh, Dehra Dun India, p. 484.

Rao, P.S., 1948. Tamarind seed (jellose pectin) and its jellying properties. *Journal of Science and Industrial Research*, 68: 89–90.

Ray, P.G. and Majumdar, S.K., 1976. Antimicrobial activity of some Indian plants. *Economic Botany*, 30(4): 317–320.

Salim, A., Simons, A., Waruhin, A. and Orwa, C., 1998. Agroforestry tree database: A tree species reference and selection guide and tree seed suppliers directory. International Council for Research in Agroforestry, PO Box 30677, Nairobi, Kenya.

Shankaracharya, N.B., 1998. Tamarind–chemistry, technology and uses: A critical appraisal. *Journal of Food Technology*, 35(3): 193–208.

Shanthi, A., 2003. Studies on variations and association in selected populations, plantations and clones in tamarind (*Tamarindus indica* Linn.). *Unpublished Ph.D. Thesis*. Bharathiar University, Coimbatore, India.

Shivanandam, V.N. and Raju, K.R.T., 1988. Correlation between some fruit characters of four tamarind (Tamarindus indica L.) types. *Mysore Journal of Agricultural Science*, 22(2): 229–231.

Singha, R.K., 1995. Biodiversity conservation through faith and tradition in India. Some case studies. *International Journal of Sustainable Development and World Ecology (UK)*, 2(4): 278–284.

Sone, Y. and Sato, K., 1994. Measurement of oligosaccharides derived from tamarind xyloglucan by competitive ELISA assay. *Bioscience Biotechnology and Biochemistry*, 58: 2295–2296.

Sozolnoki, T.W., 1985. Food and fruit trees of Gambia. Hamburg, Federal Republic of Germany: 132 Tiftung Walderhatung in Afrika.

Storrs, A.E.G., 1995. *Know Your Trees: Some Common Trees Found in Zambia*. Regional Soil Conservation Unit (RSCU), Zambia.

Swaminath, M.H., 1988. Technique of raising instant trees. *My Forest*, 24(1): 1–5.

Troup, R.S., 1921. *The Silviculture of Indian Trees, Vol 2: Leguminosae (Caesalpinieae) to Verbenaceae. 7. Tamarindus indica L.:* Clarendon Press, Oxford, p. 263–363.

Usha, K. and Singh, B., 1994. Selection of tamarind tree types for higher yield. *Crop Improvement,* 23(2): 199–202.

Usha, K. and Singh B., 1996. Influence of open and cross pollination on fruit set and retention in tamarind (*Tamarindus Indica* L). *Recent Horticulture*, 3(1): 60–61.

Von Mueller, B.F., 1881. Select extra tropical plants readily available for industrial culture or naturalisation. *Tamarindus indica* L. Thomas Richards, Government Printer, Sydney, p. 328.

Watt, G., 1893. *Tamarindus*. Linn. *Dictionary of Economic Products of India,* Vol. 6, Part 3: Silk to Tea, p. 404–409.

2013, Biodiversity in Horticultural Crops Vol. 4   *Pages 255–274*
Editor: **Professor K.V. Peter**
Published by: **DAYA PUBLISHING HOUSE, NEW DELHI**

Chapter 14

Tomato

Ravindra Mulge

Department of Vegetable Science, K.R.C. College of Horticulture, Arabhavi,
University of Horticultural Sciences, Bagalkot, Karnataka
E-mail: ravindramulge@yahoo.com

The cultivated tomato (*Solanum lycopersicum*, syn. *Lycopersicon lycopersicum* and *Lycopersicon esculentum*) is a relatively recent addition to the world's important food crops. Within the past century, it has become the most popular and widely consumed vegetable crop, with an annual world production of 1,27,920,000 tons (2007-08). Leading tomato producing countries are China, USA, India and Turkey. It is also America's most popular and pampered home garden vegetable, occupying space in more than 90 per cent of the home gardens in the United States. Its versatility in fresh or processed form has played a major role in its rapid and widespread adoption as an important food commodity. Tomato is a tender perennial that is almost universally cultivated as an annual. Despite its susceptibility to frost, the tomato can be grown outdoors successfully from the equator to as far north as Alaska. Cultivars have been developed for a variety of different environments, methods of production, and food uses. Its adaptation to fit into many diverse uses and environments is a reflection of the great wealth of genetic variability existent in the genus and the relative ease with which this diversity can be applied in breeding programs.

Botany

It is a herbaceous, usually sprawling plant with weak stem belonging to the family *Solanaceae* or nightshade family that is typically cultivated for the purpose of its fruits for human consumption. The fruit of most varieties ripens to a distinctive red colour. Tomato plants typically reach to 1 to 3 m in height, and have a weak, woody stem *i.e.* vine. The leaves are 10 to 25 cm long, pinnate, with 5 to 9 leaflets on petiole each leaflet up to 8 cm long, with a serrated margin, both the stem and leaves are densely covered with glandular-hairs. The flowers are 1 to 2 cm long, yellow with five pointed lobes as petals are united (corolla); they are borne in a cyme of 3 to 12 in cluster. It is a perennial, but grown in temperate and tropical climates as an annual.

The Root

Tomato produces a strong taproot but very often it is damaged at the time of transplanting; consequently, the lateral adventitious roots develop profusely. The adventitious roots develop at the rate of about 3 centimetres per day penetrating around 1 meter depth and attaining a length of about 1 to several meters. However, the root development largely depends upon the cultivar and the growing media and the moisture in it. The root regeneration takes place in leaf and stem cuttings. A comparative study of the cellular configurations in the meristamatic and the mature cortical cell of the primary roots has been described by Duffy (1951). The root hair emerges near the apical end of a cell where the wall is more plastic. The epidermis is unilayered with a cuticular coverage. Below the epidermis, a parenchymatous cortex is present which is demarcated with a vascular system by an endodermis which is distinct by the presence of casparian strips in the cells. The outer- most layer of the vascular system is a single layered pericycle. The xylem is exarch and the protoxylem is composed of small vessels and tracheids and xylem parenchyma whereas, the metaxylem has a vast number of large vessels, tracheids and a lesser proportion of parenchyma. The central position is occupied by the pith. The cortex and the endodermis are suppressed and they degenerate after the radial cell division (Kalloo, 1986).

The Stem

The stem is soft, brittle, and hairy when young and hard, woody and branched when mature. It is erect to semi erect. Epidermal hairs with oil glands are present at the early plant growth stage. The plants can be classified into three groups on the basis of their growth habit. The first one is determinate growth habit (self-pruning, bush) where, lateral axis terminates with inflorescence. There will be one or two nodes bearing leaves in between two successive flowering nodes. The lateral always exists to continue vegetative growth. Second is indeterminate growth habit where plant growth is continued and there will be three nodes bearing leaves between two successive flowering nodes. The lateral buds always exist to continue vegetative growth. The third one is semi determinate growth habit which is in between the determinate and indeterminate growth habits. The branching pattern is sympodial, but the first a few nodes which do not produce inflorescence may be monopodial. The side-shoots are produced from the axil of the leaves. On the outer side of transverse section of stem, there is an epidermis with hairs followed by two or three layers of collenchymatous and parenchymatous cells. The endodermis is the innermost layer of the cortex which surrounds the pericycle. It forms a single layer having the distinct characteristics of a wall. The collateral-type of vascular bundles are present which are arranged in a ring. The metaxylem is present towards the pericycle and the protoxylem towards the pith. The xylem vessels consist of wood vessels, tracheids, fibers and xylem parenchyma. The stem becomes woody due to secondary growth. After 30 to 35 days of germination, the cambium forms a continuous cylinder and produces a solid zone of secondary xylem. The secondary cambium becomes active to form the secondary xylem and phloem. The lignification starts in the secondary xylem. The radial division of the cells in the xylem, cambium, pericycle and cortex lead to secondary increase in the diameter(Kalloo, 1986).

The Leaf

The leaf is compound, alternate, petiolate, 6 to 45 cm wide, the leaflets are unequal odd pinnate; the apex is narrowed or acuminate, acute, irregularly serrate. The number of leaflets varies from four to several and the length of the leaf from 18 to 36 cm. On the leaves, hairs or trichomes are formed by the outer growth of the epidermal cells. The hairs may be unicellular or may become multicellular by subsequent cell division. Three kinds of trichome are found on the veins (Flores and Espinoz, 1977). The leaf primordium is initiated as a small protuberance from two or three outermost cell layers of the shoot apex flank. The transverse section of leaves shows the epidermal cells at the outermost region which consists of guard cells of stomata and trichomes.

Inflorescence

Inflorescence is the cyme with dichotomous or polychotomous branching. It arises from the stem opposite to the leaf. It may be lateral or terminal or may continue with the vegetative shoot. The number of flowers per cluster varies from three to several (Table 14.1). The early flowers which open in the cluster are large sized as in a typical cyme. A single plant can produce 20 or more flower clusters.

The Flower

The plant produces bright yellow flowers. The flowers are pentamerous, complete, e-bracteate, bisexual, regular and hypogynous. The pistil has two to several carpels. The anthers connate appearing in the throat of the corolla and they dehisce through side sutures present in inner side of the anthers. Usually the style is shorter than anther cone, therefore a high degree of self-pollination takes place. Microsporogenesis where in a very small anther, the hypodermal archesporial cells divide to produce the primary sporogenous cells. The latter cells continue to divide to form the microspore mother cells (MMC) which on meiotic division produce microspores (pollens). Megasporogenesis where in the archesperial cells act as megaspore mother cells which undergo a reduction division to form four megaspores of which three chalizal megaspores degenerate leaving one functional megaspore at the micropylar end. This undergoes three successive nuclear divisions to form a polygonum type of embryo sac. Later, two polar nuclei fuse to form the secondary nuclei. Fertilisation takes place after 10 to 30 hours after pollination. After fertilisation, the egg cell divides transversely twice forming four cells arranged linearly (Kalloo, 1986).

The Seed

The seed is oval, flattened buff to pale brown, 3 to 5 mm in length. It is derived from the anatropous ovule. A curved filiform embryo and an endosperm are enclosed in the seed. The embryo is surrounded by a small endosperm. The seed can be bisected to observe outer epidermis, intermediate parenchymatous tissue and inner epidermis. With the development of the seed, the outer parenchymatous tissue enlarges. In the advanced stage, the outer epidermal cells get elongated. When the seed is matured, the outer wall ruptures forming hair-like structures and the parenchymatous cells mostly disappear (Czaja, 1963). Number of seeds and 1000 seed weight vary greatly among the genotypes and are presented in Table 14.1.

The Fruit

The fruit is soft berry. Glandular hairs and glands are usually present on the small fruit which degenerate in the advanced stage. The fruits of some varieties or species do not have glandular hairs and glands. There are various shapes of fruits. The pericarp constitutes wall of the fruit. Number of locules varies from two to several (Table 14.1). Each locule has seeds and jelly like substances.

Table 14.1: Variability, Heritability and Genetic Advance for Growth, Earliness, Yield and Quality Parameters in Tomato

Characters	Mean	Range	PV	GC	PCV (%)	GCV (%)	h² (%)	GA	GAM	References
Plant height (cm)	67.95	49.87-99.65	297.31	286.30	25.50	24.90	96.30	-	50.33	Bora *et al.* (1993)
	57.14	43.80-88.40	147.96	57.75	258.91	101.06	39.03	9.78	17.12	Narendrakumar and Arya (1995)
	73.19	44.0-100.8	165.07	144.93	17.55	16.45	87.80	15.79	21.57	Sahu and Mishra (1995)
	117.37	-	2088.69	1035.55	38.94	27.42	50.00	6.37	-	Krishnaprasad and Mathurarai (1999)
	53.92	38.67-93.67	-	-	28.07	26.81	91.20	28.43	52.73	Mohanty (2003)
Number of primary	5.76	3.90-7.80	0.83	0.29	15.90	9.49	35.67	0.45	8.02	Sahu and Mishra (1995)
branches	6.86	5.06-8.46	1.51	0.27	22.60	4.06	17.99	1.79	26.09	Narendrakumar and Arya (1995)
	-	-	1.80	1.709	20.37	19.79	94.42	-	22.96	Mala and Vadivel (1999)
	6.87	4.97-13.73	-	-	32.35	30.62	89.60	4.11	59.83	Mohanty (2003)
Stem girth (cm)	1.39	1.17-1.80	0.045	0.031	-	12.61	68.00	0.30	21.58	Paranjothi and Muthikrishnan (1979)
Days to first flowering	59.54	51.00-72.00	26.28	19.50	8.61	7.42	74.19	5.32	8.94	Sahu and Mishra (1995)
	79.63	74.57-86.60	-	-	5.07	4.58	81.43	12.65	8.50	Parvindarsingh et al. (2002)
Days to 50% flowering	84.93	75.0-100.0	29.75	22.72	35.03	26.75	76.37	8.58	10.10	Pujari *et al.* (1995)
Days to first fruit set	-	-	0.018	0.016	13.78	13.11	90.50	0.249	25.69	Singh *et al.* (1988)
Days to first fruit	58.78	49.25-68.25	-	-	-	5.90	85.31	8.34	1.41	Singh et al. (1974)
maturity	164.38	150.00-183.00	-	-	8.64	8.04	86.67	25.36	15.42	Nandpuri *et al.* (1977)
	101.56	100.50-109.10	-	-	1.96	1.62	68.00	2.87	-	Reddy and Gulshanlal (1987)
	166.79	157.50-174.30	-	-	2.47	2.38	93.56	7.93	4.76	Das *et al.* (1998)
Per cent fruit set	45.65	20.90-80.60	-	-	20.10	19.71	94.30	2.68	-	Reddy and Gulshanlal (1987)
Number of fruits/cluster	2.55	2.00-3.43	0.13	0.07	5.18	2.82	54.57	0.39	15.29	Narendrakumar and Arya (1995)
	2.38	1.86-3.00	0.20	0.15	18.79	16.27	75.00	-	29.21	Bora *et al.* (1993)
Number of fruits per	39.26	18.02-70.80	475.67	457.25	55.55	54.47	96.13	-	110.01	Bora *et al.* (1993)
plant	31.54	21.00-42.67	24.15	22.08	76.53	70.01	91.48	9.26	29.35	Narendrakumar and Arya (1995)
	41.95	8.00-85.50	7036.14	6972.43	16772.6	16620.8	99.09	171.23	408.18	Pujari *et al.* (1995)

Contd...

Table 14.1–Contd...

Characters	Mean	Range	PV	GC	PCV (%)	GCV (%)	h² (%)	GA	GAM	References
Number of fruits per plant	29.28	18.66-53.80	82.07	74.07	30.94	29.39	90.25	11.44	39.09	Sahu and Mishra (1995)
	42.82	29.40-64.85	-	-	22.30	21.38	91.23	18.01	42.06	Das et al. (1998)
	45.91	22.83-75.00	-	-	31.79	30.29	90.80	27.30	59.46	Mohanty (2003)
Total yield per plant (kg)	1.81	1.30-2.50	0.166	0.141	22.09	20.44	84.93	-	39.77	Prasad and Prasad (1976)
	0.867	0.547-1.653	0.038	0.034	22.50	21.36	90.11	0.42	48.44	Bhutani et al. (1983)
	1.62	0.95-2.18	0.18	0.15	26.19	23.90	83.33	-	44.44	Bora et al. (1993)
	1.15	0.160-2.240	0.20	0.18	17.39	16.65	90.00	0.83	72.10	Pujari et al. (1995)
Average fruit weight (g)	1097.25	698.4-1708.6	63719.1	56894.9	23.01	21.74	89.29	315.54	28.75	Sahu and Mishra (1995)
	54.83	32.87-87.53	1780.93	1742.65	76.96	76.93	97.85	-	155.14	Bora et al. (1993)
	41.88	1.00-122.50	384.16	366.17	917.29	874.33	95.32	38.49	91.81	Pujari et al. (1995)
	38.43	20.53-52.20	67.59	61.37	21.00	20.38	95.25	10.45	27.19	Sahu and Mishra (1995)
	58.53	27.97-77.27	-	-	24.79	24.49	97.53	49.81	-	Parvindarsingh et al. (2002)
	51.25	29.83-92.67	-	-	38.96	37.91	94.70	38.94	75.98	Mohanty (2003)
Fruit shape index	1.15	0.65-1.77	0.045	0.036	3.91	3.13	80.00	0.35	30.40	Pujari et al. (1995)
Fruit polar diameter (cm)	5.35	4.40-6.78	-	-	-	9.15	87.57	0.93	1.73	Singh et al. (1974)
	4.21	3.10-5.80	0.37	0.20	14.38	10.58	54.15	0.46	10.50	Sahu and Mishra (1995)
	5.27	3.87-7.22	-	-	18.64	18.24	95.79	1.92	36.71	Das et al. (1998)
Fruit Equatorial diameter (cm)	6.09	4.49-8.14	-	-	-	13.44	96.87	1.68	27.58	Singh et al. (1974)
	-	-	0.581	0.458	19.99	17.75	78.84	1.238	32.47	Singh et al. (1988)
	5.34	3.10-6.68	0.66	0.57	15.25	14.07	85.15	0.97	18.17	Sahu and Mishra (1995)
Pericarp thickness (cm)	4.64	2.90-6.66	0.89	0.71	-	18.1	79.70	1.54	33.40	Padda et al. (1971)
	0.407	2.56-7.08	0.010	0.008	24.56	21.97	82.01	0.41	100.74	Bhutani et al. (1983)
	0.47	0.20-0.75	0.019	0.012	4.04	2.55	63.16	0.18	38.16	Pujari et al. (1995)
Number of locules per fruit	6.04	2.00-12.20	10.03	8.51	52.31	48.17	84.84	-	91.72	Prasad and Prasad (1976)
	3.90	2.40-6.30	0.95	0.77	24.95	22.45	80.99	1.62	41.63	Arora et al. (1982)
	3.66	2.07-5.42	0.571	0.534	20.63	19.95	93.46	0.40	10.18	Bhutani et al. (1983)

Contd...

Table 14.1– *Contd...*

Characters	Mean	Range	PV	GC	PCV (%)	GCV (%)	h² (%)	GA	GAM	References
TSS (°Brix)	7.11	6.10-7.90	0.42	0.39	9.00	8.72	92.85	-	17.58	Prasad and Prasad (1976)
	5.40	4.30-7.20	0.41	0.24	11.86	9.67	58.84	0.77	14.29	Arora *et al.* (1982)
	4.43	3.40-5.99	-	0.053	-	-	26.20	0.242	5.46	Kasrawi and Amr (1990)
	4.24	3.00-6.20	0.55	0.42	12.97	9.91	76.36	1.17	27.51	Pujari *et al.* (1995)
TSS : acid ratio	8.89	5.76-12.09	1.55	0.73	-	9.50	47.00	1.20	13.60	Padda *et al.* (1971)
Acidity (%)	0.62	0.32-1.04	0.012	0.010	-	16.10	85.60	0.197	31.90	Padda *et al.* (1971)
	1.76	0.80-2.80	0.20	0.17	25.54	23.65	85.64	0.79	45.01	Arora *et al.* (1982)
	0.402	0.316-0.569	-	0.002	-	-	61.30	0.076	18.90	Kasrawi and Amr (1990)
Fruit pH	4.64	4.11-7.20	-	0.116	-	-	78.10	0.624	13.50	Kasrawi and Amr (1990)
			0.64	0.60	21.67	21.10	94.68	-	42.27	Mala and Vadivel (1999)
Ascorbic acid	33.57	24.70-43.00	26.63	26.02	15.37	15.19	97.70	-	30.92	Prasad and Prasad (1976)
(mg/100 g)	20.72	14.60-27.80	6.99	3.99	12.76	9.65	57.21	3.12	15.03	Arora *et al.* (1982)
	-	11.52-26.08	10.78	10.78	19.39	19.39	100.00	-	39.91	Pradeepkumar and Tewari (1999)
Lycopene (mg/100 g)	-	0.89-2.16	0.097	0.096	22.59	22.45	98.80	-	46.27	Pradeepkumar and Tewari (1999)
	-	-	0.09	0.09	29.47	29.14	97.80	-	59.37	Mala and Vadivel (1999)
Fruit density	0.99	0.82-1.04	-	0.001	-	-	46.30	0.044	4.40	Kasrawi and Amr (1990)
	-	0.95-1.08	0.0008	0.0007	2.91	2.79	92.20	-	5.97	Pradeepkumar and Tewari (1999)
Number of seeds per fruit	85.936	30.00-174.33	1044.47	991.37	37.61	36.64	94.92	0.74	0.86	Bhutani *et al.* (1983)
Thousand seed weight (g)	5.20	2.00-8.10	1.52	1.47	23.72	23.29	96.20	2.44	46.99	Arora *et al.* (1982)
	3.13	1.40-8.60	-	-	42.73	42.32	98.07	2.70	-	Reddy and Gulshanlal (1987)

Reproductive Biology, Genetics and Cytogenetics

The tomato flower is normally perfect, having functional male (anthers) and female (pistil) parts. Continuous flowering facilitates crosses between cultivars which represent extremes in variation for maturity, since flowering occurs over a long period of time. Present cultivated varieties form a tight protective anther cone surrounding the stigma which greatly reduces the possibility for natural cross-fertilization. Lower movement aided by wind is sufficient to release pollen, but under greenhouse conditions, manual vibration of open flowers is required to effect pollination and fruit set. Genetic or environmental modification of stigma position can affect both fruit set and degree of cross-fertilization. Emasculation for the purpose of controlled pollination must be done approximately a day prior to anthesis or flower opening to avoid accidental self-pollination. At this time, the sepals have begun to separate and the anthers and corolla are beginning to change from light to dark yellow, characteristic of fully opened flowers. The stigma appears to be fully receptive at this stage (a day prior to anthesis), thus allowing for pollination immediately after emasculation. With favourable environmental conditions, 200 or more seeds may be obtained from a single pollination. Generally under greenhouse conditions no protection is required following emasculation to prevent uncontrolled crossing. Making controlled pollinations under field conditions may be less efficient than under greenhouse environments because hot and dry winds may cause rapid desiccation of the exposed pistil before fertilization is achieved. Cool, dry and relatively wind-free weather is preferred for better success in outdoor crossing, and protection of flowers with butter paper bags may be necessary to avoid chance crosses. Under optimal temperature and growth conditions, the tomato will complete its reproductive cycle in 95-115 days, depending upon cultivar. The first flower opens 7 to 8 weeks after seeding, and an additional 6 to 8 weeks required from flowering to ripe fruit. Seed is physiologically mature when the fruit reaches full ripeness. This makes it theoretically possible to complete three reproductive cycles a year using greenhouse facilities for off-season plantings. Self-incompatibility is commonly observed in wild relatives of tomato, and is transmitted to hybrids with *Solanum lycopersicum.* Self-incompatibility is conditioned by a single locus. Genic male sterility has also been reported frequently within the genus and many loci producing male sterility have been identified and described.

The cultivated tomato has been a favoured crop for genetic studies because of the wealth of variability within the species and the ease with which it can be manipulated. Tomato has proved to be an ideal plant for genetic studies because of its relatively simple reproductive biology, its ease of culture, and the wealth of genetic variation in cultivated and wild forms. It has been more extensively studied genetically than any other food crop except possibly maize and more than 970 genes were reported by 1979. The Tomato Genetics Cooperative, established in 1951 by Dr. C. M. Rick, University of California at Davis, provided an invaluable service to many workers in tomato genetics by coordinating gene nomenclature and mapping efforts. It is a highly self pollinated crop where flowers can easily be emasculated and pollinated and individual crosses may yield as many as several hundred seeds. Rates of natural cross-pollination in temperate zones vary from 0.5 to 4% however, much higher rates occur in Peru, presumably as a result of native insect vectors that can transfer the pollen. Rick suggested that the change from moderate to almost exclusive self pollination occurred following the introduction of the crop to Europe and was accompanied by a change in stigma position from exerted to inserted within the anther cone.

Tomato is a diploid possessing 12 chromosome pairs. Nuclear genome DNA content is of about 900 Mb. At pachytene stage, tomato chromosomes are very distinct and the length and physical features of the chromosomes are distinctive that the different chromosomes can be identified. The total length of the tomato chromosomes at the metaphase stage is about 50 μm. Haploid plants can be obtained at low frequencies by X-ray treatment of pollen grains before fertilization and also from polyembryonic seeds. Haploids are sterile, but homozygous diploids derived from haploids are highly fertile. Haploids are useful for fast screening of recessive characters. However, the success of haploids in tomato is very low and hence ploidization is rarely used. Crossing a diploid with a tetraploid genotype can produce triploids. Although 12 trivalents are expected at diakinesis, bivalents and univalents are frequently observed. Trisomics and tetrasamics, as well as other aneuploids are produced from the progeny of triploids. Tetraploids are generally obtained after colchicine treatment. Alternatively, a high frequency of tetraploids can be obtained from tissue or cell culture regenerates. Tetraploids are characterized by thicker and darker leaves, reduced inter-node length, increased stem thickness, reduced fertility and reduced fruit set and fruit quality. Consequently tetraploid tomatoes are not commercially useful. This is also true for all other non-diploid tomatoes. Triploids are the best source for trisomics, since almost all trisomics can be expected from the triploid progeny. Trisomics have been used to assign linkage groups to chromosomes (Lindhout, 2005). The addition of extra chromosomes to the diploid set has huge effect on the tomato phenotype and meiosis. It is accepted that three extra chromosomes are the limit that can be added to the diploid set. A very limited number of monosomics have been obtained. Generally, monosomics are very unstable. Tomato monosomics can be obtained by fusion of tomato and potato somatic cells followed by a series of backcrosses to the potato genome (2n=48) as recurrent parent. By DNA markers and genomic *in situ* hybridization (GISH) a complete set of monosomic lines was obtained, each carrying a different tomato chromosome in a potato background (Ali *et al.*, 2001).

Early History, Origin, Evolution and Spread of Tomatoes

Cultivated tomato is native to South America although it is domesticated in Mexico. The species suffered genetic bottlenecks as crop was carried from New World to Europe and then back to North America. Genetic diversity for correction of these short falls is not available in the immediate proximity in the process of evolution. Genetic evidence shows that the progenitors of tomatoes were herbaceous green plants with small green fruits with a centre of diversity in the high lands of Peru. These early *Solanums* diversified into the dozens of species related to tomato and are recognized as wild ancestors of tomato. *Solanum lycopersicum* was transported to Mexico where it was grown and consumed by prehistoric humans. The exact date of domestication may not be known. Evidence supports the theory that the first domesticated tomato was a little yellow fruit, ancestor of *Solanum cerasiformae* grown by the Aztecs of Central Mexico who called it xitomatl meaning "plump thing" with a navel, and later called tomatl by other Mesoamerican people. Aztec writings mention tomatoes were prepared with peppers, corn and salt, likely to be the original salsa recipe.

The word tomato comes from a word in the Nahuatl language, *tomatl*. French botanist Joseph Pitton de Tournefort provided the Latin botanical name, *Lycopersicon esculentum*, to the tomato. The Latin name *lycopersicon* means wolf peach because it was round and eaten by wolf and it was considered to be toxic due to its botanical connection to the Solanaceae or nightshade family. Aztecs and other peoples in the region used the fruit in their cooking; it was being cultivated in Southern Mexico and probably other areas by 500BC. It is thought that the Pueblo people believed that those who witnessed ingestion of tomato seeds were blessed with powers of divine. The large, lumpy tomato, a mutation from a smoother, smaller vegetable, originated and was encouraged in Mesoamerica. According to Andrew F Smith, the tomato probably originated in the highlands of the west coast of South America however, there is no evidence that the tomato was cultivated or even eaten in Peru before the Spanish arrived. Numerous wild and cultivated relatives of the tomato can still be found in a narrow, elongated mountainous region of the Andes in Peru, Ecuador and Bolivia as well as in the Galapagos Islands. These primitive relatives of the edible tomato occupy diverse environments based on latitude as well as altitude and represent large gene pool for improvement of the species.

According to Jenkins (1948), Mexico, particularly the Veracruz-Puebla area, is the centre of a varietal diversity of cultivated tomato. It might have been the source of cultivated tomatoes of the old world, on the basis of archaeological and historical evidence. The cultivated types of South America also might have been introduced from Mexico. Thus, this might have been the centre of domestication.

Domestication and cultivation of the tomato appear to have first occurred outside its centre of origin by early Indian civilizations of Mexico. The cultivated tomato is common in Peru today however; it is used as food primarily by the non-Indian population. Where it is cultivated by the native Indians of Peru, it appears to be a recent addition to their diet. Quite the contrary is true in Mexico, where the tomato is widely used by Indians and great diversity is evident in cultivars being grown. It is believed that the Spanish explorer Cortez may have been the first to transfer the small yellow tomato to Europe after he captured the Aztec city of TenochtItlan (now Mexico City) in 1521. It is also believed that Christopher Columbus, an Italian working for the Spanish monarchy, was the first European to take back the tomato, earlier in 1493. The first written account documenting the arrival of the tomato in the Old World appeared in 1554 by the Italian herbalist Pier Andrea Mattioli. The first cultivars introduced to Europe probably originated from Mexico rather than South America. These early introductions were presumably yellow, rather than red in colour, since the plant was first known in Italy as pomi d'oro or golden apple. It was also known as the love apple "pomm d'amou" in France. This appealing name did little to hasten its acceptance as food crop. In most places, the tomato was remarkably slow to gain acceptance, except an ornamental curiosity. Apparently the tomato's similarity to familiar poisonous members of the nightshade family such as mandrake and belladonna caused concern over its safety as a food. Such unfounded superstitions persisted widely, even into the twentieth century and undoubtedly had a major impact in slowing its adoption as a useful nutritious food crop. The first recorded mention of the tomato in North America was made in 1710. It was apparently brought from the Old World by early colonists but did not gain widespread acceptance, presumably because of the persisting view that its fruits were unhealthy and poisonous.

After its acceptance as safe food crop, its improvement gained momentum. The early progress of tomato breeding in the United States is poorly documented and is the best illustrated by the length of time, cultivars remained in demand and listed by seed suppliers. On this basis, the cultivars Red Cherry, Red Pear Shaped, and Trophy represented the most important, popular or persistent cultivars during the early history of the tomato in the United States. A few, if any, of these cultivars are of commercial significance today. However, they represent the foundation upon which modern-day cultivars were developed. Extensive efforts are under way to maintain old cultivars and the invaluable wild relatives which have served as the progenitors of the present-day tomato. The U.S. Seed Storage Laboratory in Fort Collins, Colorado, has the responsibility of maintaining seeds of old cultivars. Tomato introductions from other countries and from germplasm collection expeditions are maintained by the North Central Regional Plant Introduction Station at Ames, Iowa, and are available to private and public breeders working on tomato improvement.

Domestication of *S. lycopersicum* has taken place with the transition of exerted to inserted stigma, as a result of the change of allogamy to autogamy (Rick, 1976). *S. lycopersicum* var. *cerasiformae*, the wild cherry tomato, has been found in Mexico, most commonly at wild places, borders of fields, road-cuts or stream banks, where there are suitable conditions for tomato-growing. This has been found in most of the tropical places (Jenkins, 1948). Even the wild types behave as introduced plants and do not seem to have originated from this place. A large group of diversity existing in Mexico may be the mutants of the original types. A critical analysis indicates that the cultivated types might have originated from this place. The cultivated tomato, presumably, might have been carried from the slopes of the Andes into Central America and Mexico by the Indians at the time of their migration. The allozyme variations in cultivated and closely related species of tomato indicate the hypothetical sequence from var. *cerasiformae* domestication in Mexico to cultivars in Europe (Rick and Fobes, 1975). The biosystematics studies in *Solanum* and closely related species by Rick (1979) suggest that the species of tomato have been evolved via gene substitution. For a long time tomato was not consumed by human beings due to persistent superstitions of its poisonous nature. The plant belongs to the night shade family which consists of a large number of poisonous plant species. The poisonous fruit may be attributable to alkaloids and the strong odour of the leaves and the stem. The Europeans consumed it first and then the Americans followed. In Italy only yellow fruited tomato was known since 1544. The red fruited tomato might have been known in 1554. It was known as a medicinal plant since long and not as food or vegetable. In U.S.A. the plant was first grown by Thomas Jefferson in Virginia in 1781. Later it was spread to Philadelphia by the French refugees in 1789 and to Massachusetts in 1802 by an Italian painter. Since 1800, tomatoes are being used as food (Boswell, 1949). Now there has been a tremendous improvement in this crop and it has become a major source of vitamins and minerals. The details of the history of the use of tomato have been given by McCue (1952). Tomatoes that originated in the early days were of yellow fruited type with ribs. This indicates that tomato fruit including two types, contains carotenoid and lycopene. The modern round-fruited type might have been originated by hybridisation with *S. cerasiformae* and *S. pyrifome.*

Spanish Distribution

After the Spanish colonization of the Americas, the Spanish distributed the tomato throughout their colonies in the Caribbean. They also took it to the Philippines, from where it moved to southeast Asia and then the entire Asian continent. The Spanish also brought tomato to Europe. It grew easily in Mediterranean climates, and cultivation began in the 1540s. It was probably eaten shortly after it was introduced, and was certainly being used as food by the early 1600s in Spain. The earliest discovered cookbook with tomato recipes was published in Naples in 1692, though the author had apparently obtained these recipes from Spanish sources. However, in certain areas of Italy, such as Florence, the fruit was used solely as tabletop decoration before it was incorporated into the local cuisine in the late 17th or early 18th century (Anon., 2010).

Britain

Tomatoes were not grown in England until the 1590s, according to Smith. One of the earliest cultivators was John Gerard, a barber-surgeon. Gerard's *Herbal*, published in 1597 and largely plagiarized from continental sources, is also one of the earliest discussions of the tomato in England. Gerard knew that the tomato was eaten in Spain and Italy. Nonetheless, he believed that it was poisonous (tomato leaves and stems actually contain poisonous glycoalkaloids, but the fruit is safe). Gerard's views were influential, and the tomato was considered unfit for eating (though not necessarily poisonous) for many years in Britain and its North American colonies. But by the mid-1700s, tomatoes were widely eaten in Britain; and before the end of that century, the *Encyclopædia Britannica* stated that the tomato was "in daily use" in soups, broths, and as a garnish. In Victorian times, cultivation reached an industrial scale in glasshouses, most famously in Worthing. Pressure for housing land in the 1930s to 1960s saw the industry move west to Littlehampton, and to the market gardens south of

Chichester. Over the past 15 years, the British tomato industry has declined as more competitive imports from Spain and the Netherlands have reached the supermarkets (Anon., 2010).

Middle East

The tomato was introduced to cultivation in the Middle East by John Barker, British consul in Aleppo (1799 – 1825). In 1881, it is described as only eaten in the region, within the last forty years. The tomato entered Iran through two separate routes. One route was through Turkey and Armenia and the second route was through the Qajar royal family's frequent travels to France. The early name used for tomato in Iran was "Armani Badenjan" (Armenian Eggplant). The Spanish tomato dish, Paella, is called "Istanbuli Polao" (Istanbul Pilaf) by Iranians. Currently, the name used for tomato in Iran is "Gojeh Farangi" (Foreign Plum) (Anon., 2010).

North America

The earliest reference to tomatoes being grown in British North America is from 1710, when herbalist William Salmon reported seeing them in what is today South Carolina. They may have been introduced from the Caribbean. By the mid-18th century, they were cultivated on some Carolina plantations, and probably in other parts of the Southeast as well. It is possible that some people continued to think tomatoes were poisonous at this time; and in general, they were grown more as ornamental plants than as food. Thomas Jefferson, who ate tomatoes in Paris, sent some seeds back to America. Because of their longer growing season for this heat-loving crop, several states in the US Sun Belt became major tomato-producers, particularly Florida and California. In California, tomatoes are grown under irrigation for both the fresh fruit market and for canning and processing. The University of California, Davis (UC Davis) became a major center for research on the tomato. The C.M. Rick Tomato Genetics Resource Center at UC Davis is a gene bank of wild relatives, monogenic mutants and miscellaneous genetic stocks of tomato. The Center is named after the late Dr. Charles M. Rick, a pioneer in tomato genetics research. Research on processing tomatoes is also conducted by the California Tomato Research Institute in Escalon, California (Anon., 2010).

Botanical Classification

In 1753, the tomato was placed in the genus *Solanum* by Linnaeus as *Solanum lycopersicum* L. (lyco means wolf, persicum means peach, *i.e.,* "wolf-peach"). However, in 1768 Philip Miller placed it in its own genus, and he named it *Lycopersicon esculentum*. This name came into wide use but was in breach of the plant naming rules. It was decided to conserve the well-known *Lycopersicon esculentum*, making this the correct name for the tomato when it is placed in the genus *Lycopersicon*. However, genetic evidence (Peralta and Spooner, 2001) has shown that Linnaeus was correct in the placement of the tomato in the genus *Solanum*. If *Lycopersicon* is excluded from *Solanum*, *Solanum* is left as a paraphyletic taxon. The exact taxonomic placement of the tomato will be controversial for some time to come, with both names found in the literature. Two of the major reasons that some still consider separate genus for tomato are the leaf structure where tomato leaves are markedly different from any other *Solanum* and the biochemistry where many of the alkaloids common to other *Solanum* species are conspicuously absent in the tomato. The tomato can with some difficulty be crossed with a few species of diploid potato with viable offspring that are capable of reproducing. Such hybrids provide conclusive evidence of the close relationship between these genera. An international consortium of researchers from 10 countries, among them researchers from the Boyce Thompson Institute for Plant Research began sequencing the tomato genome in 2004 and is creating a database of genomic sequences and information on the tomato and related plants. The genomes of its organelles (mitochondria and chloroplast) are also expected to be published as part of the project.

The Species

The tomato genome might have been evolved from *Solanum* containing lycopene and *S. peruvianum* and *S. hirsutum* from species containing β-carotene. According to Zhukovsky (1958), all the earlier tomatoes are of *S. cerasiforme* type. He has further said that *S. lycopersicum* is divided into three sub-species: ssp. *pimpinellifolium* with vars. *eupimpinellifolium* and *racemigerum;* ssp. sub- *spontaneum* with vars. *cerasiforme, humboldtil, pyriforme, elongatum*, and *succenturiatum;* and ssp. *validum*. The species are divided into two groups: (A) Eulycopersicon: consisting of (1) *S. lycopersicum* L and (2) *S. pimpinellifolium* and (B) *Eriopersicon:* (1) *S. peruvianum* (2) *S. glandulosum;* (3) *S. hirsutum;* (4) *S. cheesmanii* (5) *S. pissisi.*

The group *Eulycopersicon* includes red- fruited species and *Eriopersicon* mostly green-fruited types. Interspecific crosses between *S. lycopersicum* and *Solanum pimpinellifolium* are easily made and very a few barriers to gene exchange. Both members of the group *Eulycopersicon* are also compatible with members of the group *Eriopersicon,* but in some cases only when the latter functions as the pollen parent. The self-incompatibility common to many of the wild species is also transmitted to interspecies hybrids, and aberrant genetic segregation is common in such wide crosses. Embryo abortion may occur in crosses of *Solanum lycopersicum* with *Solanum peruvianum*, but this barrier can be overcome by use of embryo culture.

Solanum pennellii is self-incompatible, controlled by gametophytic system of self-incompatibility. It has a wide range of genetic variation and polymorphism (Rick and Tanksley, 1981). Cytological studies were made in the hybrid of *Solanum lycopersicum* x *S. pennellii* and a low level of preferential pairing was observed indicating a close genetic affinity. Georgieva *et al.* (1968) studied in detail the relationship of *S. pennelli* with *S. hirsutum, S. hirsutum* f. *glabratum*, and *S. glandulosum* and found that *S. pennellii* is genetically closely related to *S. hirsutum. S. chmielskii* species has been described in detail by Rick *et al.* (1976). According to them, the plants are almost similar to *parviflorum* but the plants of *chmielewskii* are more robust. The characteristics of this species are the presence of anthocyanin in the upper stem; inflorescence unbranched, sometimes divided into two cymes, length 9 to 12 cm, bracteate. The basal and the upper bracts are more abundant and larger than in *S. parviflorum*. The calyx segments are 6 mm long, the corolla is strongly recurved, 20-25 mm diameter divided halfway to base into elongated triangular segments at tips. The stamen cone is 8-9 mm long, style exerted 1-2 mm. The flowers are showy; fruits yellowish green colour; morphologically closely associated with *S. pimpinellifolium*. According to Rick *et al.* (1976), *Solanum parviflorum* is also morphologically associated with *S. pimpinellifolium*. The plant is copiously branched, perennial; stem slender, internode 3-5 cm long. The leaves are broad, margin serrate or undulate, the leaves resemble those of *S. peruvianum* var. *humifusum;* inflorescence 5-8 flowered unbranched. The calyx is 5-parted at base; corolla yellow, divided into triangular lobes with narrow apices, staminal cone 6 cm long, and bottle shaped, the anther subsessile, stigma slightly exerted. The fruits are berry, globose, 10-14 mm diameter, whitish green. International corporations are focusing on identifying protein and metabolic fingerprints of tomato. Together with the DNA sequence information, this research is described as genomics, it is expected that these developments will result in an important step forward in knowledge and understanding of the structure of the tomato genome.

Endemic Galápagos tomatoes (*Solanum cheesmanii*) are of great value for cultivated tomato breeding, and therefore their conservation is of significance. Within *S. cheesmanii* there is heterogeneity for many traits. There are three forms, without taxonomic significance, of *S. cheesmanii 'short'* (one- to two-pinnate leaves, short internodes, and coastal habitats), *S. cheesmanii 'long'* (one- to two-pinnate leaves, long internodes, and inland habitats), and *S. cheesmanii* form *'minor'* (three- to four-pinnate leaves, short internodes, and coastal habitats). In a recent survey of tomato populations in the Galápagos Islands, indicated that, several populations of *S. cheesmanii* reported 30–50 years earlier had disappeared, mostly as a consequence of human activity. The total diversity (estimated with amplified fragment length polymorphisms [AFLPs]) within *S. cheesmanii* ($H_T = 0.051$) is almost as high as that for the mainland wild species *S. pimpinellifolium* ($H_T = 0.072$). *Solanum lycopersicum* 'Gal *cer*,' showed much lower diversity ($H_T = 0.014$). Comparison of AFLP fragments shared by *Solanum lycopersicum* 'Gal *cer*' with other species showed that it is closely related to weedy tomato var. *cerasiforme* and, therefore, likely of recent origin. Genetic differentiation among the three native *S. cheesmanii* forms is low ($G_{ST} = 0.235$), indicating that they share a common genetic background. Nonetheless, *S. cheesmanii* '*short*' is about twice as diverse as *S. cheesmanii* '*long*' or *S. cheesmanii* f. *minor*. Apart from the pressure of humans, some native *S. cheesmanii* populations, especially *S. cheesmanii* '*long*,' might be displaced by invasive tomato 'Gal *cer*' because they share a similar habitat (Fernando Nuez *et al.*, 2010).

Solanum Phylogeny

With an estimated 1400 species, *Solanum* is the largest genus in the Solanaceae and one of the largest genera of flowering plants. Molecular phylogenetic analyses established that the formerly segregate genera *Lycopersicon, Cyphomandra, Normania* and *Triguera* are nested within Solanum, and all species of these four genera have been transferred to *Solanum*. Conversely, molecular data confirm that *Lycianthes,* sometimes considered to belong within *Solanum*, should be maintained as a separate genus. Traditional taxonomists have recognized three subfamilies within the Solanaceae: the Solanoideae, Nolanoideae, and the Cestroideae (D'Arcy, 1979). Hunziker excludes the Nolanoideae (i.e., genus *Nolana*) from the Solanaceae and expands the number of subfamilies to six: the Solanoideae, Cestroideae, Juanulloideae, Salpiglossoideae, Schizanthoideae and Anthocercidoideae. In both of these schemes, *Solanum* is placed within the subfamily Solanoideae, characterized by flattened seeds with curved embryos (Hunziker, 1979). Within the Solanoideae, *Solanum* belongs to the large and complex tribe Solaneae, which encompasses about 50 genera in the scheme of Hunziker (Hunziker, 2001).

Molecular Markers

The high-density tomato genetic linkage map enabled a whole series of new applications. Before the advent of molecular markers only qualitative traits were mapped, usually in populations that segregated for one or two morphological markers. The advantage of DNA markers is the simultaneous analysis of hundreds of markers in any population, allowing the mapping of any gene by cosegregation of such a gene with DNA markers (Lindhout, 2002). Polymorphisms, observed between obsolete susceptible lines and modern resistant cultivars, are in many cases due to introgressed genome fragments harbouring disease-resistance genes. As a result, DNA markers may be present on the introgressed fragment, carrying the resistance gene, and hence are associated with resistance (Lindhout, 2005). An association of the RFLP marker TG 301 and the Cf-9 gene for resistance to *Cladosporium fulvum* was found (Van der Beck *et al.*, 1992). As the map position of TG 301 is known, Cf-9 could be mapped on the short arm of chromosome 1. Interspecific populations, which are obtained when susceptible cultivated varieties are crossed with resistant wild relatives, offer many opportunities for genetic disease analysis. Linkage analysis within such populations often reveals cosegregation of the resistant phenotype with the presence of specific marker bands. For instance, many DNA markers on chromosome 6 were closely linked to the *Ol-1* gene involved in resistance to *Oidium lycopersici* (Huang *et al.*, 2000). Alternatively, the map position of the *Verticillium* resistance gene *Ve* was determined through linkage analysis of interspecific recombinant inbred lines obtained from a cross between a cultivated tomato accession and a *Solanum cheesmanii* accession (Diwan *et al.*, 1999). The technique can be extended to traits controlled by many genes through Quantative Trait Loci (QTL).

International corporations are focusing on identifying protein and metabolic fingerprints of tomato. Together with the DNA sequence information, it is described as genomics, it is expected that these developments will result in an important step forward in knowledge and understanding of the structure of the tomato genome. The challenge will be how to use the information resources effectively. Tomato (*Solanum lycopersicum*) has rich genetic and genomic resources including comprehensive databases of Expressed Sequence Tags (ESTs), Bacterial Artificial Chromosome (BAC) libraries, and genetic and comparative maps which are in the process of being linked to a physical map and eventually the euchromatic genomic sequence. These resources serve as template to study genetic variation and to manipulate agricultural traits. Current genetic maps for tomato include 2,200 Restriction Fragment Length Polymorphisms (RFLPs), Cleaved Amplified Polymorphic Sequences (CAPs), and Simple Sequence Repeats (SSRs), as well as emerging genetic resources which include a comparative map with *Arabidopsis* of over 500 Conserved Orthologous Set (COS) markers (Tanksley *et al*, 1992). These maps were derived from populations, developed between wild relatives (various *Solanum* species) and cultivated varieties. This approach maximizes genetic variation and has led to the discovery and introgression of novel alleles for disease resistance (Kabelka *et al.,2004*) and fruit traits (Frary *et al.,* 2000; Fridman *et al.,* 2000) into cultivated germplasm. However, the nearly exclusive focus on wide crosses has left a void in our knowledge and ability to manipulate other traits of agricultural importance within cultivated tomato. There is a lack of molecular markers which detect nucleotide polymorphisms among elite breeding lines. With the exception of genes that were introgressed from wild species, the majority of the breeding efforts in tomato are derived from elite-by-elite intra-specific crosses, resulting in consistent improvement for yield and fruit quality (Grandillo *et al.,* 1999). Increasing molecular marker density to facilitate the evaluation and analysis of elite-by-elite breeding populations is highly desirable.

The tomato genome is comprised of approximately 950 Mb of DNA - more than 75% of which is heterochromatin and largely devoid of genes. The majority of genes are found in long contiguous stretches of gene-dense euchromatin located on the distal portions of each chromosome arm. A minimal tiling path of BAC clones will be identified through this approximately 220 Mb euchromatin. The starting point for sequencing the genome will be approximately 1500 "seed" BAC clones individually anchored to the tomato high density genetic map based on a single, common *S. lycopersicum x S. pennellii* F2 population (referred to as the F2.2000; view map on SGN). Sequencing will proceed on a BAC-by-BAC basis. Each sequenced anchor BAC will serve as a seed from which to radiate out into the minimum tiling path. Identification of the correct next BACs in the euchromatin minimum tiling path for sequencing will be based on the use of a BAC end sequence database that will be created as part of this project, as well as a fingerprint contig physical map that is currently being constructed. A subset of the sequenced BACs will be localized on pachytene chromosomes via FISH (fluorescence *in situ* hybridization) to help guide the extension of the tiling path through the euchromatic arms of each chromosome and to determine when the heterochromatin and telomeric regions have been reached on each arm. A bioinformatics portal will be created for this project that will be mirrored at several locations around the world and provide a mechanism by which researchers in different locations can develop and contribute bioinformatics tools and information to the project. The details of chromosomes assigned to different countries for sequencing and completion of the work till 2004 are furnished below as per sol genomics networks (Anon., 2004).

Genetic Variability and Divergence

The success of a breeding programme depends upon the extent and magnitude of variability existing in the germptasm. Variability may be defined as the amount of variation present among the members of a population or species for one or more characters at genotypic or phenotypic levels. Genotypic coefficient of variability (GCV) and phenotypic coefficient of variability (PCV) are derived from standard deviation divided by mean and are used to assess the extent of variation. Heritability is the transmissibility of characters from parents to offspring. Heritability in broad sense is the ratio of genotypic variance to total phenotypic variance generally expressed in percentage. Effectiveness of selection of genotypes depends on heritability. Genetic advance (GA) is the improvement over the base population that can potentially be achieved from selection. It is a function of the heritability of the trait, the amount of phenotypic variation and the selection differential that, the breeder uses. When high heritability is accompanied with high genetic advance, it indicates additive gene effects and selection may be effective; when low heritability is accompanied with low genetic advance, it indicates predominance of environmental effects and selection would be ineffective. High heritability with low genetic advance indicates the importance of non- additive gene action, while low heritability with high genetic advance indicates the importance of additive gene effects. Genetic variability has been reported for yield (Prashanth *et al.*, 2006) and quality (Prashanth *et al.*, 2007) attributes in tomato. Variability for growth, earliness, yield and yield components in tomato has been reported by several workers. The review of literature on variance and its components, heritability, genetic advance and genetic advance over mean for various characters are presented in tabular form (Table 14.1). For getting high heterosis or for recovering transgressive segregants, parents chosen for hybridisation need to be genetically diverse or distant. Multivariate analysis has been put to good use enabling quantification of degree of divergence between populations (Michener and Sokal, 1957; Morishina and Oka, 1960 and Murty and Qadri, 1966). Several methods of divergence analysis based on quantitative traits have been proposed to suit various objectives, of which, Mahalanobis's generalised distance (Mahalanobis, 1936) occupies a unique place in plant breeding. It is a very sensitive and potent biometrical tool in quantifying the degree of divergence between biological populations and also to assess the relative contribution of different components to the total divergence both at inter- and intra-cluster levels (Nair and Mukherjee, 1960; Khanna and Misra, 1977; Suyambhulingam and Jobarani, 1978 and Singh and Singh, 1980). The concept of Mahalanobis D^2 statistic is based on the technique of utilising the measurements in respect of aggregate of characters. Khanna and Misra (1977) studied 50 varieties of tomato with respect to taxonomic distance between them using the Mahalanobis's D^2 technique and grouped them into 10 clusters on the basis of intra- and inter-cluster similarity with respect to plant height, fruit number per plant, number of branches, number of locules, number of days to flowering, total soluble solids and plant height. The author found that, total soluble solids, locule number and fruit number per plant were the major contributors to the divergence. Bhattacharya *et al.* (1979) attempted to know the species differentiation in tomato based on the results of genetic divergence by D^2 analysis using 50 tomato genotypes comprising of 4 wild collections and rest of them were common cultivars belonging to different geographical areas and were evaluated for 17 characters. Genotypes were grouped into 16 clusters on the basis of relative magnitude of D^2 values. The four wild collections of which three belonging to *Solanum peruvianum* and one to *Solanum pimpinellifolium* were grouped into three distant clusters, which were highly diverse from all the other clusters comprising only the cultivars of *tomato*. The major contribution to the divergence was by fruit number per plant which helped in species differentiation. Shape index was useful for varietal differentiation.Chaurasia and Majorsingh (1998) evaluated 71 genotypes during 1993-94 and 1994-95. Based on Mahalanobis D^2 values, genotypes were grouped into 9 clusters in 1993-94 and 10 clusters in 1994-95. Fruit weight showed maximum contribution to the genetic diversity in both years followed by plant height. Considering the cluster distances and cluster means, the genotypes KS-7, DVRT-2, Antey and PS-1 were recommended as potential parents for hybridisation. Dharmatti *et al.*

(2001) studied genetic divergence for 402 summer tomato lines using multivariate analysis method. The 402 lines were grouped into 4 clusters based on similarities of D^2 values. Considerable diversity within and between the clusters were observed and the characters like tomato leaf curl virus resistance, fruit yield per plant and number of white flies per plant contributed maximum to the divergence. Peter and Rai (1976) reported that, plant height and locule number greatly contributed to diversity. Sachan and Sharma (1971) found that, plant height, number of branches and number of fruits contributed for divergence. The cultivars from widely separated localities have been usually included in the hybridisation programme, presuming the presence of genetic divergence and maximum likelihood of recovering promising segregants. As per expectations, this has not yielded very satisfactory and consistent results. Eco-geographical diversity has been regarded as a reasonable index of genetic diversity (Vavilov, 1926; Moll *et al.*, 1962 and Ram and Panwar, 1970). However, it was reported later that, there does not exist any parallelism between geographic distribution and genetic diversity (Sachan and Sharma, 1971 and Peter, 1975 in tomato). Therefore genotypes selected based on divergence studies using multivariate analysis can be used in hybridization to realise heterosis or to recover transgressive segregants. However, selection of parents based on genetic distance to get high heterosis is valid only when a relation between genetic distance and magnitude of heterosis exists (Mulge *et al.*, 2009). If the genetic distance is a true indicator of heterosis for a given trait then efficiency of breeding programme can be enhanced greatly.

References

Ali, S.N.H., Ramanna, M.S., Jacobsen, E. and Visser, R.G.F., 2001. Establishment of a complete series of a monosomic tomato chromosome addition lines in the cultivated potato using RFLP GISH analyses. *Theoretical and Applied Genetics,* 103: 687–695.

Allen Van Deynze, Kevin Stoffel, C Robin Buell, Alexander Kozik, Jia Liu, Esther van der Knaap and David Francis, 2007. Diversity in conserved genes in tomato, *BMC Genomics,* 8: 465doi: 10.1186/1471–2164–8–465.

Anonymous, 2004. Sol genomics networks: www.sgn.cornell.edu.

Anonymous, 2010. http: //en.wikipedia.org/wiki/Tomato.

Arora, S.K., Pandita, M.L. and Pratap, P.S. 1982. Genetic variability and heritability studies in tomato in relation to quality characters under high temperature conditions. *Haryana Agricultural Universities Journal of Research,* 12(4): 583–590.

Bhattacharya, M.K., Nandpuri, K.S. and Singh, S., 1979. Genetic divergence in tomato. *Acta Horticulturae,* 93: 289–300.

Bhutani, R.D., Kalloo and Pandita, M.L., 1983. Genetic variability studies for yield and physico-chemical traits of tomato (*Lycopersicon esculentum* Mill.). *Haryana Journal of Horticultural Sciences,* 12(1–2): 96–100.

Bora, G.C., Shadeque, A., Bora, L.C. and Phookan, A.K., 1993. Evaluation of some tomato genotypes for variability and bacterial wilt resistance. *Vegetable Science,* 20(1): 44–47.

Bosewell, V. R., 1949. Our vegetable Travellers. *The National Geographic Magzine.* August, Xc. 6: 145–217.

Chaurasia, S.N.S. and Majorsingh, 1998. Genetic divergence in tomato. *Indian Journal of Plant Genetic Resources,* 11(2): 203–206.

Czaja, A.T., 1963. Neue unterzuchungen an der Testa der Tomatensan, en. *Planta,* 59: 262–279.

D'Arcy, W.G., 1979. The classification of the Solanaceae. In: *The Biology and Taxonomy of the Solanaceae,* (Eds.) J.G. Hawkes, R.L. Lester and A.D. Skelding. Academic Press, London, p. 3–47.

Das, B., Hazarika, M.H. and Das, P.K., 1998. Genetic variability and correlation in fruit characters of tomato (*Lycopersicon esculantum* Mill.). *Annals of Agricultural Research,* 19(1): 77–80.

Dharmatti, P.R., Madalageri, B.B., Mannikeri, I.M., Patil, R.V. and Patil, G., 2001. Genetic divergence studies in summer tomatoes. *Karnataka Journal of Agricultural Sciences,* 14(2): 407–411.

Diwan. N., Fluhr, R.. Eshed, Y., Zamir, D. and Tanksley, S.D. 1999. Mapping of vein tomato: A gene conferring resistance to the broad-spectrum pathogen, *Verticillium dahliae* race 1. *Theoretical and Applied Genetics,* 2: 315–319.

Duffy, R.M., 1951. Comparative cellular configurations in the meristematic and mature cortical cells of the primary root of tomato. *Am. J. Bat.,* 38: 393–408.

Fernando Nuez, Jaume Prohens and José M. Blanca, 2010. Relationships, origin, and diversity of Galápagos tomatoes: Implications for the conservation of natural populations, Centro de Conservación y Mejora de la Agrodiversidad Valenciana, Universidad Politécnica de Valencia, Camino de Vera 14, 46022 Valencia, Spain (From web page)

Flores, E.M. and Espinoz, A.M., 1977. Leaf morphology of Lycopers icon esculentum. *Revista de Biologia Tropical,* 25: 289–299.

Frary, A., Nesbitt, T.C., Frary, A., Grandillo, S., van der Knaap, E., Cong, B., Liu, J.P., Meller, J., Elber, R., Alpert, K.B. and Tanksley, S.D., 2000. A quantitative trait locus key to the evolution of tomato fruit size. *Science,* 289(5476): 85–88.

Fridman, E., Pleban, T. and Zamir, D., 2000. A recombination hotspot delimits a wild–species quantitative trait locus for tomato sugar content to 484 bp within an invertase gene. *Proc. Nat. Acad. Sci. (USA),* 97: 4718–4723.

Georgieva, R., Cikova, E. and Slavov, S., 1968. The taxonomic status of the wild species *Solanum pennelli,* Corell. *Genet. Pl. Breed. Sofia,* 4: 259–269.

Grandillo, S., Zamir, D. and Tanksley, S., 1999. Genetic improvement of processing tomatoes: A 20 years perspective. *Euphytica,* 110(2): 85–97.

Huang, C.C., Cui, Y.Y., Weng, C.R., Zabel, P. and Lindhout, P., 2000. Development of diagnostic PCR markers closely linked to the tomato powdery mildew resistance gene 01–1 on chromosome 6 of tomato. *Theoretical and Applied Genetics,* 101: 918–924.

Hunziker, A.T., 1979. South American Solanaceae: A synoptic survey. In: *The Biology and Taxonomy of the Solanaceae,* (Eds.) J.G. Hawkes, R.N. Lester, and A.D. Skelding. Academic Press, London.

Hunziker, A.T., 2001. *Genera Solanacearum.* A.R.G. Ganter, Ruggell, p. 49–85.

Jenkins, J.A., 1948. The origin of cultivated tomato. *Econ. Bot.,* 2: 379–392.

Kabelka, E., Yang, W.C. and Francis, D.M., 2004. Improved tomato fruit color within an inbred backcross line derived from *Lycopersicon esculentum* and *L. hirsutum* involves the interaction of loci. *J. Am. Soc. Hortic. Sci.,* 129(2): 250–257.

Kalloo, G., 1986. *Tomato.* Allied Publishers Pvt. Ltd., India.

Kasrawi, M.A. and Amr, A.S., 1990. Genotypic variation and correlations for quality characteristics in processing tomatoes. *Journal of Genetics and Breeding*, 44: 85–90.

Khanna, K.R. and Misra, C.H., 1977. Divergence and heterosis in tomato. *SABRAO Journal*, 9(1): 43–50.

Krishnaprasad, V.S.R. and Rai, Mathura, 1999. Genetic variation, component association and direct and indirect selections in some exotic tomato germplasm. *Indian Journal of Horticulture*, 59(3): 262–266.

Lindhout, P., 2002. The perspectives of polygenic resistance in breeding for durable disease resistance. *Euphytica*, 124: 217–226.

Lindohout P., 2005. Genetics and breeding. In: *Tomatoes*, (Ed.) E. Heuvelink. Crop Production Science in Horticulture–13 CAB International, Oxfordshire, UK.

Mahalanobis, P.C., 1936. On the generalized distance in statistics. *Proceeding of the National Academy of Sciences* (India), 2: 49–55.

Mala, M. and Vadivel, E., 1999. Mean performance of tomato genotypes for leaf curl incidence. *South Indian Horticulture*, 47(1–6): 31–37.

MeCue, G.A., 1952. The history of the use of tomato: An annotated bibliography. *Ann. Rept. Mp Bot. Gard.*, 39: 299–348.

Michener, C.D. and Sokal, R.R., 1957. Quantitative approach to a problem of classification. *Evolution*, 11: 130–162.

Miller, J.C. and Tanksley, S.D., 1990. RFLP analysis of phylogenetic relationships and genetic variation in the genus *Lycopersicon. Theoretical and Applied Genetics*, 80: 43 7–448.

Mohanty, B.K., 2003. Genetic variability, correlation and path coefficient studies in tomato. *Indian Journal of Agricultural Research*, 37(1): 68–71.

Morishina, H. and Oka, H.J., 1960. The pattern of interspecific variation in the genus *Oryza*, its quantitative representation by statistical methods. *Evolution*, 14: 153–165.

Mulge, R., Mahendrakar, Praveen, Prashanth, S.J., Madalageri, M.B. and Patil, M.P., 2009. Relevance of genetic divergence in prediction of heterosis in tomato (*Solanum lycopersicon* L.). In: *International Conference on Horticulture "Horticulture For Livelihood Securing And Economic Growth"*, held at Bangalore, November 9 to 12, p. 26, 1.1–P2.

Murty, B.R. and Qadri, M.I., 1966. Analysis of divergence in some self-compatible forms of *Brassica compestris* var. Brown Sarson. *Indian Journal of Genetics and Plant Breeding*, 26: 45–58.

Nair, K.R. and Mukherjee, H.K., 1960. Classification of natural and plantation teak (*Tectona grandis*) grown at different locations in India and Burma with respect to its physical and mechanical properties. *Sankhya*, 22: 1–20.

Nandpuri, K.S., Kanwar, J.S. and Roshanlal, 1977. Variability, path analysis and discriminant function selection in tomato (*Lycopersicon esculentum* Mill.). *Haryana Journal of Horticultural Sciences*, 6(1–2): 73–78.

Narendrakumar and Arya, M.C., 1995. Association among yield components in tomato. *Madras Agricultural Journal*, 82(9,10): 536–539.

Padda, D.S., Saimbhi, M.S. and Kirtisingh, 1971. Genotypic and phenotypic variabilities and correlations in quality characters of tomato (*Lycopersicon esculentum* Mill.). *Indian Journal of Agricultural Sciences*, 41(3): 199–202.

Paranjothi, G. and Muthukrishnan, C.R., 1979. Studies on variability and association of characters in tomato. *Madras Agricultural Journal*, 66(5): 300–304.

Peralta, I.E. and Spooner, D.M., 2001. Granule-bound starch synthase (Gbssi) gene phylogeny of wild tomatoes (*Solanum* L. section *Lycopersicon* Mill. Wettst. Subsection *Lycopersicon*). *American Journal of Botany*, 88(10): 1888–1902

Peter, K.V. and Rai, B., 1976. Analysis of genetic divergence in tomato. *Indian Journal of Genetics and Plant Breeding*, 36(3): 379–383.

Peter, K.V., 1975. Genetic analysis of certain quantitative characters in tomato (*Lycopersicon esculentum* Mill.). *Ph.D. Thesis,* G.B. Pant University of Agriculture and Technology, Pantnagar.

Pradeepkumar, T. and Tewari, R.N., 1999. Studies on genetic variability for processing characters in tomato. *Indian Journal of Horticulture*, 56(4): 332–336.

Prasad, A. and Prasad, R., 1976. Studies on variability in tomato (*Lycopersicon esculentum* Mill.). *Plant Science*, 8: 45–48.

Prashanth, S.J., Mulge, Ravindra, Madalageri, M.B., Chavan, Mukesh L. and Gasti, V.D., 2007. Studies on gentic variablilty for quality characters in tomato (*Lycopersicon esculentum* Mill.). *Journal of Asian Horticulture*, 3(2): 72–74.

Prashanth, S.J., Mulge, Ravindra, Patil, M.P. and Patil, B.R., 2006. Genetic variability studies for yield characters in tomato (*Lycopersicon esculentum* Mill.). *Journal of Asian Horticulture*, 3(1): 63–64.

Pujari, C.V., Wagh, R.S. and Kale, P.N., 1995. Genetic variability and heritability in tomato. *Journal of Maharashtra Agricultural Universities*, 20(1): 15–17.

Ram, J. and Panwar, D.V.S., 1970. Interspecific divergence in rice (O*ryza sativa* L.). *Indian Journal of Genetics and Plant Breeding*, 30: 1–10.

Reddy, M.L.N. and Gulshanlal, 1987. Genetic variability and path coefficient analysis in tomato (*Lycopersicon esculentum* Mill.) under summer season. *Progressive Horticulture*, 19(3–4): 284–288.

Rick, C.M., 1976. Tomato (Family Solanaceae). In: *Evolution of Crop Plants*, (Ed.) N.W. Simmonds. London, Longman.

Rick, C.M. and Fobes, J.H., 1975. Allozyme variation in the cultivated tomato and closely related species. *Symp. On Biochem. System. Genet. and Origin of cultivated plants at* Oregon in August 1975, *Bull. Torr. Bot. Club*, 102: 376–384.

Rick, C.M. and Tanksley, S.D., 1981. Genetic variation in *Solanum penelli* comparison with 2 other sympatric tomato species. *Pl. System Evol.*, 139: 11–45

Rick, C.M., Kesicki, E., Fobes, J.F. and Holle, M., 1976. Genetic and biosystematic studies on two new siblings of *Lycopersicon* from Interandeam Peru . *Theoret. Appl. Genet.*, 47: 55–68.

Sachan, K.S. and Sharma, J.R., 1971. Multivariate analysis of genetic divergence in tomato. *Indian Journal of Genetics and Plant Breeding*, 31(1); 86–93.

Sahu, G.S. and Mishra, R.S., 1995. Genetic divergence in tomato. *Mysore Journal of Agricultural Sciences*, 29: 5–8.

Singh, H.N., Singh, R.R. and Mital, R.K., 1974. Genotypic and phenotypic variability in tomato. *Indian Journal of Agricultural Sciences*, 44(12): 807–811.

Singh, Parvindar, Singh, Surjan, Cheema, D.S. and Dhaliwal, M.S., 2002. Genetic variability and correlation study of some heat tolerant tomato genotypes. *Vegetable Science*, 29(1): 68–70.

Singh, P.K., Singh, R.K., Saha, B.C. and Rajeshkumar, 1988. Genetic variability in tomato (*Lycopersicon esculentum* Mill.). *Indian Journal of Agricultural Sciences*, 58(9): 718–720.

Singh, R.R. and Singh, H.N., 1980. Genetic divergence in tomato. *Indian Journal of Agricultural Sciences*, 50: 591–594.

Suyambhulingam, C. and Jobarani, W., 1978. Genetic divergence in medium duration rice (*Oryza sativa* L.). *Madras Agricultural Journal*, 65: 56–58.

Tanksley, S.D., Ganal, M.W., Prince, J.P., de Vicente, M.C., Bonierbale, M.W., Broun, P., Fulton, T.M., Giovanonni, J.J., Grandillo, S., Martin, G.B., Messeguer, R., Miller, J.C., Miller, L., Paterson, A.H., Pineda, O., Roder, M., Wing, R.A., Wu, W. and Young, N.D., 1992. High density molecular linkage maps of the tomato and potato genomes. *Genetics*, 132: 1141–1160.

Van der Beek, J.G., Verkerk, R., Zabel, P. and Lindhout, P., 1992. Mapping strategy for resistance genes in tomato based on RFLPs between cultivars: cf9 (resistance to *Cladosporium fulvum*) on chromosome 1. *Theoretical and Applied Genetics*, 84: 106–112.

Vavilov, N.I., 1926. Studies on the origin of cultivated plants. *Bull. Appl. Bot, Genet. and Pl. Breed.*, 16(2).

Zhukovsky, P.M., 1958. Cultivated flora of the USSR XX Vegetable plants farm. Solanaceae Tomato, Common egg plant, black night shade pepino, pepper , husk tomato, Mandrake. Gosudar stevvnnoe Izdatel Stvo sel' skohozjajsty ennoi Literatury , Muskva, Leningrade 15r 40 k: pp. 531.

2013, **Biodiversity in Horticultural Crops Vol. 4**   *Pages 275–283*
Editor: **Professor K.V. Peter**
Published by: **DAYA PUBLISHING HOUSE, NEW DELHI**

Chapter 15

Aonla

Sant Ram[1] and S.K. Sharma[2]

[1]*Scientist,* [2]*Director,*
Central Institute for Arid Horticulture, Beechwal, Bikaner – 334 006, India
E-mail: [1]*santramiari@gmail.com,* [2]*ciah@nic.in*

The aonla or Indian gooseberry [*Emblica officinalis* Gaertn.] belongs to family Euphorbiaceae and Sub-family Phyllalthoidae. The chromosome number in *Emblica* species is 2n=4x=28. It is native to tropical region of South East Asia, particularly Central and Southern India (Morton, 1960). In India, aonla is cultivated since antiquity, which is mentioned in *Ayurvedic* and religious literatures for its medicinal and nutritional importance. Besides these, the Indian gooseberry is also wildly and naturally occurring in other tropical countries mainly in Sri Lanka, Cuba, Puerto Rico, Hawaii, Florida, Iran, Iraq, Java, West Indies, Trinidad, Pakistan, Malaya and China (Benthal, 1946). In India, its commercial cultivation is practiced in Uttar Pradesh since more than 200 years ago and recently large area has been expended in other states particularly in Maharashtra, Gujarat, Rajasthan, M.P., A.P., Karnataka, Tamil Nadu, Haryana, Punjab and Himachal Pradesh.

Aonla is a well known species for its fruits as well as plant, though all parts of the tree are useful and yield a range of products. The main use is as a fresh or dried fruit. The fruit is important for consumers due to its high nutritive and therapeutic values, and also to farmers because it provides higher return even in marginal lands. Demand for aonla continues to grow due to its numerous medicinal uses and ability to produce quality dye and tannins. The popularity and nutritional value of aonla can be explained from change of the western refrain "an apple a day keeps the doctor away" to "a aonla a day keeps the doctor away". Dried fruits are used in curing haemorrhage, diarrhea, chronicdysentery, diabetes, jaundice, dyspepsia, cough etc. Aonla is a main ingredient of Chavanprash and Trifla powder which is used in curing different abnormalities. Aonla has a wide range of domestic and industrial uses.

The aonla has vast potential and wider adaptability to grow in variable range of soil and climatic conditions. However, the most of the traditional varieties *viz.* Banarasi, Chakaiya and Francis are originated from chance seedlings or wildly grown population of aonla, which exhibits a number of variations in plant population. These genetical variations in respect to desirable traits are not well understood and fully exploited by the fruit breeders in crop improvement of aonla. No efforts have been made to study inheritance of variable characters. It is also not determined for particular quantitative and qualitative characters which are governed by single or polygene factors. Due to lack of such basic knowledge on genetical inheritance of available genetic resources, it is desirable to develop any high yielding hybrid varieties with high fruit quality and tolerance to biotic and abiotic stress conditions. On the other hand aonla is a cross pollinated, monoecious, entomophilous and fruit species having very less number of female flowers per determinate shoot, which contains more than 450 flowers. The natural crossing due to honey bee and wind pollination resulted in much variations within the genotypes. Besides, the emasculation is the major limiting factor to carry out the breeding programme in aonla. The poor success of fruit set are also affected by dry atmosphere, temperature, low humidity and dormancy of fruit buds which is very difficult to visualize at early stage of fruit setting in the month of March-April. Natural regeneration of aonla is by seed, owing to a lot of variation in existing natural population with respect to tree vigour, fruiting behavior, yield and quality characters. The natural variability of *Emblica officinalis* is available in the forest areas of Vindhyan hills, lower hills of Uttrakhand and Himachal Pradesh, Chhattisgarh, Jharkhand, M.P. Rajasthan and Bihar, whereas *Phyllanthus acidus* is found in North-Eastern hill region (Mizoram). Among natural growing plants, they differ not only in production but also in characters like, coloured fruit, cluster bearing etc. Therefore, there is need to collect these diversities *in-situ* as well as *ex-situ* for further utilization in crop improvement programme. In this direction, the efforts are being made by N.D.U.A.T., Faizabad, Anand Agriculture University, Anand and Central Institute for Arid Horticulture, Bikaner for collection, conservation and evaluation of aonla genotypes for various objectives. Various efforts have been made to develop the superior varieties from the locally and commercially cultivated varieties in different aonla growing belts. A number of promising genotypes *viz.* NA-10, NA-7, NA-6, Kanchan, and Krishna have been selected from the chance seedlings by clonal selection. These genotypes are different in their growth behaviour, flowering, fruiting and fruit quality.

However, the genetical inheritance of different desirable traits expressed in each selected variety is not well understood due to wide range of genetic base originated from seedling propagated varieties. No efforts are still made to study the genetic inheritance controlling the desirable traits in particular varieties. It is very difficult to explain which desirable traits are transmitted from which genetical sources, whether it is controlled by single/polygene/multiple gene factors. A plant breeder enables to determine all these genetical factors of desirable traits in variable varieties. It is also needed to study the extent of out crossing; intensity of natural pollination through honey bee, wind and hand pollination (Bajpai, 1968).

The extent of natural variability and cross-compatibility are very much required for determination of inheritance and development of varieties through hybridization. The cause of poor fruit set seems to be the high number of staminate flowers or less number of female flowers or improper pollination. However, the percentage of fruit set can be increased from 18 to 28 by hand pollination (Bajpai, 1968). It has been reported that in a few of the isolated plants of Banarasi trees, there is no fruit set even after profuse flowering, indicating presence of self- incompatibility.

Botany

The genus *Phyllanthus* belongs to family Euphorbiacae. *Phyllanthus* is derived from two Greek words: *Phullon*, meaning a leaf, and *anthos*, a flower, referring to the bearing of flowers on the axils of leaves. In general *Phyllanthus* species, including aonla resemble legumes, as feathery leaves, which are identical to pinnate leaves. *Phyllanthus* is a large, pantropically distributed genus that contains about 700 species, mostly shrubs and some herbs or trees. It bears two types of shoots and on the basis of growth characteristics these have been categorized as long or indeterminate and short or determinate. Systematic cytogenetic studies have not been carried out for aonla. The basic chromosome number of aonla is X=7 (Amal and Raghvan, 1957; Van Schaik-van Banning, 1991). The cultivated forms of aonla have been identified as polyploid (2n=28), while a variation from 2n=23 to 104 has also been reported (Amal and Raghvan, 1957; Van Schaik-van Banning, 1991). Aonla is a small to medium sized, much branched tree, usually grows to 10-20 m in height, but sometimes reaching even up to 30 m. Aonla is a deciduous plant, shedding its branchlets as well as its leaves. Indeterminates are longer and continue to put forth new growth in this season. The determinate shoots appearing on the nodes of the indeterminate shoots and their number at each node may vary from 3-5 different cultivars. These determinate shoots bear small sized leaves (9-14 x 2-4 mm) which are compound type. Flowering in aonla takes place during March-April on newly emerged determinate shoots under hot arid conditions. Size of determinate shoot of aonla is 4.5-114 cm. (Shukla and Singh, 2008). The fertilized ovary of aonla, unlike those of other plant, remains dormant for 3-5 months and resumes growth following division in endosperm and zygote nuclei in the month of August. Despite their variability, almost all *Phyllanthus* species express a specific type of growth called "phyllanthoid branching" in which the vertical stems bear deciduous, floriferous (flower-bearing), plagiotropic (horizontal or oblique) stems. The leaves on the main (vertical) axes are reduced to scales called "cataphylls", while leaves on the other axes develop normally. *Phyllanthus* is distributed in all tropical and subtropical regions on Earth. Leaf flower is the condition of flowers with leaflets for all *Phyllanthus* species. The aonla plant bears separate staminate and pistillate flowers. It is a monoecious species found generally in most genotypes of aonla. In determinate shoots, first a few proximal nodes are without leaves. Succeeding nodes are with green but reduced leaves, followed by nodes with male flowers. The numbers of male flowers per determinate shoot ranged between 327.50 to 501.75 and 4.75 to 9.69 range noticed for female flowers (Bala *et al.*, 2009). Fruit is fleshy drupe. They are globose, round to oblate in shape. It is pale green, changing to light yellow or brick-red (rarely) when mature. It consists of 3 sub-dehiscent, 2 seeded, crustaceous endocarp enclosed in a thick fleshy mesocarp. Endocarp consists usually of 6 trigonous having three prominent longitudinal angles.

The aonla or Indian gooseberry (*Emblia officinalis* Gaertn.) is one of the most important fruits of arid tropics of India which is valued for its high medicinal and nutritional property and significant as a health care drug tree. However, the potential of commercially grown cultivars is highly affected by variable environments. In addition to that, the genetical behavior of different desirable genotypes/ cultivars also varied due to out crossing/open or natural pollination. However, the commercially grown varieties are not genetically uniform and show stable genetical characters. There is need to exploit hybrid vigour by involving distinct and desirable progenies. Most of the traditional varieties *viz.*, Chakaiya and Francis originated from the chance seedlings or wildly grown aonla which exhibited wide range of variations in plant population. It is difficult to determine particular quantitative and qualitative characters which are governed by single/polygene factors. On the other hand, aonla is a cross pollinated and monoecious fruit species having very less number of female flowers per determinate shoot. The prime objective of plant breeding programme is to improve yield attributes, but yield is a complex character and a combined result of a number of component traits. Yield is controlled by polygenetic characters and influenced greatly by the environmental fluctuations. The identification of donor parents for important characters, assessment of genetic variables and diversity in available germplasm are prerequisites as breeding tool to commercial and successful breeding programme.

Varieties of *Emblica officinalis*

So far, the improvement of aonla was done through selection only. The earlier varieties of aonla such as Banarasi, Francis and Chakaiya have some limitations. Banarasi is a very shy bearer, Francis is affected by fruit necrosis and Chakaiya bears small sized fibrous fruits. Pathak *et al.* (1993) made some varietal improvement in aonla at NDUAT, Faizabad. Some varieties have also been developed by AAU, Gujarat. Characteristics of the important cultivars are as below:

Banarasi

Tree has upright growing habit with three branchlets per node. Fruits are large, triangular and slight conical at apex, skin thin, smooth, semi translucent, whitish green to straw yellow in colour; segment six, easily separated, flesh whitish green, nearly fibreless and soft. It has less number of female flowers and has self-incompatibility, hence it is shy bearing. It matures early, keeping quality poor; sometimes segment gets splitted during preserve making.

Chakaiya

This is late maturing cultivar, fruits small in size (10.34cm²) flattened, skin smooth, greenish in color, strips six not distinct, thin and flush prolific bearer, average number of female flowers per branchlet varies 7.86-8.50.

Francis (Hathijhool)

This is a mid season cultivar of aonla, fruit large in size (14.38 cm²), greenish in color, flattened, oral thick at upper side and thin to basin, flush fibreless and soft superior susceptible to necrosis. It is a prolific bearer and average number of female flowers per branchlet is 4.84.

NA-6

It is a chance seedling selection from Chakaiya. The size of fruit is medium (13.22 cm²), yellowish in colour, fibrous and prolific in bearing, average number of female flowers per branchlet is 10.83.

NA-7

It is a selection of chance seedling of Francis. The size of fruit is large (15.23cm²), free from necrosis and fibreless. It is more precocious and prolific bearer and seems to be an outstanding selection of aonla. Average number of female flowers per branchlet is 9.71.

NA-10

This is a selection from the seedling of Banarasi which is early in bearing, medium fruit size (13.5 cm²), round skin, brown in colour and prolific bearer.

NA-4 (Kanchan)

It is a chance seedling of Chakaiya. Tree is tall with spreading growth habit. Fruit is small to medium, flattened oblong, skin is smooth, light green, segment six and difficult to separate, flesh

Figure 15.1: Aonla Germination-Seedlings

Figure 15.2: Bearing Habit

Figure 15.3: Chakaiya

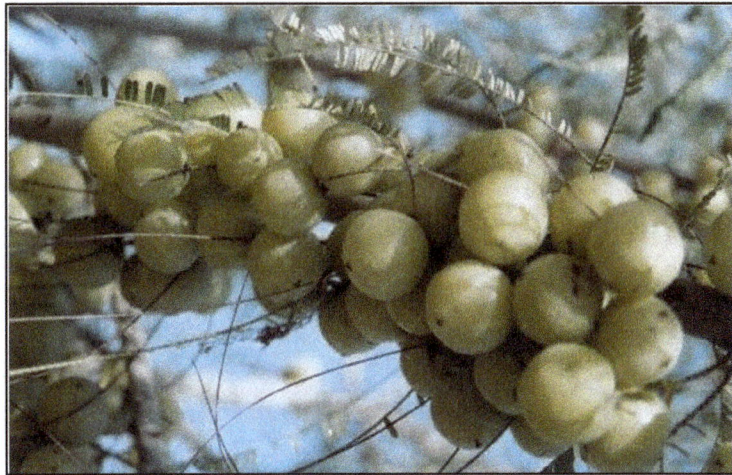

Figure 15.4: Aonla NA-6

fibrous and hard, keeping quality good and ideally suitable for preparation of pickles. It is a late maturing cultivar and free from fruit necrosis, TSS 9.5 per cent, acidity 2.11 per cent and vitamin C 549.20 mg/100 g.

NA-5 (Krishna)

This is a chance seedling of Banarasi from Pratapgarh district of Uttar Pradesh. Tree is semi tall with spreading growth habit, fruit large, triangular, skin smooth, whitish green to apricot yellow with red spot on exposed surface.

Figure 15.5: Aonla NA-7

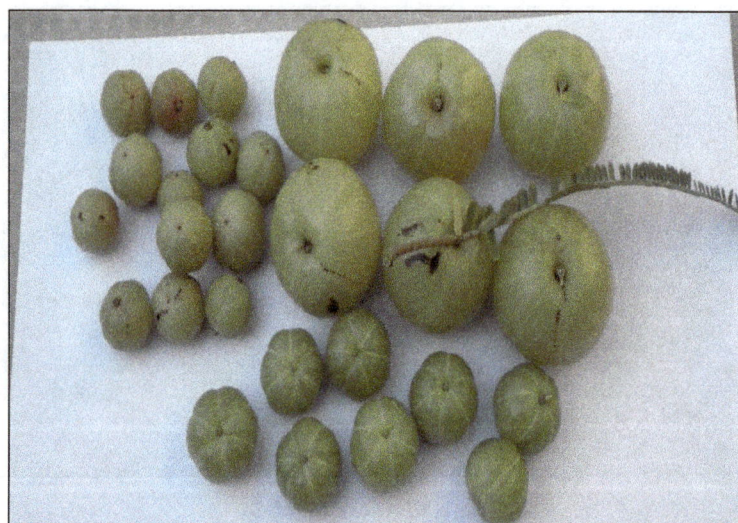

Figure 15.6: NA-7 Variability in Fruit

NA-8

It is a selection from open pollinated seedlings of Chakaiya. Trees are tall with upright growth habit. Leaves are small, oblong with acute apex. It is a moderate bearer having 3.5 female flowers per branchlet. Fruits are small and flattened round, skin slightly rough, thick and light green in colour. Flesh is fibrous and hard, keeping quality good. It is susceptible to fruit necrosis.

NA-9

It is a seedling selection from open pollinated strain of cultivar Banarasi. Trees are tall and semi spreading growth habit, fruit large, flattened round, skin smooth, light green in colour, flesh soft and

Figure 15.7: Aonla Goma Aishwariya

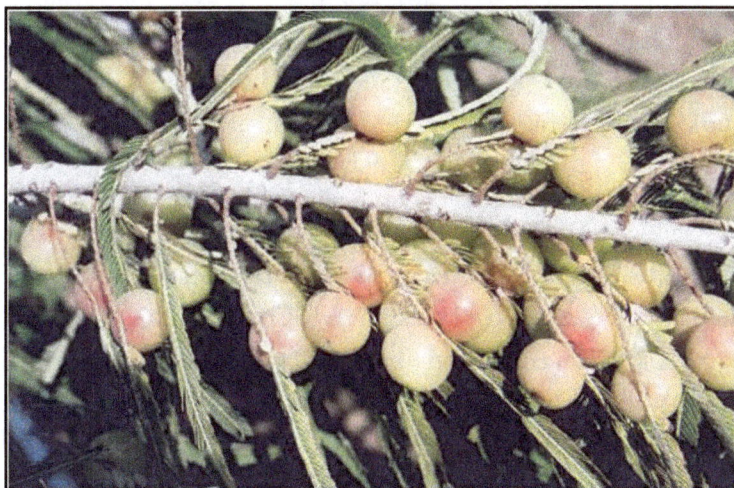

Figure 15.8: Aonla NA-5 (Krishna)

moderately fibrous, shy bearer and early maturing, having moderate keeping quality. It is susceptible to fruit necrosis. T.S.S. 11.50 per cent, acidity 2.52 per cent and Vitamin C 499 mg/100 g.

Goma Aishwariya

A selection from plus tree and released from CIAH, Bikaner, it is an early and drought tolerant cultivar. The average yield potential is 102.9 kg/tree. It has low fiber content and suitable for processing and export.

Anand-1

Tree is small with upright growth habit. It is a moderate bearer having 1-2 female flowers per branchlet. Fruits are small, skin slightly rough and thick. Flesh is fibrous and hard.

Anand-2

Tree is small with upright growth habit. It is a moderate bearer having 2-2.5 female flowers per branchlet. Fruits are small to medium, skin rough and keeping quality medium.

Lakshmi-52

Seedling selection of Francis from the farmers field in district Pratapgarh (U.P.), tree is semi erect type, prolific bearer with big size of fruit. During early period of fruit growth, the colour is light pink, which disappears on full development. This variety should be popularized for commercial cultivation.

BSR-1

This is a seedling selection of Tamil Nadu. It is self fruitful, yields round the year, small size fruits, fibrous and low moisture content, heavy bearing and thus in very high demand by *Ayurvedic* industries.

References

Amal, E.K.J. and Raghvan, R.S., 1957. Physiology and vitamin C in *Emblica officinalis*. Proc. *Ind. Acad. Sci.*, 47: 312–314.

Bajpai, P.N., 1968. Studies on flowering and fruit development in aonla (*E. officinalis* Gaertn.). *Hort. Adv.*, 7: 38–67.

Bajpai, P.N., 1957. Blosson biology and fruit set in aonla (*Phyllanthus emblica*). *Indian J. Hort.*, 14: 95–102.

Bajpai, P.N., 1968. Studies on flowering and fruit development in aonla (*Emblica officinalis* Gaertn.). *Hort. Adv.*, 7: 38–67.

Bala, S., Ram, S. and Prasad, J., 2009. Studies on variability and genetic diversity in selected aonla genotypes. *Indian J. Hort.*, 66(4): 433–437.

Benthal, A.P., 1946. *Trees of Calcutta and its Neighbourhood*. Thacker Spink and C. Ltd. Calcutta.

Morton, J.E., 1960. The emblic (*Phylanthus emblica* L.). *Econ. Bot.*, 14: 119–127.

Pathak, R.K., Srivastava, A.K., Dwivedi, R. and Singh, H.K., 1993. *Aonla Descriptor*. Department of Horticulture, N.D.U.A.T. Kumarganj, Faizabad.

Shukla, A.K. and Singh, D., 2008. Genetic variability of aonla.(*Emblica officinalis* Gaertn.). *The Hort. J.*, 21(1): 11–13.

Van Schaik-van Banning, A.J.J., 1991. *Phyllanthus emblica* L. In: *Plant Resource of South East Asia No. 3: Dye and Tannin-Producing Plant*. (Eds.) R.H.M.J. Lemmens and N. Wulijarni-Soejipto. PUDOC, Wageningen, pp. 105–108.

2013, Biodiversity in Horticultural Crops Vol. 4   *Pages 285–300*
Editor: Professor K.V. Peter
Published by: DAYA PUBLISHING HOUSE, NEW DELHI

Chapter 16

Bael

S.K. Sharma, R.S. Singh* and A.K. Singh

Central Institute for Arid Horticulture,
Bikaner – 334 006, Rajasthan
E-mail: rssingh1@yahoo.com

Bael (*Aegle marmelos* Correa.) also known as *Shri Phal, Baelpatra,* Bengal quince, Beel etc. is an important indigenous fruit of India. Bael fruit is known from pre-historic times. In India, bael is being grown throughout the country.It is found growing in sub-tropical, tropical, arid and semi-arid regions of the country. The bael tree has found mention in mythological treatises. It is mainly planted near temples of Lord Shiva. According to Hindu customs, the leaves of the tree are considered sacred and offered to the Lord Shiva. The fruits form an essential ingredient of holy offerings to the God and Goddess through holy pyre (havan). In history, the mention of bael tree is traced to Vedic times (C 2000 B.C. – C 800 B.C.) in the *Yajur veda.* Om Prakash (1961) recorded mentions of bael in early Buddhist and Jain literature (C 800 B.C. – C 325 B.C.) describing various methods of ripening of the bael fruit along with some other fruits. In the 'Ramayana' period, the bael fruit was known and its trees were reported growing in 'Chitrakuta' hills and 'Panchvati'. In the 'Upavana Vinod' a Sanskrit treatise on silviculture and in the 'Brihat Samhita' mention is made of bael fruit as the legend goes, in the forest, Lord Rama performed religious rites by offering various fruits including bael. Bael fruit is portrayed in painting of Ajanta Caves along with other fruits (Om Prakash, 1961). Like many other species of Rutaceae family, bael also has fragrant flowers. It is believed that this tree acts as an indicator plant for tracing of underground water (Singh and Roy, 1984).

It is a nutritious and medicinal fruit plant, which is the most suitable to grow in water scarce areas of the country. Its fruit is highly nutritive and rich in riboflavin, vitamin A, carbohydrate, etc. Various chemical constituents' *viz.,* alkaloids, coumarins and steroids were isolated and identified from different parts of bael plants. Marmelosin is probably the therapeutically active factor of bael fruit and is known as the panacea of stomach ailments. The aroma component of bael fruits was studied by Tokitoma *et al.* (1982). All the parts of tree, whether stem, bark, root, leaves or fruit at different maturity stages have some use or the other. It is also used in the preparation of several Ayurvedic medicines since ancient times (Sharma, *et al.*, 2007; Rai and Dwivedi, 1992). The pulp of fresh ripe fruits is eaten and used for preparation of value added products like jam, squash, powder, nectar, toffee etc. The physico- chemical studies of different bael cultivar fruits were carried out by Ram and Singh (2003). They have laxative properties and good for heart and brain. Fruit pulp is used for preparing *sherbet* and syrup. The *sherbet* and marmalade prepared from fruit pulp are prescribed for diarrhea and dysentery. It is said that the ripe fruit is a tonic, an astringent and laxative, whereas unripe fruit is an astringent and digestive. Moreover, the fruits available mostly in the markets are harvested from the trees either naturally growing in the forest areas or from seedling plantations. The mature fruits can be stored up to 10-15 days under normal conditions and at low temperature (9° C) up to three months.

Nutritive Value and Uses

The bael fruit is mentioned in the Indian Pharmacopoeia. They are also valued in *Ayurvedic* medicines. The peripheral part just within the rind is fleshy and thick, and has a pleasant resinous odour. The walls separating the chambers have a light yellow tint which becomes yellowish brown on exposure and have slightly acrid bitter taste. The chambers are full of amber or honey coloured viscous, very sticky or glutinous, translucent pulp, slightly sweet and feebly aromatic. The gummy substance surrounding the seeds serves as a good adhesive and is added to water-paints to improve strength and brilliancy. The shell of the hard fruits is fashioned into pill and snuff boxes, sometimes decorated with gold and silver. It is more abundant in young fruits. The gum is used for the stabilization of drilling fluids. The stem also contains a gum similar to gum arabic. The leaves are mildly flavored, and hence chewed as mouth freshener. Fruits have mucilaginous substance. Gum is also obtained from stem and seed locules. The use of bael gum was reported as early as in 1961 by Haskar and Kendurkar. The gum is used to prepare adhesive, waterproofing and oil emulsion coating. The wood is used for making small agricultural implements. The wood pulp is used for manufacturing wrapping paper. The dried twigs are used as firewood. The leaves are also used as fodder for feeding of sheep and goats. The leaves contain 0.6 per cent of essential oil, mostly composed of limonene. The use of bael leaf powder as a protectant against storage pests of paddy was reported by Prakash *et al.* (1983). An antifungal activity of essential oil of bael is reported by Jain (1977). The antimicrobial and anti-helmintic properties as well as chemical properties and medicinal importance of bael seeds have also been studied by Singh *et al.* (1983) and Pal *et al.* (1993), respectively.

The importance of bael fruit lies in its curative properties, which make the tree one of the most useful medicinal plants of India (Kirtikar and Basu, 1935). Its medicinal properties have been dealt with in the 'Charaka Samhita', an early medical treatise in Sanskrit. All the parts of the tree–stem, bark, root, leaf, flower, seed oil or fruits of any stage of maturity and ripening–are used in various *Ayurvedic* medicines. Bael fruit was in use from time immemorial in traditional medicine for relieving constipation, diarrhea, dysentery, peptic ulcer and respiratory infections (Sharma *et al.*, 2007). The important medicinal properties of bael are antidiabetic, antimicrobial, anti-inflammatory, antipyretic, analgesic, cardioprotective, antispermogenic, anticancer and radioprotective.

The demand of bael fruit is not much higher as other major fruits like mango, banana, guava, apple etc. The ripe fruit is eaten fresh and has a more demand from them, who wants it for therapeutic use. Sometimes the pulp diluted with water and added with requisite amount of sugar and tamarind, forms a delicious cooling drink. The bael pulp is used as a base of various fruit products like squash, jam, slab, toffee, powder, nectar, RTS, etc. A firm jelly is made from the combined pulp of bael and guava. The bael fruit toffee is prepared by combining the pulp with sugar, glucose, skimmed milk powder and hydrogenated fat. Green bael fruits are used for preparing *murabba* (preserve). The leaves, roots, flower, fruit and other parts are known to have medicinal properties especially for stomach disorders and ailments of digestive system. Bael fruit pulp dressed with palm sugar, eaten as breakfast, is a common practice in Indonesia. The green bael fruit slices often are dried and stored for future use. The unripe or half ripe fruit is regarded as astringent, digestive stomachic and good for heart and brain (Kirtikar and Basu, 1935). The fruit is used in chronic diarrhea and dysentery, and is said to act as tonic for heart and brain. Bael fruit pulp is dried to make powder, which can be stored for long period (Verma and Ahmed,1958). It is useful adjuvant, as it helps to remove constipation which hinders the healing ulcerated surfaces of intestine. Besides the fruits, the root is an ingredient of *'Dashmool'* (ten roots), used as Ayurvedic medicine. The roots as well as bark are used in the form of a decoction as a remedy in melancholia, intermittent fever and palpitation of heart. Root has anti-amoebic and hypoglycemic properties. The young leaves and shoots are used as fodder for cattle, sheep and goats. The leaves are bitter and used as febrifuge. Poultice made of leaves is used for ophthalimia and ulcers. Fresh leaves are also used as remedy for 'dropsy, and beriberi associated with weakness of heart. The astringent rind of ripe fruits and bark are employed in dyeing and tanning. The timber is commonly used for making pestles of oil and sugar mills, for posts, shafts, axles and naves of cart. The shells of smaller fruits are used as snuff boxes. A yellow dye is extracted from the rind of unripe fruit. The diluted leaf juice is used for catarrh. The alkaloid aegeline present in the leaves are efficacious in asthma. Utilization and storage of bael fruit were reported by Roy and Singh (1979, 1981).

Various chemical constituents *viz,* alkaloids, coumarins, steroid etc. have been isolated and identified from the different parts of the plants. Bael fruit is one of the most nutritious fruits. Analysis of the fruit gave the following values: 61.5g moisture, 1.8 g protein, 0.39 g fat, 31.8 g carbohydrates, 1.7 g minerals, 55 mg carotene, 0.13 mg thiamine, 1.19 mg riboflavin, 1.1mg niacin and 8.0 mg vitamin C per 100 g of edible portion (Gopalan *et. al.,* 1985).No other fruit has such a high content of riboflavin. Tannic acid is only phenolic substance detected from bael fruits. The fruit contains allo-imperatorin, marmelosin identical with imperatorin and β-sitosterol. The fruit yields two per cent dried, water soluble gum. Hydrolysis of the gum gave: galactose, 20.4, arabinose, 10.7, D-galacturonic acid, 25.2 per cent and traces of rhamnose (Haskar and Kendulkar, 1961). The presence of Aegeline, an alkaloid, has been reported in the leaves of bael fruit (Chatterjee and Roy, 1957). Analysis of the leaves gave the following values (dry basis): crude protein, 15.13, ether extract, 1.54, crude fibre, 16.45, N-free extract, 52.83, ash, 14.05, calcium, 5.93 and phosphorus, 0.69 per cent. Marmelosin is most probably the therapeutically active compound present in fruit (0.03-0.37 per cent) and varies according to variety and locality (Dixit and Dutt, 1932). The mature bark contains marmesin, auropotins, umbelliferone, lupeol and skimmianine (Chatterjee and Roy, 1959).The bael fruit mucilage on hydrolysis shows the presence of three reducing sugars, galactose, arabinose and rhamnose (Parikh *et al.,* 1958). The wood contains a furoquinoline alkaloid, dictamnine, marmesin and neutral compound. The seed yields oil 34.4 per cent on dry basis and the fatty acid composition of oil as follows: palmitic, 16.6; stearic, 8.8; oleic 30.5; linoleic, 30.0; and linolenic, 8.1 per cent. Bael seed contains 62 per cent protein and 3 per cent each carbohydrate and ash (Banerjee and Maiti, 1980).

Origin and Distribution

There are no systematic plantations of bael trees in our country, hence, exact data on acreage and production are not available. However, in recent years, concerted efforts have been made for collection of elite genotypes of bael from all over the country and their evaluation and establishment of germplasm block at ICAR Institutes/Regional stations and State Agricultural Universities. The bael is grown in

India and in neighboring countries namely Nepal, Sri Lanka, Pakistan, Bangladesh, Myanmar, Thailand and most of the South East Asian countries. In India, it is distributed throughout the country, but concentrated area under bael is in eastern parts of the Gangetic plains and nearby areas particularly in Uttar Pradesh, Bihar, M.P. Chhatisgarh, Jharkhand, West Bengal and Orissa. Its trees are also available in wild state in sub-Himalayan tracts from Rajasthan to West Bengal, central and southern India. In Gujarat, bael trees are found growing naturally in the forest with great diversity. Most of the genotypes available in Gujarat have small size fruits (Singh *et al.,* 2008). Looking to its potential in waste land and dry areas of the country, plantations of improved cultivars are being developed. Some progressive farmers have also started planting bael varieties either in the form of small orchards or as boundary plantation of other fruits. Apart from systematic orchards, bael trees are also planted in nutritional gardens, parks, temple gardens and roadsides for various purposes.

The *bael* tree is very hardy, it can thrive well even in saline, alkaline, and stony soils having pH range from 5 to10, where many other fruit trees fail to establish. The extent of hardiness of bael plants under Thar desert have been observed that the plant even after being buried under sand for 2-3 months are capable of rejuvenating itself. It can tolerate salinity up to 9 dsm^{-1}. However, well-drained sandy loam soil is ideally suitable for *bael* orchards. *Bael* plants can grow easily in wasteland and in sandy soils of arid ecosystem, having low fertility status and poor moisture holding capacity. Marked reductions in the contents of leaf NPK and Ca were observed in response to increase in salinity and sodicity levels in the soil in which plants were grown. Salinity caused significant increase in leaf Mg, while sodicity decreased it. Leaf Na was at toxic levels in both saline and sodic soils. Bael plants tolerate salinity and sodicity (Shukla and Singh, 1996a). Seed germination and seedling growth were influenced by sodic soils. Delayed and poor seed germination and reduced plant growth were observed in response to increased sodicity (Shukla and Singh, 1996b).

The bael tree grows successfully and produces higher yield in sub-tropical climate where summer is hot and winter mild. It has wide range of adaptability to adverse soil and climate. Further, the seedling *bael* tree is reported cold hardy grown up to an elevation of 1200 m asl and is not damaged by temperature as low as –7° C. Under arid conditions of western part of Rajasthan, the leaves and tender twigs of budded *bael* cultivars and young seedlings were affected by low temperature/frost during winter. However, plants tolerate high temperature and drought by shedding leaves during summer.

Botanical Description

The genus *Aegle* belongs to family Rutaceae. Other members of Rutaceae are *Citrus, Casimiroa, Clausena, Eremocitrus, Limonia, Feroniella, Fortunella, Poncirus, Triphasia* etc. The generic name is of Greek origin and the species *marmelos* is of Portuguese origin. The chromosome number is x=9 and 2n=36. The tree is medium to tall, deciduous, slow growing and 5-10 m in height. The bark is furrowed and corky yellowish-brown in colour. Its leaves are green aromatic, trifoliate; branches have spines, and trunk is strong and stout. Misra *et al.* (1999) reported great variations in leaf characters, development pattern and shoot growth of 8 selected genotypes of bael under *Tarai* conditions of Uttarakhand. The bisexual flowers are nearly 2 cm wide and borne in clusters and greenish white to white in colour, appear in axil of leaves and pedicilate. The calyx is shallow with 5 short, broad teeth and pubescent outside. There are 5 petals, which are oblong oval, blunt, thick, pale greenish white, dotted with glands. Stamens are numerous, sometimes coherent in bundle. The ovary is oblong ovoid, slightly tapering, the axis being wide; cells are numerous 8-20, small and arranged in a circle, with numerous ovules in each cell. Fruit is hard-shelled berry, globose, pericarp smooth and round to oblong in shape, greyish yellow in colour. Variation in morphological characters of fruit have been reported by Vishal Nath *et al.* (2003) in germplasm collected from East Central India. Seeds are numerous and arranged in seed cavity/locules. Mucilage is present in seed locules. The testa is white with wooly hairs. The embryo has large cotyledons (Roy and Singh, 1979, Singh *et al.*, 2009).

The branches are unusual with long, straight spines. The bark is shallowly furrowed and corky. The bisexual flowers are nearly 2 cm wide and borne in clusters, sweet scented and greenish white. The calyx is shallow with 5 short, broad teeth, pubescent outside. There are 5 petals which are oblong oval, blunt, thick, pale greenish white, dotted with glands. Stamens are numerous, sometimes coherent in bundles. The ovary is oblong ovoid, slightly tapering, the axis being wide; cells are numerous 8-20, small and arranged in a circle, with numerous ovules in each cell. The ripe fruits are woody, large, spherical, up to 23 cm in diameter, oblong or pear shaped, with a more or less smooth or slightly tuberculate surface. The fruit is usually globose with a pericarp nearly smooth, greyish yellow, 1.6-2.7 mm thick, hard and filled with soft, yellow and orange, very fragrant and pleasantly flavoured pulp. Botanically, the fruit is berry with hard pericarp. The number of cells in the fruit, arranged in a circle, is equal to the number of cells in ovary. Seeds are numerous, compressed and arranged in closely packed tiers in the cell surrounded by very tenacious, slimy, transparent mucilage, which becomes hard when dry. The testa is white with wooly hairs and the embryo has large cotyledons and a short superior radicle. Pollination is usually done by honeybees. The nectar secreting disc found beneath the ovary is the main source of attraction for insects (Srivastava and Singh, 2000; Singh *et al.*, 2008).

Flower bud emergence, flowering duration, time of anthesis, dehiscence of anther, stigma receptivity and pollen viability vary according to variety and locality (Srivastava and Singh, 2000). Variation in inflorescence and floral morphology were observed in bael genotypes by Jaiswal and Misra (1996) under *Tarai* conditions. Flower bud emergence and flowering in bael take place during the month of May under arid conditions. Size and shape of floral organs in terms of bud size, flower size, petal size etc. of the varieties evaluated at CHES, Godhra under rain fed conditions of semi-arid ecosystem according to Singh *et. al.* (2008) are given in Tables 16.1 and 16.2.

Table 16.1: Flowering Behavior of Bael Varieties

Genotypes	Flower Bud Emergence			Flowering Period		
	Start	Peak Period	End	Start	Peak Period	End
CISHB-1	20 April	24-29 May	16 June	22 May	8-12 June	18June
CISHB-2	30 April	23-28 May	20 June	24 May	7-14 June	22 June
NB-5	3 May	24-29 May	22 June	26 May	10-15 June	24 June
NB-7	2 May	25-30 May	15 June	22 May	7-12 June	16 June
NB-9	5 May	25-30 May	23 June	15 May	10-18 June	24 June
NB-16	7 May	24-29 May	26 June	14 May	10-20 June	25June
NB-17	30 April	23-28 May	20 June	24 May	7-14 June	22 June
Pant Aparna	5 May	24-29 May	16 June	15 May	9-14 June	18 June
Pant Sujata	1 May	25-30 May	15 June	22 May	7-12 June	16 June
Pant Shivani	2 May	25-29 May	20 June	23 May	10-15 June	17 June
Pant Urvashi	8 May	25-30 May	23 June	15 May	10-18 June	24 June
Goma Yashi	21 April	16-23 May	19 June	12 May	9-16 June	14 June

Table 16.2: Morphomatrics of Floral Organs of Different Bael Varieties

Genotypes	Flower Size (mm)		Bud Size (mm)		Petal Size (mm)		Pedicel Size (mm)	
	Length	Width	Length	Width	Length	Width	Length	Width
CISHB-1	14.00	35.00	11.00	7.00	15.00	7.00	07.50	2.00
CISHB-2	19.00	26.00	13.00	9.50	13.00	8.00	06.50	2.00
NB-5	14.15	29.00	11.50	8.00	18.00	7.50	10.50	2.50
NB-7	18.00	35.00	13.00	9.50	19.00	9.00	05.50	2.50
NB-9	12.00	25.00	11.00	8.00	11.00	7.00	04.00	2.00
NB-16	14.00	28.00	11.00	8.00	12.00	7.00	04.50	2.00
NB-17	18.00	34.00	13.00	9.50	18.00	9.00	05.00	2.50
Pant Aparna	16.00	30.00	10.00	9.00	18.00	10.00	10.00	2.50
Pant Sujata	15.15	29.00	12.50	8.00	17.00	7.50	09.50	2.50
Pant Shivani	15.00	30.00	11.00	7.50	18.50	8.00	09.50	2.00
Pant Urvashi	15.00	26.00	11.00	8.50	16.00	9.00	7.00	2.50
Goma Yashi	12.00	22.00	12.00	8.00	15.00	9.50	6.50	2.00

Genetic Diversity

Bael plant grows naturally in Uttar Pradesh, Bihar, Jharkhand, Madhya Pradesh, Orissa, West Bengal and Chhattisgarh with large genetic variability, which should be exploited. In Uttar Pradesh, Deoria, Basti, Gorakhpur, Gonda, Faizabad, Sultanpur, Jaunpur, Pratapgarh, Mirzapur, Allahabad, Lucknow, Etawah, Agra, etc. are the districts where large number of promising genotypes are either growing naturally or planted near the houses. There is a fast genetic erosion in wild bael genotypes, therefore, its conservation has become necessary (Srivastava *et al.,* 1998). Rai and Dwivedi (1992) reported vivid account of bael genetic diversity available in India. The variability in bael germplasm was observed in identified types at different locations (Rai *et al.,* 1991). Apart from the tree morphological characters, wide variability exist in fruit size and shape, bearing habit, flesh colour, texture, fibre content, sugar content, mucilage content, etc. In Jaunpur area of UP, very old naturally growing bael plants are available. Some types have more number of seeds, gum locules and thick pericarp (Misra *et al.,* 2000). However, some selections were made at NDUA&T, Faizabad and GBPUA&T, Pantnagar and CISH, Lucknow, which are gaining popularity for commercial cultivation. At Central Institute for Arid Horticulture, Bikaner also, collection of bael germplasm has been done, which are under evaluation (Table 16.3). Besides this, some germplasm were also collected from nursery/farmers' field and maintained in the field repository at CAZRI, Jodhpur; CISH, Lucknow, CCS HAU, RRS, Bawal; NDUAT, Faizabad, GBPUAT, Pant nagar and TNAU, Aruppokottai for conservation and evaluation. Lal (2002) evaluated 12 genotypes from Jaipur (Rajasthan) and found that 8 genotypes produced fruits of excellent quality under semi-arid conditions. In Chomu area of Jaipur, fruit samples from seedling plants were collected during 2009. Variation was observed in fruiting, size, quality of fruits and two genotypes were found promising. Fruit cracking was also observed in bael trees grown in Sikar district.

Table 16.3: Status of Bael Germplasm at CIAH, Bikaner

Genotypes/Cultivars	Source of Collection
N.B.-5, N.B.7, NB.-9,NB-16,NB-17	NDUAT, Faizabad.
Pant Aparna, Pant Urvashi, Pant Shivani, Pant Sujata	GPBUAT, Pant Nagar
Dhara Road	Jodhpur
Seedling	Udaipur
Seedling	Jodhpur/Ajmer
Goma Yashi	CHES, Godhra, Gujarat

Variation in morphological and physico-chemical traits of fruits were reported by Pandey *et al.* (2008) from samples collected from U.P. and M.P. The number of fruits per tree varied from 56-695, fruit weight (0.38-2.84 kg) and fruit yield per tree (75.74-450.08 kg/tree) among the different accessions collected. The fruit characters *viz.*, fruit length (7.93-18.60 cm), fruit circumference (28.20-61.67 cm), number of seeds per fruit(33.0-95.0),number of seed sacs per fruit (10.0-19.0), skull weight per fruit (0.10-0.69 kg), skull thickness (1.25-3.32 mm), seed weight (3.74-10.48 g), pulp weight (0.16-1.76 kg), TSS content (28.67-46.33°B), acidity (0.22-0.56 per cent), vitamin 'C' content (6.25-17.71 mg/100g pulp), tannin content (1.68-4.68 per cent) and total carotenoids (0.90-1.72mg/100g pulp) were evaluated among the 40 accessions. The fruit drop and cracking are major problems in bael growing areas under rain fed conditions. These disorders can be regulated by using micronutrients, plant growth regulators and soil amendments (Saini *et al.,* 2004).

The efforts for varietal improvement in bael are less in our country so far. Jaisawal and Misra (1998) reported that fruit set, retention and yield of fruits in bael germplasm varied under *Tarai* conditions. The available promising types are through the clonal selection only. Some basic works on floral biology, fruit set, fruit drop and fruit retention and pollen viability have been done on seven years old plants of NB-4, NB-5, NB-7 and NB-9 at NDUAT, Faizabad. It was observed that the flower bud emergence in bael cultivars started from third week of May and continued up to last week of May, it was the earliest in NB-9 followed by NB-4, NB-7 and the latest in NB-5. The flower bud took 15-22 days from flower bud emergence to anthesis. The maximum flowers open between 6-7 AM followed by 5-6 AM and 7-8 AM in all the bael types under study. The maximum flowers dehisced between 6.30-7.30 AM. The stigma receptivity was the highest with NB-9 closely followed by NB-5 and the lowest with NB-7. The stigma receptivity was maximum on the day of anthesis. The pollen viability was the highest in NB-7 followed by NB-5, NB-4 and the lowest in NB-9. The fruit retention was the highest in NB-9 closely followed by NB-5 and the lowest in NB-7 (Srivastava and Singh, 2000).

Metroglyph grouping and association analysis on physical characters of bael fruit were reported by Ghosh and Gayen (1990). Analysis of genetic divergence in seventeen bael germplasm using Mahalanobis D2 statistics indicated existence of substantial genetic diversity. The genotypes were grouped into 3 clusters, which included one solitary group (PB-3). The clustering pattern of genotypes was random and not parallel to geographic distribution. The characters *viz.*, ascorbic acid content, fruit weight, fruit length, number and weight of seeds per fruit, fibre content and petiole length contributed maximum to genetic divergence. These characters may be used in selecting diverse parents in hybridization and intercrossing between clusters to develop superior clones of bael with most desirable traits (Deepti Rai, *et al.,* 2002).

Varietal Wealth

Among local types, Ayodhya, Kagzi, Etawah, Gonda and Mirzapuri bael are popular in Uttar Pradesh due to their good yield potential and quality of fruits. Out of the existing extensive biodiversity of bael in India, two distinct types *viz.* the small fruiting type having acrid pulp and high seed and mucilage and the large fruiting with thin skin skull, less seeds and mucilage and more sweet pulp are broadly identified. In recent past, some promising varieties of bael were developed through clonal selection at G.B. Pant University of Agriculture and Technology, Pantnagar. Pant Bael-3 and Pant Bael-1 produced vigorous growth and the tallest trees while Pant Bael-9 (Pant Aparna) and Pant Bael-12 (Pant Sujata) produced dwarf trees. Pant Bael-10 and Pant Bael-1 produced higher number of fruits per tree while Pant Bael-5 (Pant Shivani) followed by Pant Bael –12 (Pant Sujata) produced the highest yield in terms of fruit weight per tree. In general, Pant Shivani, Pant Aparna and Pant Sujata were better for physico-chemical characters of fruits as compared to other clones.

Cultivars NB-5 and NB-9 are performing well under arid conditions of Western Rajasthan. The vegetative growth parameters and physico-chemical characters of some bael genotypes at NDUAT, Faizabad (Anon., 2008) were recorded and presented in Table 16.4. The bael varieties are characterized on morphological and physiochemical basis (Singh *et al.,* 2009). Dhara Road, a prolific bearing, erect growing, small size fruits, is a local genotype of bael suitable for arid areas (Singh and Vishal Nath, 1999). Tree size is very tall and number of fruits/plant is higher than the other local genotypes. Recently, CISH-B-1 and CISH-B-2 selections of bael have been developed at Central Institute for Sub-tropical Horticulture, Lucknow (Pandey *et al.,* 2001). In addition to this, a collection *viz.* NB-16 and NB-17 from Pratapgarh and Sultanpur (U.P.) is also found promising (Singh *et al.,* 2009)

Table 16.4: Evaluation of Bael Genotypes

Characters	Genotypes			
	NB- 4	*NB- 5*	*NB -7*	*NB -9*
Plant height (m)	3.80	4.12	4.38	4.30
Av. Plant Spread (m²)	5.30	6.26	6.00	5.20
Fruit Wt. (kg)	0.75	1.10	2.30	1.40
Fruit length (cm)	22.75	27.27	27.50	26.50
Fruit Width (cm)	27.20	31.00	29.00	23.00
No. of seeds/fruit	60.00	61.00	72.00	84.00
TSS (per cent)	32.00	34.00	38.00	35.00
Acidity (per cent)	0.15	0.10	0.26	0.23
Ascorbic Acid (mg/100 g of pulp)	19.02	21.40	25.10	24.00
Yield (fruits/tree)	30.00	46.00	19.00	41.00

Description of Varieties

The description of varieties of bael under dry land conditions of semi-arid and arid region is given below:

Narendra Bael-5

The plants are small with semi-spreading growth habit, precocious and prolific in bearing. The budded plants start fruiting in the 4th year. The average fruit yield of six-year old plant is 28.78 kg. The

Figure 16.1: Genetic Diversity in Bael Germplasm Collected from Jharkhand

Figure 16.2: Diversity in Bael Germplasm Collected from Gujarat

Figure 16.3: Genetic Diversity of Bael Germplasm Collected from U.P. and M.P.

Figure 16.4: Bearing Tree of Bael cv. N.B-5

Figure 16.5: Variability in Bael Genotypes

Figure 16.6: Five Year Old Bael Tree c.v NB-9

Figure 16.7: Bael Variety Goma Yashi

fruits are medium in size (12.50 cm X 11.50 cm), round, with smooth surface and very thin rind (0.16-0.17cm), straw yellow at maturity, low in mucilage, moderately fibrous and an attractive yellow pulp, with low seed content. Excellent in taste and flavour, the fruits have 33°Brix total soluble solids in pulp and 48° Brix in mucilage and ascorbic acid (18.63 mg/100g) of edible portion. The fruit weight ranged from 0.8-1.0 kg under rain fed conditions of semi-arid ecosystem during sixth year of orchard life. Number of fruits per plant is more than NB-9. Fruit cracking has also been observed under hot arid environment. Its taste is good and can be used as fresh as well as for processed products.

Narendra Bael-7

Plants are tall and semi-spreading. They are sparse in bearing with large size fruit. The budded plants start fruiting in the 4th year. The average fruit yield of six-years old plant is 32.10 kg (6th year). The fruits are medium in size (18.25cm x22.50 cm), round and with smooth surface and very thick rind, yellow at maturity, low in mucilage and fibrous, an attractive yellow pulp and with low seed content. Fruits are good in taste and flavour, having 30° Brix total soluble solids, slight acidity and 19.78 mg/100 g ascorbic acid. It is highly suitable for processing.

Narendra Bael-9

The plants are semi-vigorous and spreading having compact canopy. The variety is precocious and prolific bearer. The average fruit yield of a six-year old plant is 56 kg. Fruits are medium to large in size (16.00 cm x 13.50cm), roundish-oblong, with smooth surface and thick rind (0.31cm), light yellow at maturity, average in mucilage, moderately fibrous, slightly golden-yellow pulp with low seed content. The fruits are good to taste containing 38° Brix total soluble solids, slightly acidic and ascorbic acid 19.20 mg/100 g of edible portion. The cracking in fruits has been noticed under hot arid conditions. It can be used as fresh as well as processed into several value added products. Its keeping quality for storage is very good.

Pant Aparna

Its trees are dwarf with drooping foliage, almost thornless, precocious and heavy-bearer. The leaves are large, dark green and pear shaped. Fruit has globose shape with average size of fruit 13.00 cm x 12.00 cm and weight of 0.8-1.25 kg. Fruit pulp is yellow and rind is thin. TSS 35 per cent, titratable acidity 0.67 per cent and ascorbic acid 17.15 mg/100 g of pulp. Mucilage, seed and fibre are low. Mucilage and seeds are enclosed in separate segments. Flavour and taste are very good. Yield during 6th year is 40.25 kg/plant.

Pant Shivani

It is an early to mid-season maturing variety. Trees are tall, vigorous, dense, upright growth, precocious and heavy-bearer. Fruit shape is ovoid, oblong and the size being 18.50 cm x 15.00 cm. Fruit weight ranges from 2 to 2.4 kg. Colour of fruit is lemon-yellow and its storage quality is good. Rind is medium-thin, pulp is lemon-yellow with pleasant flavour and mucilage, seeds and fibre are low to medium. Taste is very good. It has 69 per cent pulp, TSS 36°Brix, total titratable acidity 0.47 per cent and ascorbic acid 19.55 mg/100 g of flesh.

Pant Sujata

It is an early mid-season variety but has problem of fruit splitting. So for it has not been reported under rainfed conditions of hot semi-arid ecosystem of western India. Trees are medium-dwarf with drooping and spreading foliage, dense, precocious and heavy bearer. Thorns are stout and bigger. Fruit is globose shaped, depressed at both ends with average size of 14.50 cm x 13.50 cm and weight varied from 1.12 to1.40 kg under rainfed conditions of hot semi arid ecosystem of western India. Fruit and pulp are light yellow. Storage life is better, rind is thin, and seeds, mucilage and fibre are low. Its flavour is pleasant and taste is very good. Flesh is 72 per cent, TSS 32° Brix, acidity 0.44 per cent and ascorbic acid 17.10 mg/100 g of flesh.

Pant Urvashi

It is a mid-season variety. Trees are tall, vigorous, dense, upright growing, precocious and heavy bearer. Fruit is ovoid-oblong with average size of 14.50 cm x 17.20 cm and fruit weight ranges 1.5-2.50 kg. Fruit is yellow, rind is medium to thin and pulp is light yellow. Fruit has 62.35 per cent pulp with pleasant flavour. Seeds and mucilage are medium, fibre content low, TSS 33° Brix, titratable acidity 0.49 per cent and ascorbic acid 17.15 mg/100g pulp.

CISH Bael-1

It is early maturing variety. The plants are semi-tall and having spreading growth habit. The budded plants start fruiting in the 4th year. The average fruit yield of six-year old plant is 42.64 kg. The fruits are medium in size (16.50 cm x 12.00 cm), oval-oblong, with smooth surface, yellow at maturity, low in mucilage and fibrous, an attractive yellow pulp, with high seed content. Excellent in taste and flavour, the fruits have 30-34° Brix total soluble solids in pulp and 43° Brix in mucilage. The fruit weight varies from 0.8 to 1.40 kg under rain fed conditions of western India.

CISH Bael-2

The plants are dwarf and spreading and this variety is developed at CISH, Lucknow. The average fruit yield of six-year old plant is 38.45 kg. The fruits are medium in size (16.00 cm x 14.00 cm), rounded, with smooth surface, yellow at maturity and the rind is thick, low in mucilage and fibrous, an attractive yellow pulp, with low seed content. Good in taste and flavour, the fruits have 31° Brix total soluble solids and titratable acidity (0.41 per cent). The fruit weight ranges from 1.7 to 2.6 kg/fruit.

Goma Yashi

It is early maturing variety, semi-spreading, drooping growth habit, less lateral branches growth (20-35 cm annually), small tree relatively low stature, spineless, central leaf size (13.20x6.70 cm), lateral leaf size (9.20x5.30cm), ovate having acute apex. This variety is developed at CHES, Godhra Gujarat and is highly suitable for high density planting (Singh *et al.,* 2011). Initiation of flowering (last week of April), end of flowering (last week of June), bud size (12 mm x 8 mm), flower size (15 mm x29 mm), petals (4-5), anthesis (6 A.M -8 A.M), dehiscence period (6.3 A.M 0-8.30 A.M), yield (51.00 kg 6thyear), fruit size (13.00 cm x 12.50 cm), fruit girth (41-45 cm), shell thickness (0.17cm), seed weight (25-30g), fibre weight (40.24-51.20g), shell weight (180-210g), fruit weight (1.00-1.62 kg), locules in cross section (13-15), TSS pulp (35-39°B), TSS mucilage (41-43°B), titratable acidity (0.26-0.32 per cent) and vitamin C (22.00 mg/100 g pulp) are characters.

Future Thrust Areas

A rich genetic diversity is available throughout the country especially in the states of U.P., Bihar, Uttarakhand, Jharkhand, Chhatisgarh, M.P., etc. which should be exploited for selection of better genotypes. It is a medicinal fruit tree and naturally grown in forest areas in addition to orchards. Fruit drop and cracking of *bael* fruit are the major problems of cultivation in different parts of country. To overcome this problem, suitable genotypes should be identified with high yield potential and better fruit quality. Development of low seeded variety is necessary to lure this fruit. More emphasis on post harvest technology should be given for value added and export oriented processed products. Establishment of small scale processing units should be promoted. Screening of genotypes/cultivars for problematic soils is also essential. Standardization of production technology and medicinal importance of *bael* fruit should be explored. There is need to bring out documents related to collection, conservation, characterization and utilization from various sites.

The collected material should be utilized properly for improvement. Efforts should be made in this direction to utilize available genetic resources. The resistance for some traits found at some place needs rigorous testing before it is used in breeding programme. Bael based farming models should be developed for higher yield and proper land utilization.

References

Anonymous, 2008. Annual Report, AICRP on AZF, CIAH, Bikaner.

Banerjee, N. and Maiti, M. 1980. Some physico-chemical studies on a protein rich fruit seed (*Aegle marmelos* Correa). *J. Indian Chem. Soc.,* 57(9): 945-46.

Chatterjee, A. and Roy, S.K. 1957. Agelenine: a new alkaloid of leaves of *Aegle marmelos.* Sci. Culture,23:106-107

Chatterjee, A. and Roy, S.K.,1959. Chemistry of extractives from hardwood. 1. Constituents of hard wood of *Aegle marmelos. J. Indian Chem. Soc.,*36:267-69

Deepti, Rai., Misra, K.K. and Singh, V.P. 2002. Analysis of genetic divergence in bael (*Aegle marmelos* Correa) germplasm. *Prog. Hort.*, 34(1): 35-38.

Dixit, B.B.L. and Dutt, S. 1932. The constituent of Marmelosin. *J. Indian Chem.Soc.*,9:271-79

Gopalan, C. B. M., Ramashastri, B.V. and Balasubramanium, S. C. 1985. Nutritive value of Indian foods, ICMR, Hyderabad.

Ghosh, D.K. and Gayen, P. 1990. Metroglyph grouping and association analysis on physical characters of bael fruits (*Aegle marmelos*). *Prog. Hort.*, 22(1-4): 6-11.

Haskar, C.N. and Kendurkar, S.G. 1961. Uses of bael gum. *Printindia*, 11(1): 135.

Jain, N.K., 1977. Antifungal activities of *Aegle marmelos* Correa, *Indian J. Microbiology*, 17(1): 51.

Jaiswal, H.R. and Misra, K.K. 1996. Clonal variation in inflorescence and floral morphology of bael (*Aegle marmelos* Correa). *Prog. Hort.*, 28(3-4): 119-123.

Jaisawal, H.R. and Misra, K.K. 1998. Studies on fruit set, retention and yield of bael (*Aegle marmelos* Correa) clones under Tarai conditions. *Prog. Hort.*, 30(3-4): 164-167.

Kirtikar, K.R. and Basu,B.D. 1935. Indian Medicinal Plants,vol.-I published by L.M.Basu, Allahabad.

Lal, G. 2002. Evaluation of bael (*Aegle marmelos* Correa) germplasm in semi-arid regions of Rajasthan. *Curr. Agric.*, 26 (1-2): 127-129.

Misra, K.K., Singh, R. and Jaiswal, H.R.1999. Studies on leaf characters, development pattern and shoot growth in bael genotypes. *Prog. Hort.*, 31(3-4): 144-150.

Misra, K. K., Singh, R. and Jaiswal, H.R. 2000. Performance of bael (*Aegle marmelos*) genotypes under foothills regions of Uttar Pradesh. *Indian J.Agric. Sci.*, 70(10): 682-683.

Om Prakash 1961. Food and drinks in Ancient India, Munshi Ram Manohar Lal Book Sellers and Publishers, Delhi.

Pal, P.K., Bhattacharya, D.K. and Ghosh, S. 1993. Chemical examination of seeds of *Aegle marmelos* (Bael fruits) and its medicinal importance. *J. Oil Technologists Association* India, 25(3): 55-59.

Parikh,V.M., Ingle,T.R. and Bhide, B.V.1958. Studies in carbohydrates. Bael fruit (*Aegle marmelos*) mucilage. *J. Indian Chem. Soc.*,35:125-29

Pandey, D., Shukla,S.K. and Akhilesh Kumar 2008. Variability in bael (*Aegle marmelos Correa.*) germplasm collected from Uttar Pradesh and Madhya Pradesh. *Jour. Tropical Forestry*,24:31-36.

Pandey, D., Pathak, R. K. and Pramanik, P. K. 2001. *Shri Phal Bael. Udyan Rashmi* (In Hindi), 2(2):13-18.

Prakash, A., Pasalu, I.C. and Mathur, K.C. 1983. Use of bael (*Aegle marmelos*) leaf powder as protectant against storage pests of Paddy. *Agril. Res. J. Kerala*, 21(2): 79-80.

Rai, M. and Dwivedi, R. 1992. Bael (*Aegle marmelos* Correa) diversity in India. In: *Underutilized fruits of medicinal value, IBPGR Newsletter*, 10:13-14.

Rai, M., Dwivedi, R. and Gupta, P. N. 1991. Variability and potentials of identified germplasm of bael (*Aegle marmelos* correa.). *Indian Jour. Pl. Gen. Resources*, 4(2): 86-92.

Ram, D. and Singh, I.S. 2003. Physico-chemical studies on bael (*Aegle marmelos* Correa) fruits. *Prog. Hort.*, 35(2): 199-201.

Roy, S.K. and Singh, R.N. 1979. Preliminary studies on storage of *bael* fruits (*Aegle marmelos*). *Prog. Hort.*, 11:3

Roy, S.K. and Singh, R.N. 1981. Studies on ripening of bael fruits (*Aegle marmelos*). *Punjab. Hort.* J., 21(122): 74-82.

Saini, R.S., Surender Singh and Deswal, R.P.S., 2004. Effect of micronutrients, plant growth regulators and soil amendments on fruit drop, cracking, yield and quality of bael (*Aegle marmelos* Correa) under rainfed conditions. *Indian J. Hort.*, 61(2): 175-176

Sharma,P.C., Bhatia,Vivek, Bansal, N. and Sharma Archana 2007. A review on bael tree. *Natural Product Radiance,* 6(2):171-178.

Shukla, S.K. and Singh, G.N 1996 a. Seed germination and growth of bael (*Aegle marmelos* Correa) seedlings as influenced by soil sodicity, *J. Tropical Forestry*, 12(4): 242-245.

Shukla, S.K. and Singh, G.N. 1996 b. Nutrient imbalances in bael (*Aegle marmelos* Correa) leaves as induced by salinity and sodicity, *Indian J. Plant Physiol.*, 1(4): 293-294.

Singh, A.K., Singh, S.,Singh,R.S., Bagle,B.G. and Sharma, B. D. 2011. The Bael- fruit for dryland, Tech. Bull.No. 38, CIAH, CHES, Vejalpur, Godhra, p.46.

Singh, A.K., Singh, S., Joshi, H.K., Bagle, B.G. and More, T.A. 2008. Evaluation of bael genotypes for growth behavior and floral traits under semi-arid ecosystem of western India.*The Hort. J.,* 21(30):140-142

Singh, K.V., Bhatt, S.K. and Sthapak, J.K. 1983. Antimicrobial and anthelmintic properties of the seeds of *Aegle marmelos. Fitoterapia,* 54(6): 261-264.

Singh, R. N. and Roy, S. K. 1984. *The Bael.* I.C.A.R., New Delhi, p.25.

Singh, R. S. and Vishal Nath 1999. *Anupayogi Bhoomi mein – Upayogi Bael Vriksh Lagayein. Unnat Krishi,* 38 (3):3-8.

Srivastava, A.K. Singh, H.K and Singh I. S. 1998. Genetic diversity in Bael (*Aegle marmelos* Correa) *Indian Horticulture, cover page,II.*

Srivastava, K.K. and Singh, H.K. 2000. Floral biology of bael (*Aegle marmelos*) cultivars. *Indian J. Agriculture Sciences,* 70(11): 797-798.

Singh, H.K.; Srivastava, A.K., Prasad, J. and Dwivedi, R. 2009. Descriptor of bael (*Aegle marmelos* Correa.) AICRP on Arid Zone fruits, NDUA&T, Faizabad, p.23.

Tokitoma, Y., Shimono, Y., Kobayashi, A. and Yamanishi, T. 1982. Aroma component of bael fruits (*Aegle marmelos* Correa.). *Agril. Biological Chemistry,* 46(7): 1873-1877.

Verma, J.P. and Ahmed, S. 1958. Drying of bael (*Aegle marmelos* Correa.) pulp. *Indian Food Packer,* 12(7): 7-9.

Vishal Nath, Pandey, D. and Das, Bikas 2003. Diversity of Bael (*Aegle marmelos* Correa) in East Central India, *Indian Jour. Plant Genetic Res.,* 16: 85-89.

2013, Biodiversity in Horticultural Crops Vol. 4
Editor: Professor K.V. Peter
Published by: DAYA PUBLISHING HOUSE, NEW DELHI

Pages 301–334

Chapter 17

Citrus

T.K. Hazarika

Department of Horticulture, Aromatic and Medicinal Plants
School of Earth Sciences and Natural Resources Management,
Mizoram University, Aizawl – 796 004, Mizoram
E-mail: tridip28@gmail.com

Citrus is one of the most important commercial and nutritional fruits of the tropical and subtropical regions of the world which has gained a separate dimension due to its gigantic industrial expansion the world over. *Citrus* belongs to the family Rutaceae, which comprises 140 genera and 1300 species throughout the world (Singh and Rajam, 2009). It is the third most important fruit crop in the world after apple and banana and accounts for the production of about 100 million tonnes with an area of cultivation spread over a massive 7.2 million hectares (FAO, 2001) and the production has reached 122 million tonnes (FAO, 2008). It is a long lived perennial crop and is grown in more than 100 countries across the world (Saunt, 1990). Brazil and the US are the leading producers and produce 42 per cent of the world's total *Citrus* production (FAO, 2001). The major economically important species of the family Rutaceae are- *Citrus sinensis* (L.) Osbeck (Sweet orange), *C. reticulata* Blanco (mandarin), *C. grandis* (L.) Osbeck (pummelo), *C. paradisi* Mac. f. (grapefruit), *C. limon* (L.) Burm. f. (lemon), *C. aurantifolia* (Christm) Swing. (lime) and *C. aurantium* (L.) (sour orange). Sweet orange alone accounts for 75 per cent of world total *Citrus* production followed by mandarin, grapefruit and lemon. India, having a varied range of climatic conditions and being one of the eight Vavilovian centres of crop plants origin and diversity have given rise to a wide range of variability in *Citrus* and related genera (Singh and Singh, 2001). The diverse geographical regions characterized by varying temperatures and rainfall conditions ranked India sixth amongst the various *Citrus* producing countries in the world. In India, the commercial *Citrus* cultivars are the mandarin (*C. reticulata* Blanco), sweet orange (*C. sinensis* (L.) Osbeck and acid lime (*C. aurantifolia* Swingle).

Origin and Evolution of Citrus

Different hypothesis were formulated on the history and geographical origin of *Citrus*. It is believed that all the species belonging to *Citrus* and its related genera originated in the tropical and subtropical regions of the South-East Asia–North Eastern India, China, the Indo-Chinese peninsula and the Malay Archipelago and then spread to other continents (Webber, 1967; Chapot, 1975). Tolkowsky (1938) considered that the mountainous regions of Southern China and North East India are the centres of origin, while Tanaka (1954), reported that *Citrus* may have originated in north-eastern India and Burma; China should be considered only as secondary centre of distribution. Calabrese (1998) indicated that the primordial genetic nucleus of *Citrus* originated in China and the *Citrus* slowly passed from its original location to other oriental regions and from there followed the paths of civilization. Chinese literature written as early as 2200 BC showed evidence of *Citrus* cultivation in China. The first indication regarding the presence of *Citrus* in China was contained in the book 'Tribute of Yu' (2205-2197 BC). During the Chou dynasty (1027-256 BC) there was report of two types of *Citrus*, the *Chu* (Kumquat) and the *Yu* (Pummelo). Similarly, during the Han dynasty (202 BC-220 AD) other *Citrus* species *viz. Cheng* (sour orange), *Lu Chu* (kumquat) and *Huang Kan* (yellow mandarin) were mentioned (Nicolosi, 2007). In India, the mention of various *Citrus* fruits has been found in Ayurvedic works of Dhanwantri (10th BC), Charak and Susruta Samhita written in 600 to 500 BC. The species described in chapters 27 and 46 were *Citrus medica, C. jambhiri, C. arantifolia, C. limetta, C. grandis* and *C. reticulata*. Hooker (1875) thought that not less than 78 species of the family Rutaceae are native to India. Dymock *et al.* (1890) quoting Hindu *Materia Medica* mentioned 8 *Citrus* types and 22 varieties belonging to them.

The sour orange (*Citrus aurantium* L.) is believed to be native to South-east Asia, possibly India, while the sweet orange (*Citrus sinensis* (L.) Osbeck) originated in southern China and possibly as far south as Indonesia (Webber, 1967). The pummelo (*C. grandis* (L.) Osbeck) is of tropical origin; according to Hodgson (1967), it seems reasonably certain that pummelo is indigenous to the Malayan and East Indian archipelagos. The grapefruit (*C. paradisi* Mac. f.) appears to have originated as a mutation or hybrid of the shaddock in the West Indies, perhaps Barbados. Hirai *et al.* (1986) reported that cultivated mandarins (*C. reticulata*) have three origins: India, China and Japan. The exact area of origin of the lemon (*C. limon*) is still uncertain. Tolkowsky (1938) and others suggested that the lemon is native to India. Webber, (1967) considered Southern China and probably Upper Burma to be the native home of the lemon. The limes *viz. Citrus aurantifolia* is native to the Malaysian region of South-Western Asia, while *C. latifolia* probably originated in the east and then spread to Persia, and then to Tahiti, possibly via Brazil and Australia, and finally to California. *Citrus limettoides* Tan., sweet lime, is native to north eastern India where it is known as *Mitha nimboo*. The citron (*C. medica* L.) is probably native to India. It is well accepted by all authorities that it was the first *Citrus* fruit known to Europeans, but there are different opinions about the exact period and the steps by which it was first brought from its native land (Nicolosi, 2007).

Taxonomy and Systematics

Citrus belongs to the Geraniales order, the Rutaceae family and the Aurantioideae subfamily. Rutaceae is one of the 12 families in the Geraniineae suborder, and the Aurantioideae subfamily- one of the 7 belonging to the Rutaceae family (Engler, 1931) is rather numerous and comprises the commercial *Citrus* species and also several important related genera. Aurantioideae, the Orange subfamily, is subdivided by Swingle into two tribes: *Clauseneae* with five genera and *Citreae* with twenty eight genera including *Citrus* and related genera, *i.e. Fortunella, Poncirus, Eremocitrus, Microcitrus* and *Clymenia* (Nicolosi, 2007).

Currently two different classification systems of *Citrus* are commonly accepted for the *Citrus* taxonomy, the Swingle system (Swingle, 1943; Swingle and Reece, 1967) and that of Tanaka (1954; 1961). Swingle (1943) recognized 16 valid species under two subgenera - *Citrus* and *Papeda*, 10 under subgenus *Citrus* and 6 under subgenus *Papeda* and did not accord species status to budsports or natural hybrids. The sub genus *Citrus* is characterized by pulp vesicles are nearly free from oil droplets and never contain acrid oil. Petioles with narrow wings or wingless or if broadly winged, are subcordate and never 3/4th as broad as the leaf blade, flowers large and fragrant, stamens cohering in bundles. The sub genus *Citrus* comprises of 10 species *viz. Citrus tachibana, C. medica, C. limon, C. reticulata, C. indica, C. grandis, C. paradisi, C. aurantifolia, C. sinensis* and *C. aurantium*. On the other hand, the sub-genus *Papeda* is characterized with pulp vesicles containing numerous droplets of acrid oil. Petioles are long and very broadly winged but not cordate, often nearly as broad as the leaf blade. Stamens are usually free. Six species are included under the sub genus *Papeda* are: *C. ichangensis, C. latipes, C. micrantha, C. celebica, C. macroptera* and *C. hystrix*. On the other hand, Tanaka proposed his system of classification of *Citrus* in a more elaborative fashion. He included almost all sorts of geographic and morphological variability of various *Citrus* plants and proposed an exhaustive list of species (Naqvi, 2001). In 1954, in '*Species Problems in Citrus*', Tanaka published a classification of the genus *Citrus* in which two subgenera, *Archicitrus* and *Metacitrus,* eight sections, thirteen subsections, eight groups, two subgroups, two microgroups and 145 species were distinguished. In 1961, he added two new subsections, another group and 12 new species to his system, taking the total to 157 species (Nicolosi, 2007). In 1977, he again increased the number of species up to 162 mostly represented variants and hybrids.

Table 17.1: The Genus *Citrus*–A Summary

Species	Common Name	Year Named	Probable Origin	Probable Native Habitat	Seed Reproduction
C. medica	Citron	1753	True Species	India	Sexual
C. aurantium	Sour orange	1753	Hybrid	China	Nucellar
C. sinensis	Sweet orange	1757	Hybrid	China	Nucellar
C. maxima	Pummelo	1765	True species	China	Sexual
C. limon	Lemon	1766	Hybrid	India	Partly sexual
C. reticulata	Mandarin	1837	True species	China	Variable
C. aurantifolia	Lime	1913	Hybrid	Malaya	Partly sexual
C. paradisi	Grapefruit	1930	Hybrid	Barbados	Nucellar
C. tachibana	Tachibana	1924	Unknown	Japan	Sexual
C. indica	Indian wild orange	1931	Unknown	India	Sexual
C. hystrix	Mauritus papeda	1813	Unknown	S.E. Asia	Sexual
C. macroptera	Malesian papeda	1860	Unknown	S.E. Asia	Sexual
C. celebica	Celebes papeda	1898	Unknown	Celebes	Sexual
C. ichangensis	Ichang papeda	1913	Unknown	China	Sexual
C. micrantha	Papeda	1915	Unknown	Philippines	Sexual
C. latipes	Khasi papeda	1928	Unknown	N.E. India	Sexual

Source: Krueger and Navarro, 2007.

Tanaka's classification is more complex as compared to that of Swingle's because of the far greater number of species included in each subgenus. Tanaka very often has given rank of species to cultigens, chance seedlings and apomicts *etc*. On the other hand, Swingle's classification was easier to understand although it does not provide an exhaustive description of *Citrus* systematic. It was very concise, but not free from such inclusions by admitting *C. paradisi* and *C. limon* as species and ignoring the most distinctive forms which are probably more valid than his *C. indica* or *C. tachibana*. Even so, Swingle's system is the most used, with some addition and deletions and adoption from Tanaka's system (Nicolosi, 2007).

Subsequent classifications on *Citrus* proposed by various other botanists (Bhattacharya and Dutta, 1956; Singh and Nath, 1969) are mere modifications of either Swingle's or Tanaka's system. Hooker (1875), Bonavia (1880-90), Lushington (1910) and Singh (1967) attempted to study and classify Indian *Citrus* from botanical as well as horticultural perspectives. Scora (1975) and Barrett and Rhodes (1976) suggested that there are only three 'basic' true species of *Citrus* within the subgenus *Citrus* as defined by Swingle: Citron (*C. medica* L.), Mandarin (*C. reticulata* Blanco) and Pummelo (*C. maxima* L. Osbeck). Other cultivated *Citrus* species within the subgenus *Citrus* are believed to be hybrids derived from these true species, species of the subgenus *Papeda*, or closely related genera. This idea has recently been supported by data derived from molecular markers (Federici *et al.*, 1998; Nicolosi *et al.*, 2000; Barkley *et al.*, 2006). Recently Mabberley (1998, 2004) circumscribed *Citrus* in a broader sense by merging three of the closely related genera *viz. Fortunella* Swingle, *EremoCitrus* Swingle and *MicroCitrus* Swingle within it. Systematics of *Citrus* thus continue to be an intriguing topic for further research and discussion.

Genetic Diversity of Citrus in India

India enjoys a remarkable position in the "*Citrus belt of the world*" due to her rich wealth of *Citrus* genetic resources, both wild and cultivated (Nair and Nayar, 1997). A vast reservoir of *Citrus* diversity exists in different parts of India in wild, semi wild as well as cultivated forms. A large number of *Citrus* species/progenitors of commercial *Citrus* fruits are believed to have originated in India (Bhattacharya and Dutta, 1951, 1956; Dutta, 1958; Singh, 1967; Singh and Singh 1967, 1968 and Singh, 1977). Many of these species are wild (Bhattacharya and Dutta, 1956). As early as 1950 Vavilov reported the occurrence of sweet orange (*C. sinensis* Osbeck), mandarin (*C. reticulata* Blanco), citron (*C. medica* L.), sourlime (*C. aurantifolia* Blanco), Jenerutenga (*C. nobilis* lour), Rangpur lime (*C. limonia* Osbeck) and lemon (*C. limon* Burrn.) both in cultivated and wild form in the NEH region of India. Tanaka (1958) believed that sweet oranges originated in India in addition to many other *Citrus* species. Some of the indigenous and wild mandarin types are found in South India, whereas Hill lemon (Galgal) *C. pseudolimon* and Attani (*C. rugulosa*) are prevalent in the foot hills of Himalayas in the north-west part of the country (Verma *et al.*, 1999; Singh and Singh, 2001). According to Singh (1981), NEH region and parts of North Western India are the best locations for collecting primitive germplasm of *Citrus*. Whereas Malik *et al.* (2006) reported that, north-eastern Himalayan region and foothills of the central and western Himalayan tracts are rich sources of *Citrus* genetic diversity. *Citrus* is almost universally present in various forms in the entire north-eastern India. The favourable climatic conditions aiding in easy hybridization amongst different species and genera have brought in numerous forms of *Citrus* growing wild and semi wild condition in the region (Hore *et al.*, 1997; Singh *et al.*, 2001). The wide distribution of "Soh Nairiang" a wild sweet orange, wild Indian mandarin (*C. indica* Tanaka) (Bhattacharya and Dutta 1951, 1956), *C. assamensis*, *C. ichangensis* Swingle, *C. latipes* Tanaka and *C. macroptera* Mont in various parts of NE region of India growing up to an elevation of 2000 m above msl (Ghosh, 1977) are reported.

Table 17.2: Citrus Germplasms Maintained at Different Field Gene Banks of India

Sl.No.	Locations	Number of Accessions
1.	NRC for *Citrus*, Nagpur, Maharashtra	169
2.	ICAR, NEH Region, Umiam, Barapani, Meghalaya	92
3.	RFRS (PAU), Abohar, Punjab	132
4.	Dr. PDKV, Akola, Maharashtra	18
5.	HRS (RAU), Birouli, Bihar	14
6.	RRS(IIHR), Chethali, Karnataka	126
7.	IIHR, Hessarghatta, Banglore, Karnataka	76
8.	RFRS (Dr. PDKV), Katol, Maharashtra	33
9.	PAU, Ludhiana	76
10.	HRS (Marathwada Agric. Univ.), Parbhani, Maharashtra	68
11.	HRS (TNAU), Periyakulam, Tamil Nadu	20
12.	Mahatma Phule Krishi Vidyapeeth, Rahuri, Maharashtra	66
13.	HRRS(YSPUHF), Dhaulakuan, Sirmour, H.P.	11
14.	FRS(MPKV), Srirampur, Rahuri, Maharashtra	34
15.	*Citrus* Research Station, AAU, Tinsukia, Assam	36
16.	S.V. College of Agriculture, Tirupati, A.P.	115
17.	HRS(TNAU), Yerud, Tamil Nadu	41

Source: Singh and Singh, 2006.

In North West India, kinnow mandarin is grown commercially, whereas sweet oranges and lemons are confined to a limited extent. In the foothills of Western Himalaya hill, Galgal (*C. pseudolimon*) and Athanni, Chawanni (*C. regulosa*) are commonly seen. Besides this, Malta type sweet orange (*C. sinensis*) was observed to grow at comparatively higher altitudes of Uttarakhand. In Poanta valley of H.P. and Doon valley of Uttarakhand, a large extent of variability in pummelo and galgal were seen. Similarly, Pathankot area of H.P. is famous for local mandarins (Butwal) cultivation. Besides mandarin, many rough lemon strains were reported to grow in this region (Singh and Singh, 2006). According to Verma *et al.* (1999), presence of Jambhiri, Jambhira, Khatta, Chukh, Madkakaree showed possible variability of *Citrus* belonging to different groups. They also reported a wild relative of *Citrus*, *Glycomis pentophylla* from Bajpur Khatima and from Majhere area of Nainital district of Uttarakhand. During 1999-2001 NRCC led three exploration missions in collaboration with NBPGR in North West India and collected 57, 43 and 28 accessions respectively from these areas. During these explorations, two new types (probable hybrids) were reported from Nainital area of Uttarakhand. It seems to be a cross between Malta and Pummelo. Different species of *Citrus viz. C. pseudolimon, C. jambhiri, C. limonia, C. karna, C. reticulata, C. rugulosa, C. limmetoides, C. sinensis, C. medica, C. grandis, C. limon* and *C. aurantifolia* were reported to grow on commercial scale in different parts of NW India. The exploration showed that the Nainital, Chamoli, Bageshwar, Almora, Champawat and Pithorgarh areas of Uttarakhand have tremendous diversity of rare and cultivated *Citrus* species (Singh and Singh, 2006).

In West India not much diversity is reported except acid lime. Total of 13 collections were reported from different locations like Kheda, Anand, Mehsaan, Gandhinagar and Ahmadabad area of Gujarat. These areas were potential areas for Acid lime (Singh and Singh, 2006).

North east region of India is considered the natural home of many *Citrus* species (Ghosh, 1977; Govind and Yadav, 1999) and is reservoir of various *Citrus* species including mandarin orange (Tanaka, 1958; Bhattacharya and Dutta, 1956). Bhattacharya and Dutta (1956) collected a total of 56 accessions of *Citrus* from this region. During 1980-1995, ICAR, Barapani, Meghalaya collected 92 *Citrus* accessions from NEH region and planted in field gene bank. Kaul (1981) reported that there are 17 *Citrus* species with 52 variety/types of *Citrus* occurring in NEH region. Besides two commercial species *viz.* Khasi mandarin (*C. reticulata*) and Assam lemon (*C. limon*), many inedible and edible species like *C. indica, C. ichangensis, C. macroptera* and *C. latipes* are found growing wild and semi wild in this region. Different strains of citron (*C. medica*), sour pummelo (*C. megaloxycarpa*), rough lemon (*C. jambhiri*) and sour orange (*C. aurantium*) are also reported to grow semi wild in this region (Verma and Ghosh, 1979). Presence of three wild types of sweet orange (*C. sinensis*) *viz.* Soh bitara, Soh nairiang and Tasi in Meghalaya and Arunachal Pradesh provided a strong evidence that most of the *Citrus* species originated in this region. Hore *et al.* (1997) found maximum diversity of *C. reticulata, C. limonia* and *C. grandis* in western parts of Aizawl district, Mizoram and Jampui hills area as well as north Tripura bordering Mizoram. During 2000-2002, NRCC led 4 exploration missions to collect *Citrus* genetic diversity from this region and they collected 16, 79, 40 and 12 accessions respectively from this region. During the exploration work, germplasms from different groups *viz.* mandarin, rough lemon, sweet orange, citron, pummelo and acid lime and species *viz. C. pseudolimon, C. jambhiri, C. karna, C. reticulate, C. macroptera, C. sinensis, C. medica, C. grandis, C. limon, C. aurantifolia, C. indica* and *Poncirus trifoliate* were collected from different parts of NE region. Three endangered species *viz. C. latipes, C. assamensis* and *C. megaloxycarpa* were also collected from Muktapur area of Jaintia hills, Meghalaya. Similarly, three probable hybrids were also reported from near Nokrek Biosphere reserve area of Meghalaya.

In Central India, besides the three commercial *Citrus* species, mandarin, sweet orange and acid lime, other *Citrus* species *viz.* rough lemon, rangpur lime, sweet lime and lemon are also grown. Nagpur mandarin is commercially grown in this region. Variation in genetic diversity of mosambi (*C. sinensis*) was also observed. Acid lime is also grown in Akola, Nanded, Srigonda and Parbhani areas of Maharashtra (Singh and Singh, 2006).

In South India, acid lime (*C. aurantifolia*) is commercially grown in areas like Nellore, Gundur area of A.P. and Periyakulam area of Tamil Nadu, Coorg mandarin in Coorg region of Karnataka and Sweet orange cv. Sathguti in Anantpur area of A.P. Other commercially important indigenous types in this area are Gajanimma (*C. pennivesculata* Tan), Kichli (*C. maderaspatana* Tan) and some wild mandarin types *viz.* Kodaithuli, Billikichili, Nakoor lemon and Mole Puli (Sour orange type) (Singh and Singh, 2006).

Species of Citrus

Citrus Species of Horticultural Importance

Mandarin Group

Citrus reticulata

The tree is usually thorny, with slender twigs, broad-or slender-lanceolate leaves having minute, rounded teeth, and narrowly-winged petioles. Flowers are borne singly or a few together in the leaf axils. Fruits oblate, peel bright-orange or red-orange when ripe, loose, separating easily from the segments, usually 10-14 segments. Seeds small, pointed at one end, green inside. *C. reticulata* tends to be the hardiest of the common *Citrus* species and can withstand short periods of –10 °C, but temperatures not falling below –2°C are required for successful cultivation.

C. unshiu

A seedless and easy-peeling *Citrus* species, it is also known as cold hardy mandarin, satsuma mandarin or christmas orange. Trees are evergreen small to medium sized, often with a weeping aspect is due to the weight of the mandarins at the tips of the branches. Leaves are generally wider than other mandarin and tangerine leaves. Fruit is sweet and usually seedless, about the size of other mandarin oranges or smaller than an orange. The thin, leathery skin dotted with large and prominent oil glands, which is lightly attached around the fruit, enabling it to be peeled very easily in comparison to other *Citrus* fruits. It has particularly delicate flesh, which cannot withstand the effects of careless handling. The flesh is orange, sweet, juicy and not very acidic.

C. deliciosa

It is commonly known as willow leaf mandarin or Mediterranean mandarin. Tree is of medium size, broadly spreading and pendant. Leaves are narrow and lanceolate, suggestive of a willow's leaves, and fragrant. Flesh is light orange, tender, juicy and sweet. Fruit medium, moderately oblate, frequently slightly lobed; base sometimes even, but usually with low collared and strongly furrowed neck; apex depressed and commonly slightly wrinkled; areole lacking; small navel-like structure fairly common. Segments 10 to 12, very loosely adherent; axis hollow. Flesh colour light orange; tender; juicy; flavour sweet; have distinctive aroma. Seeds numerous, small, round, plump, and highly polyembryonic, with light green cotyledons. Rind thin, not leathery, loosely adherent; surface smooth and glossy with large, deep coloured oil glands; colour yellowish-orange at maturity.

C. nobilis

Commonly known as King orange, a small tree 4 to 5 or rarely to 6 meters high, the branches spreading or pendulous, the crown somewhat globose. Leaves 11 cm long, with narrowly winged petiole. Flowers terminal, solitary or paired, 3.2 cm across, stamens 20-27. Fruit is large, deep orange, easily peelable. Fruits compressed globose, rind deep orange, rough, segments 10 to 13, separating easily, medium sweet and slightly acidic.

Citrus reshni Hort. ex Tan.

This species is the *chota* or *billi kichili* of India and the Cleopatra mandarin of the United States. Considered to be native to India, *C. reshni* is increasingly popular as a rootstock in the United States and elsewhere. It is an attractive ornamental and bears fruit the year round. Tree is attractive, round-topped, symmetrical, and thornless, with small, dark-green leaves. Fruit is orange-red, small, oblate, and highly depressed at the apex, with thin, somewhat rough rind. Flesh soft, juicy, flavour is somewhat acid. Seeds are small, polyembryonic.

C. medurensis

Commonly called the calamondin, golden lime, panama orange, chinese orange, acid orange, calamonding, or calamandarin. In North India, it is popularly called hazaara because of bearing a very large number of very small orange shaped fruits around the whole periphery plant. The plant is characterized by wing-like appendages on the leaf stalks and white or purplish flowers. Leaves are broadly oval, dark green above and paler below, petioles narrowly winged, articulated. Flowers are small, white fragrant, borne singly or in pairs. Fruits are small, 2.5 to 3 cm wide, sub-globose or oblate to spherical or oblate to spherical, bright orange with depressed apex and juicy, weighing 20-30 g, seeded with orange coloured flesh, acidic in taste, but the peel is sweet and edible. Peel orange coloured, smooth, very thin. Calamandin is commonly grown in China, Taiwan, Japan, Phillipines and northern India.

C. madaraspatana

Commonly known as Kitchli sour orange, Guntur sour orange or Vadlapudi sour orange, it is of commercial importance in South India, principally in the Guntur district, where it is grown on a somewhat extensive scale. Fruit medium-sized, depressed globose to broadly obovoid; sometimes slightly necked; colour yellowish orange; seedy; rind rough, somewhat warty, and of medium thickness and adherence. Core semi-hollow at maturity. Flesh pale orange-coloured; flavour pleasant at full maturity, with slightly bitter aftertaste and musky aroma. Prior to maturity, flesh sharply acid, cotyledons light green.

C. tangarina

The tangerine is closely related to the mandarin orange. Fruits are smaller than most oranges and are usually much easier to peel and to split into segments, taste is often less sour, or tart, than that of orange. Tangerines are the most commonly peeled and eaten out of hand. The fresh fruit is also used in salads, desserts and main dishes. The peel is dried and used in Sichuan cuisine. Fresh tangerine juice and frozen juice concentrate are commonly available in the United States. The number of seeds in each segment (carpel) varies greatly.

Table 17.3: Commercial *Citrus* Varieties of India and their Distribution

Citrus Group	Cultivars	Distribution
Mandarin orange (*C. reticulata* Blanco)	Nagpur mandarin, Khasi mandarin, Darjeeling mandarin, Sikkim mandarin, Coorg mandarin, Kinnow mandarin	Maharashtra, Madhya Pradesh, North eastern States, Darjeeling district of West Bengal, Sikkim, Coorg region of Karnataka, Punjab, Rajasthan, Haryana
Sweet orange [*C. sinensis* (L.) Osbeck]	Mosambi, Sathgudi, Jaffa, Valencia, Hamlin, Malta,	Maharashtra, Andhra Pradesh, Punjab, Rajasthan
Acid lime (*C. aurantifolia* Swingle)	Kagzi lime	Andhra Pradesh, Tamil Nadu, Karnataka, Maharashtra
	Jai Devi, Vikram, Pramalini, Sai Sarbati, Kagzi Kala and Tenali	Tamil Nadu, Maharashtra, Delhi, Andhra Pradesh, Uttar Pradesh
Lemon (*C. limon* Burm.f.)	Hill lemon, Galgal, Assam lemon, Eureka, Baramasia, Lisbon, Pant lemon 1,2,3	Uttar Pradesh, Himachal Pradesh, Assam and NE states, Karnataka, UP
	Eureka, Hill Galga	Gujarat, Andhra Pradesh, Uttar Pradesh
	Baramashi, Nepali oblong, Italian lemon, Lisbon lemon, Eureka lemon and Seville	Karnataka
Grapefruit (*C. paradisi* Macf.)	Saharanpur special	Uttar Pradesh
Pummelo [*C. grandis* (L.) Osbeck]	Red fleshed/white fleshed	Andhra Pradesh, NEH Region and Uttar Pradesh

Source: Singh, 2001.

Orange Group

C. sinensis

Tree medium-sized; twigs are angled when young, usually with slender, somewhat flexible, rather blunt spines in the axils of leaves; leaves medium-sized, pointed at apex, rounded at base;

petioles narrowly winged, articulated both with the twig at base and with leaf blade at the tip; flowers in small racemes or singly in axils of the leaves, medium-sized, 5 petals and 20-25 stamens. Fruits subglobose, oval or flattened globose; peel thin, tight, not bitter, central axis solid.

C. aurantium

Tree medium sized, dark green and distinctively scented leaves. Flowers large, fragrant, fruits medium, orange red colour at ripening, rind easily peelable, flesh orange coloured, very sour in taste or sometimes bitter, seeds numerous, sour oranges are usually not eaten fresh. Their importance lies in the oil that can be extracted from the flowers, leaves, seeds and rind. This oil gives its typical orange-like flavour to spices, sweets, liqueurs etc. Some varieties of sour orange impart their aroma to expensive perfumes, soaps and after-shaves. The main constituent in the aroma of many soft drinks is from sour orange. In addition to the many food uses what really makes the sour orange important for the whole *Citrus* industry is its suitability for being used as rootstock. Being resistant to many plant diseases and compatible with the majority of commercial *Citrus* types, makes it invaluable.

Pummelo-Grapefruit Group

C. grandis (Syn: C. maxima)

The Pummelo is the biggest *Citrus* fruit. Tree is 5-15 m tall. Some forms are dwarfed. Young branchlets are angular and often densely hairy, spines present on branchlets, old limbs and trunk. Leaves alternate, ovate, ovate-oblong, or elliptic, 5-20 cm long, 2-12 cm wide, leathery, dull-green, glossy above, dull and minutely hairy beneath, petiole broadly winged to occasionally nearly wingless. Flowers are fragrant, yellowish-white, 1.5-3.5 cm long, somewhat hairy on the outside. Fruit ranges from nearly round to oblate or pear-shaped; 10-30 cm wide; the peel, easily removable, greenish-yellow or pale-yellow colour, 1.25-2 cm thick. Pulp varies from greenish-yellow or pale-yellow to pink or red; 11 to 18 segments, very juicy to fairly dry. Flavour varies from mildly sweet and bland to sub acid or rather acid.

C. paradisi

This fruit is believed to have originated as a natural cross between sweet orange (*C. sinesis*) and Shadock (*C. grandis*). Evergreen trees usually grow to around 5–6 m tall, although they can reach 13–15 m. Leaves dark green, long (up to 150 mm) and thin. It produces 5 cm white four-petaled flowers. Fruit is yellow-orange and largely an oblate spheroid; it ranges in from 10–15 cm. The flesh is segmented and acidic, varying in colour, which include white, pink and red pulps of varying sweetness.

Acid Group

C. limon (L.) Burm. f.- Lemon

Tree is vigorous, upright-spreading and open. Most varieties are comparatively thorny, with relatively short and slender spines. Flowers occur in clusters produced throughout the year, large and purple-tinged in bud and on the lower surface of the petals. New shoot growth is purple-tinted. Although more resistant than the citron and limes to cold and heat, lemon is much more sensitive than the other *Citrus* fruits of major importance and hence its commercial culture is restricted to subtropical regions of mild winter temperatures. Relatively equable growing-season temperatures are advantageous in that they seem to emphasize the ever-flowering tendency and are favourable for fruit setting.

Jambhiri–Rough Lemon

Trees 3-6 m in height and usually have sharp thorns on the twigs. Alternate leaves, reddish when young, become dark-green above, light-green below at maturity; oblong, elliptic or long-ovate, 6.25-

11.25 cm long, finely toothed, with slender wings on the petioles. Mildly fragrant flowers solitary or 2 or more clustered in leaf axils. Buds are reddish. Flowers have 4 or 5 petals, 2 cm long, white on the upper surface, purplish beneath and 20-40 more or less united stamens with yellow anthers. Fruit is oval with a nipple-like protuberance at the apex; 7 -12 cm long; peel rough and yellow, aromatic, dotted with oil glands, 6-10 mm thick. Pulp pale-yellow, 8 to 10 segments, juicy, acidic. Fruits have a few seeds, elliptic or ovate, pointed, smooth, 9.5 mm long.

C. aurantifolia Acid Lime

Trees about 5 m height and if not pruned become shrub like, rarely grow straight, with many branches that often originate quite far down on the trunk. Its branches spread and are irregular, with short, stiff twigs, small leaves, and many small, sharp thorns. Leaves pale green; ovate 2.5 – 9 cm long, resembling orange leaves, small white flowers in clusters, 2.5 cm in diameter, yellowish white with light purple tinge on the margins. Fruits are 3 to 4 cm in diameter, oval to nearly globular in shape, often with a small apical nipple; skin thin, turns yellow on ripening. Pulp is tender, juicy, yellowish green in colour and decidedly acidic. Juice is as sour as lemon juice, but more aromatic. Limes exceed lemons in both acid and sugar content.

C. limettioides Tan.- Sweet lime (syn. C. lumia Risso et Poit.)

Also called *mitha limbu, mitha nimbu,* or *mitha nebu* in India (*mitha* meaning "sweet"), it is often confused with the sweet lemon, *C. limetta* Tan. The tree, its foliage, and form and size of the the fruit resembles the Tahiti lime (*C. latifolia*). Leaves serrated, petioles nearly wingless. Flowers borne singly in leaf axils or in terminal clusters of 2 to 10; fruits may be solitary or in bunches of 2 to 5. oblong, ovoid or nearly round, with rounded base and small nipple at apex, occasionally slightly ribbed; peel aromatic, greenish to orange-yellow when ripe, smooth, with conspicuous oil glands; pulp pale-yellow, 10 segments, tender, very juicy, non-acidic faintly bitter.

C. medica - Citron

Believed to be the first cultivated *Citrus*, records in literature indicate that citron was planted in gardens of the Mediterranean region as early as 4000 BC. Citron is a shrub or small tree reaching 2.4-4.5 m high with stiff branches and stiff twigs and short or long spines in the leaf axils. Leaflets are evergreen, lemon-scented, ovate-lanceolate or ovate elliptic, 6.25-18 cm long; leathery, with short, wingless or nearly wingless petioles; flower buds are large and white or purplish; fragrant flowers about 4 cm wide, in short clusters; 4 - 5 petalled, often pinkish or purplish on the outside. Fruit is highly fragrant, oblong, obovoid or oval, occasionally pyriform, but highly variable; various shapes, smooth or rough fruits; size varies 9- 30 cm long. Peel is yellow when fully ripe; very thick, fleshy; pulp pale-yellow or greenish, 14 or 15 segments, firm, juice less, acidic or sweet; seeds numerous, monoembryonic.

C. karna - Karna (Khatta)

Also known as Karna nimbu, Karna Khatta, Khatta, Soh-sarkar, Karna is an old fruit from Maharashtra, India. It is suspected to be a cross between sour orange and lemon. The commercial importance of karna is use as rootstock, second only to rough lemon. Tree is vigorous, medium to large in size, 4.5-6 m in height, upright-spreading. Leaves ovate or ovate oblong, 6.5-9.5 cm long, 4.5-5.5 cm broad; margin serrulate, articulated, petioles prominently winged; flowers tinged with red or purple. New growth purple-tinted. Fruits round to oval, medium to medium large, 9-12 cm long, 8-11 cm in diameter, orange coloured, usually with broad and prominent nipple, rind rough and irregular, moderately thick, firm, colour golden yellow to deep orange, core open at maturity, Flesh colour dull orange; moderately juicy, seeds many, moderately polyembryonic.

C. latifolia Persian Lime

Also known as Tahiti lime or Bears lime. Tree is moderately vigorous, medium to large, 4.5-6 m, nearly thornless, widespread, drooping branches. Leaves broad-lanceolate, winged petioles; young shoots are purplish. Flowers borne off and on during the year and are slightly purple-tinged. Fruit oval, obovate, oblong or short-elliptical, usually rounded at the base, occasionally ribbed or with a short neck; apex rounded with brief nipple; 4-6.25 cm wide, 5-7.5 cm high; often with slightly nippled ends, colour is usually quite green, yellow as it reaches full ripeness. Peel is vivid green until ripe when it becomes pale-yellow; segments 10, acidic; usually seedless, rarely with one or a few seeds.

C. limetta Sweet Lime

It is believed to be a native of India, commonly known as sweet limetta, Mediterranean sweet lemon, sweet lemon and sweet lime. It is a small tree up to 8 m. irregularly branched and relatively smooth, brownish-grey bark. It possesses about 1.5 to 7.5 cm long numerous thorns. Petioles narrow, distinctly winged, 8 to 29 mm long. It has leaflets rather than leaves, which are obovate and 5.5 to 17 cm long, 2.8 to 8 cm wide; apex of leaflet is acuminate, base rounded. Flowers white in bud and in bloom, 2 to 3 cm wide. Fruit is light yellow at maturity; rind is white and about 5 mm thick. Pulp is greenish and juice is sweet rather than acidic and delicious.

C. limonia Rangpur Lime

It is also known as *lemandarin* or mandarin lime. It is a hybrid between the mandarin orange and lemon. Tree is medium, often confused with tangerine or other *Citrus*. Fruit is polyembryonic and usually reproduces true to seed and commonly used as rootstock. Spreading and drooping branches have dull green foliage with an occasional purple tint on new growth. Numbers and size of thorns vary from tree to tree with some trees being almost thornless. Fruit rind is orange to reddish orange with minutely pitted moderately loose skin with oil glands and a lime like aroma, highly acidic and very juicy, number of segments 8 to 10 with numerous seeds and slightly hollow centre.

Wild, Semi-Wild Species and Related Genera of Citrus

Citrus indica Tanaka

C. indica Tanaka is one of the most primitive species of *Citrus* and perhaps the progenitor of cultivated *Citrus* (Singh 1981, Malik *et al.*, 2006). It bears flowers and fruits more profusely in dense forests than those found in the periphery of human habitants. Fruits are small (15-20 g), spherical, surface deep orange red to almost scarlet, smooth, juicy, inedible and mostly used as medicines. One important characteristic feature of this species is that its flowering time starts from September and lasts till January, a period when generally all other *Citrus* species undergo rest period due to severe low winter temperature. Due to this reason, the purity of the species is still maintained. It has no value as a commercial fruit. *Citrus indica* has been observed to be hardy and to be free from pests and diseases in its natural habitat (Malik *et al.*, 2006).

C. latipes Tanaka; Soh Shyrkhoit

Commonly called "Khasi papeda", Soh Shyrkhoit is an endemic species of Khasi hills of NE India. Tree is medium to tall, densely foliaged and winged petiole. Flowers borne in small axillary racemes with 5-7 very small flowers. Fruits borne singly and resemble those of *C. ichangensis* except for having a thicker peel, of which the inner layer is chalky white just below the outer green layer. Fruits are big in size with large and bold seeds. The seeds are also smaller and more numerous than those of *C. ichangensis* and are arranged 5-7 in each segment. It is expected to have some valuable gene of cold tolerance or resistance, since it is well adapted to cooler areas.

C. macroptera Montr. Satkara or Hatkara

Evergreen forests of N. E. India and moist deciduous forests of the North Himalayas and Assam are the areas of diversity of this species (Nair and Nayar, 1997). Locally it is known as Satkara, or Hatkara, Sat refers to multiple of seven as the fruit contains fourteen segments. Hatkara means multiple of eight as some fruits contains sixteen segments per fruit. Plants are very tall, densely foliaged, thorny and hardy. Petiole is thick and larger than the blade. Fruits are very hard and look like small bael fruit with storage life of about two months. Now it is considered as one of the endangered *Citrus* species.

C. ichangensis Swingle Ichang papeda

It is a slow-growing species, which has characteristic lemon-scented foliage and flowers. It differs greatly from other species in the subgenus *Papeda* in having large flowers and large very thick seeds. Trees of medium size (6.6 m), commonly thorny, petiolar wing and leaves most closely resemble to those of *Citrus latipes*. Fruits are medium in size, smooth, oval, spherical, or flattened in shape, ripening to yellow or orange. Fruit contains many large monoembryonic seeds. Juice is bitter or sour; some fruits lack juice entirely and are instead filled with a mass of pith and seeds. The plant has been successfully employed in evolving cold resistant hybrids such as ichandrins, ichang lemon and yuzu (Swingle and Reece 1967).

Citrus assamensis (Dutta and Bhattacharya); Ada Jamir or Ginger Lime

It is found in Khasi Hills region of Meghalaya and North Cachar area of Assam of North East India. It is a fairly tall tree (up to 4 m) of medium vigour with an open growth habit with medium length thorns. Flower and leaf buds have purplish colour. Leaves are elliptical, 8 × 4 cm in size with crenulated edges and have a faint ginger odour when crushed. Fully-grown flowers have white petals. Fruit is ovoid, yellow-green when ripe with a smooth skin and a vestigial nipple. Fruit size is 8 × 7 cm and has 11-13 segments with firm, yellow flesh and about a dozen seeds. Rind is thick, tightly adherent with a white albedo. Juice is very sour.

Citrus megaloxycarpa Lushaigton; Sour Pummelo

This is one of the indigenous *Citrus* species of Mizoram, Assam, Tripura and Meghalaya of N.E. India. It is rated as the sourest *Citrus* fruit in Indian folklore. It is a medium sized hardy and vigorous tree up to 4.7 m with upright and spreading, round and compact crown; leaflets broadly elliptic to ovate oblong, serrate, obtuse, broadly cuneate base, scented; petiole 0.6-1.5 cm long, narrowly winged, distinctly articulate, flowers highly fragrant, in clusters of 4-12, mostly terminal, creamy white. Fruits oblong ovate, lemon yellow, rough, coarse, apex rounded to slightly nipples, base slightly depressed, rind 1.8-2.3 cm thick, strongly adhering, pulp light yellow and coarse, vesicles spindle to cylindrical shape, loosely packed and seeds many.

Related Genera

Poncirus

Known as Japanese Bitter orange or Trifoliate orange, it is believed as the hardiest close relative of *Citrus* and a native of China and Korea. While other *Citrus* species are evergreen and unifoliate, it is deciduous and trifoliate. *Poncirus trifoliata* is recognized by large 3–5 cm thorns and its deciduous leaves with three or rarely five leaflets, typically with the middle leaflet 3–5 cm and the two side leaflets 2–3 cm. Flowers are white, with pink stamens, 3–5 cm in diameter, fruits green, ripening to yellow, 3–4 cm in diameter. They are very bitter, are not edible fresh. *P. trifoliata* is a valuable species within *Citrus* because it possesses genes conferring many agriculturally important traits not discovered in other *Citrus*. As a primary rootstock and rootstock breeding parent for *Citrus*, it exhibits resistance to *Citrus* tristeza virus (CTV), *Phytophthora* root rot, and *Citrus* nematode (CN). It also confers tolerance to low temperature and has dwarfing effect and other beneficial attributes.

Fortunella Kumquat

Kumquats or *cumquats* have been called "the little gems of the *Citrus* family". The common name, cumquat means "gold orange" in China. Tree is slow-growing, shrubby, compact, 2.4-4.5 m tall, branches light-green and angled when young, thornless or with a few spines. Leaves alternate, lanceolate, 3.25-8.6 cm long, finely toothed from the apex to middle, dark-green, glossy above, lighter beneath. Sweetly fragrant, 5-parted, white flowers are borne singly or 2- 4 in leaf axils. Edible fruit closely resembles that of the sweet orange but it is much smaller and ovular, being the size and shape of an olive; peel is golden-yellow to reddish-orange, with large, conspicuous oil glands, fleshy, thick, tightly clinging; pulp is scant, 3 to 6 segments, not very juicy, acid to sub acid; contains small, pointed seeds or sometimes none.

Hybrids of Citrus

Intergeneric and Intragerneric Hybrids

Intergeneric Hybrids

Poncirus Hybrids

Citrange

It is a hybrid of the sweet orange and the trifoliate orange. Fruit is usually much larger and more orange-like in appearance. In general, the citranges exhibit some degree of intermediacy between the parental species. The influence of the trifoliate orange is strongly marked in the citranges by the trifoliate nature of their leaves, the acidity and bitterness of their fruits and the cold-hardiness of the trees. The influence of the sweet orange is shown, in the evergreen nature of the trees, though a few are semi-deciduous and in their greater vigour. Of great horticultural importance in connection with their use as rootstocks is the fact that with a few exceptions they come remarkably true from seed. Since they exhibit some of the most desirable features of the trifoliate orange combined with the greater vigour and wider range of soil adaptation of the sweet orange, some of them have achieved importance as rootstocks.

Citrangequat

It is a trigeneric hybrid of a citrange and kumquat. Combining the cold-hardiness of the kumquat and trifoliate orange, the citrangequats appear to be more cold-resistant than the citranges or the calamondin and kumquat, for they succeed in regions too cold for these fruits. Tree is vigorous, erect, thorny or thornless, with mostly trifoliate leaves; highly cold-resistant. Fruits are bitter in taste, but are considered edible by some at the peak of their maturity. Fruit resembles the oval kumquat, mostly very acidic.

Citrangedin

A hybrid between citrange (*Poncirus trifoliata*) and calamandin.

Citrangor

A hybrid between Trifoliate orange (*Poncirus trifoliata*) and sweet orange (*C. sinensis*). Crown is compact or dense, not weeping, petiole glabrous, long, wings narrow, adjoining the blade, leaflets one or three, margin bluntly toothed, rachis wings absent, fruit broader than long or as broad as long or longer than broad, rind light green to orange or red orange, rind texture rough, firmness leathery, navel absent, flesh orange or yellow, taste sour.

Cicitrange

A back cross hybrid between citrange and *Poncirus trifoliata*.

Citrumelo

-*Citrus paradisi* Macf. × *Citrus trifoliata* (L.) Raf.- It is a cross of grapefruit and trifoliate orange. Swingle Citrumeolo is a hybrid of Duncan grapefruit and *Poncirus trifoliata*. Phytophthora foot rot and nematode tolerance of Swingle citrumelo makes it suitable for replant sites. Trees are vigorous, large and produce intermediate to high yields. Swingle citrumelo rootstock produces fruit with high juice and soluble solids content and mid-range acidity. It is a superior rootstock for grapefruit producing high yields of large, excellent quality fruit with high juice content.

Citrandarin

Citrandarin is a cold-hardy intergeneric hybrid of *Citrus trifoliata* (Trifoliate Orange) and *Citrus reticulata* (Mandarin). It can tolerate down to –12°C for short duration depending upon cultivar and individual seedling variation.

Citremon

Citrus limon (L.) Burm. × *Poncirus trifoliata* L. C-1499 is one popular cultivar of Citremon.

Citradia

A hybrid between sour orange (*Citrus aurantium*) X Trifoliate orange (*Poncirus trifoliata*). Crown compact or dense, not weeping. Petiole glabrous, length short or long, wings medium, adjoining the blade. Leaflets one, margin crenate/crenulate, Fruit as broad as long or longer than broad, rind yellow to orange, rind texture slightly rough, firmness leathery, navel absent, flesh orange, taste sour. These hybrids greatly resemble citranges but seem to be even more vigorous and cold-resistant.

Citrumquat

A hybrid between Trifoliate orange (*P. trifoliata*) and Kumquat (*Fortunella japonica* or *F. margarita*)

Fortunella Hybrids

Hybrids of Fortunella include the following:

Limequat

Limequats are Mexican lime x Kumquat hybrids. Trees are vigorous, evergreen, the single leaflets having narrowly-winged petioles; nearly spineless or with a few short thorns; more cold-tolerant than the lime but not as hardy as the Kumquat; very resistant to wither tip. Fruit is like the Mexican lime. There are three named cultivars: Eustis, Lakeland, and Tavares. All limequats are more cold-resistant than the lime parent but much less than Kumquats. None of the limequats has achieved commercial importance for the fruit, but Eustis and Lakeland are grown to some extent as ornamentals

Indio Mandarinquat

Mandarin x Kumquat hybrid (*Citrus reticulata* Blanco X *Citrus japonica* Thunb. sp.). Orange bell-shaped fruit much larger than a typical Kumquat. The sweet peel is eaten along with the tart flesh for a unique flavour combination. This variety usually blooms during the summer months and produces abundant crops of fruit that stay on the tree during winter months. Indio looks like a giant Kumquat, with similar tangy-sweet flesh and edible rind.

Nippon Mandarinquat

The name "Orangequat" originally given to this class of hybrid is misleading since its parentage involves a mandarin rather than an orange. This variety originated from a cross between the Satsuma mandarin (*C. unshiu*) and Meiwa kumquat (*Citrus japonica* Thunb. 'Meiwa'). It is an attractive ornamental and the fruit makes excellent marmalade. Tree is slow-growing, medium-small, spreading; foliage dark green. Fruit is small (smaller than Satsuma but larger than Kumquat), broadly oval to obovate; orange-coloured; rind relatively thick and spongy; flavour mild and pulp acidic. Matures early but holds well on tree for several months.

Calamondin

Calamondin is a cross between the mandarin (*Citrus reticulata*) var. *austera* Swingle) and oval Kumquat (*Citrus japonica* Thunb. 'Nagami'). It is itself a parent in many other hybrids, Faustrimedin among them (*Citrus × microcarpa* Bunge × *Citrus australasica* F. Muell.). Calamondin has inherited more qualities from the mandarin than Kumquat. Whereas the rind of Kumquats is usually edible, the rind of Calamondin resembles those of the genus *Citrus* in being often too sour for consumption especially in the larger fruits (4 - 6 cm in diameter). The inner parts of the flower (pedicel and ovary) are also more like those of the *Citrus*. Calamondin is increasingly popular as a potted house plant.

Citrangequat

Combining the cold-hardiness of the Kumquat and trifoliate orange, the citrangequats appear to be more cold-resistant than the citranges or the Calamondin and Kumquat. Tree is vigorous, erect, thorny or thornless, with mostly trifoliate leaves; highly cold-resistant. Fruit resembles the oval Kumquat, mostly very acidic.

Procimequat

A hybrid between Limequat x Hong Kong Kumquat (*Fortunella hindsii*). Crown open, not weeping. Petiole glabrous, length short, wings narrow, adjoining the blade. Leaflets one, margin crenate/crenulate, Fruit broader than long, rind yellow, yellow-orange, orange or red-orange, rind texture slightly rough, flesh yellow, taste sour.

Sunquat

A hybrid of lemon x kumquat (*Citrus japonica* Thunb. sp. × *Citrus limon*). The fruit usually known as Lemonquat is in fact a Sunquat, having an edible rind. It is most likely a type of mandarinquat, possibly a hybrid of Clementine and a Meiwa kumquat. The whole fruit with peel is edible and has no off-flavours. Sunquat is said to be excellent for marmalade. The cultivated variety 'Rio Grande Valley' is sold as a lemonquat.

Yuzuquat

A trigeneric hybrid between Yuzu lemon and 'Nagami' kumquat. The fruit is often used as a lemon substitute and is very seedy.

Intrageneric Hybrids

Tangor

The tangor (*C. reticulata* × *C. sinensis*) is a hybrid of mandarin orange (tangerine, *Citrus reticulata*) and the sweet orange (*Citrus sinensis*). The name "tangor" is a portmanteau using the "tang" of tangerine and the "or" of "orange". The popular varieties of Tangor are: King ("King of Siam"; formerly *Citrus nobili*), Murcott ("Honey Murcott"; "Murcott Honey Orange"; "Red"; "Big Red"), Ortanique (originally found in Jamaica, the name comes from the words "orange", "tangerine", and "unique"), Temple ("Magnet" of Japan) and Umatilla (misnomer "Umatilla tangelo").

Tangelo

The tangelo (*C. reticulata* × *C. maxima* or *C. paradisi*), is a hybrid of tangerine and pummelo or grapefruit. Widely known as honeybells, tangelos have the size of an adult fist, have a tangerine taste and are juicy. They generally have loose skin and are easier to peel than oranges, readily distinguished from them by a characteristic "nipple" at the stem. The Orlando and minneola are two popular varieties of Tangelo. The early maturing Orlando tangelo is noted for its juiciness, mild and sweet flavour, large size and flat-round shape with a characteristic knob. Most Minneola tangelos are characterized by a stem-end neck, which tends to make the fruit appear bell-shaped. The fruit is usually fairly large.

Lemonime

The lemonime (*Citrus limon*) × (*Citrus aurantiifolia*) fruit is larger than a regular lime. The plant is resistant to wither tip and *Citrus* scab. The most important variety is the Perrine of Florida. Other varieties include: 'Fourny' (Corsica), 'Oscar' (Italy), 'La Valette' (Malta), 'Mohtasseb' (Morocco), 'Khangi' (Nepal) and 'AK' (Turkey).

Lemmonnage

Citrus limon x *Citrus sinensis*-, a hybrid of lemon and orange.

Lemandarin

Citrus limon × *Citrus reticulata*. Lemandarin or Mandarin lemon or Mandarin lime belongs to a group of several closely related types of *Citrus*. They resemble the mandarin in appearance but taste more like limes. It is originated in China, where it is called the Canton lime or Canton lemon. It has three well known varieties: Rangpur, Otaheite and Kusaie.

Cultivars of Citrus

Mandarins

Mandarins 'the loose skinned orange' include a group of *Citrus* fruits which are easy-peeling with nice rind colour. Mandarin, its hybrids and relatives together form the largest group in the genus *Citrus*. The number of commercially important mandarin varieties almost equals the number of all other horticultural *Citrus* types combined. The grouping of all mandarins into 3 species by Swingle earlier seemed too arbitrary and the division into 36 mandarin species and their subtypes by the Japanese author Tanaka seemed too elaborate. The system followed by Robert Willard Hodgson in his book 'Horticultural varieties of *Citrus*' divides mandarins into four species. *Deliciosa* (Mediterranean mandarin), *nobilis* (King type) and *unshiu* (satsumas) formed species of their own because of their origin, distribution and characteristics and all other mandarin types (*common mandarin, clementines, tangerines, tangors, tangelos and their hybrids*) were grouped under *reticulata*.

Satsuma Mandarin Group

This mandarin group is one of the famous mandarins of the world and commonly grown in Japan. Some of the important cultivars include 'Miyagawa, Okitsu, Seto, Miho,Kuno, Matsuyama, Obawase and Iseki. They are a few early ripening varieties while Owari, Silverhill, 'Dobashi Beni' and Kimbrough are late ripening varieties.

Willow Leaf Mandarin Group

Also known as Mediterranean mandarin, in comparison with other mandarins, the most distinctive characteristics of this species are the small size and narrow-lanceolate form of leaves; the mild and pleasantly aromatic flavour of the juice; the distinctive nature and fragrance of the rind oil and the plump and almost spherical seeds. Another distinctive characteristic not confined to this mandarin is the spreading-drooping habit of growth. A few of the cultivars under this group are: Avana, Chios, Emperor, Montegrina, Natal and Willowleaf.

King Mandarin Group

Fruits are the largest of the mandarins and oblate to depressed globose. Rind thick, moderately adherent but peelable; surface moderately smooth to rough and warty. Deep yellowish-orange to orange at maturity. King of USA, Kunembo of Japan, Japon, Guanxi Shagan, Huang Yen Man Chieh, Huangyan Bendi-guangju, Shagan and Som-Chuck are common cultivars.

Common Mandarin

There are numerous cultivars in this group and some of the important cultivars are described below:

Khasi Mandarin

It is a commercial cultivar of mandarin in North-Eastern India. Trees are medium to large with erect habit, dense foliage, both thorny and thorn less. Fruits depressed, globose, oblate to ovate, orange yellow to bright orange, surface smooth, glossy, rind thick to moderate, adherence slight, base even, occasionally short-necked, segments moderate in number, generally 8-10; abundantly juicy, juice orange coloured with peculiar sour sweet blend; seeds 9-25, polyembryonic, possesses very good post harvest life.

Kinnow Mandarin

It is believed to be a first generation hybrid between King mandarin (*C. nobilis*) and willow leaf mandarin (*C. deliciosa*). First introduced in Punjab, it performed well and now gained commercial significance in Kinnow region of Punjab. Plants are medium to large, erect, symmetrical, dense foliage with a few scattered spines; leaves broadly lanceolate. Fruits medium, globose to oblate base flattened, deep orange yellow coloured on ripening; rind medium to thin, adherent to segments, number of segments 9-10, easily peelable, very juicy; seeds 12-25; fruit maturity in mid January; somewhat irregular bearer.

Coorg Mandarin

The most important commercial variety in South India, it is particularly grown on large scale in Coorg and Wynad tracts and Chickmagalur district of Karnataka. Trees are very vigorous and upright with compact foliage, sparingly spinous. Fruits medium to large, bright orange in colour, oblate to globose in shape, base necked or depressed, rind thin to medium thick, easily peeled, segments 8-12; juice abundant with deep chrome colour and attractive flavour; seeds 14-30, matures during February-March.

Nagpur Mandarin

Among all commercial mandarin types of India, it is the most important cultivar and its comercial cultivation is confined to Satpura hills (Vidarbha region), Nagpur of central India. Trees are vigorous, spineless with compact foliage; leaves narrowly lanceolate. Fruits medium, sub globose, cadmium coloured, surface smooth, base slightly drawn out with glandular furrows; rind thin, loosely adherent; easily peelable, segments 10- 12, juice abundant, saffron coloured; seeds 4-10. It matures during January- February

Darjeeling and Sikkim Mandarin

Its cultivation is mainly concentrated in Darjeeling district of West Bengal and Sikkim of North East India, fruits globose to oblate, orange yellow in colour, juice moderate, orange in colour.

Ponkan

Ponkan is probably the most widely-grown mandarin in the world, being heavily-grown in China, India and Brazil. Rind medium-thick, fairly loosely adherent; surface relatively smooth but pebbled, with prominent, sunken oil glands; orange-coloured at maturity, flesh colour orange; tender and melting, juicy; flavour mild, pleasant and aromatic.

Clementine

Clementine is a hybrid of the Mediterranean mandarin and a sweet orange. It is an early maturing monoembryonic cultivar with good quality fruits. Fruit size is variable, ranging from medium-small to medium. Well-known clementine cultivars include: Clementines Marisol, Clemenules, Oroval, Clementines Algerian, Nour, Sidi Aissa, Monreal and Carte Noir, Corsica, Fina and Caffin, Bruno, Esbal, Fina Sodea, Orogrande and Ragheb.

Dancy

Fruits are oblate to pear-shaped and of medium size. Peel is deep orange-red to red, smooth, glossy at first but lumpy and fluted later, thin, leathery and tough. Pulp dark-orange, of fine quality, richly flavoured. This cultivar is grown in many parts of the world and is also known as: Bijou, Christmas, Lady, Moragne, Obeni-mikan, Trimble, Welschart and Weshart.

Beauty

Tree is vigorous, medium to large, upright-spreading, virtually thornless, with dense foliage of medium-sized, broadly lanceolate leaves. Fruits are medium in size, oblate, Rind is thin, firm, but easily removed; surface smooth and glossy; colour orange-red at maturity. Flesh is orange-coloured, tender, juicy and sprightly flavoured.

Sweet Orange

This group is commonly known as tight skinned orange. The increased demand for fresh juice and the development of very early and very late maturing varieties have contributed to its present popularity and its year-round availability. Some popular cultivars under this group are described below:

Ambersweet

It is a complex *Citrus* hybrid. A mid-season sweet orange was crossed with a clementine x Orlando tangelo hybrid. It can be peeled more easily than other oranges. The fruit resembles those of navel orange in size and appearance more than other types and have good orange rind and juice colour at maturity. Trees are moderately vigorous and upright in shape. Foliage is usually fairly dense. Young shoots have small thorns.

Berna

Tree is slow-growing, compact, medium-small, very productive. Fruits are medium-small, seed a few or none. Rind medium-thick, firm; surface finely pebbled. Flesh well-coloured; moderately juicy; flavour sweet. Late in maturity (end of March).

Jaffa

Also known as 'Florida Jaffa' or 'Jaffa Blood Orange' was introduced to Florida in 1883 from Palestine. Fruits are small to medium, commercially seedless (0-6 seeds), with a thin, smooth peel. Peel colour is yellow. The flesh is melting in texture and of very high quality, producing a thick, nectar-like juice. It usually achieves maturity by Christmas time. Flecks of pigment sometimes occur in the fruit during cooler winter condition.

Hamlin

Hamlin is an early variety; the first fruit reaches maturity in October and is seedless, juicy and very productive. Hamlin is also known as Norris.

Pineapple

The fruits are medium large, somewhat flattened on both ends, with a moderately thick, smooth peel that develops good orange colour under cool night conditions. Juice colour and quality are very good. It usually contains 15-25 seeds. It succeeds on sour orange and rough lemon rootstocks. If the crop is allowed to remain too long on the tree, it may induce alternate-bearing. It is the favourite midseason orange, sweet with flavour resembles pineapple and good for processing.

Shamouti

The peel is thick but comes off easily in segments. The fruit is juicy, with a distinct flavour and of good quality. Fruits are large, oval to ellipsoid with rounded apex, deep orange in colour, nearly seedless and sweet.

Valencia

The fruits are medium to large and commercially seedless (0-6 seeds). The peel is thin and the pulp is tender and very juicy. Valencia is a late variety and keeps well on the tree. It can be picked until late spring, or even early summer, which prolongs the season and increases productivity. Fruits are of high quality but in the heat and without cool nights colour break does not occur and the fruits remain greenish in the tropics even when fully ripe. Campbell nucellar, Cutter, Delta, Dom Joao, Frost nucellar, 'Harward Late, Midknight, Olinda, Rhode Red, Cotidian and Valencia Late are a few of the popular varieties of Valencia.

Navel Oranges

Navel oranges syn. Bahia oranges get their name from the navel-like protrusion at the lower end of the fruit. It is actually a secondary embryo inside the same fruit. Navel oranges have no functioning pollen. They do not cross-pollinate with other citruses and produce seedless fruit. Navel oranges are propagated from budwood and recently by cloning. Cloning is the most cost-effective way of propagation especially in developing countries. Many varieties grow somewhat slower and smaller than other orange varieties.

Washington Naval

It is the second most important orange variety in the world after Valencia. It is the leading variety in Brazil, California, Paraguay, South Africa, Australia and Japan. Fruit is medium to large, yellow, exceptionally delicious, round and seedless, rich in flavour and has a slightly pebbled orange rind, easily peeled, segments are seedless, loose, very juicy, very sweet, texture is firm but melting. The sugar to acid balance is on the sugary side but had enough acids to make a pleasing *Citrus* flavour. It is early in maturity.

Mosambi

Fruits are medium to large, slightly oblate to globose, colour light yellow to pale orange, rind medium, tight, difficult to peel; flesh firm, juicy with straw yellow colour, seeds 15-25.

Malta Blood Red

Fruits medium to large, ellipsoid to spherical, cadmium yellow in colour with scarlet blush, apex broad, rind thin to medium, tight, flesh blood red when fully ripe, seeds many.

Sathgudi

Fruits spherical to globose, attractive orange yellow in colour, medium to large, segments 10-12, juicy, sweet with acid tinge, moderate flavour, seeds 10-20

Lemon

Eureka

Tree is spreading, moderately vigorous, sparsely foliaged, less thorny, more prone to wind blemish, sunburn and frost damage than other varieties, less cold-resistant than other varieties. Flowers are tinged with pink. Fruits have a smooth medium thin rind, high juice content, high acid level, low number of seeds and good flavour.

Lisbon

In comparison with Eureka, Lisbon is more cold-resistant, more productive and vigorous. It has denser foliage and the fruit inside the canopy are better protected from sun, wind and cold. Trees are also much thornier. Fruits have a less pronounced nipple and slightly rougher rind texture. In suitable conditions, Lisbon outyields Eureka by 20-25 per cent due to bigger size of tree and fruit.

Volkamer

It was first thought of as a variant of mandarin lime. More recently, it is identified as a cross of lemon and sour orange. The attractively dense foliage makes it an excellent ornamental tree. Fruits are lemon-shaped, wide and with a rough, bright reddish rind. Flesh and juice are yellow-reddish in colour. Fruit has a few or no seeds, tastes slightly bitter and has a pleasantly fresh taste and aroma. It can be used in cooking instead of lemon. It is used as rootstock because of its resistance to many diseases.

Galgal

It is an important local item of trade in the Himachal Pradesh, Jammu and Kashmir and Punjab of India. Tree vigorous, upright or spreading but irregular and open, with stout branches, numerous thick spines; leaves large and dull-green. Flowers large, purple-tinged and produced in spring only. New shoot growth is purple-tinted, fruit medium-large to large, oblong to ellipsoid, usually with short blunt-pointed nipple, sometimes depressed and flat. Rind medium-thick; surface usually smooth, tightly adherent; colour pale to golden yellow. Segments about 10, flesh colour pale yellow, moderately juicy; flavour very sour and with trace of bitterness, seeds numerous and large.

Assam Lemon

Fruits elliptic to oblong-obovate, lemon yellow, medium to large, long, apex nippled and base rounded, rind medium thick, axis hollow, segments 11-13, pulp light yellow or whitish, juicy, acidic, normally seedless or occasionally a few seeds.

Baramasi

Fruits ovoid to globose in shape, medium sized, light yellow in colour, rind medium to thick, flesh firm, acidic with good flavour, seedless or occasionally a few seeds.

Lucknow Seedless

Fruits oblong, lemon yellow, smooth, apex nippled, base rounded, rind thin, axis hollow, segments 10-13, pulp light yellow and coarse, juicy, flavour good and sour, seeds absent to a few. Fruits ripen from November to January.

Kagzi Kalan

Fruits medium, spherical, yellow with apex slightly nippled, base rounded, rind thin, smooth, flesh acidic, light yellow, juicy, 8-13 seeds. Susceptible to tristeza, canker and die back diseases.

Pant Lemon 1

Fruits medium, 80-100 g, round and smooth, rind thin, juicy, tolerant to tristeza, canker and die back.

Limes

West Indian Lime

West Indian lime (Key lime, Mexican lime, Common lime) is the common round lime grown especially in Mexico, Florida and West Indies. In India, it is known as Kagzi lime. Colour dark green when immature to yellowish green when ripe. Fruits small, round obovate or short elliptical, round base, sometimes with short neck, apex round, rind thin, tight, flesh colour greenish yellow, highly acidic, seeds small and moderate in number.

Pramalini

Bears fruits in clusters of 3-7 and yields 30 per cent more than normal Kagzi lime. Fruits have 57 per cent juice.

Vikram

Fruits bear in clusters of 5-10 and produce some off season fruits during September, May and June. It gives 30-32 per cent higher yields than local strains.

Selection 49

Prolific bearer, fruits larger sized, juice abundant, Fruits have a tendency of bearing summer crop and show tolerance to canker, tristeza and leaf miner.

Seedless Lime

It is a selection from Kagzi lime. Fruits oblong, skin thin, primrose coloured, prolific bearer, yields double than normal but late.

Tahiti Lime (*C. latifolia*)

Tahiti lime (Bears lime, Persian lime) is the elongated or ovate seedless lime, often with beautiful spring green flesh and juice. It is less acidic and often juicier than the West Indian lime. Trees are thornless and bigger than Mexican lime. Leaves are bigger and darker. Fruits are often more oval in shape. It is slightly more tolerant to cold than the Mexican lime but still prefers warmer areas than lemon. The extremely rare seeds are strongly monoembryonic. Most seeds are flat, have shrivelled look and rarely germinate.

Indian Lime

Indian lime (Palestine sweet lime) is a sweet lime of Indian origin often thought to be a hybrid of limetta and lime. It has lesser sugar than sweet orange but almost no citric acid and therefore tastes sweeter. Tree has big shiny leaves that are often cupped or rolled. Fruit is large and usually has a pronounced nipple, colour turns orange-yellow at maturity. It is one of the a few limes that can be enjoyed fresh. In India the fruits are also cooked whole and eaten as a dessert or preserved. Following two are the varieties of Sweet lime.

Mitha Chickna

Fruits spherical, globose, yellow, glossy surface with oil glands, rind thin, leathery, flesh medium coarse, juicy, sweet, seeds a few.

Mitha Tora

Fruits large with depressed apex and necked back, lemon yellow coloured, rind tough and thick with oil glands, flesh yellowish white, juicy, sweet and well flavoured.

Rongpur Lime

It is indigenous to India. Trees evergreen, spreading in growth habit, highly productive, rind and pulp orange coloured, thin and readily separates from pulp. It is commonly used as rootstock.

Pummelo

Tahitian

A typical pummelo but with a thin peel of greenish-yellow colour and the flesh has a greenish shade as well, but may become amber in full maturity. Fruit is juicy and taste is sweet and has a flavour of melon or lime. Tahitian pummelo is often used in the development of new cultivars of both pummelo and grapefruit.

Siamese Sweet

It is an acidless and sweet-tasting pummelo from Thailand. Fruit is oblate to broad ovoid; pulp white, with large, crisp, non-juicy sacs easily separating from each other; mild-flavoured but faintly bitter. Tree is dwarf with drooping branches and hairy new growth.

Sweetie

Sweetie is one of the newest pummelo hybrids. Its parents are Siamese Sweet and a variant of Marsh grapefruit. Flesh is pale yellow and taste is mild and sweet. This is one of the pummelos that retains some green on its rind even when ripe.

Chandler

It is a cross between the Siamese Sweet (white, acidless) and Siamese Pink (acid) pummelos. Fruit is almost perfect globe, medium size to very big with a smooth peel that sometimes has a pinkish tinge. Pulp is pink to medium red, fine grained, tender and fairly juicy. Segment walls are thin, flavour very good; subacid, about 12 per cent sugar and seedy.

Honey

Fruits are seedless and sweet, flesh semi-transparent, colour light green to lemon yellow, Fruit weight (½ - 2 kg), main harvest is from mid September to mid October, but the fruit is available from August to February. 100 ml juice contains 9.2-9.9 g sugar, 0.73-1.01 g acid, 49-52 mg Vitamin C. One hundred gram fruit contains 0.6g fat, 0.7g protein and 57 Kcal energy.

Cocktail

It is a cross between Siamese Sweet pummelo and Frua mandarin. Trees are large and vigorous. Fruit size varies from size of orange to size of grapefruit, thinner peeled than most pummelos, fruit has seeds and is very juicy, peel, flesh and juice are mandarin-coloured, flavour is pleasant and sub-acid, matures in early winter and the fruits hold well on the tree.

Crop Improvement in Citrus

Tissue Culture and Biotechnology for Genetic Improvement of *Citrus*

Plant tissue-culture techniques are part of a large group of strategies and technologies, ranging through molecular genetics, recombinant DNA studies, genome characterization, gene-transfer techniques, aseptic growth of cells, tissues, organs, and *in vitro* regeneration of plants. Tissue culture has been exploited to create genetic variability from which crop plants can be improved, to improve the state of health of the planted material and to increase the number of desirable germplasm available to the plant breeder. Tissue culture and micropropagation protocols were described for a number of *Citrus* species and explant sources (Grinblat 1972, Chaturvedi and Mitra 1974, Barlass and Skene 1982, Edriss and Burger 1984, Duran-Vila *et al.*, 1989). The environmental conditions and composition of culture media are known to be crucial for the growth of tissue cultures (Kobayashi *et al.*, 1985,

Duran-Vila *et al.*, 1992, Randall 1994). Impact of MS and Murashige and Tucker (MT) media with various kinds and concentrations of cytokinins on the tissue culture of *Citrus reticulata* revealed that MT medium supplemented with (0.5 mg l^{-1}) BA gave the greatest percentage of shoots and root numbers (Te-Chato and Nudoung, 1998).

Regeneration of different species of *Citrus* was already investigated using MS medium supplemented with BA (3 mg l^{-1}) or with BA (1 mg l^{-1})) (Pena *et al.*, 1995a, b, 1997; Cervera *et al.*, 1998a, b, 2000; Ghorbel *et al.*, 1999: Pena and Navarro 1999; Dominguez *et al.*, 2000;). Kaneyoshi *et al.* (1994) used MS medium supplemented with BA (5 mg l^{-1}) and NAA (0.1 mg l^{-1}) for regeneration of *Poncirus trifoliata* Rad. Maximum shoot regeneration response (70 per cent) from callus was observed on MS medium supplemented with BA (3 mg l^{-1}) followed by MS medium supplemented with BA (2 mg l^{-1}) + NAA (0.2 mg l^{-1}) (54 per cent). In direct shoot regeneration higher shoot regeneration response (83 per cent) was observed on MS medium supplemented with BA (3 mg l^{-1}) followed by MS medium supplemented with BA (0.5 mg l^{-1}). Costa *et al.* (2002) reported shoot regeneration at BA from (0.5–4 mg l^{-1}) with the best at (2 mg l^{-1}) for *Citrus paradisi* (Macf) epicotyl explants. Te-chato and Nudoung, (1998) reported that BA (0.5 mg L–l) gave the best result (75 per cent) of shooting response in *Citrus reticulata* Blanco cv Shogun from different explants of *in vitro* raised seedlings.

Sim *et al.* (1989) cultured cotyledon, epicotyls, shoot tips, nodal explants and leaves from 2-4 months old seedlings and 15 years old grafted cv. Blanco and grown in MS medium with 2 per cent sucrose and BA, NAA, 2, 4-D, IBA or a mixture and they obtained regenerated shoots from seedling epicotyls, leaf, shoot tip and nodal stem segments, while from the 15 years old plants regeneration obtained only from shoot tips and nodal stem segments. They did not get organogenesis with mature leaf explants. Hazarika *et al.* (1995) developed an efficient acclimatization protocol involving direct culturing of 4-week old *in vitro* grown proliferated shoots in soilrite topped over sterile FYM. The shoots rooted in 7-15 days depending upon species. Parthasarathy and Nagaraju (1996) cultured the shoot explants obtained from *in vitro* germinated embryos of seven *Citrus* species *viz. C. reticulata, C. medurensis, C. reshni, C. volkemerina, C. taiwanica, C. sinensis* cvs. Malta and Mosambi and *C. limon* on MS and B5 media supplemented with BAP and reported MS was the best medium for proliferation of shoots in all species except *C. reshni*. Barua *et al.* (1996) reported *ex vitro* rooting and simultaneous acclimatisation of microshoots of *C. assamensis, C. latipes* and *C. indica* proliferated *in vitro* in soilrite. Dass *et al.* (1997) studied the effect of different rootstocks *viz.* rough lemon, troyer citrange, and carrizo citrange of different ages. They were grafted with shoot tip scion of Nagpur mandarin. The success of *in vitro* micro-grafting was dependent on the type of rootstocks used. Overall success of *in vitro* shoot tip grafting was more in troyer citrange followed by Carrizo and rough lemon.

Miah *et al.* (2002) achieved somatic embryogenesis and plantlet regeneration in callus cultures of nucellus derived from undeveloped ovule of immature fruits of *Citrus macroptera* by using modified MS medium supplemented with malt. Calli produced in malt supplemented MS medium were embryogenic in nature. After transfer of this embryogenic callus in hormone free MS medium, somatic embryos were developed. Independent plantlets were developed from these somatic embryos by further subculture in the same medium.

Ali and Mirza (2006) studied the effect of various concentrations and combinations of 2,4-D, BA and NAA on regeneration of *Citrus jambhiri* Lush. explants. Optimal callus induction response was observed on MS medium supplemented with 2,4-D (1.5 mg l^{-1}) from all types of explants, with stem explants showing the highest response (92 per cent). Maximum shoot regeneration response (70 per cent) from callus was observed on MS medium supplemented with BA (3 mg l^{-1}). Direct shoot regeneration was the highest in stem segment explants on MS medium with BA (3 mg l⁻¹). MS medium supplemented with NAA (0.5 mg l⁻¹) provided 70 per cent rooting response.

Jajoo (2010) developed an efficient and highly reproducible plant regeneration protocol from nucellar embryo of *Citrus limonia* by using MS medium. BA (2.22 mM) induced the highest number of multiple shoots as 18.26 shoots per explant. On transfer of individual shoots to root inducing MS medium supplied with auxins, IBA (2.46 ìM) and BA (1.11 ìM) proved to be the best combination. This media resulted in 78.80 per cent rooting and produced plantlets with an average of 5.53 roots/shoot.

Pérez-Tornerov *et al.* (2010) investigated the influence of basal medium and different plant growth regulators on micropropagation of nodal explants from mature trees of lemon cultivars. Several combinations of BA and GA were used to optimise the proliferation phase. The best results were obtained with (2 mg l⁻¹) BA and (1 or 2 mg l⁻¹) GA. Explants length was shorter with the higher BA concentrations and, in all genotypes, shoot length was greater with (2 mg l⁻¹) GA. The best results for productivity (number of shoots × the average shoot length) were obtained with (2 mg l⁻¹) BA and (2 mg l⁻¹) GA, although explants with chlorosis and narrow leaves were observed. The highest rooting percentages were obtained on media containing (3 mg l⁻¹) IBA alone or IBA in combination with 1 mg l⁻¹ IAA and on these media, the highest number of roots was produced. The average root length was affected significantly by the IBA and IAA concentrations. Root length was greater when only (3 mg l⁻¹) IBA was used and explants had a better appearance, with greener and larger leaves.

Biochemical and Molecular Markers

Markers are used over the years for classification of plants. Markers are any trait(s) of an organism that can be identified with confidence and relative ease, and can be followed in a mapping population (Bhat *et al.,* 2010). They can be defined as heritable entities associated with the economically important trait under the control of polygenes (Beckman and Soller, 1986). Molecular markers are any kind of molecule indicating existence of a chemical or a physical process. Molecular markers include biochemical constituents (*e.g.* secondary metabolites in plants) and macromolecules (e.g. proteins and deoxyribo-nucleic acid) (Joshi *et al.,* 1999). These macromolecules show easily detectable differences among different strains of a species or among different species. Strauss *et al.* (1992) distinguished the molecular markers into two classes. Bio-chemical molecular markers are derived from the chemical products of gene expression *i.e.* protein based markers and molecular genetic markers derived from direct analysis of polymorphism in DNA sequences *i.e.* DNA based markers.

Isozyme analysis is a useful technique for determining the genetic variability, identification of nucellar and zygotic seedlings and identification of somatic hybrids and studies in phylogeny. In *Citrus* more than 20 isozyme loci have been genetically characterized. Isozyme analysis is increasingly being used to distinguish nucellar and zygotic seedlings (Bhat *et al.,* 2010). The identification of intergeneric and interspecific somatic hybrids obtained by protoplast fusion has also been done using isozyme analysis (Grosser and Gmitter, 1990; Tusa *et al.,* 1990). In *Citrus,* cultivars within the same species could have arisen through mutation and given species status and isozyme profiles have shown that sweet orange and grapefruit have similarities (Hirai and Kajiura, 1987; Roose, 1988).

DNA markers based polymerase chain reaction (PCR) are becoming increasingly popular for various uses for which isozymes are used besides the construction of linkage maps and the marker assisted selection. RAPD has been successfully used to measure the genetic diversity (Sawazaki *et al.,* 1997; Coletta *et al.,* 1998) and to understand the genetic origin of cultivars (Russo *et al.,* 1998). Fang *et al.* (1997) found that acidless trait in pummelo 2240 is controlled by a single recessive gene called *acitric.*

Several molecular markers, particularly polymerase chain reaction (PCR)- based DNA markers have now been utilized in inferring systematic and phylogenetic relationships and identication of cultivars and crop relatives in several economically important plant groups (Weising *et al.,* 2005). In *Citrus*, molecular phylogeny at various taxonomic levels was examined in several earlier studies through application of isozymes (Herrero *et al.,* 1996), RAPD and PCR- RFLP (Federici *et al.,* 1998; Abkenar *et al.,* 2004), RAPD, SCAR and PCR- RFLP (Nicolosi *et al.,* 2000), AFLP (Liang *et al.,* 2007; Pang *et al.,* 2007), SSR (Barkley *et al.,* 2006), ISSR (Shahsavar *et al.,* 2007) and sequence data analysis of non-coding chloroplast DNA (cpDNA) regions (Chase *et al.,* 1999; Araujo *et al.,* 2003; Morton *et al.,* 2003). The trnL (UAA)-trnF (GAA) intergenic spacer of cpDNA has also been reported as a potential region for systematics and phylogenetic studies in *Citrus* (Araujo *et al.,* 2003). Yingzhi *et al.* (2007) utilised trnL-trnF intergenic spacer for inferring phylogeny in wild as well as cultivated mandarins in China, while Jung *et al.* (2005) evaluated the potential of the trnL-trnF region in analyzing the phylogeny and systematic relationships in Korean *Citrus*. Recently Jena *et al.* (2009) studied the systematic and phylogenetic relationships in Indian *Citrus* using PCR-RFLP of two intergenic spacer regions (rbcL-ORF106 and trnD-trnT) and sequence data analysis of trnL-trnF intergenic spacer of cpDNA and PCR-RFLP and trnL-trnF data supported the recognition of *C. maxima, C. medica* and *C. reticulata* as the basal species of edible *Citrus*. The separation of *Citrus maxima, C. medica* and *C. reticulata* in distinct groups or subclusters by using PCR-RFLP and trnL-trnF sequence support their distinctiveness as the true basal species of edible *Citrus* as supported by recent morphological, biochemical and molecular studies in *Citrus*(Scora, 1975; Barrett and Rhodes, 1976; Nicolosi *et al.,* 2000; Araujo *et al.,* 2003; Mabberley, 2004; Liang *et al.,* 2007; Pang *et al.,* 2007). Similarly, Jena *et al.* (2009) by using trnL-trnF sequence reported closer affinities among the above three taxa.

Nicolsi *et al.* (2000) investigated *Citrus* phylogeny by using RAPD, SCAR and cpDNA markers indicated that *Fortunella* is phylogenetically close to *Citrus* while the other three related genera *Poncirus, MicroCitrus* and *EremoCitrus.*are distant from *Citrus* and from each other. Different phylogenetic relationships were revealed with cpDNA data. In another study, Malik *et al.* (2006) studied the morphological characterization of leaves, fruits and seeds in *C. indica and C. macroptera* and reported presence of sizable variability within accessions of these two *Citrus* species. Marak and Laskar (2010) analysed the polymorphism in ISSR amplicons and revealed diverse genetic relationship between *Citrus indica* and other *Citrus* species. In a consensus UPGMA dendrogram, the *C. indica* samples were clustered together with 99 per cent bootstrap support. *C. reticulata, C. sinensis* and *C. aurantifolia* formed a cluster with 67 per cent bootstrap separation. *C. macroptera* and *C. maxima* samples formed separate clusters with respectively 100 per cent and 98 per cent bootstrap supports. Principal Components Analysis projected *C. indica* to be more closely related to *C. aurantifolia* than to the other *Citrus* species of the study.

Genetic Engineering in Citrus Improvement

Several techniques such as polyethylene glycol (PEG)-mediated direct uptake of DNA by protoplast (Kobayashi and Uchimaya, 1989), particle bombardment (Yao *et al.,* 1996) and *Agrobacterium*-mediated transformations (Hidaka and Omura, 1993) were developed and used with various *Citrus* spp. However, the latter transformation system is now the most commonly used method because it has been proven most successful with higher transformation efficiencies resulting in the production of transgenic plants (Peña *et al.,* 2007; Singh and Rajam, 2009; Yu *et al.,* 2002).

Protoplast transformation is mostly used with commercially important *Citrus* genotypes which are either seedless or contain very a few seeds, which are required in most *Agrobacterium*-mediated transformation procedures (Fleming *et al.,* 2000). Regeneration using this system has been used with many *Citrus* species, including lemons [*C. limon* (L.) Burm. F.], limes [*C. aurantifolia* (Cristm.) Swingle], mandarins (*C. reticulata* Blanco), grapefruits (*C. paradise* Macf.), sweet orange (*C. sinensis* Osbeck) and sour orange (*C. aurantium* L.). Although, limited success has previously been reported using protoplast transformation with sweet orange, rough lemon (*C. jambhiri* Lush.) and 'Ponkan' mandarin (Hidaka and Omura, 1993; Kobayashi and Uchimaya, 1989; Vardi *et al.,* 1990). Fleming *et al.* (2000) have reported success in recovering transgenic sweet orange plantlets by an optimized version of this method.

The Agrobacterium mediated transformation uses the ability of the *Agrobacterium*-plant interaction to transfer and integrate genetic information into the plant's genome. *Agrobacterium*-mediated transformation experiments were carried out with numerous hybrids and species of *Citrus,* such as grapefruit, sour orange, sweet orange, trifoliate orange (*Poncirus trifoliata* Raf.), 'Carrizo' citrange, 'Mexican' lime, 'Swingle' citrumelo (*C. paradisi* x *P. trifoliata*), 'Cleopatra' mandarin, and alemow (*C. macrophylla* Wester) (Dominguez *et al.,* 2000; Ghorbel *et al.,* 2000; Gutierrez-E *et al.,* 1997; Luth and Moore, 1999; Molinari *et al.,* 2004; Moore *et al.,* 1992; Peña *et al.,* 2004, 2007).

Genetic Engineering in *Citrus* Disease Control

Several studies have transformed sequences from a variety of economically important viruses into different *Citrus* types to attempt to produce resistant plants. One of such viral diseases is caused by *Citrus tristeza virus* (CTV). Severe strains of CTV can dramatically reduce production and in some instances lead to tree death in a relatively short period of time (Moreno *et al.,* 2008). Transforming the major CP (p25) into 'Mexican' lime had two types of response to viral challenge. In replicate plants, propagated from the same line (*i.e.* genetically identical), 10 to 33 per cent were resistant to CTV while the rest developed typical symptoms, despite a significant delay in virus accumulation (Domínguez *et al.,* 2002). Similar results were obtained in 'Duncan' grapefruit when translatable and untranslatable versions of the major CP were transformed (Febres *et al.,* 2003, 2008). The use of the 3' region of the p23 and the contiguous 3'-untranslated region (UTR), either as a hairpin or as single copy, has also been transformed into 'Duncan', 'Flame', 'Marsh', and 'Ruby Red' grapefruit and alemow plants with similar results as described above in which some plants derived from a particular line were fully resistant and others were not (Ananthakrishnan *et al.,* 2007; Batuman *et al.,* 2006; Febres *et al.,* 2008).

In another strategy to control *Citrus* canker, a *hrpN* gene derived from *Erwinia amylovora* was transformed into 'Hamlin' sweet orange plants. The *hrpN* encodes a harpin protein that elicits the hypersensitive response (HR) and systemic acquired resistance (SAR) in plants. The *hrpN* gene was inserted in a construct made up of *gst1,* a pathogen-inducible promoter, a signal peptide for protein secretion to the apoplast. Several of the *hrpN* transgenic lines showed reduction in their susceptibility to *Citrus* canker as compared to wild type plants and one line in particular displayed very high resistance to the pathogen (up to 79 per cent reduction in disease severity) (Barbosa-Mendes *et al.,* 2009).

Resistance to another important viral disease, *Citrus psorosis virus* (CPsV) was reported in transgenic sweet orange plants transformed with intron-hairpin constructs (ihp) corresponding to the viral CP, the 54K or the 24K genes (Reyes *et al.,* 2011).

References

Abkenar, A.A., Isshiki, S. and Tashino, Y., 2004. Phylogenetic relationships in the true *Citrus* fruit trees revealed by PCR-RFLP analysis of cp DNA. *Sci. Hortic.* **102**: 233–242.

Ali, S. and Mirza, B. 2006. Micropropagation of rough lemon (*Citrus jambhiri* Lush.): Effect of explant type and hormone concentration. *Acta Bot. Croat.* **65**: 137–146.

Ananthakrishnan G.; Orbovic V.; Pasquali G.; Calovic M. and Grosser J. W. 2007. Transfer of *Citrus Tristeza Virus* (CTV)-Derived Resistance Candidate Sequences to Four Grapefruit Cultivars through *Agrobacterium*-Mediated Genetic Transformation. *In Vitro Cellular and Developmental Biology-Plant,* **43**: 593-601

Araujo, E.F., Queiroz, L.P. and Machado, M.A., 2003. What is *Citrus*? Taxonomic implications from a study of cp-DNA evolution in the tribe Citreae (Rutaceae subfamily Aurantioideae). *Org. Divers. Evol.* **3**: 55–62.

Barbosa-Mendes J. M.; Mourao F. D. A.; Bergamin A.; Harakava R.; Beer S. V. and Mendes B. M. J. 2009. Genetic Transformation of *Citrus sinensis* cv. Hamlin with *Hrpn* Gene from *Erwinia amylovora* and Evaluation of the Transgenic Lines for Resistance to *Citrus* Canker. *Scientia Horticulturae,*.**122**: 109-115

Barkley, N.A., Roose, M.L., Krueger, R.R. and Federici, C.T., 2006. Assessing genetic diversity and population structure in a *Citrus* germplasm collection utilizing simple sequence repeat markers (SSRs). *Theory Appl. Genet.* **112**: 1519–1531.

Barlass, M. and Skene, K. G. M., 1982: *In vitro* plantlet formation from *Citrus* species and hybrids. *Scientia Hort.* **17**: 333–341.

Barrett, H.C. and Rhodes, A.M., 1976. A numerical taxonomic study of afnity relationships in cultivated *Citrus* and its close relatives. *Syst. Bot.* **1**:105–136.

Barua, A., Nagaraju, V. and Parthasarthy, V.A. 1996. *Ann. Pl. Physiol.* **16**: 124-128.

Batuman O.; Mawassi M. and Bar-Joseph M. 2006. Transgenes Consisting of a dsRNA of an RNAi Suppressor Plus the 3' UTR Provide Resistance to *Citrus Tristeza Virus* Sequences in *Nicotiana benthamiana* but Not in *Citrus*. *Virus genes,* **33**:319-327

Beckman, J. S. and Soller, M., 1986. Restriction fragment length polymorphism and genetic improvement of agricultural species. *Euphytica* **3**:111-24.

Bhat, Z.A., Dhillon, W.S., Rashid, R., Bhat, J.A. Dar, W.A.and Ganaie, M.Y.2010. The role of Molecular Markers in Improvement of Fruit Crops. *Not. Sci. Biol.* **2**:22-30.

Bhattacharya, S.C. and Dutta, S. 1956. Classification of *Citrus* fruits of Assam. Sc. Monogr. 20. ICAR, New Delhi, pp 110.

Bhattacharya, S.C. and Dutta, S. 1951. *Citrus* varieties of Assam. *Indian J. Genetics* **11**: 57-62.

Bonavia E. 1880-90. The cultivated oranges and lemons etc. of India and Ceylon. WH Allen and Co. London, pp 384.

Calabrese, F. 1998. *La Favolosa Storia degli Agrumi.* L'Epos Societa Editrice, Palermo.

Cervera, M., Juarez, J., Navarro, A., Pina, J. A., Duran-Vila, N., Navarro, L. and Pena, L., 1998a: Genetic transformation and regeneration of mature tissues of woody fruit plants bypassing the juvenile stage. *Transgenic Res.* **7**: 51–59.

Cervera, M., Ortega, C., Navarro, A., Navarro, L. and Pena, L., 2000: Generation of Transgenic *Citrus* plants with tolerance-to-salinity gene *HAL2* from yeast. *J. Hort. Sci. Biotechnol.* **75**: 26–30.

Cervera, M., Pina, J. A., Juarez, J., Navarro, L. and Pena, L., 1998b: A*grobacterium*-mediated transformation of citrange: factors affecting transformation and regeneration. *Plant Cell Rep.* **18**: 271–278.

Chapot, H. 1975. The *Citrus* plant. In: *Citrus* Technical Monograph. No. 4. Ciba- Geigy Agrochemicals, pp. 6-13.

Chase, M.W., Morton, C.M. and Kallunki, J.A., 1999. Phylogenetic relationships of Rutaceae: a cladistic analysis of the subfamilies using evidence from rbcL and atpB sequence variation. *Am. J. Bot.* **8**:1191–1199.

Chaturvedi, H.C. and Mitra, G., 1974: Clonal propagation of *Citrus* from somatic callus cultures. *Hort. Sci.* **9**:118–120.

Coletta, F.H.D., Machado, M. A., Targon, M.C.P.N. Moreira, M.C.P.G. and Pompeu, J.J. 1998. *Euphytica,* **102**: 133-139.

Costa, M. G. C., Otoni, W. C. and Moor, G. A., 2002. An elevation of factors affecting the efficieny of *Agrobacterium*-mediated transformation of *Citrus paradisi* (Macf.) and production of transgenic plants containing carotenoid biosynthetic genes. *Plant Cell Rep.* **21**:365–373.

Dass, H.C., Srivastava, A.K. Lalan Ram and Shyam Singh. 1998. *Indian J. Agric. Sci.* **68**: 692-694.

Domínguez A.; de Mendoza A. H.; Guerri J.; Cambra M.; Navarro L.; Moreno P. and Peña L. 2002. Pathogen-Derived Resistance to *Citrus Tristeza Virus* (CTV) in Transgenic Mexican Lime (*Citrus aurantifolia* (Christ.) Swing.) Plants Expressing Its *P25;* Coat Protein Gene. *Molecular Breeding,*10: 1-10

Dominguez A.; Guerri J.; Cambra M.; Navarro L.; Moreno P. and Peña L. 2000. Efficient Production of Transgenic *Citrus* Plants Expressing the Coat Protein Gene of *Citrus Tristeza Virus. Plant Cell Reports,* **19**: 427-433

Duran-Vila, N., Gogorcenna, Y., Ortega, V., Ortiz, J. and Navarro, L., 1992: Morphogenesis and tissue culture of sweet orange (*Citrus sinensis* (L.) Osb. Effect of temperature and photosynthetic radiations. *Plant Cell Tis. Org. Cult.* **29**: 11–18.

Duran-Vila, N., Ortega, V. and Navarro, L., 1989: Morphogenesis and tissue cultures of three *Citrus* species. *Plant Cell Tiss. Org. Cult.* **16**: 123–133.

Dutta, S. 1958. Origin and history of *Citrus* fruits of Assam. *Indian J. Hort.* **15**: 146-153.

Dymock, W., Warden, C.J.H. and Hooker, D. 1890. Pharmacographia Indica. Periodical experts. Delhi, pp. 268-277.

Edriss, M. H. and Burger, D. W., 1984: *In vitro* propagation of troyer citrange from epicotyl segments. *Scientia Hort.* **23**: 159–162.

Engler, A. 1931. Rutaceae. In: Engler A. and Prantl, K. (eds.) *Die naturlichen Pflanzenfamilien,* 2nd ed. Engelmann, Leipzig, pp.19, 187-359.

Fang, D.Q. Federici, C.T. and Roose, M.L. 1997. *Genome,* **40**: 841-850.

FAO, 2001. http.//apps.fao.org/lim500/nph-wrap.pl.

FAO, 2008. *Food and Agriculture Organization. FAOSTAT.* Statistical database http://faostat.fao.org

Febres V. J.; Lee, R. F. and Moore G. A. 2008. Transgenic Resistance to *Citrus Tristeza Virus* in Grapefruit. *Plant Cell Reports,* **27**: 93-104

Febres V. J.; Niblett C. L.; Lee R. F. and Moore G. A. 2003. Characterization of Grapefruit Plants (*Citrus paradisi* Macf.) Transformed with *Citrus Tristeza Closterovirus* Genes. *Plant Cell Reports,* **21**: 421-428

Federici, C.T. Fang, D. Q., Scora, R. W. and Roose, M.L., 1998. Phylogenetic relationship within the genus *Citrus*(Rutaceae) and related genera as revealed by RFLP and RAPD analysis. *Theory Appl. Genet.* **96**:812-822.

Fleming G. H.; Olivares-Fuster O.; Del-Bosco S. F. and Grosser J. W. 2000. An Alternative Method for the Genetic Transformation of Sweet Orange. *In Vitro Cellular and Developmental Biology Plant,* **36**:450-455

Ghorbel R.; Domínguez A.; Navarro L. and Peña L. 2000. High Efficiency Genetic Transformation of Sour Orange (*Citrus aurantium*) and Production of Transgenic Trees Containing the Coat Protein Gene of *Citrus Tristeza Virus. Tree Physiology,*Vol.**20**: 1183-1189.

Ghorbel, R., Juarez, J., Navarro, L., and Pena, L., 1999: Green fluorescent protein as a screenable marker to increase the efficiency of generating transgenic woody fruit plants. *Theor. Appl. Genet.* **99**: 350–358.

Ghosh, S. P. 1977. *Citrus* industry of north east India. *Punjab Hort J.,* 13-21.

Govind, S. and Yadav, D.S. 1999. Genetic Resources of *Citrus* in North Eastern Hill region of India. In: Singh S, Ghosh SP (eds), Hi-Tech *Citrus* management. ISC, ICAR, NRCC. pp. 38-46.

Grinblat, V., 1972. Differentiation of *Citrus* stem *in vitro. J. Amer. Soc. Hort. Sci.* **97**:599–603.

Grosser, J. W. and Gmitter, G.G.Jr. 1990. *Plant Breeding Rev.* **8**: 339-397.

Gutierrez-E M. A.; Luth D. and Moore G. A. 1997. Factors Affecting *Agrobacterium*-Mediated Transformation in *Citrus* and Production of Sour Orange (*Citrus aurantium* L.) Plants Expressing the Coat Protein Gene of *Citrus Tristeza Virus. Plant Cell Reports,* **16**: 745-753.

Hazarika, B. N., Nagaraju, V. And Parthasarathy, V. A. 1995. *Abstr. Nat. Symp., Role of Plant Biotech. Improving Agric. Challenges and Opp.,* 23-25 March, Jaipur, India.

Herrero, R., Asins, M.J., Carbonell, E.A. and Navarro, L., 1996. Genetic diversity in the orange subfamily Aurantioideae I. Intraspecies and Intragenus genetic variability. *Theory Appl. Genet.* **92**: 599–609.

Hidaka T. and Omura M. 1993. *Agrobacterium*-Mediated Transformation and Regeneration of *Citrus Spp.* From Suspension Cells. *Journal of the Japanese Society for Horticultural Science,* **62**: 371-376.

Hirai, M. and Kajiura, I. 1987. I Jap. J. Breed., **37**: 337-388.

Hirai, M., Kozaki, I. and Kajiura, I. 1986. Isozyme analysis and phylogenic relationship of *Citrus. Japan Journal of Breeding.* **36**: 377-389.

Hodgson, R.W. 1967. Horticultural varieties of *Citrus.* In: Reuther, W., Batchelor, L. D. and Webber, H.D. (eds.) *The Citrus Industry,* 2nd edn. University of California Press, California, pp. 431-591.

Hooker, J. D. 1872. Rutaceae. *Flora of British India.* Reeve and Co. London. pp. 484-517

Hore, D.K., Govind, S. and Singh, I.P. 1997. Collecting of *Citrus* germplasm from Mizoram and Tripura hills of India. *Plant Genetic Resources Newsletter* (IPGRI/FAO) **110**: 57-59.

http://cdn.intechopen.com/pdfs/18815/InTech-*Citrus*_transformation_challenges_and_ prospects.pdf

http://hortportal.org

http://www.britannica.com/EBchecked/topic/341295/lime.

http://www.fruitipedia.com

http://www.tradeswindfruits.com

http://www. wikipedia.org

http://www. hort.purdue.edu

http://www. users.kymp.net/citruspages

Jajoo, A. 2010. *In vitro* Propagation of *Citrus limonia* Osbeck Through Nucellar Embryo Culture. *Curr. Res. J. Biol. Sci.* **2**: 6-8.

Jena, S.N., Kumar, S. and Nair, N. K. 2009. Molecular pylogeny in Indian *Citrus* L. (Rutaceae) inferred through PCR-RFLP and trnL-trnF sequence data on chloroplast DNA. *Scientia Horticulturae.* **119**: 403-416.

Joshi, S. P., P. K. Ranjekar and V. S. Gupta 1999. Molecular markers in plant genome analysis. *Curr. Sci.* **77**:230-40.

Jung, Y.H., Kwon, H.M., Kang, S.H., Kang, J.H. and Kim, S.C., 2005. Investigation of phylogenetic relationships within the genus *Citrus* (Rutaceae) and related species in Korea using plastid trnL/trnF sequences. *Sci. Hort.* **104**:179–188.

Kaneyoshi, J., Kobayashi, S., Nakamura, Y., Shigemoto, N. and Doi, Y., 1994: A simple and efficient gene transfer system to trifoliate orange (*Ponicirus trifoliata* Raf.). *Plant Cell Rep.* **13**: 541–545.

Kaul, G.L. 1981. Development of Horticulture in North Eastern region – A case study of North Cachar district, NEC Secretariat, Shillong, Meghalaya.

Kobayashi S. and Uchimaya H. 1989. Expression and Integration of a Foreign Gene in Orange (*Citrus sinensis* Osb.) Protoplasts by Direct DNA Transfer. *Japanese Journal of Genetics,* **64**: 91-97

Kobayashi, S., Ikeda, I. and Uchimiya, H., 1985. Conditions for high frequency embryogenesis from orange (*Citrus sinensis*) protoplasts. *Plant Cell Tiss. Org. Cult.* **4**:249–259.

Krueger, R.R. and Novarro, L. 2007. *Citrus* germplasm resources. In: *Citrus*, genetics, Breeding and Biotechnology, Khan, I (ed.). CAB International, UK.

Liang, G., Xiong, G., Guo, Q., He, Q. and Li, X., 2007. AFLP analysis and the taxonomy of *Citrus*. *Acta Hort. (ISHS),***760**: 137–142.

Lushington A. W.1910. The genus *Citrus. Indian Forester.* **36**: 323-353.

Luth D. and Moore G. A. 1999. Transgenic Grapefruit Plants Obtained by *Agrobacterium tumefaciens*-Mediated Transformation. *Plant Cell, Tissue and Organ Culture,* **57**: 219-222.

Mabberley, D. J. 1998. Australian Citrerae with notes on other Aurantoideae (Rutaceae) *Telopea* **7**: 333-344.

Mabberley, D. J., 2004. *Citrus* (Rutaceae): A review of recent advances in etymology, systematics and medical applications. *Blumea,* **49**: 481-498.

Malik, S.K., Chaudhury, R., Dhariwal, O.P. and Kalia, R.K. 2006. Collection and characterization of *Citrus indica* Tanaka and *C. macroptera* Montr. : wild endangered species of north-eastern India. *Genet Resources Crop Evol.,* **53**: 1485-1493.

Marak, C. and Laskar, M. A. 2010. Analysis of phonetic relationship between *Citrus indica* Tanaka and a few commercially important *Citrus* species by ISSR marker. *Scientia Horticulturae*, **124**: 345-348.

Miah, M. N.; Islam Sahina and Hadiuzzaman Syed 2002. Regeneration of Plantlets Through Somatic Embryogenesis from Nucellus Tissue of *Citrus macroptera* Mont. var. *anammensis* ('Sat Kara'). *Plant Tissue Cult.* **12** : 167-172.

Molinari H.; Bespalhok J. C.; Kobayashi A. K.; Pereira L. F. P. and Vieira L. G. E. 2004. Agrobacterium Tumefaciens-Mediated Transformation of Swingle Citrumelo (*Citrus paradisi* Macf.× *Poncirus trifoliata* L. Raf.) Using Thin Epicotyl Sections. *Scientia Horticulturae,* **99**: 379-385

Moore G. A.; Jacono C. C.; Neidigh J. L.; Lawrence S. D. and Cline K. 1992. *Agrobacterium*-Mediated Transformation of *Citrus* Stem Segments and Regeneration of Transgenic Plants. *Plant Cell Reports,* **11**: 238-242

Moreno P.; Ambros S.; Albiach-Marti M. R.; Guerri J. and Pena L. 2008. *Citrus Tristeza Virus*:A Pathogen That Changed the Course of the *Citrus* Industry. *Molecular Plant Pathology,* **9**: 251-268.

Morton, C.M., Grant, M. and Blackman, S., 2003. Phylogenetic relationships of Auratioideae inferred from chloroplast DNA sequence data. *Am. J. Bot.* **90**:1463–1469.

Nair, K.N. and Nayar, M. P. 1997. Rutaceae. In: Hajra PK, Nair VJ, Daniel P (eds) Flora of India, vol iv, Botanical survey of India, Calcutta, pp 259-408.

Naqvi, N. S. 2001. Morphology and taxonomy of *Citrus*. In: *Citrus* (Singh, S. and Naqvi, S.A.M.H.) International Book Distributing Co. Lucknow, India. Pp.69-80.

Nicolosi, E. 2007. Origin and taxonomy. In: *Citrus* Genetics, breeding and biotechnology (ed. I. Khan). CAB International.UK.

Nicolosi, E; Deng, Z. N.; Gentile, A; La malfa, S; Ciontinella, G. and Tribulato, E. 2000. *Citrus* phylogeny and genetic origin of important species as investigated by molecular markers. *Theor. Appl. Genet.* **100**: 1155-1166.

Pang, X.M., Hu, C.G. and Deng, X. X., 2007. Phylogenetic relationships within *Citrus* and its related genera as inferred from AFLP markers. *Genet. Resour. Crop Evol.* **54**: 429–436.

Parthasarthy, V.A. and Nagaraju, V. 1996. *Ind. J. Hort.* **53**: 171-174.

Peña L.; Cervera M.; Fagoaga C.; Romero J.; Juárez J.; Pina J. A. and Navarro L. 2007. *Citrus. Biotechnology in Agriculture and Forrestry.* **60**:35-50.

Peña L.; Perez R. M.; Cervera M.; Juarez J. A. and Navarro L. 2004. Early Events in *Agrobacterium*-Mediated Genetic Transformation of *Citrus* Explants. *Annals of Botany,* **94**: 67-74

Pena, L., Cervera, M., Juarez, J., Navarro, A., Pina, J. A., Duran-Vila, N. and Navarro, L., 1995a: *Agrobacterium*-mediated transformation of sweet orange and regeneration of transgenic plants. *Plant Cell Rep.* **14**: 616–619.

Pena, L., Cervera, M., Juarez, J., Ortega, C., Pina, J. A., Duran-Vila, N. and Navarro, L., 1995b: High efficiency *Agrobacterium-mediated* transformation and regeneration of *Citrus. Plant Sci.* **104**: 183–191.

Pena, L. and Navarro, L., 1999. Transgenic *Citrus. Biotechnology in agriculture and forestry.* **44**: 39–54.

Pena, L., Cervera, M., Juarez, J., Navarro, A., Pina, J. A. and Navarro, L., 1997: Genetic transformation of lime (*Citrus aurantifolia* Swing.): factors affecting transformation and regeneration. *Plant Cell Rep.* **16**: 731–737.

Pérez-Tornero, O.; Tallon, C.I. and Porras, I. 2010. An efficient protocol for micropropagation of lemon (*Citrus limon*) from mature nodal segments. *Plant Cell Tiss. Org. Cult.* **100**: 263-271.

Randall, P. Niedz., 1994: Growth of embryogenic sweet orange callus on media varying in the ratio of nitrate to ammonium nitrogen. *Plant Cell Tiss. Org. Cult.* **39**: 1–5.

Reyes C. A.; De Francesco A.; Pena E. J.; Costa N.; Plata M. I.; Sendin L.; Castagnaro A. P. and Garcia M. L. 2011. Resistance to *Citrus Psorosis Virus* in Transgenic Sweet Orange Plants Is Triggered by Coat Protein-Rna Silencing. *Journal of Biotechnology,* **151**: 151-158

Roose, M. L. 1988. *Proc. VI Intl. Citrus Symp.*,**1**: 155-165.

Russo, M.P., Rosa, M.L., Astuto, A. and Reupero, G.R. 1998. *Adv. Horti. Sci.* **12**: 85-88.

Saunt, J. 1990. *Citrus varieties of the world.* Sinclair International Ltd. Pp 126.

Sawazaki, H. E., Muller, G.W. and Sodek, L. 1997. *Revista Brasileira de Biologia.* **57**: 337-342.

Scora, R.W. 1975. On the history and origin of *Citrus*. Bull. Torrey Bot. Club **102**: 369–375.

Shahsavar, A.R., Izadpanah, K., Tafazoli, E. and Tabatabaei, B.E.S., 2007. Characterization of *Citrus* germplasm including unknown variants by inter-simple sequence repeat (ISSR) markers. *Sci. Hort.* **112**:310–314.

Sim, G. E., Goh, C. J. and Loh, C. S. 1989. *Plant Sci.* **59**: 203-210.

Singh I.P. and Singh S. 2006. Exploration, collection and characterisation of *Citrus* germplasm- A review. *Agric. Rev.* 27: 79-90.

Singh, B. 1981. Establishment of First Gene sanctuary for *Citrus* in Garo hills. Concept Pub Co, New Delhi, pp. 182.

Singh, D. and Singh, R. 1967. Taxonomic status of Indian *Citrus* rootstocks. *Proc. Int. Symp. Subtrop. Hort.* 424-428.

Singh, D. and Singh, R. 1968. Taxonomy of Indian rough lemon. *Indian J. Agric. Sci.* **38**:25-37.

Singh, I.P. and Singh, A. 2001. *Citrus* gemplasm and its utility. In: *Citrus* (Singh, S. and Naqvi, S.A.M.H.) International Book Distributing Co.Lucknow, India. Pp.45-66.

Singh, I.P., Singh, S., Singh, K. and Srivastava R. 2001. Exploration and collection of *Citrus* germplasm from NEH region (Meghalaya) of India. *Indian J Plant Genet Resour,* **14**: 70-73.

Singh, R. 1967. A key of the *Citrus* fruits. *Indian J. Hort.,* **4**: 71-83.

Singh, R. and Nath N. 1969. Practical approach to the classification of *Citrus*. In : Proc. Intl *Citrus* Symp, Champman HD (ed), 1: 435-440.

Singh, R., 1977. Indigenous *Citrus* germplasm in India. In: Fruit Breeding in India. (Nijjar, G.S. ed.) Oxford IBH Publ. Co., New Delhi, pp. 39-49.

Singh, S. 2001. *Citrus* Industry in India. In: *Citrus* (Singh, S. and Naqvi, S.A.M.H.) International Book Distributing Co., Lucknow, India. Pp.3-41.

Singh, S. and Rajam, M. V. 2009. *Citrus* biotechnology: Achievements, limitations and future directions. *Physiol. Mol. Biol. Plants.* **15**: 3-22.

Strauss, S. H., J. Bonsquet, V. D. Hipkins and Hong, Y.P., 1992. Biochemical and molecular genetic markers in biosystematic studies of forest trees. *New Forests* **6**:125-158.

Swingle W.T. 1943. The botany of *Citrus* and its relatives of the orange subfamily Aurantioidae of the family Rutaceae. In: Webber H.J. and Batcheler L.D. (eds) The *Citrus* Industry, vol 1, University of California, Berkley, pp. 129-474.

Swingle W.T. and Reece P.C. 1967. The botany of *Citrus* and its wild relatives. In: Reuther W, Batchelor L.D. and Webber H.J. (eds), The *Citrus* industry. University of California Press, Berkley, pp 190-340.

Tanaka, T. 1954. Species problem in *Citrus* (Revisio aurantiacearum, IX). *Japan Society Prom. Sci.,* Veno, Tokyo

Tanaka, T. 1958. The origin and dispersal of *Citrus* fruits having their centre of origin in India. *Indian J Hort.,* **15**: 101-115.

Tanaka, T. 1961. Contribution to the knowledge of *Citrus* classification. *Reports Citrologia.* 107-114.

Tanaka, T. 1977. Fundamental discussion of *Citrus* classification. *Stud. Citrol.* **14**:1-6.

Te-Chato, S. and Nudoung, S., 1998: Tissue culture of *Citrus reticulata* Blanco cv. Shogun and gene transformation by *Agrobacteria*. Proc. Abstracts IMT-GT UNITE Conf, Songkxhla.

Tolkowsky, S. 1938. Hespirides: A History of the Culture and use of *Citrus* Fruits. John Bale, Sons and Curnow Ltd., London.

Tusa, N., Grosser, J.W. and Gmitter, Jr. G.G. 1990. *J. Amer. Soc. Horti. Sci.,* **115**: 1043-1046.

Vardi A.; Bleichman S. and Aviv D. 1990. Genetic Transformation of *Citrus* Protoplasts and Regeneration of Transgenic Plants. *Plant Science,* **69**: 199-206.

Vavilov, N.I. 1950. Phytogeographic basis of Plant Breeding. *Chron. Bot.* Pp. 13-54.

Verma, A. N. and Ghosh S.P. 1979. *Citrus* germplasm collection of NE hills and their evaluation. *Indian J Hort.,* **36**: 2-4.

Verma, S. K. *et al.,* 1999. In: Hi tech *Citrus* management (Singh, S. and Ghosh, S.P. eds.) ISC, NRCC, Nagpur, pp. 54-620.

Webber, H.J. 1967. History and development of the *Citrus* industry. In: Reuther, W., Batchelor, L. D. and Webber, H.J. (eds.) *The Citrus Industry,* 2nd edn. University of California Press, California, pp. 1-39.

Weising, K., Nybom, H., Wolff, K. and Kahl, G., 2005. DNA Fingerprinting in Plants: Principles, Methods and Applications, 2nd ed. Taylor and Francis group, Boca Raton, FL.

Yao J. L.; Wu J. H.; Gleave A. P. and Morris B. A. M. 1996. Transformation of *Citrus* Embryogenic Cells Using Particle Bombardment and Production of Transgenic Embryos. *Plant Science,* **113**: 175-183

Yingzhi, L., Yunjiang, C., Nengguo, T. and Xiuxin, D., 2007. Phylogenetic analysis of mandarin land races, wild mandarins, and related species in China using nuclear LEAFY second intron and plastid trnL-trnF sequence. *J. Am. Soc. Hort. Sci.* **132**: 796–806.

Yu C. H.; Huang S.; Chen C. X.; Deng Z. N.; Ling P. and Gmitter F. G. 2002. Factors Affecting *Agrobacterium*-Mediated Transformation and Regeneration of Sweet Orange and Citrange. *Plant Cell Tissue and Organ Culture,* **71**: 147-155.

2013, Biodiversity in Horticultural Crops Vol. 4   *Pages 335–353*
Editor: Professor K.V. Peter
Published by: DAYA PUBLISHING HOUSE, NEW DELHI

Chapter 18

Okra in Thailand

Amnuai Adthalungrong
Senior Agricultural Researcher,
Horticulture Research Institute, Department of Agriculture, Thailand
E-mail: amnuai.th@gmail.com

Okra is one of the most important indigenous vegetables in Thailand. Besides commercial scale production for export or domestic sale, okra is generally grown at backyard for household consumption as well. Nevertheless, a proportion of a domestic consumption of okra is considered insignificant because it is taken as a side dish. In Thailand, okra is commonly known as Kra-Jeab-Kheaw. However, other names are also used in certain local parts such as Gra-Tad (Samudsakron and Samudprakran province), Ma-Keao-Mon (central region), Ma-Keao-Muen (northern) and Toa-Lea (northeastern).

Okra pod contains high nutrition values especially vitamins A and C, folate, and potassium. The dietary fibres are also high but it is low in fat and calories (Ebadi, 2002). Protein and oil are found in high contents in seed kernel while the crude fibre is detected in seed coat (Udayasekhara Rao, 1985). There are 20 per cent protein and 14 per cent oil in dry matter of seed (Charrier, 1984; Kittipakron *et al.*, 2001). Gum and pectin are also high. These two polymers are the core structure of the mucilage (Suttipolpiboon, ny), which is found in fresh as well as dried pod wall, but not in seed (Kumar *et al.*, 2010).

Apart from being consumed as a vegetable, okra is currently used as a medicinal herb. A number of reports disclosed the roles of okra on health protection and prevention from many diseases such as diabetes mellitus (Amin, 2011), sore throat, and urinary infection (Li, 2008). In addition, okra has been considered as a health promoting vegetable since it contains abundant amounts of antioxidant, hypolipidemic, antiulcer, and antimicrobial agents, etc. (Lim, 2012).

Thailand started to export okra in 1981. However, the export amounts had not been considered an important economic scale until ten years later. In 1991, Thailand exported 2,582 metric tons of okra, which were ten times higher than the export quantity of the previous year. Japan has been the biggest market for okra from Thailand, since it shares over 95 per cent of total export quantity. Okra is shipped as fresh or chilled and frozen vegetable.

The popularity of okra consumption in Japan started after 1960 when the cultivars of okra were satisfactorily developed until pod quality met the consumer requirements. However, the researches on okra cultivar improvement are undertaken continually till specific pod characteristics are obtained *i.e.* five angles and short tip, thick flesh, green to dark green color and line-up seeds in young pod (Figure 18.1a).

Japan imports okra for one-third of the country's total consumption quantity. The major sources are from Thailand and Philippine which export okra mostly from September to May which is the cold season in Japan because Japanese growers are unable to produce enough okra under the cold weather. The major production areas in Japan are Kagoshima, Kochi, Okinawa and Gunma.

Japanese market is punctilious to the pod quality in particular. The qualities of pod for Japanese market are fresh young pod, 7-12 cm in length, erect pod with five angles, green to dark green color all over the pod without defects from insects or diseases. The satisfied package is 100 gram or ten pods per pack (Figure 18.1b), except some local markets such as Osaka which prefers 50 gram per pack.

a: Japanese Cultivar   b: Okra Packet for Export

Figure 18.1: Pod Characteristics of Japanese Okra and Okra Packet for Export

Export and Production of Okra in Thailand

The export quantities of okra in Thailand are shown in Figure 18.2. The tendency continuously increased during 1991-1994 but decreased during 1995-2000 and started to fluctuate after the year 2000. The lowest export quantity, only 1,787.92 metric tons, was recorded in 2009. This was a consequence from lower quality products which did not meet the Japanese market standard criteria and the spread of yellow vein mosaic disease. Excessive residue levels (over maximum residue limits, MRLs) in the products due to an overuse of chemicals were the major cause of rejection as well.

The important production areas of okra are in central and western regions of Thailand. Ratchaburi, Nakhon Pathom, Kanchanaburi, Nonthaburi, Singburi and Angthong provinces are among the leading provinces for okra production. In general, the contract farming system has been applied between growers and export companies. The system normally is established by the government extension agencies and the network is extended by export companies. In the terms and conditions of most

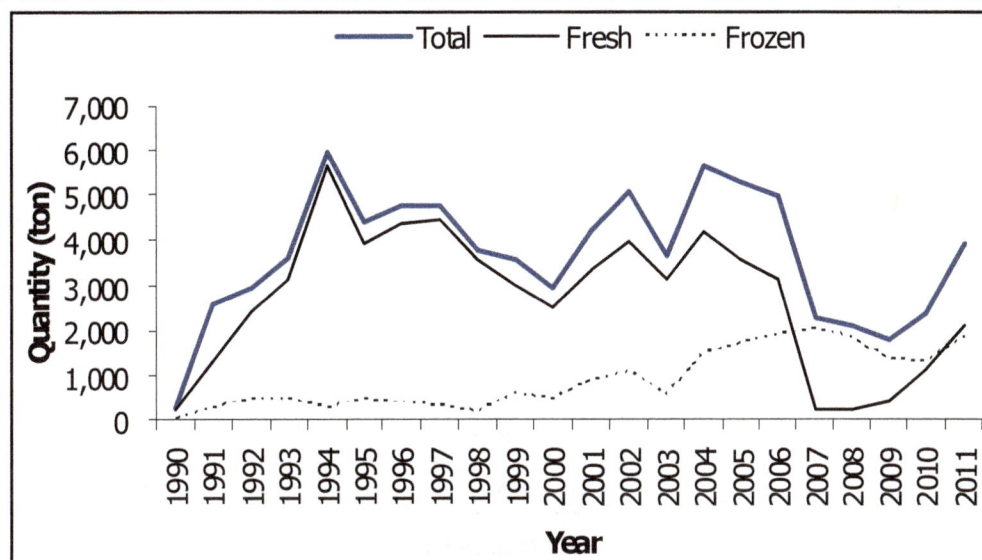

Figure 18.2: Okra Export during 1990-2011
Source: Department of Customs

contracts, the companies have to support necessary agricultural inputs including seeds, pesticides, fertilizer, etc. and buy fresh products from the growers.

Okra growers for export mostly possess only 0.16-0.32 hectare of production area due to the limitation of labor intensive farming processes. Usually, the laborers are limited only to family members. However, most growers have good skills and experiences as well as are ready to learn the new technology and to take the advices from export company or government agency. In some areas, farmers have established the group to assist one another with transferring good practices for farming, harvesting the products and negotiating with the company. The group leaders sometimes collect and sell the seeds to their members.

Okra Cultivars

Previously, the cultivars of okra grown for export to Japan belonged to Japanese cultivars like Star Light, Early Five or Green Star. For decades, a number of agencies put a lot of efforts to improve okra cultivars for export. For instance, Kasetsart University released OK 1, 2, 4 and 5, Phichit Horticulture Research Center released PC 014, PC 003, PC 03 and PC 01 (Suttipolpiboon, 2012) and The Royal Hybridization Project released Morakot F1 (Lavapaurya *et al.,* 1997). Nevertheless, the cultivars used were still strict to Japanese cultivars. Consequently, Thailand needed to import at least 2 tons of okra seeds from Japan each year during 1991-1997. This costs about 0.13-0.10 millions US$ annually (The calculation was made based on 30 Baht/US$).

When the yellow vein mosaic disease (YVMD) spread to the production areas, it has become the serious problem for okra production in Thailand since 1998. Therefore, the YVMD resistant cultivars are needed. Unfortunately, the Japanese cultivars and the improved cultivars mentioned earlier are susceptible to the disease. Hence, the resistant cultivars must be imported from India. In Figure 18.3, the overall data suggested that over 95 per cent of total quantities and values of imported okra seeds

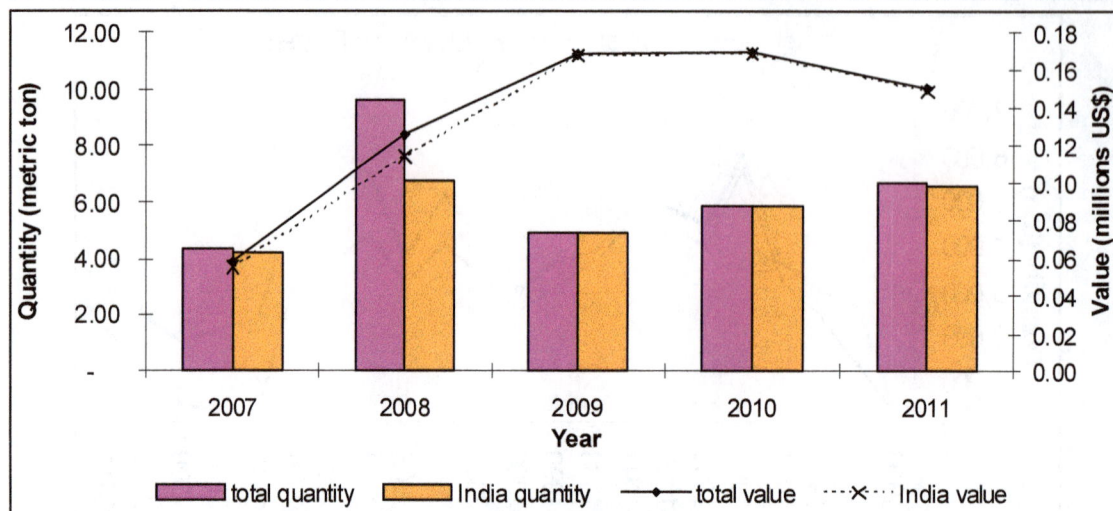

Figure 18.3: Statistic of Imported Okra Seed Quantities and Values During 2007-2011
Sources: **Department of Agriculture**

were from India. During 2007-2011, the average quantities of okra seeds imported from India were 4,738.07 tons annually in Thailand, corresponding to approximately US$ 0.11 millions per year. Nowadays, the Thai government agencies and certain companies have launched the projects of cultivar improvement in okra for YVMD resistant capability. The developed cultivars have been released to farmers for trials.

Botany and Horticulture

Okra is classified as a member of the genus *Abelmoschus*, family *Malvaceae*. In 1990, the International Okra Workshop held at National Bureau of Plant Genetic Resources reclassified the genus *Abelmoschus* into nine species (IBPGR, 1991) as given in Table 18.1.

Table 18.1: Classification Adopted by IBPGR, 1991

Sl.No.	Species
1.	*A. moschatus* Medikus- subsp. *moschatus* var *moschatus*, - subsp. *moschatus* var *betulifolius* (Mast) Hochr, - subsp *biakensis* (Hochr) Borss, - subsp *tuberosus* (Span) Borss
2.	*A. manihot* (L) Medikus
3.	*A. tetraphyllus* (Roxb.ex Hornem.) R. Graham var. *Tetraphyllus* var. pungens (Roxb ex Hornem)
4.	*A. esculentus* (L) Moench
5.	*A. tuberculatus* Pal & Singh
6.	*A. ficulneus* (L) W & A.ex. Wight
7.	*A. crinitus* Wall
8.	*A. angulosus* Wall ex. W, & A
9.	*A. caillei* (A. Chev) Stevels

Only three species namely *A. esculentus* (L) Moench, *A. manihot* (L) Medikus and *A. moschatus* Medikus are known as cultivated species (IBPGR, 1991). *A. esculentus* (L) Moench is cultivated in most tropical and subtropical areas including Africa, America, Mediterranean (Small, 2012), Sri Lanka, India, and Southeast Asia. It is commonly named as okra, gombo, or lady's finger but different names are more familiar in certain areas such as bhindi (India) and Kra Jeab Khiew (Thailand). The origin of *A. esculentus* remains unclear but most people tend to believe that *A. esculentus* is an amphiploid of *A. tuberculatus* (2n=58) and *A. culneus* (2n=72) (Bishtan and Bhat, 2007). Therefore, the origins of okra are probably from North India (a habitat of *A. tuberculatus*) and East Africa: north Egypt or Ethiopia (habitats of *A. culneus*) (Hamon and van Sloten, 1995; Lim, 2012). Chromosome number (2n) of okra is found in the range of 66-144. The most frequently observed chromosome number is 130 (DOB, 2001) or between 108 and 144 (Charrler 1984). For the *A. manihot* (L) Medikus, it is cultivated as leafy vegetable in Far East and Papua New Guinea. Finally, *A. moschatus* Medikus is grown in Africa, Asia and America for leaf vegetables as well. Besides, the seed extract can be used as perfume or condiment (Charrler, 1984).

In Thailand, Chomchalow *et al.* (1987) found 78 accessions of okra which were collected from western, northeastern and northern parts. These included *A. moschatus* Medikus (30 accessions), *A. manihot* (L) Medikus (35 accessions), *A. esculentus* (L) Moench (7 accessions) and interspecific hybrids (6 accessions). The six interspecific hybrids were *A. esculentus* (L) Moench x *A. manihot* (L) Medikus (1 accession), *A. manihot* (L) Medikus x *A. moschatus* Medikus (2 accessions) and the last three were probably *A. manihot* (L) Medikus x *A. esculentus* (L) Moench or *A. moschatus* Medikus (Chomchalow, 1987). Moreover, *A. crinitus* Wall and *A hainanensis* Hu were mentioned by Chomchalow *et al.* (1987) in northern Thailand.

Morphology

Okra is cultivated as an annual crop by seed. The roots of okra are the taproot system. Primary root grows vertically in soil for about 20-60 cm in depth and allows lateral roots and root hairs to branch out from primary root. The stem of okra is robust, erect, and green color. Sometimes red and purple patches might be observed and most slightly pubescent. Stem becomes woody at maturity and usually branches out from the base (2-5 twigs/plant). Plant height varies from 0.5 to 4.0 m (DOB, 2001) but often reaches 60 to 180 cm, depending on the cultivar and cultivated environment. The growth of okra was classified into indeterminate type which means growth and reproductive stages of okra plant can occur synchronously at certain plant ages. The pod growth pattern is considered as a single sigmoid curve.

Simple leaf is palmatifid to palmatisect (Charrier, 1984). The arrangement of leaf is alternate and there is only one leaf per node. Normally there are 3-5 primary veins which arise from the petiole. The base is cordate, margins toothed (serrate), and apex acute. Leaf surface is hispid since it is covered with long stiff hairs or bristles at both sides. The ventral side of lamina has deeper color than the dorsal side of the leaf.

Okra flower is axillary and solitary flower with yellow color. The diameter is 4-7 cm when completely bloom. The colors of young pod can be light to deep green, white or burgundy. There are two types of pod, unridged and ridged (normally 5-8 ridges). The mature pod is 5-25 cm in length, 1.5-2.5 cm or larger in diameter.

Seed is relatively large and heavy: 50-60 g/100 seeds, about 5 mm in diameter, round to reniform shape. Seed surface is rather glabrous or sometimes hirsute or scattered simple short hairs on the surface, light green to gray color.

Floral Biology

Okra is classified as a short-day plant (Charrler, 1984; Rashid and Singh, 2000; George, 2009) but some cultivars fall into the day-neutral type (Rithichai *et al.,* 2004). The short-day group requires the period of day light which is shorter than 12.15 hrs per day to blossom while the day-neutral group needs a critical day length of 12.30 hrs or longer. However, it is noted that there is an interaction effect between the period of light exposure and daytime temperature on flowering (Oyolu, 1977). The number of short day directly affects the quantity of flower and total number of seeds per plant. However, the effects of those factors vary depending on cultivars. Nwoke (1980) reported that the number of flower buds increased following the increase of short day numbers from 5 to 20. Likewise, the fruit set was better and seed developed greater in the longer period of short days.

Okra flower usually begins to blossom at 34 to 67 days after sowing. The first one appears at the 3rd to 19th nodes but generally at the 5th to 7th nodes (Hamon and van Sloten 1989). Flower takes about 22 days to fully develop (Bisht and Bhat, 2007) or takes about 13-16 days to develop from 0.8-1.0 cm in flower size to bloom, depending on the cultivars (Adthalungrong, 2009).

Okra flower comprises 8-10 linear to lanceolate epicalyx segments that will fall off in mature pod. Calyx appears light green or green which longitudinally breaks up from the tip to base when it starts blooming. There are five petals which have yellow color with dark red color at both sides of the base or inside each petal (Figure 18.4). The fully blooming size is about 4-7 cm in diameter (Adthalungrong, 2009).

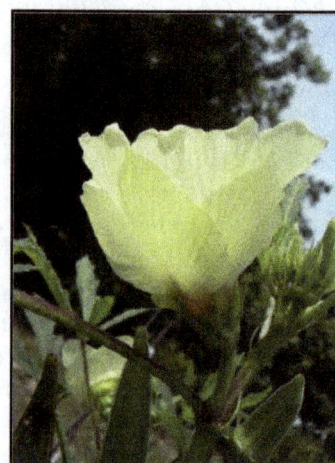

| The Break Up of Calyx | Dark Red Color at Both Sides of the Base | Dark Red Color Inside the Base |

Figure 18.4: Okra Flower Structures

Flower of okra is complete and perfect type (male and female reproductive structures exist in the same flower). The filaments combine into the staminal column where anthers are arranged in concentric superimposed circles (5-6 anthers per circle). Inside the staminal column, there is a female reproductive structure called the style. The ovary type is superior ovary or perigynous. There are 5-9 lopes of dark red stigmas, about 5-7 mm in diameter (Figure 18.5).

The anther resembles bivalvia (two shells attached) and is adjacent to the staminal column. On the flowering day, the anther may start to dehisce around 6 AM or earlier although the flower still remains unopened. Interestingly, the anther dehisces with an uncertain pattern *i.e.* It may start to split up from one or both side and sometimes from the central. The anther contains about 100 pollen grains, about 0.10-0.15 mm in diameter. Pollen grain appears spherical, porate and echinate, yellow and is covered by sticky substance (Figure 18.5).

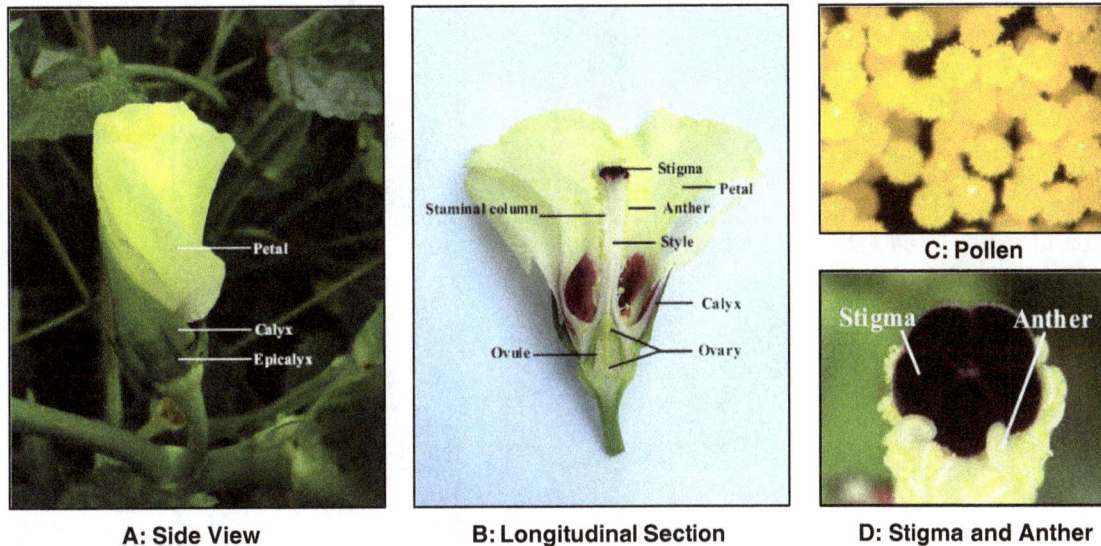

A: Side View   **B: Longitudinal Section**   **D: Stigma and Anther**

Figure 18.5: Anatomy of Okra Flower

The pistil is composed of 5-9 carpels. Five carpels are most common. The carpels combine together and the number of carpels is equal to the number of locules. The base of each carpel joins together and develops to the ovary. The upper parts of the ovary are located by the style (2-3 cm in length) and stigma. Style is a female structure which links between ovary and stigma. The appearance of the stigma is like velvet. It comprises short hairy structure with dark red color. Those hairs are extended and lose the fertilization capability after receptive period. The ovules are orderly lined up at the axile placentation (Figure 18.5).

For some reasons, the most anthesis is observed between 8 and 9 AM. Then, the flower continually opens up until 3-4 PM before wilting (Figure 18.6). Pollination can be observed before flower opening. The incident takes place when the anthers that are arranged surrounding the stigma dehisce and release pollen grains into the stigma. Abiotic pollinators such as win, water and rain are usually uninvolved in okra pollination due to the limitation of big pollen size and weight. The stigma receptivity is perfect on the day of flowering. But receptive capability becomes less efficient before and after flowering (DOB, 2001). Adthalungrong (2009) reported that the appropriate pollination time was during 7.00-9.00 AM since high numbers of pod and satisfied seed yields were achieved when pollination was induced at that time. The high fertility rate was observed at 7 AM (Hamon *et al.,* 1990; Adthalungrong, 2009).

Okra is classified as a self pollinated plant or autogamy (Charrier, 1984; Hamon and Koechlin a, b, 1991; Adthalungrong, 2009; George, 2009; McCormack, 2010). However, there is a chance of cross

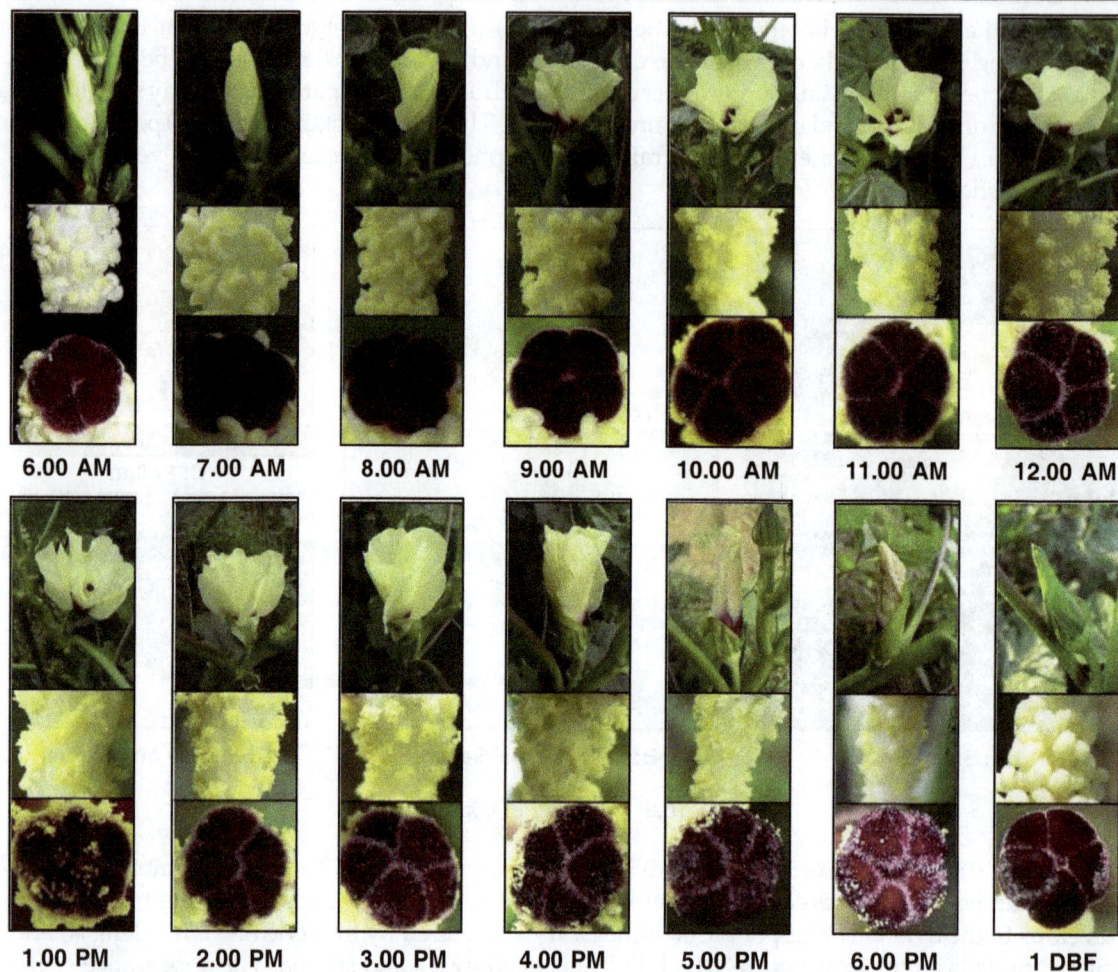

Figure 18.6: Development of Okra Flower from 6.00 AM to 6.00 PM and 1 DBF

pollination by insect as well (McLaurin, 2000; Sukprakarn *et al.,* 2005; George, 2009). Flower structures, particularly the distance between anther and stigma, insect types and movement habits have major roles in okra cross pollination (Hamon and Koechlin b, 1991). Honey bees, *Apis mellifera* and *A. cerana*, are known as possible causes (Tanda, 1985) and around 4-19 per cent out crossing are arisen by those bees (Shoba and Mariappan, 2007).

Pod and Seed Development

After fertilization, pod grows up rapidly in the first 1-13 days after flowering (DAF) and the growth rate will decrease after that. Devadas *et al.* (1998) reported the weight of pod which increased after fertilization and reach the maximum weight at 9 days. Color of young pod is green to dark green or depends on the cultivar. Green pod is pallid and turns to yellow green during 21-27 DAF (Figure 18.7).

Figure 18.7: Development of Okra Pod at 3-45 Days After Flowering (DAF)

The split of pod is first noticed at the tip and the pod is easy to tear off at 27 DAF. After that pod color turns brown and dry at 29-45 DAF. The immoderate split of pod is observed at 37-45 DAF, depending on the cultivars. This causes seed loss or damage (Figure 18.7).

Young seed of okra is white. It grows and develops after fertilization. The size of seed is almost unchanged 13 DAF. During this time, seed color is changed from white to yellow in 21-25 days. Subsequently, it turns deep yellow, brown and black within another 2-5 days (Figure 18.8). However, the phase of seed development is not synchronous which can be noticed by the mixed seed color in the same pod at 25-27 DAF. When the stationary phase of seed growth is reached, it is the phase for seed dry weight accumulation (Devadas *et al.,* 1998; Anitha, 2001).

After seed color is completely black, seed size becomes smaller due to the loss of moisture content and seed will be completely dry eventually. The first harvesting time for seed production is suggested at 35-37 DAF. To prevent seed loss and seed damage, harvesting interval should not be more than 10

Figure 18.8: Development of Okra Seed during 3-27 DAF

days to avoid the splitting of pod which leads to seed drop. Besides, seed quality will be spontaneously lower when harvesting interval is too long (Adthalungrong, 2009).

Yellow Vein Mosaic Disease and Breeding for Resistance in Thailand

Situation of Yellow Vein Mosaic Disease

Yellow vein mosaic disease is known as okra vein yellowing disease in Thailand. It is caused by okra yellow vein virus (OYVV). The disease was first discovered in 1995 and has become a serious problem of okra production since 1998. Disease pathogen is the geminivirus group which belongs to the Genus *Begomovirus*. It is transmitted by an insect vector, the tobacco whitefly (*Bemisia tabaci*) (Kittipakron *et al.,* 1999; Adthalungrong *et al.,* 2011; Mukhopadhyay, 2011). Disease spread is related to the vector activity. In addition, disease spread and vector populations are positively correlated to warm weather (Mukhopadhyay, 2011).

The OYVV has round shape with polyhedral, usually in pairs. Its approximate size is 18 x 30 nm. The virus transmitting capability of an insect was 26 per cent, after acquisition and inoculation feeding for at least 24 hr each time. Mechanical method and seed are not capable transmission routes (Kittipakron *et al.,* 1999).

Symptoms of the disease are mosaic, yellow leaf vein surrounding by green tissue or totally yellow/cream leaf color, yellow shoot, curling leaf and top and yellow pod (Figure 18.9). Seedling and young plant express severe symptoms, stunted or undeveloped and bare very a few deformed and small fruits. In the serious infected areas, disease symptoms are observed in susceptible cultivars after 18 days of sowing (Adthlungrong, *et al.,* 2011)

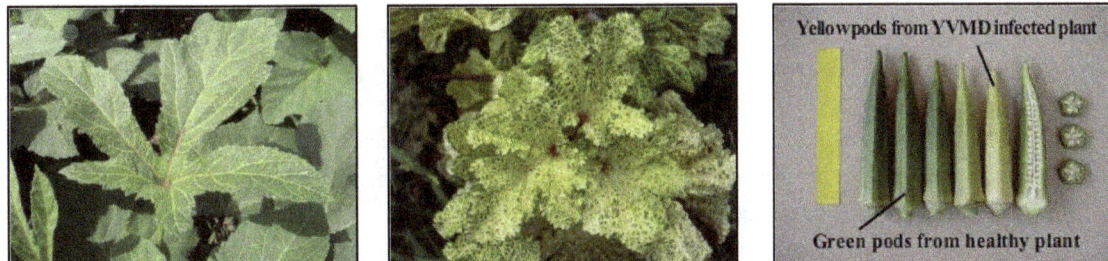

Figure 18.9: Severity of Yellow Vein Mosaic Disease Symptoms on Leaf and Pod

The leaf turns yellow due to decrease of chlorophyll content (Atiri and Ibidapo, 1989; Sarma *et al.,* 1995). This affects the growth and yield but the degree of severity varies with different stages of plant growth when being infected. Sastry and Singh (1975) reported that yields of okra were 93.80, 83.63 and 49.36 per cent lower when infection occurred at 35, 50 and 65 days after germination, respectively.

Different virus isolates cause different symptoms, disease severities and time duration after plant infection until the disease is first observed (Figure 18.10). On the other hand, different resistant cultivars also respond dissimilarly to the same pathogen (Srithongchai and Adthalungrong, 2010). Furthermore, the virus pathogen has developed itself and breakdown the resistance capability of its host. For this reason, the resistant cultivars usually function efficiently for only 2-3 years (Adthlungrong, *et al.,* 2011).

OYVV-PC   **OYVV-KB1**   **OYVV-KB2**

Figure 18.10: Disease Symptoms on Okra cv. Phichit 03 by OYVV-PC (Phichit province), OYVV-KB1 and OYVV-KB2 (Kanchanaburi province)

The host plants for YVMD are in a wide range and symptoms vary with different hosts. Kittipakron *et al.* (2001) found 21 plant species belonging to 7 families which were the disease hosts, as given in Table 18.2. Moreover, zinnia, gerbera, morning glory (kangkong), and numerous kinds of weeds such as para cress, billy goat weed, false daisy and paragrass are the host plants (Srithongchai, 2009)

Table 18.2: Host Plants of OYVV Transmission by Tobacco White Fly

Common Name	Scientific Name	Disease Symptom
Okra, Lady finger	*Abelmoschus esculentus* (L.) Moench	Mosaic, vein yellowing, yellow pod.
Kang pla khao (THAI, Ang Thong, Chiang Mai)	*Phyllanthus reticulatus* Poir	Yellow mosaic, leaf distortion.
Ivy gourd, Scarlet gourd	*Coccinia indica* Wight & Arn	Yellow mosaic.
Cucumber	*Cucumis sativus* L.	Mosaic, leaf curl.
Angled gourd, Loofah	*Luffa acutangula* (L.) Roxb.	Leaf curl, enation.
Cotton	*Gossypium barbadense* L. cv. Pima	Leaf curl, enation.
Vinca	*Catharanthus roseus* (L.) G. Don	Crinkle top leaf, leaf curl, mottle
Chilli	*Capsicum annuum* L.	Crinkle top leaf, leaf curl, mottle
Mung Bean	*Vigna radiata* (L.) R.	Crinkle top leaf, leaf curl, mottle
French Bean, Bush Bean	*Phaseolus vulgaris* L.	Crinkle top leaf, leaf curl, mottle
Soya bean	*Glycine max* (L.) Merrill.	Vein clearing and white spot on inoculated leaf.
Tomato	*Lycopersicon esculentum* Mill.	Yellow leaf roll, witches-broom and stunting.
Yellow berried nightshade	*Solanum xanthocarpum* Schrad. & Wendl.	Mosaic, leaf curl.
Egg plant	*Solanum melongena* L.	Mosaic, leaf curl.
Sweet potato	*Impomoea batatas* (L.) Poir.	Leaf curl, mottle.
Cumin	*Cuminum cyminum* L.	Pale mosaic.
Tobacco	*Nicotiana tabacum* L. cv. White Burley	Leaf roll, yellowing, stunting.
Tobacco	*Nicotiana benthamiana* L.	Yellow leaf roll, stunting.
Tobacco	*Nicotiana glutinosa* L.	Yellow leaf roll, stunting.
Hollyhock	*Althaea rosea* Cav.	Crinkle leaf, enation, stunting.
Brazil Jute, Malachra	*Malachra capitata* L.	Vein yellowing, crinkle leaf, enation.

Modified from: Kittipakron *et al.* (2001)

There are many ways to control the YVMD. Elimination of the host plants, crop rotation, selection of area and time for planting are common strategies to decrease the disease. In addition, control of tobacco white fly should be done simultaneously by an application of proper chemical substances (such as carbosulfan and imidacloprid) or natural organic substances (such as neem extract and petroleum oil). However, the use of plants which possess genetic resistance against plant virus is the simple and effective method to control the disease.

YVMD Resistance Gene and Heritability

The sources of YVMD resistance gene were reported in various plants *e.g. Abelmoschus manihot*, the wild species from Ghana (Sharma and Dhillon, 1983; Nerkar, 1991), some West African cultivated okras (*Abelmoschus esculentus* (L.) Moench) (Salehuzzaman, 1986), and certain mutant okras (Phadvibulya *et al.,* 2004; 2009; Boonsirichai *et al.,* 2009).

The tolerance character is also claimed in the cultivated okras in Bangladeshat (Ali *et al.,* 2000). Moreover, leaf attributes including leaf angle, leaf shape, number of leaves/plant, leaf area, leaf area index and specific leaf weight were important characteristics against the whitefly associated with virus transmission (Amandeep and Neelima, 2008).

In Thailand, even if the wild species *A. moschatus* Medikus, *A. manihot* (L) Medikus and interspecific hybrid were found in western, northeastern and northern regions (Chomchalow *et al.,* 1987; Chomchalow, 1987), there are no research on the disease and insect resistance of those species. Most researches solely focus on the mutant okras which possess the YVMD resistance.

The inheritance of YVMD resistance varies widely. It probably occurs via single dominant gene (Jambhale and Nerkar, 1981), single dominant or more major gene(s) along with minor genes (Arora *et al.,* 2008), two complimentary dominant genes (Dhankhar *et al.,* 2005; Pullaiah *et al.,* 1998), or incomplete dominant resistance gene (Boonsirichai *et al.,* 2009). Ali *et al.* (2000) quoted that the tolerance inherits via a few major genes with incomplete gene action.

Creation of Base Population

Base populations for YVMD resistance of okra in Thailand are mostly derived from Indian cultivars which have been proved to control the disease. Hybridization between the resistance cultivar and the F1 of Japanese cultivar was also created to improve the agronomic characters particularly plant height, pod quality, time of harvest and yield (Adthalungrong *et al.,* 2008).

Moreover, the mutation technique was done by Thailand Institute of Nuclear Technology for the new source of YVMD resistance. Gamma rays were applied to okra seeds at 400 and 600 Gy doses. Those seeds were planted and self fertilized to build up the population (Phadvibulya *et al.,* 2004 and 2009).

Methodology for YVMD Resistance Selection

Various breeding methods for self pollinated plant *i.e.* the pedigree method, pure line selection, mass selection and backcross, have been propelled to improve the crop plant. The methods can be applied to base population to acquire pure lines which can be used as parent line in hybrid or as open pollinated cultivar. Single or combined methods may be assigned based on the efficiency and effectiveness of the methods to yield the desired characteristics.

For the YVMD resistance selection in the field, artificial virus inoculums and susceptible cultivars are required throughout the crop to assure the reliable selection. Susceptible cultivar was planted together with selected lines and also at the borders (Figure 18.11a). Virus transmission by insect was investigated in the screen house (Figure 18.11b). The whitefly inoculation (5 insects/plant) was done by feeding on the disease plant prior to the test plant, each for 48 hr. The test plant was used at the seedling stage, 4-6 days after sowing, and 30 plants/line. Then, the resistance was morphologically examined every week and ELISA (enzyme-linked immunosorbent assay) test was applied (Srithongchai and Adthalungrong, 2010).

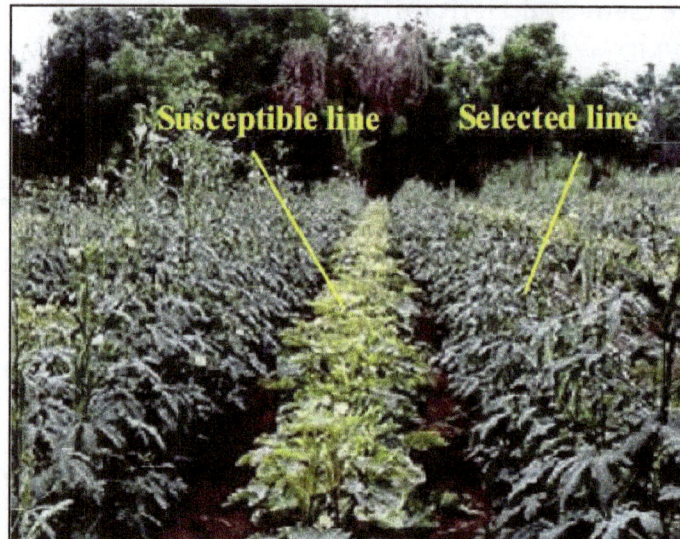

a: Planting Susceptible and Selection Lines

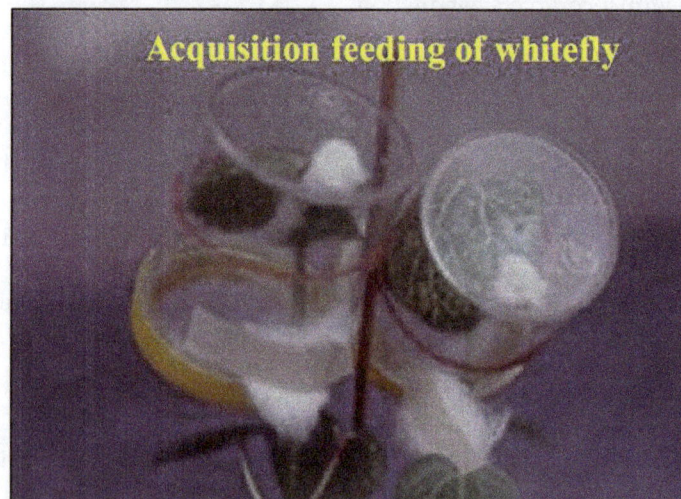

b: Insect Transmission

Figure 18.11: Field and Screen House Selection

Advance in Breeding of YVMD Resistant Okra in Thailand

After the YVMD became a serious problem in Thailand, the groups of government and private sectors were established to resolve the problem. Three government sectors including Department of Agriculture (DOA), Department of Agriculture and Extension (DOAE), and Maejo University, in collaboration with some seed companies and okra exporters, have searched for suitable cultivars with YVMD resistance and other desired attributes to substitute the Japanese cultivars for export.

For this reason, exotic cultivars were introduced from India, Taiwan, and some other countries to test for the disease resistance as well as their agricultural performances. As a result, most of appropriate cultivars were from India. Therefore, Thailand needs to import the resistant cultivars from India since then. However, Indian okra cultivars provide too long, thin and light green pod which are not desirable. Another major disadvantage is that the plants usually are too high which cause the plant easily broken down by wind. In addition, YVMD resistance does not last long since virus pathogen continually develops itself. Therefore, the improvement of okra has been done uninterruptedly by many agencies.

DOA has a high concern about YVMD since 1997. A research project launched by Plant Pathology Research Group, Plant Protection Research and Development Office and Phichit Horticultural Research Center, Horticulture Research Institute are set up. However, the project engrossed in breeding program two years later and the research goal at the beginning was to obtain the open pollinated resistant cultivar. The pedigree selection method was applied with specific Indian cultivars, both in the screen house and in the field, and three promising lines were achieved as the project outcome. A few years later in 2006-2007, those lines were tested at farmer fields in Kanchanaburi. Unfortunately, they provided good yield and resisted to YVMD only at the first year but the resistance declined at the second year. The resistance instability also happened to commercial cultivars grown as control (Adthalungrong and Buara, 2008).

Afterwards, the research has emphasized on an improvement of the disease resistance and agronomic characteristics by hybridization. Study on virus isolates and the interaction with the cultivars were carried out. Currently, researches on a comparison of eight promising lines that resist to the Kanchanaburi virus isolate are on going and selected lines will be released to farmers soon. However, the improvement of YVMD resistance and agronomic characteristics must be done continually. Therefore, a new base population of okra was created at Phichit Agricultural Research and Development Center (previous name: Phichit Horticultural Research Center).

Apart from the cultivar improvement by DOA, researches on breeding by other agencies like seed companies and okra exporters distribute seeds to farmers or contractors. East West Seeds Co. Ltd. has a project on YVMD resistance development and used to release "Green Star 691" variety to the market. However, the variety failed to succeed because of the unsatisfied pod size. Their pod size is about 5-25 g/pod while the standard size is about 10 g/pod. Besides, the pod color is light green which is undesirable. Currently, the better improved cultivar called "Green Star 695" has been marketed.

The Uniseeds Co. Ltd., in collaboration with National Science and Technology Development Agency, has developed the YVMD resistant F1 hybrid okra. Six F1 hybrid cultivars revealed good performance and high resistance to YVMD when testing at the farmer field in 2007 (BIOTEC, 2012).

Maejo University also created an okra breeding project using a conventional method to develop OP cultivars. They now released the okra cultivars named Maejo 49 and 70 to market. Lampang Rajabhat University also has a number of researches on okra breeding.

Thailand Institute of Nuclear Technology (TINT) conducted an okra breeding project using a mutation breeding method. Okra seeds are exposed to gamma rays to induce the mutation and the selection of okra variety is undertaken by a pedigree selection method. TINT successfully obtained one resistant line (Phadvibulya *et al.,* 2009). The resistance characteristic is controlled by incomplete dominant gene and it is unstable (Boonsirichai *et al.,* 2009). Although the research output is not very successful, the method can be another strategy to gain the new source of YVMD resistant okra.

Acknowledgement

The author would like to express his high gratitude to all those who co-operate with the okra breeding project: Mrs. Kruapan Kittipakorn who initiated okra selection for YVMD resistance and provided the author all needed materials, Dr. Wanphen Srithongchai, a virologist who fully dedicates in virus disease part, Mrs. Valailak Phadvibulya who has assisted and worked together from the beginning of the project.

References

Adthalungrong, A and Buara, P., 2008. On farm testing of yellow vein disease resistant okra cultivars. Research and development on plants and gricultural technology 2007. Department of Agriculture, Ministry of Agriculture and Cooperatives, Thailand. 103. (in Thai).

Adthalungrong, A., Choodee, K. and Tayamanon, P., 2008. Comparison of okra promising lines derived from the hybridization of Japanese and Indian okra. Research and development on plants and gricultural technology 2007. Department of Agriculture, Ministry of Agriculture and Cooperatives, Thailand. 104. (in Thai)

Adthalungrong, A., 2009. Floral biology and appropriate pollination time of okra (*Abelmoschus esculentus* (L.) Moench). *Agri. Sci. J.,* 40 (3) (Suppl): 63–66 (in Thai).

Adthalungrong, A., Choodee, K. and Tsai, Wen–shi, 2011. Yellow vein mosaic disease inflicts severe damage on okra in Thailand. *AVRDC Feedback from the Field* 11: 1–2.

Ali, M., Hossain, M.Z. and Sarker, N.C., 2000. Inheritance of Yellow Vein Mosaic Virus (YVMV) tolerance in a cultivar of okra (*Abelmoschus esculentus* (L.) Moench). *Euphytica,* 111: 205–209.

Amandeep, K. and Neelima, A., 2008. Comparative studies on leaf attributes of yellow vein mosaic virus resistant and susceptible okra genotypes. *Plant Disease Research,* 23(1): 79–83.

Amin, I.M., 2011. Nutritional Properties of *Abelmoschus esculentus* as Remedy to Manage Diabetes Mellitus: A Literature Review. 2011 *International Conference on Biomedical Engineering and Technology* (ICBET 2011) 4 to 5 June 2011 Kuala Lumpur, Malaysia. 50–54.

Anitha, P., Sadhankumar, P.G. and Rajan, S., 2001. Effect of maturity stages on seed quality in okra. *Vegetable Science* 28 (1): 76–77.

Arora, D., Jindal, S.K. and Singh, K., 2008. Genetics of resistance to yellow vein mosaic virus in inter-varietal crosses of okra (*Abelmoschus esculentus* L. Moench). *Breeding and Genetics* 40(2): 93–103.

Atiri, G.I. and Ibidapo, B., 1989 Effect of combined and single infections of mosaic and leaf curl viruses on okra (*Hibiscus esculentus*) growth and yield. *J. of Agri. Sci.,* 112(3): 413–418.

BIOTEC. ny. Breeding for virus disease resistance in okra F1 hybrid. 2 p. Access Apr. 20, 2012. from http://www.biotec.or.th/th/images/stories/IU/document/5/20.pdf (in Thai)

Bisht, I.S. and Bhat, K.V., 2007. Okra (*Abelmoschus spp.*). 147–183 pp. In: *Genetic Resources, Chromosome Engineering, and Crop Improvement, Vol. 3: Vegetable Crops*, (Eds.) Ram J. Singh. CRC Press Taylor & Francis Group, Boca Raton, Florida.

Boonsirichai, K., Phadvibulya, V., Adthalungrong, A., Srithongchai, W., Puripanyavanich, V. and Chookaew, S., 2009. Genetics of the radiation-induced yellow vein mosaic disease resistance mutation in okra. In: *Induced Plant Mutations in the Genomics Era*, (Ed.) Q.Y. Shu. Food and Agriculture Organization of the United Nations, Rome, 352–354 pp.

Charrier, A., 1984. Genetic Resources of the genus *Abelmoschus* (OKRA). International Board for Plant Genetic Resources, Rome, 61 p.

Chomchalow, N., 1987. Natural interspecific hybrids of *Abelmoschus* in Thailand. Present in the fifth academic seminar of genetic at Prince of Songkla University, 13–15 May 1987. 12 p. (in Thai).

Chomchalow, N., Hamon, S., Chantaraprasong, C. and Chomchalow, S., 1987. *Abelmoschus* germplasm collection in Thailand. Present in the fifth academic seminar of genetic at Prince of Songkla University, 13–15 May 1987. 8 p. (in Thai).

Department of Biotechnology (DOB), 2001. Biology of Okra. Department of Biotechnology, Ministry of Science and Technology, India, 25 p. Access Apr. 20, 2012. from http: //dbtbiosafety.nic.in/guidelines/okra.pdf

Devadas, V.S., Rani, T.G., Kuriakose, K.J. and Nair, S.R., 1998. A note on fruit and seed development in okra (*Abelmoschus esculentus* (L.) moench). *Vegetable Science,* 25(2): 187–189.

Dhankhar, S.K., Dhankar, B.S. and Yadava, R.K., 2005. Inheritance of resistance to yellow vein mosaic virus in an interspecific cross of okra (*Abelmoschus esculentus*). *Indian Journal of Agricultural Sciences,* 75(2): 87–89.

Ebadi, M., 2002. *Pharmacodynamic Basis of Herbal Medicine.* CRC Press LLC, Boca Raton, Florida. 760 p.

George, Raymond A.T., 2009. *Vegetable Seed Production,* 3rd edn. CABI, MPG Books Group, Bodmin. 320 p.

Hamon, S. and Sloten, D.H. van, 1989. Characterization and Evaluation of Okra. In: *The Use of Plant Genetic Resources*, (Eds.) A.H.D. Brown, O.H. Frankel, D.R. Marshall and J.T. Williams. Cambridge University Press, 173–196 pp.

Hamon, S. and Koechlin, J., 1990a. The reproductive biology of okra. 1. Study of the breeding system in four *Abelmoschus* species. *Euphytica,* 53: 41–48.

Hamon, S. and Koechlin, J., 1990b. The reproductive biology of okra. 2. Self-fertilization kinetics in the cultivated okra (*Abelmoschus esculentus*), and consequences for breeding. *Euphytica,* 53: 49–55.

Hamon, S., Charrier, A., Koechlin, J. and Sloten, D.H. van, 1990. Potential contributions to okra breeding through the study of their genetic resources. 77–83 pp. In: International Crop Network Series 5, Report of an International Workshop on Okra Genetic Resources. International Board for Plant Genetic Resources, Rome.

Hamon, S. and Sloten, D. H. van, 1995. Okra, *Abelmoschus esculentirs, A. caillei, A. manihot, A. rnosclintus* (*Malvaceae*). In: *Evolution of Crop Plants,* 2nd Edn. (Eds.) J. Smartt and N.W. Simmonds. Longman Singapore Publishers, 350–357 pp.

IBPGR, 1991. International Crop Network Series 5, Report of an International Workshop on Okra Genetic Resources. International Board for Plant Genetic Resources, Rome, 133 p.

Jambhale, N.D. and Nerkar, Y.S., 1981. Inheritance of resistance to okra yellow vein mosaic disease in interspecific crosses of Abelmoschus. *Theoretical and Applied Genetics,* 60(5): 313–316.

Kittipakron, K., Adthalungrong, A. and Chiemsombat, P., 1999. Okra vein yellowing disease. *The Journal of Thailand Phytopathological Society*, 15–14(1–2): 16–30 (in Thai).

Kittipakron, K., Gajanandana, O., Kladpan, S., Adthalungrong, A., Phadvibulya, V. and Jomthaisong, J., 2001. Situation of okra vein yellowing disease. Academic seminar on "Okra for exported" at Maejo University, Oct. 26, 2001. 5 p. (in Thai).

Kumar, S., Dagnoko, S., Haougui, A., Ratnadass, A., Pasternak, D. and Kouame, C., 2010. Okra (*Abelmoschus spp.*) in West and Central Africa: Potential and progress on its improvement. *African Journal of Agricultural Research*, 5(25): 3590–3598.

Lavapaurya, T., Niumkead, P., Chotikachamroon, C. and Pao-in, K., 1997. Synthesizing 5 angled-okra F1 hybrid single cross variety Morakot F1. In: *Proceedings of the 35th Kasetsart University Annual Conference: Plant, Agricultural Extension and Communication and Agro-Industry*, 3–5 Feb. (in Thai), 277–284 pp.

Li, Thomas S.C., 2008. *Vegetables and Fruits: Nutritional and Therapeutic Values*. CRC press, Taylor & Francis Group, Boca Raton, FL. 304 p.

Lim, T.K., 2012. *Edible Medicinal and Non-Medicinal Plants, Volume 3: Fruits*. Springer Science+Business Media. 898 p.

McCormack, J.H., 2010. Principles and practices of isolation distances for seed crops: an organic seed production manual for seed growers in the Mid-Atlantic and Southern U.S. 21 p. Access Apr. 20, 2012. from www.savingourseeds.org/pubs/isolation_distances_ver_1pt8.pdf.

Mukhopadhyay, S., 2011. *Plant Virus: Vector Epidemiology and Management*. Science Publishers, Enû eld. 520 p.

Nerkar, Y.S., 1991. The use of related species in transferring disease and pest resistance gene to okra. 110–113 pp. In: IBGPR. 1991 International Crop Network Series 5. Report of an International Workshop on Okra Genetic Resources. International Board for Plant Genetic Resources, Rome.

Nwoke, F.I.O., 1980. Effect of Number of Photoperiodic Cycles on Flowering and Fruiting in Early and Late Varieties of Okra (*Abelmoschus esculentus* (L.) Moench.). *Journal of Experimental Botany*, 31(125): 1657–1664.

Oyolu, C., 1977. Variability in photoperiodic response in okra (*Hibicus esculentus* L.). *Acta Hort.,* 53: 207–216.

Phadvibulya, V., Adthalungrong, A., Kittipakron, K., Puripanyavanich, V. and Lavapaurya, T., 2004. Induce mutation for resistance to yellow vein mosaic virus in okra. In: Genetic improvement of under–utilized and neglected crops in low income food deficit countries through irradiation and related techniques. Proceedings of a final Research Coordination Meeting, in Pretoria, South Africa, 19–23 May 2003. Plant Breeding and Genetics Section International Atomic Energy Agency, Vienna, 155–173 pp.

Phadvibulya, V., Boonsirichai, K., Adthalungrong, A. and Srithongchai, W., 2009. Selection for resistance to yellow vein mosaic virus disease of okra by induced mutation. 349–351 pp. In:

Induced Plant Mutations in the Genomics Era. yellow vein mosaic disease resistance mutation in okra. 352–354 pp. In: Induced plant mutations in the genomics era. Edited by Q.Y. Shu. Food and Agriculture Organization of the United Nations, Rome.

Pullaiah, N., Bhupal Reddy, T., Moses, G.J., Reddy, B.M. and Reddy, D. Rajaram, 1998. Inheritance of resistance to yellow vein mosaic virus in okra (*Abelmoschus esculentus* (L.) Moench). *The Indian Journal of Genetics and Plant Breeding,* 58(3): 349–352.

Rashid, M.A. and Singh, D.H., 2000. A manual on vegetable seed production in Bangladesh. AVRDC–USAID–Bangladesh Project, Dhaka. 125 p.

Ritchai, P., Fujime, Y., Sukprakarn, S., Terabayashi, S., Okuda, N. and Date, S., 2004. Flowering response to photoperiod in okra (*Abelmoschus esculentus*). *Journal of the Japanese Society of Agricultural Technology Management,* 11(1): 33–41.

Salehuzzaman, M., 1986. Nature of resistance to yellow vein mosaic virus in okra (of West Africa and Bangladesh). In *Proceedings of the 11th Annual Bangladesh Science Conference.* Section 2. Dhaka (Bangladesh). BAAS, 9–10 pp.

Sarma, U.C., Bhagabati, K.N. and Sarkar, C.R., 1995 Effect of yellow vein mosaic virus infection on some chemical constituents of bhendi (*Abelmoschus esculentus* (L.) Moench). *Indian Journal of Virology,* 11(1): 81–83.

Sastry, K.S.M. and Singh, S.J., 1975. Effect of yellow-vein mosaic virus infection on growth and yield of okra crop. *Indian Phytopathology,* 27(3): 294–297.

Sharma, B.R. and Dhillon, T.S., 1983. Genetics of resistance to yellow vein mosaic virus in interspecific crosses of okra (*Abelmoschus* species). *Genetica Agraria,* 37(3–4): 267–275.

Shoba, K. and Mariappan, S., 2007. Heterosis studies in okra (*Abelmoschus esculentus* (L.) moench) for some important biometrical traits. *Acta Hort.,* 752: 437–440.

Small, E., 2012. *Top 100 Exotic Food Plants.* CRC press, Taylor & Francis Group, Boca Raton, FL. 708 pp.

Srithongchai, W., 2009. Yellow Vein Disease. In: *Handbook of Vegetable Disease.* Plant Protection Research and Development Office, Department of Agriculture, Bangkok (in Thai), 89–90 pp.

Srithongchai, W. and Adthlungrong, A., 2011. Relationship between Okra yellow vein virus and okra varieties in planting areas. 2225–2235 pp. Access Apr. 20, 2012 from http://it.doa.go.th/refs/files/1743_2553.pdf (in Thai)

Sukprakarn, S., Juntakool, S., Huang, R. and Kalb, T., 2005. Saving your own vegetable seeds a guide for farmers. AVRDC publication number 05–647. AVRDC–The World Vegetable Center, Shanhua, Taiwan. 25 p.

Suttipolpiboon, S., 2012. Exported Okra. 5 p. Accesses Apr. 20, 2012. from www.eto.ku.ac.th/neweto/e-book/plant/herb_gar/krajeab.pdf. (in Thai)

Tanda, A.S., 1985. Floral biology, pollen dispersal, and foraging behaviour of honeybees in okra (*Abelmoschus esculentus*). *Journal of Apicultural Research* 24(4): 225–227.

Udayasekhara Rao, P. 1985. Chemical composition and biological evaluation of Okra (*Hibiscus esculentus*) seeds and their kernels. *Plant Foods for Human Nutrition,* 35: 389–396.

2013, Biodiversity in Horticultural Crops Vol. 4
Editor: Professor K.V. Peter
Published by: DAYA PUBLISHING HOUSE, NEW DELHI

Pages 355–386

Chapter 19

Orchids in Arunachal Pradesh

Krishna S. Tomar and Sunil Kumar

College of Horticulture and Forestry,
Central Agricultural University, Pasighat, Arunachal Pradesh – 791 102, India
E-mail: tomarhorti@rediffmail.com

The earth is home to a rich and diverse array of living organisms, whose genetic diversity and relationship with one another and with their physical environment constitute our planet's biodiversity. This biodiversity is the natural biological capital of the earth, and presents important opportunities for all nations. It provides goods and services essential to support human livelihoods and aspirations and enables societies to adapt to changing needs and circumstances. In addition, biodiversity maintains the ecological balance necessary for planetary and human survival.

Arunachal-the land of rising sun, extends over an area of 83,740 sq km and lies between 26°28' and 29°30' North latitude and 90°30' and 97°30' East longitude. It is bounded in north by Mc Mohan Line, in the east by China and Myanmar, in the south by the states of Nagaland and Assam and in the west by Bhutan. Arunachal Pradesh has generally a hilly terrain characterized by hill ridges and valleys. The elevation ranges from 100 m in case of Siwalik formations from plains of Assam to the Himalayas ranging up to 7750 m along the Tibet-China border. Arunachal Pradesh is considered to be luxuriant in floral diversity and has been recognized as the 25th biodiversity hotspot in the world (Chowdhery, 1999). It is also among the 200 globally important eco regions (Olson and Dinerstein, 1998).

The climatic conditions vary from place to place in the state due to rapid changes in its topography. Generally December and January are the coldest months and the temperature begins to rise from the month of March and continues up to July-August. The monsoon rains occur during the months of February-March.

Floral Diversty in Arunachal Pradesh

State contains 40 per cent of the floral and faunal species of India, many of which are endemic to the region. Of the 8,000 reported species of flowering plants from the Himalayas, 5,000 species are found in the north eastern region. Apart from large number of timber species, there are innumerable varieties and kinds of orchids, medicinal plants, ferns, bamboos, canes, wild relatives of large number of cultivated plants, and even plants of biological curiosities, such as parasites, saprophytes, etc. Other important groups of plants are Rhododendrons, Hedychiums and oaks etc. Orchids form a dominant group of plants with their attractive and unique blooms. There are more than 600 species of orchids, 52 species of rhododendron, 18 species of Hedychium, 16 species of oak, 18 species of canes and 45 species of bamboo in addition to large number of medicinal and aromatic plants. About one third of the 105 species of Indian bamboo are found in Arunachal Pradesh.

The wide ranging altitudinal variations from the foothills to Himalayan mountains, the mighty rivers flowing down to the delightful valleys; high rainfall and high humidity; soil conditions etc., have blessed Arunachal with a rich and very diverse flora and fauna. Keeping in view the varied flora and fauna of Arunachal, 11 protected areas have been established in the state covering an estimated area of 9245.98 sq km including a sanctuary for orchids at Sessa in West Kameng district.

Four major types of Orchid Zones may be classified within the state according to their occurrence in different vegetation types.

1. Tropical forest: 100-1000 m
2. Subtropical forest: 1001-1800 m
   i. Mixed wet forest
   ii. Mixed dry pine forest

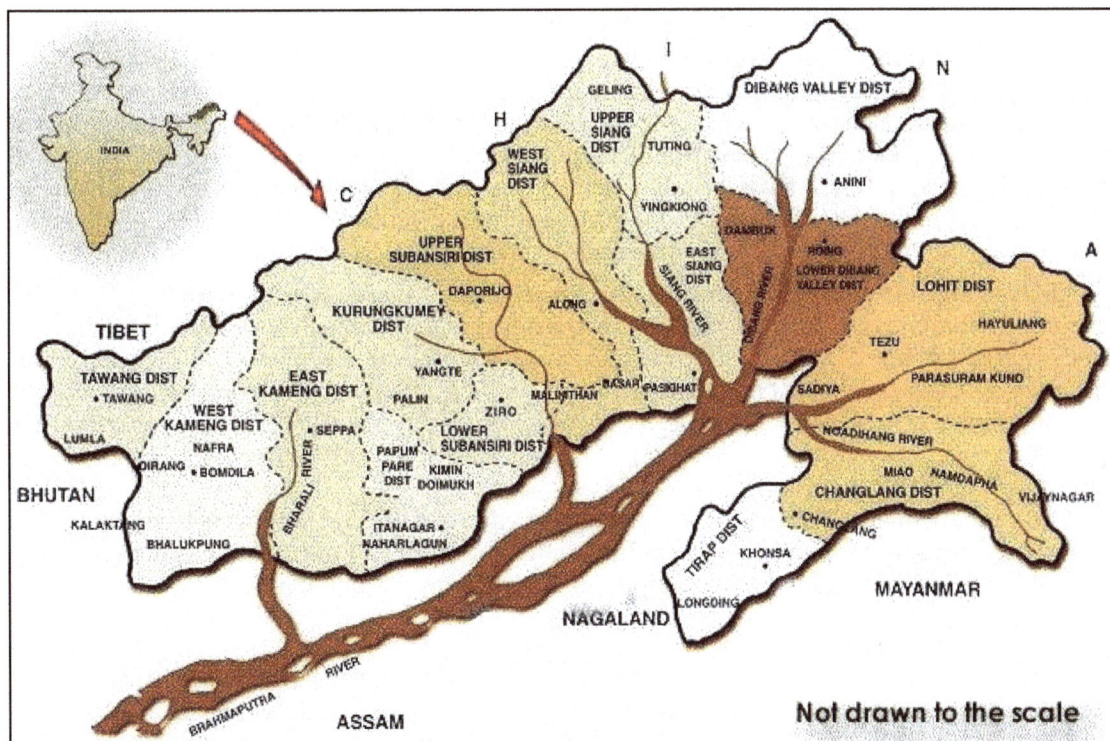

Figure 19.1: Map of Arunachal Pradesh

3. Temperate forest: 1801-3500 m
    i. Broad leaved forest
    ii. Coniferous forest
4. Alpine zone: 3501-5000 m

Orchids Diversity in Arunachal Pradesh

Orchids are highly evolved group of plants occupying a top position in the plant kingdom. They have a diverse habit, bizarre and curious flower structure with brilliant colours adding beauty to the land and the landscape. The very fact that there are about 20,000 species distributed all over the world speaks of their diversity and varied nature having a realm of their own.

In India, there are about 1,150 species distributed in various states. However, Arunachal Pradesh alone accounts for 601 species – almost 52 per cent of the total species known in India Arunachal's rich and colorful orchids find a place of pride. Out of about a thousand species of orchids in India, over 600 are to be found in Arunachal alone. Hence this state can rightly be called the "*Orchid Paradise*" of our country. These are colorful, spectacular and some bear exotic names such as Sita-Pushpa and Draupadi-Pushpa believed to have been worn by Sita and Draupadi for ornamentation. Many of these orchids are rare, endangered and highly ornamental with long-lasting flower qualities. Amongst the orchids as many as 150 species are ornamental and commercially important. Orchids are considered as "*gems*" in the field of floriculture. Orchids are in fact proud possessions of the hobbyists, sought after by the nurserymen and are a symbol of royalty in many countries. The tourists with special crave for orchids would love to visit the natural habitat of the orchids, which could be commercially exploited.

In the orchid family, there are members with primitive character and fashion on the one hand and with tremendous specialization, modification, adaptation, fanciful and catching the eyes with brilliant colours and shapes, on the other. Hence, the entire orchid family with about 601 species in Arunachal Pradesh has been classified into six-sub families, 17 tribes, 24 sub-tribes and 111 genera. Of these 148 species in 52 genera are terrestrials including saprophytes and 377 species in 61 genera are epiphytes including lithophytes. Almost those *Apostasias* are the most primitive with the least specialization, while *Vandaceous* orchids are highly evolved and advanced in nature.

Orchids in Arunachal Pradesh can also be classified into 140 species of terrestrial orchids with 15 saprophytes and about 340 epiphytes found in the different forest types. The prominent species are: *Cymbidium ensifolium, C, grandiflorum, Coelogyne corymbosa. Dendrobium aphylla, D. fimbriatum* var. *occulatum, D. densiflorum, Calanthe masuca, Phaius flavus, Paphiopedilum fairrieanum, P.venustum, Renanthera imschootiana, Vanda coenulea*, etc. *Rhynchostylis retusa* is the state flower of Arunachal Pradesh.

Habit and Habitat

Orchids have diverse habits with variously modified vegetative and floral structures, forms and colour with varying requirements of light, shade, temperature and nutrition within a particular environment. They may be Terrestrials, Epiphytes, Saprophytes and Lithophytes (growing on rocks).

Terrestrials (Growing on Ground)

The leafy orchids grown on the ground in humus rich soil are called terrestrials. There are about 132 species of leafy terrestrials known in Arunachal Pradesh. Some of the ornamental genera are *Acanthephippium, Anoectochilus, Calanthe, Phaius, Paphiopedilum,* etc, amongst them, *Arundina graminifolia*, the "bamboo orchid" is normally seen grown in the open sunny areas amongst grassy patches in the foothills. These plants look like reeds and produce attractive flowers in light pink with deep purple lip.

In the thickets of the forest floor, occur *Anoectochilus* and *Geodorum*, popularly referred to as the "Jewel orchids" group having beautifully designed velvety leaves. They are rare in occurrence however. *Acanthephiplum species* on the other hand, grow under heavy shades of trees on the ground or rocks and have curiously saccate yellow tinged purple flowers found in tropical and sub-tropical wet humid forests. Similarly, the species of *Phaius* with comparatively larger and attractive flowers in yellow, pink and brown shades are found in these forests. There are about 14 species of *Calanthe* found in various forest types. They are one of the best known orchids under cultivation too. One of the species *i.e. Calanthe masuca* has the distinction of having been used in the first ever hybrid orchid produced in the year 1856 in London and is also found in tropical and sub-tropical forests of Arunachal Pradesh.

Epiphytes (Growing on Tree Trunks)

Majority of the orchids grow on tree trunks and called epiphytes. They are either pendulous, erect growing, sympodial or monopodial, but always having aerial roots. In Arunachal Pradesh, there are 377 species in 61 genera recorded so far. Major genera amongst them are *Aerides. Arachnis, Bulbophyllum, Coelogyne, Dendrobium, Cymbidium, Eria, Cleisostoma, Rhynchostylist, Vanda* etc. Most of the epiphytic orchids are ornamentals. In the tropical valleys of Arunachal Pradesh, one finds cascades of colorful flower-spikes of *Rhynchostylis retusa. Aerides odorata, A. williamsii, A. rosea, Cymbidium pendulum, C. aloifolium, Dendrobium aphylla, D. nobile, D. moschatum, D. fimbriatum* etc. loaded on tree trunks during spring time which add beauty to the surrounding wood. Some orchids like; *Cymbidium ensifolium, Aerides odoratum* etc. fill the air with pleasant aroma during the season. As one travels through the sub-tropical hill forests, bunches of "pineapple-orchids" the *Dendrobium densiflorum, giganteum, grandiflorum, eburnium, mastersii, Vanda coerulea, Renanthera imschootiana* (*Red Vanda*), *Coelogyne* etc. greet the on lookers. The pink flowers of *Anthogonium gracile* on the cut ends of rocks and edges carpet the exposed areas adding beauty to the lanscape. It is however, significant to note that most of the epiphytic orchids are endangered in the wild as their very existence depends upon the trees, which are being exploited for various needs of man.

Saprophytes (Leafless Orchids and Growing on Decaying Organic Matter)

Some orchids grow on decaying organic matters on the forest floor having no leaves and without chlorophyll, but produce variously colored flowers. While passing through the thickets of these forests, one may find strikingly white coloured mushroom – like bodies about 11/2 tall group of plants the Epipogium and *Stereosandra species*, dull-brown coloured *Gastrodia species* and *Eulophia-zollingeri*, the giant 1–2 m tall golden yellow flowering spikes of *Galeola species*, etc. which thrill the onlooker. However these enigmatic plants cannot thrive under cultivation. There are 16 such saprophytic species recorded so far in Arunachal Pradesh.

Morphology

The orchid plant structure is basically of two types:

1. *Monopodial* – Such orchids do not have rhizomes or pseudobulbs and grow from a single vegetative apex *e.g. Vanilla, Vanda, Rhynchostylis* etc.

2. *Sympodial* – Such orchids grow from a number of vegetative apices sitiuated on the rhizome which is a modified stem, creeping or sometimes under ground and in most of the cases produce more or less erect stem like structures. These sometimes swell into reserve organs and are known as pseudobulbs.

Roots

The roots of most monopodials become considerably flattened as they creep over the surface. In the sympodials orchids, the roots are normally produced from the rhizome. The roots of most orchids are cylindrical, often threadlike, at times branching and frequently elongated. In the terrestrials orchids, these are provided with root hairs.

Pseudobulbs

In order to combat the extreme drought conditions specially faced by the epiphytic orchids growing higher on tree tops and branches, the stem in certain species is modified into special storage organs called 'Pseudobulbs'. These are of various sizes ranging from round to ovoid, ellipsoid, fusiform, cylindrical and clavate. These are very useful devices for storing of food and water to enable them to withstand periods of drought. These stems contain a large number of cells which absorb and store moisture during the rainy season.

Leaves

In most of the orchids, the leaves are arranged opposite or alternately on the pseudostem or may be grouped at the apex of a pseudobulb. The number of leaves present on the apex of a pseudobulb is often treated as an important taxonomic character to group species of a genus. The orchid leaves are of two main types – plicate and conduplicate. Plicates are thin, membranous and parallel ribbed as in *Cypripedium, Calanthe* etc., while the conduplicates are thick, fleshy or coriaceous.

Inflorescence

The most common type of inflorescence met within orchids is raceme but sometimes it may be in the form of an umbel, spike and cyme.

Flowers

Although orchids are the highest evolved among the monocotyledons, the flowers are comparatively simple within seven floral parts - three sepals, three petals and the column or Gynostemium. The lip or lebellum is the showy part of the flower. It is a modified petal and not a separate structure as is commonly believed by a layman. The lip has considerable variation in shape. The column is the reproductive part of the orchid flower and is distinct from other plants. It contains the male (staminate) and female (pistillate) organs of the flower. Striking variations are found in orchid flowers which make them attractive.

Economic Uses of Orchids

Ornamental Uses

Variously shaped, coloured, large or small, long lasting, scented or unscented orchid flowers are a constant source of attraction and delight. More than tens of thousands of hybrids of a large number of orchid genera are now under cultivation for their flowers and today orchid growing is a multi million dollar industry.

Medicinal Uses

The list of orchids and their medicinal importance are given below in Table 19.1.

Table 19.1: Medicinal and Other Uses of Orchids

Orchid Species	Plant Part Used	Ailment
Acampae papillosa	Roots	Rheumatism
Anthogonium gracile	Tuber	Joining broken articles and as glue
Calanthe triplicata	Root extract	Diarrhoea and teeth cavities
Cleisostoma williamsonii	Whole plant	Bone fractures
Coelogyne punctulata	Dried pseudobulbs	Healing of cuts
Cymbidium aloifolium	Entire plant	Purgative, emetic
Cymbidium ensifolium	Decoction of roots	Gonorrhoea
Dactylorhiza hatagirea	Tubers	Astringent, expectorent
Dendrobium hookerianum	Flower	To impart bright yellow colour to the cloth
Dendrobium ovatun	Plant juice	Stomach ache and laxative
Dendrobium nobile	Seed powder	Wound healing
Eria pannea	Decoction of roots	Bone -ache
Eulophia nuda	Tubers	Blood purifier, vermifuge
Satyrium nepalense	Tubers	Malaria, dysentry
Vanda spathulata	Plant juice	Bile

Spiritual Uses

In Arunachal Pradesh, orchids are a part and parcel of tribal life and culture. The orchids like *Dendrobium hookerianum* (Lishang Momdang), *D.nobile* (Bomzang momdang) and *D. gibsonii* are normally cultivated in Gompas symbolizing the sanctity of the place and their use for holy worship. Similarly *Vanda coerulea* (Blue Vanda), popularly known as "Rangpu" by Wangchu tribals of Tirap district is invariably associated with their festival dances for decoration, similar to that of *Rhynchostylis retusa* – "Kopu-phul" used in Assam.

Table 19.2: Threatened Orchids in Arunachal Pradesh

Sl.No.	Scientific Name	Status	Distribution Sites and Altitudes
1.	Bulleyia yunnanensis	Rare	-
2.	Calanthe mannii	Rare	Khasi hills.
3.	Cymbidium eburneum	Vulnerable	Endemic to Eastern Himalaya and N.E. India. 1000-1500 m.
4.	Cymbidium hookerianum	Vulnerable	1700-2500 m
5.	Diplomeris hirsuta	Vulnerable	Kameng. 1500-2000 m.
6.	Diplomeris pulchella	Vulnerable	Tirap district.
7.	Paphiopedilum wardii	Endangered	Lohit.
8.	Pholidota wattii	Rare	Endemic to North-eastern India.
9.	Vanda coerulea	Rare	1300-2000 m.
10.	Paphiopedilum fairrieanum	Endangered	Endemic to the Eastern Himalayas.

Source: Red Data Book Plants of India (Nayar and Sastry 1987-88).

Table 19.3: Genera and Species of Orchids Found in India and Arunachal Pradesh

Sl.No.	Genera	No. of Species Found in India	No. of Species Found in Arunachal Pradesh	Sl.No.	Genera	No. of Species Found in India	No. of Species Found in Arunachal Pradesh
1.	Acampae	6	4	36.	Dactylorhiza	1	1
2.	Acanthephippium	3	2	37.	Dendrobium	91	47
3.	Acrochaene	1	1	38.	Didymoplexis	2	1
4.	Aerides	10	5	39.	Diglyphosa	1	1
5.	Agrostophyllum	4	3	40.	Diphylax	2	1
6.	Ania	1	1	41.	Diplomeris	2	2
7.	Anoectochilus	16	7	42.	Epigeneium	5	5
8.	Anthogonium	1	1	43.	Epipogium	6	4
9.	Aorchis	2	1	44.	Eria	51	31
10.	Aphyllorchis	6	2	45.	Erythrorchis	1	1
11.	Apostasia	3	1	46.	Esmeralda	2	2
12.	Arachnis	2	2	47.	Eulophia	32	6
13.	Arundina	1	1	48.	Flickingeria	7	4
14.	Ascocentrum	4	4	49.	Galeola	6	5
15.	Biermannia	2	2	50.	Gastrochilus	12	9
16.	Brachycorythis	8	2	51.	Gastrodia	5	1
17.	Bulbophyllum	100	61	52.	Geodorum	1	1
18.	Bulleyia	1	1	53.	Goodyera	17	14
19.	Calanthe	23	17	54.	Gymnadenia	2	1
20.	Camarotis	2	1	55.	Habenaria	66	9
21.	Cephalanthera	3	2	56.	Herminium	14	6
22.	Ceratostylis	1	1	57.	Herpysma	1	1
23.	Chamaegastrodia	1	1	58.	Hetaeria	6	1
24.	Cheirostylis	9	6	59.	Katherinea	1	1
25.	Chiloschista	4	1	60.	Kingidium	2	2
26.	Chrysoglossum	6	2	61.	Liparis	45	19
27.	Cleisocentron	1	1	62.	Listera	11	1
28.	Cleisostoma	19	9	63.	Luisia	16	5
29.	Coelogyne	43	25	64.	Malaxis	19	6
30.	Collabium	1	1	65.	Micropera	3	3
31.	Corymborkis	1	1	66.	Mischobulbum	1	1
32.	Cremastra	1	1	67.	Monomeria	1	1
33.	Cryptochilus	2	2	68.	Myrmechis	1	1
34.	Cymbidium	23	19	69.	Neogyna	1	1
35.	Cypripedium	4	1	70.	Neotainiopsis	1	1

Contd...

Table 19.3–*Contd...*

Sl.No.	Genera	No. of Species Found in India	No. of Species Found in Arunachal Pradesh	Sl.No.	Genera	No. of Species Found in India	No. of Species Found in Arunachal Pradesh
71.	*Neottia*	3	1	97.	*Rhynchostylis*	1	1
72.	*Nephelaphyllum*	3	1	98.	*Ritaia*	1	1
73.	*Nervilia*	14	6	99.	*Robiquetia*	4	2
74.	*Oberonia*	50	15	100.	*Saccolabiopsis*	1	1
75.	*Oreorchis*	5	1	101.	*Sarcoglyphis*	1	1
76.	*Ornithochilus*	1	1	102.	*Satyrium*	2	1
77.	*Otochilus*	4	4	103.	*Schoenorchis*	6	2
78.	*Pachystoma*	2	1	104.	*Smitinandia*	1	1
79.	*Panisea*	3	2	105.	*Spathoglottis*	3	3
80.	*Paphiopedilum*	9	3	106.	*Spiranthes*	2	1
81.	*Papilionanthe*	4	3	107.	*Stereochilus*	2	1
82.	*Pecteilis*	4	1	108.	*Stereosandra*	1	1
83.	*Peristylus*	28	5	109.	*Sunipia*	10	7
84.	*Phaius*	7	5	110.	*Taeniophyllum*	7	2
85.	*Phalaenopsis*	7	2	111.	*Tainia*	9	3
86.	*Pholidota*	10	7	112.	*Thelasis*	4	2
87.	*Phreatia*	2	1	113.	*Thrixspermum*	9	3
88.	*Physurus*	2	1	114.	*Thunia*	2	1
89.	*Platanthera*	12	6	115.	*Trias*	3	3
90.	*Pleione*	8	4	116.	*Trichotosia*	3	3
91.	*Podochilus*	4	2	117.	*Tropidia*	3	3
92.	*Polystachya*	1	1	118.	*Tylostylis*	1	1
93.	*Pomatocalpa*	8	3	119.	*Uncifera*	3	2
94.	*Porpax*	7	1	120.	*Vanda*	12	9
95.	*Pteroceras*	5	1	121.	*Vandopsis*	1	1
96.	*Renanthera*	1	1	122.	*Zeuxine*	15	5

Orchid Research Centre at Tipi

Realising the importance of orchids, Govt. of Arunachal Pradesh established Orchid Research Centre at Tipi on way to Bomdila in West Kameng District to promote orchid conservation, research and development. Over the past 27 years, the center has been developed as a center of excellence and a place of tourist attraction with central orchidarium containing a number of exotic species and hybrids of the state. There area number of orchid bouses, natural orchidarium and botanic garden with curious and rare orchids like *Paphiopedilum venustum*, *P. spicerianum*, *Vanda coerulea*, *Coelogyne*, *Renanthera imschootiana* etc. The idyllic location surrounded by lush green forests with rising hills transversed by the River Kameng makes the center a paradise and tourist spot worth visiting for

Table 19.4: Orchid Species Distributed in Arunachal Pradesh

Sl.No.	Botanical Name	Altitudes	Area of Distribution	Classification	Flowering and Fruiting
1.	Acampae multiflora	300–1200 m	Subansiri, Tirap	Epiphyte	June–September
2.	Acampae ochracea	200–1000 m	Kameng, Tirap	Epiphyte	November–February
3.	Acampae papillosa	500–900 m	Subansiri, Tirap	Epiphyte	October–February
4.	Acampae rigida	500–1500 m	Kameng, Subansiri	Epiphyte	May–July
5.	Acanthephippium striatum	500–1000 m	Kameng, Lohit, Siang, Subansiri, Tirap	Terrestrial	July–October
6.	Acanthephippium sylhetense	1000–1500 m	Siang	Terrestrial	April–June
7.	Acrochaene punctata	1000–1500 m	Subansiri	Epiphyte	October–January
8.	Aerides falcata	1000–1500 m	Kameng, Subansiri	Epiphyte	July–September
9.	Aerides multiflora	100–1200 m	Kameng, Subansiri	Epiphyte	May–September
10.	Aerides odorata	100–1000 m	Kameng, Lohit, Siang, Subansiri	Epiphyte	June–August
11.	Aerides rosea	1500–1800 m	Subansiri	Epiphyte	August–September
12.	Aerides williamsii	500–1000 m	Kameng, Subansiri	Epiphyte	September–November
13.	Agrostophyllum brevipes	1000–1800 m	Kameng, Lohit, Siang, Subansiri, Tirap	Epiphyte	August–November
14.	Agrostophyllum callosum	1000–2000 m	Kameng, Subansiri	Epiphyte	June–September
15.	Agrostophyllum khasianum	500–800 m	Kameng, Lohit, Siang, Subansiri, Tirap	Epiphyte	August–October
16.	Agrostophyllum myrianthum	200–500 m	Kameng, Subansiri	Epiphyte	August–October
17.	Ania hookeriana	500–1000 m	Kameng, Subansiri	Terrestrial	March–May
18.	Anoectochilus brevilabris	1000–1500 m	Kameng, Lohit, Siang, Subansiri, Tirap	Terrestrial	April–August
19.	Anoectochilus elwesii	1000–1500 m	Kameng	Terrestrial	July–September
20.	Anoectochilus grandiflorus	1000–1500 m	Kameng	Terrestrial	July–September
21.	Anoectochilus lanceolatus	1000–2000 m	Kameng, Lohit, Siang, Subansiri	Terrestrial	July–October
22.	Anoectochilus roxburghii	1000–1500 m	Kameng	Terrestrial	July–October
23.	Anoectochilus sikkimensis	1200–1800 m	Kameng, Lohit, Siang, Subansiri, Tirap	Terrestrial	August–October
24.	Anoectochilus tortus	1000–1500 m	Kameng	Terrestrial	December–February
25.	Anthogonium gracile	900–2500 m	Kameng, Lohit, Siang, Subansiri, Tirap	Terrestrial	June–October
26.	Aorchis roborovskii	About 3000 m	Lohit	Terrestrial	August–September
27.	Aphyllorchis alpina	1800–2000 m	Kameng	Terrestrial	July–September

Contd...

Table 19.4—*Contd...*

Sl No.	Botanical Name	Altitudes	Area of Distribution	Classification	Flowering and Fruiting
28.	*Aphyllorchis montana*	2000–3500 m	Subansiri	Saprophytic	July–September
29.	*Apostasia odorata*	200–1000 m	Kameng	Terrestrial	June–August
30.	*Arachnis flos-aeris*	100–300 m	Kameng, Lohit	Epiphyte	April–July
31.	*Arachnis labrosa*	800–1100 m	Kameng	Epiphyte	August–November
32.	*Arundina graminifolia*	300–1000 m	Kameng, Lohit, Siang, Subansiri, Tirap	Terrestrial	March–October
33.	*Ascocentrum ampullaceum*	300–1000 m	Kameng	Epiphyte	March–April
34.	*Ascocentrum curvifolium*	300–1000 m	Kameng, Lohit, Siang, Subansiri, Tirap	Epiphyte	March–May
35.	*Ascocentrum himalaicum*	800–1200 m	Kameng	Epiphyte	May–July
36.	*Ascocentrum semiteretifolium*	100–300 m	Kameng	Epiphyte	March–April
37.	*Biermannia bimaculata*	500–1000 m	Kameng, Subansiri	Epiphyte	April–September
38.	*Biermannia jainiana*	500–1100 m	Kameng	Pendulous Epiphyte	May–June
39.	*Brachycorythis iantha*	1500–2000 m	Kameng	Terrestrial	November–January
40.	*Brachycorythis obcordata*	500–1500 m	Kameng, Lohit, Siang, Subansiri, Tirap	Terrestrial	July–September
41.	*Bulbophyllum acutiflorum*	1000–2000 m	Kameng	Creeping Epiphyte	July–September
42.	*Bulbophyllum affine*	1200–2000 m	Changlang, Kameng, Lohit, Siang, Subansiri, Tirap	Creeping Epiphyte	June–August
43.	*Bulbophyllum amplifolium*	1000–1300 m	Arunachal Pradesh	Creeping Epiphyte	October–December
44.	*Bulbophyllum andersonii*	1400–1800 m	Subansiri	Creeping Epiphyte	September–November
45.	*Bulbophyllum bisetum*	1500–2000 m	Kameng	Creeping Epiphyte	September–November
46.	*Bulbophyllum brienianum*	1000–1100 m	Kameng	Epiphyte	April–December
47.	*Bulbophyllum capillipes*	1500–2000 m	Kameng, Subansiri, Tirap	Epiphyte	July–September
48.	*Bulbophyllum careyanum*	500–1500 m	Kameng, Lohit, Siang, Subansiri, Tirap	Creeping Epiphyte	October–December
49.	*Bulbophyllum cariniflorum*	1800–2000 m	Kameng, Siang, Subansiri	Epiphyte	July–September
50.	*Bulbophyllum caudatum*	1000–1500 m	Kameng	Creeping Epiphyte	June–August
51.	*Bulbophyllum cauliflorum*	1100–3000 m	Kameng, Lohit, Siang, Subansiri, Tirap	Epiphyte	May–November
52.	*Bulbophyllum clarkeanum*	500–1500 m	Kameng	Creeping Epiphyte	June–August
53.	*Bulbophyllum cornucervi*	700–1000 m	Kameng	Epiphyte	July–September
54.	*Bulbophyllum crassipes*	100–500 m	Kameng	Creeping Epiphyte	September–October

Contd...

Table 19.4–*Contd...*

Sl.No.	Botanical Name	Altitudes	Area of Distribution	Classification	Flowering and Fruiting
55.	*Bulbophyllum cupreum*	1000–1500 m	Kameng	Creeping Epiphyte	October–December
56.	*Bulbophyllum cylindraceum*	1500–3500 m	Kameng, Lohit, Siang, Subansiri, Tirap	Epiphyte	September–November
57.	*Bulbophyllum delitescens*	500–1200 m	Kameng, Subansiri	Epiphyte	July–October
58.	*Bulbophyllum devangiriensis*	1000–1500 m	Changlang, Kameng, Subansiri, Tirap	Creeping Epiphyte	August–October
59.	*Bulbophyllum ebulbum*	Upto 500 m	Kameng, Tirap	Creeping Epiphyte	May–June
60.	*Bulbophyllum elatum*	900–1500 m	Kameng	Epiphyte	May–July
61.	*Bulbophyllum emarginatum*	1800–2200 m	Kameng	Creeping Epiphyte	September–November
62.	*Bulbophyllum eublepharum*	1800–2000 m	Changlang, Kameng, Lohit, Siang, Subansiri, Tirap	Epiphyte	August–October
63.	*Bulbophyllum gamblei*	200–2000 m	Kameng	Creeping Epiphyte	July–September
64.	*Bulbophyllum griffithii*	1500–2000 m	Kameng, Lohit, Siang, Subansiri, Tirap	Epiphyte	July–September
65.	*Bulbophyllum guttulatum*	1000–2000 m	Kameng	Epiphyte	August–October
66.	*Bulbophyllum gymnopus*	1200–1600 m	Kameng, Subansiri	Creeping Epiphyte	November–February
67.	*Bulbophyllum hastatum*	1800–2200 m	Lohit, Siang, Tirap	Creeping Epiphyte	May–July
68.	*Bulbophyllum helenae*	1000–2000 m	Kameng, Subansiri	Creeping Epiphyte	May–August
69.	*Bulbophyllum hookeri*	1200–1500 m	Arunachal Pradesh	Epiphyte	September–December
70.	*Bulbophyllum hymenanthum*	2000–3000 m	Kameng, Lohit, Siang, Subansiri, Tirap	Creeping Epiphyte	May–June
71.	*Bulbophyllum khasianum*	1000–2000 m	Kameng, Lohit, Siang, Subansiri, Tirap	Epiphyte	September–November
72.	*Bulbophyllum leopardinum*	2000–2500 m	Kameng, Lohit, Siang, Subansiri, Tirap	Epiphyte	October–January
73.	*Bulbophyllum lobbii*	1000–1500 m	Arunachal Pradesh	Creeping Epiphyte	July–September
74.	*Bulbophyllum macraei*	800–1800 m	Kameng	Epiphyte	August–September
75.	*Bulbophyllum obrienianum*	500–1000 m	Kameng	Epiphyte	April–June
76.	*Bulbophyllum odoratissimum*	1000–2000 m	Kameng, Lohit, Siang, Subansiri, Tirap	Creeping Epiphyte	May–September
77.	*Bulbophyllum ornatissimum*	1500–2500 m	Kameng, Siang	Creeping Epiphyte	June–October
78.	*Bulbophyllum parviflorum*	1500–2000 m	Kameng	Creeping Epiphyte	June–August
79.	*Bulbophyllum penicillium*	900–1100 m	Kameng	Epiphyte	November–December
80.	*Bulbophyllum pituliferum*	200–800 m	Kameng	Epiphyte	April–July
81.	*Bulbophyllum polyrhizum*	600–1100 m	Kameng	Creeping Epiphyte	December–February

Contd...

Table 19.4—*Contd...*

Sl.No.	Botanical Name	Altitudes	Area of Distribution	Classification	Flowering and Fruiting
82.	*Bulbophyllum protractum*	500–1000 m	Kameng, Lohit, Siang, Subansiri, Tirap	Creeping Epiphyte	June–August
83.	*Bulbophyllum pulchrum*	1500–2000 m	Kameng	Creeping Epiphyte	October–December
84.	*Bulbophyllum repens*	800–1000 m	Kameng	Creeping Epiphyte	November–December
85.	*Bulbophyllum reptans* var. *reptans*	1600–2000 m	Kameng,Siang, Subansiri, Tirap	Creeping Epiphyte	October–January
	var. *subracemosa*	1500–2000 m	Kameng, Lohit, Siang, Subansiri, Tirap	Creeping Epiphyte	March–May
86.	*Bulbophyllum retusiusculum*	2000–3500 m	Kameng, Tawang	Creeping Epiphyte	August–September
87.	*Bulbophyllum rolfei*	1500–2000 m	Kameng, Lohit, Siang, Subansiri, Tirap	Creeping Epiphyte	August–October
88.	*Bulbophyllum sarcophyllum* var. *minor*	200–1000 m	Kameng	Creeping Epiphyte	August–October
	Bulbophyllum sarcophyllum var. *sarcophyllum*	1000–1500 m	Kameng, Tirap	Creeping Epiphyte	June–October
89.	*Bulbophyllum scabratum*	500–1500 m	Kameng, Tirap	Epiphyte	April–July
90.	*Bulbophyllum secundum*	–	Kameng, Lohit, Siang, Subansiri, Tirap	Epiphyte	June–September
91.	*Bulbophyllum sikkimense*	200–1000 m	Kameng	Creeping Epiphyte	May–July
92.	*Bulbophyllum spathulatum*	200–500 m	Kameng, Tirap	Creeping Epiphyte	September–November
93.	*Bulbophyllum striatum*	1000–2000 m	Kameng	Creeping Epiphyte	October–December
94.	*Bulbophyllum thomsoni*	100–1000 m	Kameng, Lohit, Siang, Subansiri, Tirap	Creeping Epiphyte	April–May
95.	*Bulbophyllum tortuosum*	200–500 m	Arunachal Pradesh	Epiphyte	March–May
96.	*Bulbophyllum trichocephalum*	500–900 m	Changlang	Creeping Epiphyte	April–June
97.	*Bulbophyllum triste*	800–1000 m	Kameng, Lohit, Siang, Subansiri, Tirap	Creeping Epiphyte	March–May
98.	*Bulbophyllum viridiflorum*	1500–2500 m	Changlang	Epiphyte	October–November
99.	*Bulbophyllum wallichii*	1200–2000 m	Kameng	Epiphyte	October–December
100.	*Bulbophyllum xylophyllum*	500–800 m	Changlang	Epiphyte	September–November
101.	*Bulleyia yunnanensis*	1600–2000 m	Subansiri	Lithophytic herbs	April–June
102.	*Calanthe alismaefolia*	1000–1200 m	Kameng, Lohit, Siang, Subansiri, Tirap	Terrestrial	May–July
103.	*Calanthe alpina*	1000–3000 m	Kameng, Lohit, Siang	Terrestrial	July–September
104.	*Calanthe angusta*	200–500 m	Kameng, Siang	Terrestrial	April–June

Contd...

Table 19.4—Contd...

Sl.No.	Botanical Name	Altitudes	Area of Distribution	Classification	Flowering and Fruiting
105.	*Calanthe brevicornu* var. *brevicornu*	2000–3000 m	Kameng, Subansiri	Terrestrial	May–July
	Calanthe brevicornu var. *wattii*	1800–3000 m	Kameng, Lohit, Siang, Subansiri, Tirap	Terrestrial	May–July
106.	*Calanthe densiflora*	1200–1800 m	Siang	Terrestrial	September–February
107.	*Calanthe gracilis*	1500–2000 m	Subansiri	Terrestrial	September–October
108.	*Calanthe griffithii*	800–1000 m	Kameng, Lohit, Siang, Subansiri, Tirap	Terrestrial	August–October
109.	*Calanthe herbacea*	1000–2500 m	Kameng, Lohit, Siang, Subansiri, Tirap	Terrestrial	June–August
110.	*Calanthe masuca*	100–1000 m	Kameng, Lohit, Siang, Subansiri, Tirap	Terrestrial	August–September
111.	*Calanthe plantaginea*	800–1500 m	Kameng, Lohit, Siang, Subansiri, Tirap	Terrestrial	July–September
112.	*Calanthe puberula*	2000–2500 m	Subansiri	Terrestrial	July–October
113.	*Calanthe tricarinata*	1500–2000 m	Subansiri	Terrestrial	August–October
114.	*Calanthe triplicata*	1000–2000 m	Kameng	Terrestrial	May–August
115.	*Calanthe trulliformis*	1000–1500 m	Kameng	Terrestrial	July–September
116.	*Camarotis mannii*	300–600 m	Kameng, Subansiri	Epiphytic	June–July
117.	*Cephalanthera damasonium*	1500–1700 m	Kameng	Terrestrial	June–August
118.	*Cephalanthera longifolia*	1400–1800 m	Kameng, Lohit, Siang, Subansiri, Tirap	Terrestrial	July–September
119.	*Ceratostylis subulata*	500–1000 m	Siang	Epiphytic	May–August
120.	*Chamaegastrodia shikokiana*	500–1000 m	Kameng	Saprophytic	May–June
121.	*Cheirostylis griffithii*	500–1500 m	Kameng, Lohit, Siang, Subansiri, Tirap	Terrestrial	December–January
122.	*Cheirostylis moniliformis*	1500–2000 m	Lohit	Terrestrial	December–January
123.	*Cheirostylis munnacampensis*	800–1200 m	Kameng	Terrestrial	May–July
124.	*Cheirostylis pusilla*	500–1000 m	Changlang	Terrestrial	June–August
125.	*Cheirostylis sessanica*	1000–1200 m	Kameng	Terrestrial	March–May
126.	*Cheirostylis tippica*	100–300 m	Kameng	Terrestrial	January–April
127.	*Chiloschista parishii*	100–1000 m	Kameng	Epiphytic	May–July
128.	*Chrysoglossum erraticum*	1000–1300 m	Kameng	Terrestrial	June–August
129.	*Chrysoglossum robinsonii*	800–1200 m	Kameng	Terrestrial	April–June
130.	*Cleisocentron trichromum*	300–2000 m	Kameng, Tirap	Epiphytic	June–September

Contd...

Table 19.4—*Contd...*

Sl.No.	Botanical Name	Altitudes	Area of Distribution	Classification	Flowering and Fruiting
131.	*Cleisostoma appendiculatum*	100–1000 m	Kameng	Epiphytic	August–October
132.	*Cleisostoma aspersum*	100–1000 m	Kameng, Lohit, Siang, Subansiri, Tirap	Epiphytic	May–August
133.	*Cleisostoma discolour*	500–1200 m	Kameng	Epiphytic	June–July
134.	*Cleisostoma filiforme*	50–100 m	Subansiri	Epiphytic	April–June
135.	*Cleisostoma paniculatum*	800–1800 m	Lohit	Epiphytic	September–February
136.	*Cleisostoma racemiferum*	800–1800 m	Kameng	Epiphytic	July–September
137.	*Cleisostoma simondii*	100–1000 m	Kameng, Lohit, Siang, Subansiri, Tirap	Epiphytic	August–October
138.	*Cleisostoma subulatum*	500–1000 m	Siang	Epiphytic	May–August
139.	*Cleisostoma williamsonii*	500–1000 m	Kameng	Epiphytic	April–June
140.	*Coelogyne arunachalensis*	500–800 m	Lower Subansiri	Epiphytic	October–February
141.	*Coelogyne corymbosa*	500–2000 m	Kameng, Lohit, Siang, Subansiri, Tirap	Epiphytic	October–December
142.	*Coelogyne cristata*	1500–2000 m	Kameng, Lohit, Siang, Subansiri, Tirap	Epiphytic	January–May
143.	*Coelogyne fimbriata*	1000–2000 m	Subansiri	Creeping Epiphyte	August–November
144.	*Coelogyne flaccida*	1000–1500 m	Kameng, Siang, Subansiri, Tirap	Epiphytic	March–June
145.	*Coelogyne flavida*	1200–1800 m	Siang, Subansiri	Epiphytic	May–August
146.	*Coelogyne fuliginosa*	1000–1600 m	Kameng, Siang, Subansiri, Tirap	Creeping Epiphyte	September–December
147.	*Coelogyne fuscescens var. brunnea*	1200–2000 m	Kameng	Lithophytic	November–February
	Coelogyne fuscescens var. fuscescens	600–1500 m	Kameng, Lohit, Siang, Subansiri, Tirap	Lithophytic	September–January
148.	*Coelogyne griffithii*	1000–2000 m	Kameng, Lohit, Siang, Subansiri, Tirap	Epiphytic	April–August
149.	*Coelogyne holochila*	500–1500 m	Kameng	Epiphytic	April–July
150.	*Coelogyne longipes*	1500–2500 m	Kameng, Lohit, Siang, Subansiri, Tirap	Epiphytic	April–July
151.	*Coelogyne micrantha*	1200–1800 m	Subansiri	Epiphytic	September–January
152.	*Coelogyne nitida*	1200–1800 m	Siang, Subansiri	Epiphytic	March–August
153.	*Coelogyne occultata*	1000–1200 m	Kameng, Lohit, Siang, Subansiri, Tirap	Epiphytic	May–August
154.	*Coelogyne ovalis*	800–1800 m	Kameng, Lohit, Siang, Subansiri, Tirap	Epiphytic	August–December
155.	*Coelogyne prolifera*	1200–1800 m	Siang, Subansiri	Epiphytic	May–June

Contd...

Table 19.4—Contd...

Sl.No.	Botanical Name	Altitudes	Area of Distribution	Classification	Flowering and Fruiting
156.	Coelogyne punctulata	1500–2500 m	Kameng, Lohit, Siang, Subansiri, Tirap	Epiphytic	September–March
157.	Coelogyne radicosa	1000–1200 m	Siang	Epiphytic	November–February
158.	Coelogyne raizadae	1800–3000 m	Kameng	Epiphytic	March–September
159.	Coelogyne rigida	1500–2000 m	Kameng, Siang, Tirap	Epiphytic	June–October
160.	Coelogyne schultesii	1500–2000 m	Kameng	Epiphytic	March–June
161.	Coelogyne stricta	500–1500 m	Kameng	Epiphytic	March–September
162.	Coelogyne suaveolens	1500–2000 m	Kameng, Lohit, Siang, Subansiri, Tirap	Epiphytic	May–September
163.	Coelogyne viscosa	1200–2000 m	Kameng, Lohit, Siang, Subansiri, Tirap	Epiphytic	December–June
164.	Collabium chinense	1000–1200 m	Kameng	Terrestrial	July–October
165.	Corymborkis veratrifolia	500–1000 m	Kameng, Lohit, Siang, Subansiri, Tirap	Terrestrial	March–August
166.	Cremastra appendiculata	1000–2500 m	Kameng	Terrestrial	April–June
167.	Cryptochilus lutea	1500–2000 m	Kameng, Lohit, Siang, Subansiri, Tirap	Epiphytic	May–August
168.	Cryptochilus sanguineus	1400–1800 m	Kameng, Lohit, Siang, Subansiri, Tirap	Epiphytic	June–September
169.	Cymbidium aloifolium	500–1000 m	Kameng, Lohit, Siang, Subansiri, Tirap	Epiphytic	April–August
170.	Cymbidium bicolor	100–800 m	Kameng, Lohit, Siang, Subansiri, Tirap	Epiphytic	May–August
171.	Cymbidium cochleare	1200–2000 m	Kameng, Lohit, Siang, Subansiri, Tirap	Epiphytic	August–Dcember
172.	Cymbidium cyperifolium	1000–1800 m	Kameng, Lohit, Siang, Subansiri, Tirap	Terrestrial	November–March
173.	Cymbidium dayanum	500–1000 m	Subansiri	Epiphytic or Terrestrial	March–July
174.	Cymbidium devonianum	1500–2000 m	Kameng, Lohit, Siang, Subansiri, Tirap	Epiphytic	May–August
175.	Cymbidium eburneum	1000–1500 m	Tirap	Epiphytic	March–May
176.	Cymbidium elegans	1500–2500 m	Kameng, Lohit, Siang, Subansiri, Tirap	Epiphytic	September–December
177.	Cymbidium ensifolium	1000–2000 m	Kameng	Terrestrial	April–July
178.	Cymbidium gammieanum	1000–1200 m	Kameng	Erect	December–February
179.	Cymbidium goeringii	1500–1800 m	Kameng	Terrestrial	November–February
180.	Cymbidium iridioides	1200–1800 m	Kameng, Lohit, Siang, Subansiri, Tirap	Epiphytic	October–November
181.	Cymbidium lancifolium	1800–2500 m	Kameng, Lohit, Siang, Subansiri, Tirap	Terrestrial	May–July
182.	Cymbidium lowianum	800–1000 m	Kameng	Epiphytic	February–June

Contd...

Table 19.4—*Contd...*

Sl.No.	Botanical Name	Altitudes	Area of Distribution	Classification	Flowering and Fruiting
183.	*Cymbidium macrorhizon*	1200–2000 m	Kameng	Saprophytic	May–August
184.	*Cymbidium mastersii*	1500–2200 m	Kameng, Lohit, Siang, Subansiri, Tirap	Epiphytic	October–December
185.	*Cymbidium parishii*	1000–1500 m	Tirap	Epiphytic	June–August
186.	*Cymbidium sinense*	300–700 m	Kameng, Subansiri	Terrestrial	October–December
187.	*Cypripedium macranthon*	2500–3500 m	Kameng	Terrestrial	July–August
188.	*Dactylorhiza hatagirea*	2500–3500 m	Kameng, Subansiri	Terrestrial	June–September
189.	*Dendrobium acinaciforme*	200–500 m	Kameng, Lohit, Siang, Subansiri, Tirap	Epiphytic	August–September
190.	*Dendrobium aduncum*	500–1000 m	Kameng, Lohit, Siang, Subansiri	Epiphytic	June–September
191.	*Dendrobium amoenum*	1000–1500 m	Kameng, Lohit, Siang, Subansiri, Tirap	Epiphytic	June–September
192.	*Dendrobium anceps*	500–1000 m	Kameng, Lohit, Siang, Subansiri, Tirap	Epiphytic	April–July
193.	*Dendrobium aphyllum*	200–1000 m	Kameng, Lohit, Siang, Subansiri, Tirap	Epiphytic	April–June
194.	*Dendrobium bicameratum*	1200–1800 m	Kameng	Epiphytic	August–October
195.	*Dendrobium candidum*	2000–2500 m	Kameng, Lohit, Siang, Subansiri, Tirap	Epiphytic	May–August
196.	*Dendrobium cathcartii*	300–1200 m	Kameng, Lohit, Siang, Subansiri, Tirap	Epiphytic	April–July
197.	*Dendrobium chrysanthum*	1200–1800 m	Kameng, Lohit, Siang, Subansiri, Tirap	Epiphytic	April–May
198.	*Dendrobium chrysotoxum*	300–1000 m	Kameng, Lohit, Siang, Subansiri, Tirap	Epiphytic	April–June
199.	*Dendrobium crepidatum*	500–1000 m	Kameng, Lohit, Siang, Subansiri, Tirap	Epiphytic	March–May
200.	*Dendrobium cretaceum*	1200–1800 m	Kameng, Lohit, Siang, Subansiri, Tirap	Epiphytic	May–July
201	*Dendrobium cumulatum*	100–1000 m	Kameng	Epiphytic	May–August
202.	*Dendrobium densiflorum*	500–1500 m	Kameng	Epiphytic	April–July
203.	*Dendrobium denudans*	500–1500 m	Kameng, Lohit, Siang	Epiphytic	April–July
204.	*Dendrobium devonianum*	1000–1500 m	Kameng, Lohit, Siang, Subansiri, Tirap	Epiphytic	March–June
205.	*Dendrobium eriaeflorum*	600–1000 m	Kameng, Lohit, Siang, Subansiri, Tirap	Epiphytic	November–January
206.	*Dendrobium falconeri*	500–1200 m	Kameng, Lohit, Siang	Epiphytic	October–January
207.	*Dendrobium farmeri*	500–1200 m	Kameng, Lohit, Siang, Subansiri, Tirap	Epiphytic	April–July

Contd...

Table 19.4—*Contd...*

Sl.No.	Botanical Name	Altitudes	Area of Distribution	Classification	Flowering and Fruiting
208.	*Dendrobium fimbriatum* var. *fimbriatum*	500–1500 m	Kameng, Subansiri	Epiphytic	March–May
	Dendrobium fimbriatum var. *oculatum*	500–1500 m	Kameng, Lohit, Subansiri	Epiphytic	March–June
209.	*Dendrobium formosum*	100–1000 m	Kameng	Epiphytic	May–July
210.	*Dendrobium gibsonii*	800–1500 m	Kameng, Lohit, Siang, Subansiri, Tirap	Epiphytic	July–August
211.	*Dendrobium heterocarpum*	1000–1200 m	Kameng	Epiphytic	April–May
212.	*Dendrobium hookerianum*	500–1200 m	Kameng, Lohit, Siang, Subansiri, Tirap	Epiphytic	October–November
213.	*Dendrobium jenkinsii*	500–1000 m	Kameng, Subansiri	Epiphytic	April–June
214.	*Dendrobium kentrophyllum*	200–400 m	Tirap	Epiphytic	July–August
215.	*Dendrobium lindleyi*	500–1000 m	Kameng, Lohit, Siang, Subansiri, Tirap	Epiphytic	April–June
216.	*Dendrobium lituiflorum*	500–1500 m	Kameng, Lohit, Siang, Subansiri, Tirap	Epiphytic	September–November
217.	*Dendrobium longicornu*	1000–1500 m	Kameng, Lohit, Siang, Subansiri, Tirap	Epiphytic	September–November
218.	*Dendrobium mannii*	100–500 m	Kameng	Epiphytic	November–December
219.	*Dendrobium monticola*	500–1000 m	Kameng	Epiphytic	April–June
220.	*Dendrobium moschatum*	500–1000 m	Kameng, Lohit, Siang, Subansiri, Tirap	Epiphytic	April–June
221.	*Dendrobium nareshbahadurii*	600–700 m	Kameng	Epiphytic	October–November
222.	*Dendrobium nobile*	1000–1500 m	Kameng, Lohit, Siang, Subansiri, Tirap	Epiphytic	April–June
223.	*Dendrobium palpebrae*	100–1000 m	Kameng, Lohit, Siang, Subansiri, Tirap	Epiphytic	May–July
224.	*Dendrobium parciflorum*	100–500 m	Kameng, Siang, Subansiri	Epiphytic	May–August
225.	*Dendrobium parishii*	1200–1500 m	Kameng, Lohit, Siang, Subansiri	Epiphytic	September–October
226.	*Dendrobium pauciflorum*	1500–2000 m	Kameng	Epiphytic	June–August
227.	*Dendrobium pendulum*	800–1200 m	Lohit, Tirap	Epiphytic	April–June
228.	*Dendrobium porphyrochilum*	1000–2000 m	Kameng, Siang	Epiphytic	April–July
229.	*Dendrobium primulinum*	200–1000 m	Kameng, Lohit, Siang, Subansiri	Epiphytic	April–June
230.	*Dendrobium pulchellum*	Upto 500 m	Kameng	Epiphytic	June–August
231.	*Dendrobium stuposum*	Upto 500 m	Kameng, Siang, Subansiri	Epiphytic	June–August
232.	*Dendrobium sulcatum*	100–1000 m	Kameng, Lohit, Siang, Subansiri, Tirap	Epiphytic	June–August

Contd...

Table 19.4—Contd...

Sl.No.	Botanical Name	Altitudes	Area of Distribution	Classification	Flowering and Fruiting
233.	*Dendrobium terminale*	1200–1800 m	Kameng	Epiphytic	July–January
234.	*Dendrobium transparens*	500–1200 m	Kameng, Lohit, Siang, Subansiri, Tirap	Epiphytic	June–August
235.	*Dendrobium wardianum*	1000–1200 m	Kameng	Epiphytic	August–October
236.	*Didymoplexis pallens*	500–1000 m	Kameng	Terrestrial	May–July
237.	*Diglyphosa latifolia*	1200–1500 m	Kameng, Lohit, Siang, Subansiri, Tirap	Terrestrial	September–November
238.	*Diphylax urceolata*	500–1200 m	Kameng	Terrestrial	April–August
239.	*Diplomeris hirsuta*	500–1500 m	Kameng, Lohit, Siang	Terrestrial	June–August
240.	*Diplomeris pulchella*	100–1000 m	Kameng, Lohit, Siang, Subansiri, Tirap	Terrestrial	July–October
241.	*Epigeneium amplum*	1500–1800 m	Kameng, Lohit, Tirap	Epiphytic	September–November
242.	*Epigeneium chapaense*	500–1000 m	Changlang, Lohit, Siang	Epiphytic	November–December
243.	*Epigeneium fargesii*	500–1500 m	Kameng, Lohit, Tirap	Epiphytic	September–October
244.	*Epigeneium fuscescens*	1500–2000 m	Kameng, Tirap	Epiphytic	September–November
245.	*Epigeneium rotundatum*	2000–2500 m	Kameng, Siang, Tirap	Epiphytic	April–June
246.	*Epipogium africanus*	Upto 200 m	Kameng	Saprophytic	January–February
247.	*Epipogium indicum*	300–600 m	Subansiri	Saprophytic	August–September
248.	*Epipogium roseum*	500–1200 m	Kameng, Subansiri	Saprophytic	October–December
249.	*Epipogium sessanum*	Upto 1100 m	Kameng	Saprophytic	June–August
250.	*Eria acervata*	500–1000 m	Kameng, Lohit, Siang, Subansiri, Tirap	Epiphytic	May–July
251.	*Eria amica*	500–1000 m	Kameng, Siang, Subansiri, Tirap	Epiphytic	March–May
252.	*Eria bambusifolia*	1000–2000 m	Kameng, Lohit, Siang, Subansiri, Tirap	Epiphytic	October–January
253.	*Eria biflora*	100–1000 m	Kameng, Subansiri	Epiphytic	August–October
254.	*Eria bipunctata*	1000–1500 m	Subansiri	Epiphytic	September–October
255.	*Eria bractescens*	200–1000 m	Subansiri	Epiphytic	April–June
256.	*Eria carinata*	1200–1800 m	Kameng, Lohit, Siang, Subansiri, Tirap	Epiphytic	July–November
257.	*Eria clausa*	1000–1500 m	Kameng, Lohit, Siang, Subansiri, Tirap	Epiphytic	March–May
258.	*Eria clavicaulis*	200–1000 m	Kameng, Lohit, Siang, Subansiri, Tirap	Epiphytic	September–October
259.	*Eria connata*	1100–1500 m	Kameng	Epiphytic	July–September

Contd...

Table 19.4—*Contd...*

Sl.No.	Botanical Name	Altitudes	Area of Distribution	Classification	Flowering and Fruiting
260.	Eria coronaria	1000–1500 m	Kameng, Lohit, Siang, Subansiri, Tirap	Epiphytic	November–February
261.	Eria cristata	500–1000 m	Siang, Subansiri, Tirap	Epiphytic	February–April
262.	Eria excavata	1500–2000 m	Kameng, Lohit, Siang, Subansiri, Tirap	Epiphytic	June–July
263.	Eria ferruginea	1200–1600 m	Kameng, Lohit, Siang, Subansiri, Tirap	Epiphytic	May–September
264.	Eria globulifera	1000–1500 m	Subansiri, Tirap	Epiphytic	November–December
265.	Eria graminifolia	1500–2000 m	Kameng, Lohit, Siang, Subansiri, Tirap	Epiphytic	July–September
266.	Eria javanica	500–1000 m	Kameng, Lohit, Siang, Subansiri, Tirap	Epiphytic	July–September
267.	Eria jengingensis	650–850 m	Siang	Epiphytic	March–June
268.	Eria lohitensis	1500–1600 m	Lohit	Epiphytic	May–July
269.	Eria muscicola	1200–1800 m	Kameng, Lohit, Siang, Subansiri, Tirap	Epiphytic	July–August
270.	Eria paniculata	1000–1500 m	Kameng, Lohit, Siang, Subansiri, Tirap	Epiphytic	December–March
271.	Eria pannea	100–1000 m	Kameng, Lohit, Siang, Subansiri, Tirap	Epiphytic	May–July
272.	Eria pubescens	100–1000 m	Siang, Subansiri, Tirap	Epiphytic	March–July
273.	Eria pudica	500–1200 m	Kameng, Subansiri	Epiphytic	April–July
274.	Eria pumila	200–500 m	Kameng, Lohit, Siang, Subansiri, Tirap	Epiphytic	January–March
275.	Eria pusilla	1000–1200 m	Kameng, Subansiri	Epiphytic	March–June
276.	Eria sharmae	450–500 m	Subansiri	Epiphytic	February–April
277.	Eria spicata	1000–1500 m	Kameng, Lohit, Siang, Subansiri, Tirap	Epiphytic	August–December
278.	Eria stricta	100–1000 m	Kameng, Lohit, Siang, Subansiri, Tirap	Epiphytic	November–January
279.	Eria tomentosa	100–1000 m	Kameng, Lohit, Subansiri,	Epiphytic	September–November
280.	Eria vittata	1200–2500 m	Kameng	Epiphytic	March–May
281.	Erythrorchis ochobiensis	200–600 m	Tirap	Saprophytic	May–July
282.	Esmeralda cathcartii	1000–1200 m	Kameng, Lohit, Siang, Subansiri, Tirap	Terrestrial	March–May
283.	Esmeralda clarkei	1000–1500 m	Kameng	Terrestrial	September–November
284.	Eulophia bicallosa	500–1200 m	Kameng, Lohit, Siang, Subansiri, Tirap	Terrestrial	March–May
285.	Eulophia dabia	100–500 m	Kameng	Terrestrial	March–May
286.	Eulophia hormusjii	1400–2000 m	Kameng	Terrestrial	May–July

Contd...

Table 19.4—Contd...

Sl.No.	Botanical Name	Altitudes	Area of Distribution	Classification	Flowering and Fruiting
287.	Eulophia nuda	100–500 m	Kameng	Terrestrial	April–July
288.	Eulophia sanguinea	300–800 m	Kameng, Lohit, Siang, Subansiri, Tirap	Terrestrial	May–June
289.	Eulophia zollingeri	500–1800 m	Kameng	Saprophytic	May–June
290.	Flickingeria bancana	Around 1100 m	Kameng	Epiphytic	May–August
291.	Flickingeria fugax	500–1500 m	Kameng, Lohit, Siang, Subansiri	Epiphytic	August–October
292.	Flickingeria macraei	200–1000 m	Kameng, Siang, Subansiri	Epiphytic	May–July
293.	Flickingeria ritaeana	1000–1200 m	Tirap	Epiphytic	August–October
294.	Galeola altissima	500–1000 m	Tirap	Saprophytic	April–August
295.	Galeola falconeri	1500–2500 m	Kameng, Lohit, Siang, Subansiri, Tirap	Saprophytic	August–November
296.	Galeola javanica	1000–1200 m	Kameng	Saprophytic	June–August
297.	Galeola lindleyana	1000–1500 m	Kameng, Lohit, Siang, Subansiri, Tirap	Saprophytic	August–November
298.	Galeola nudifolia	300–500 m	Subansiri	Saprophytic	April–May
299.	Gastrochilus acutifolius	800–1200 m	Kameng, Siang, Subansiri	Epiphytic	October–December
300.	Gastrochilus affinis	1000–1500 m	Kameng, Lohit, Subansiri, Tirap	Epiphytic	June–August
301.	Gastrochilus arunachalensis	100–500 m	Kameng	Epiphytic	November–January
302.	Gastrochilus calceolaris	500–1000 m	Tirap	Epiphytic	March–June
303.	Gastrochilus dasypogon	100–1000 m	Kameng, Siang, Subansiri, Tirap	Epiphytic	November–January
304.	Gastrochilus distichus	1200–1800 m	Subansiri, Tirap	Epiphytic	September–November
305.	Gastrochilus inconspicuus	500–1000 m	Kameng, Subansiri	Epiphytic	March–May
306.	Gastrochilus intermidius	500–1200 m	Kameng, Subansiri, Tirap	Epiphytic	July–September
307.	Gastrochilus crassilabris	NA	NA	Epiphytic	NA
308.	Gastrodia mishmensis	Upto 1500 m	Kameng	Saprophytic	August–September
309.	Geodorum densiflorum	200–800 m	Kameng, Lohit, Siang, Subansiri, Tirap	Terrestrial	June–July
310.	Goodyera biflora	1000–2000 m	Kameng	Terrestrial	July–September
311.	Goodyera foliosa	200–1000 m	Kameng, Lohit, Siang, Subansiri, Tirap	Terrestrial	September–November
312.	Goodyera fumata	1000–1400 m	Kameng	Terrestrial	March–April
313.	Goodyera fusca	1500–2500 m	Kameng	Terrestrial	July–September

Contd...

Table 19.4—*Contd...*

Sl.No.	Botanical Name	Altitudes	Area of Distribution	Classification	Flowering and Fruiting
314.	*Goodyera grandis*	500–1000 m	Siang	Terrestrial	August–September
315.	*Goodyera hemsleyana*	1500–2500 m	Kameng, Subansiri	Terrestrial	July–August
316.	*Goodyera hispida*	100–500 m	Siang	Terrestrial	August–October
317.	*Goodyera macrantha*	500–1200 m	Kameng	Terrestrial	July–September
318.	*Goodyera procera*	500–800 m	Kameng, Lohit, Siang, Subansiri, Tirap	Terrestrial	May–July
319.	*Goodyera recurva*	2000–3000 m	Kameng	Terrestrial	April–June
320.	*Goodyera repens*	1000–1200 m	Kameng	Terrestrial	July–September
321.	*Goodyera schlechtendaliana*	1500–2500 m	Kameng, Lohit, Siang, Tirap	Terrestrial	September–December
322.	*Goodyera viridiflora*	500–1500 m	Kameng, Lohit, Siang, Subansiri, Tirap	Terrestrial	September–November
323.	*Goodyera vittata*	200–4000 m	Kameng, Lohit, Siang, Subansiri, Tirap	Terrestrial	July–September
324.	*Gymnadenia orchidis*	3000–4000 m	Kameng, Lohit, Siang	–	June–August
325.	*Habenaria aculifera*	1200–1800 m	Kameng	Terrestrial	July–December
326.	*Habenaria arietina*	1500–3000 m	Kameng, Lohit, Siang, Subansiri, Tirap	Terrestrial	July–September
327.	*Habenaria cumminsiana*	3000–4000 m	Tawang	Terrestrial	September–November
328.	*Habenaria dentata*	1000–2500 m	Kameng, Siang, Subansiri	Terrestrial	April–August
329.	*Habenaria digitata*	1000–1500 m	Kameng	Terrestrial	July–September
330.	*Habenaria ensifolia*	1500–2000 m	Kameng, Siang	Terrestrial	May–August
331.	*Habenaria pectinata*	1500–1800 m	Kameng, Lohit, Siang, Subansiri, Tirap	Terrestrial	July–August
332.	*Habenaria seshagiriana*	800–2000 m	Kameng, Lohit, Subansiri, Tirap	Terrestrial	August–December
333.	*Habenaria stenopetala*	200–1500 m	Kameng, Lohit, Siang, Subansiri, Tirap	Terrestrial	August–October
334.	*Herminium haridasanii*	At 3000 m	Lohit	Terrestrial	August–September
335.	*Herminium jaffreyanum*	2200–3500 m	Kameng, Subansiri	Terrestrial	August–October
336.	*Herminium josephi*	3500–4000 m	Kameng	Terrestrial	June–August
337.	*Herminium lanceum*	1500–2000 m	Kameng	Terrestrial	July–October
338.	*Herminium longilobatum*	3000–3800 m	Kameng, Lohit	Epiphytic	September–October
339.	*Herpysma longicaulis*	1200–1800 m	Kameng, Lohit, Siang, Subansiri, Tirap	Terrestrial	July–September
340.	*Hetaeria rubens*	600–1000 m	Kameng, Lohit, Siang, Subansiri, Tirap	Terrestrial	August–October

Contd...

Table 19.4.–Contd...

Sl.No.	Botanical Name	Altitudes	Area of Distribution	Classification	Flowering and Fruiting
341.	*Katherinea navicularis*	1000–2000 m	Kameng	Epiphytic	July–October
342.	*Kingidium delicosum*	500–1000 m	Kameng	Epiphytic	June–August
343.	*Kingidium taenialis*	1000–2000 m	Kameng, Siang, Subansiri	Epiphytic	May–July
344.	*Liparis assamica*	100–1000 m	Kameng, Lohit, Siang, Subansiri, Tirap	Epiphytic	October–November
345.	*Liparis bistriata*	1200–1500 m	Kameng, Lohit, Siang, Subansiri, Tirap	Epiphytic	July–August
346.	*Liparis bootanensis*	1200–1500 m	Kameng, Lohit, Siang, Subansiri, Tirap	Epiphytic	July–August
347.	*Liparis caespitosa*	1500–2500 m	Tirap	Epiphytic	August–October
348.	*Liparis cathcartii*	1600–2500 m	Kameng, Lohit, Siang, Subansiri, Tirap	Terrestrial	July–August
349.	*Liparis cordifolia*	500–1000 m	Kameng, Subansiri	Terrestrial	October–November
350.	*Liparis delicatula*	1500–1800 m	Kameng, Lohit	Epiphytic	August–September
351.	*Liparis distans*	500–1000 m	Kameng, Lohit, Siang, Subansiri, Tirap	–	August–October
352.	*Liparis elliptica*	1200–2000 m	Kameng	Epiphytic	August–December
353.	*Liparis longipes*	400–1500 m	Kameng, Subansiri	Epiphytic	October–December
354.	*Liparis luteola*	1000–1500 m	Kameng, Lohit, Siang, Subansiri, Tirap	Epiphytic	May–September
355.	*Liparis mannii*	500–1000 m	Kameng	Epiphytic	Nvember–January
356.	*Liparis nervosa*	1000–1500 m	Kameng, Lohit, Siang, Subansiri, Tirap	Terrestrial	June–October
357.	*Liparis paradoxa*	100–500 m	Kameng, Lohit, Siang, Subansiri	Terrestrial	July–September
358.	*Liparis plantaginea*	200–1000 m	Kameng, Lohit, Siang, Subansiri, Tirap	Epiphytic	June–August
359.	*Liparis resupinata*	1500–2000 m	Kameng, Lohit, Siang, Subansiri, Tirap	Epiphytic	October–December
360.	*Liparis stricklandiana*	1500–1800 m	Kameng, Lohit, Siang, Subansiri, Tirap	Terrestrial	October–December
361.	*Liparis viridiflora*	600–1500 m	Kameng, Lohit, Siang, Subansiri, Tirap	Epiphytic	November–December
362.	*Liparis wrayii*	500–1000 m	Tirap	Terrestrial	August–October
363.	*Listera divaricata*	3000–4000 m	Kameng, Lohit	Terrestrial	August–September
364.	*Luisia brachystachys*	1000–1200 m	Kameng	Epiphytic	July–September
365.	*Luisia filiformis*	300–1000 m	Kameng	Epiphytic	April–June
366.	*Luisia psyche*	500–1000 m	Kameng, Lohit, Subansiri	Epiphytic	August–October
367.	*Luisia trichorhiza*	100–1000 m	Kameng, Siang, Subansiri, Tirap	Epiphytic	March–May

Contd...

Table 19.4—Contd...

Sl.No.	Botanical Name	Altitudes	Area of Distribution	Classification	Flowering and Fruiting
368.	*Luisia zeylanica*	500–1000 m	Kameng, Lohit, Siang, Subansiri, Tirap	Terrestrial	May–July
369.	*Malaxis acuminata* var. *acuminata*	500–1500 m	Kameng, Lohit, Siang, Subansiri, Tirap	Terrestrial	June–September
	Malaxis acuminata var. *biloba*	500–1000 m	Kameng	Terrestrial	July–September
370.	*Malaxis aphylla*	200–1000 m	Kameng	–	June–October
371.	*Malaxis josephiana*	500–1000 m	Kameng, Tirap	Terrestrial	May–August
372.	*Malaxis khasiana*	200–1000 m	Kameng, Tirap	Terrestrial	July–October
373.	*Malaxis latifolia*	200–1000 m	Kameng, Siang	Terrestrial	August–October
374.	*Malaxis mucifera*	1500–2500 m	Kameng	Terrestrial	April–July
375.	*Micropera mannii*	200–1000 m	Kameng	Epiphytic	June–July
376.	*Micropera obtusa*	100–1000 m	Kameng	Epiphytic	July–August
377.	*Micropera rostratum*	500–1500 m	Changlang	Epiphytic	July–September
378.	*Mischobulbum wrayanum*	200–500 m	Tirap	Terrestrial	July–September
379.	*Monomeria barbata*	1200 m	Kameng, Subansiri	Epiphytic	February–March
380.	*Myrmechis pumila*	1000–1200 m	Kameng, Lohit, Siang, Subansiri, Tirap	Terrestrial	June–August
381.	*Neogyna gardneriana*	1000–1500 m	Kameng	Epiphytic	June–September
382.	*Neotainiopsis barbata*	Upto 1100 m	Lohit	Epiphytic	September–October
383.	*Neottia acuminata*	2500–3000 m	Kameng	Saprophytic	August–September
384.	*Nephelaphyllum sikkimensis*	1000–1200 m	Kameng	Terrestrial	August–October
385.	*Nervilia aragoana*	100–1000 m	Kameng, Siang	Terrestrial	July–September
386.	*Nervilia gammieana*	1000–1200 m	Kameng, Siang	Terrestrial	May–July
387.	*Nervilia hookeriana*	1000–1500 m	Kameng	Terrestrial	August–October
388.	*Nervilia infundibulifolia*	1200–1500 m	Kameng	–	June–July
389.	*Nervilia juliana*	200–500 m	Kameng, Siang	Terrestrial	May–July
390.	*Nervilia macroglossa*	500–1200 m	Kameng	Terrestrial	July–September
391.	*Oberonia acaulis*	1000–1200 m	Kameng	Epiphytic	July–September
392.	*Oberonia anthropophora*	1000–1200 m	Kameng	Epiphytic	July–September
393.	*Oberonia auriculata*	100–1000 m		Epiphytic	September–November

Contd...

Table 19.4–*Contd...*

Sl.No.	Botanical Name	Altitudes	Area of Distribution	Classification	Flowering and Fruiting
394.	Oberonia caulescens	100–1000 m	Kameng	Epiphytic	July–September
395.	Oberonia emarginata	1600–2000 m	Kameng, Lohit, Siang, Subansiri, Tirap	Epiphytic	August–November
396.	Oberonia ensiformis	500–1000 m	Kameng, Siang, Subansiri	Epiphytic	October–December
397.	Oberonia falcata	1800–2200 m	Kameng, Lohit, Siang, Subansiri, Tirap	Epiphytic	July–September
398.	Oberonia falconeri	500–1800 m	Kameng, Siang, Subansiri, Tirap	Epiphytic	September–December
399.	Oberonia iridifolia	100–1000 m	Kameng, Lohit, Siang, Subansiri, Tirap	Epiphytic	September–December
400.	Oberonia jenkinsiana	200–500 m	Kameng, Lohit, Siang, Subansiri, Tirap	Epiphytic	August–October
401.	Oberonia maxima	1800–2200 m	Kameng	Epiphytic	June–October
402.	Oberonia obcordata	1600–1800 m	Kameng	Epiphytic	May–July
403.	Oberonia pachyrachis	800–1500 m	Kameng, Lohit, Siang, Subansiri, Tirap	Epiphytic	November–March
404.	Oberonia pyrulifera	300–800 m	Kameng, Lohit, Siang, Subansiri, Tirap	Epiphytic	November–January
405.	Oberonia rufilabris	100–1500 m	Kameng	Epiphytic	August–October
406.	Oreorchis micrantha	upto 2500 m	Kameng	Terrestrial	May–July
407.	Ornithochilus difformis	1400–2000 m	Kameng, Lohit, Siang, Subansiri, Tirap	Epiphytic	July–September
408.	Otochilus albus	1000–2000 m	Kameng, Lohit, Siang, Subansiri, Tirap	Epiphytic	May–September
409.	Otochilus fuscus	1200–1600 m	Kameng, Lohit, Siang, Subansiri, Tirap	Epiphytic	December–January
410.	Otochilus lancilabius	1000–1800 m	Arunachal Pradesh	Epiphytic	October–January
411.	Otochilus porrectus	1500–2000 m	Kameng, Lohit, Siang, Subansiri, Tirap	Epiphytic	October–January
412.	Pachystoma senile	1000–1500 m	Kameng, Tirap	Terrestrial	June–August
413.	Panisea demissa	1000–1500 m	Kameng	Epiphytic	November–December
414.	Panisea tricallosa	1200–1500 m	Kameng	Epiphytic	April–June
415.	Paphiopedilum fairieanum	1500–1800 m	Kameng	Terrestrial	October–February
416.	Paphiopedilum venustum	–	Kameng	Terrestrial	December–February
417.	Paphiopedilum	–	Lohit	Terrestrial	December–February
418.	Papilionanthe subulata	1500–2000 m	Kameng	Epiphytic	April–June
419.	Papilionanthe teres	50–100 m	Kameng, Subansiri, Tirap	Epiphytic	April–May
420.	Papilionanthe uniflora	1500–2000 m	Subansiri	Epiphytic	September–October

Contd...

Table 19.4—Contd...

Sl.No.	Botanical Name	Altitudes	Area of Distribution	Classification	Flowering and Fruiting
421.	*Pecteilis susannae*	Upto 1200 m	Kameng	Terrestrial	August–October
422.	*Peristylus affinis*	1000–2500 m	Arunachal Pradesh	Terrestrial	July–October
423.	*Peristylus fallax*	200–1000 m	Kameng, Lohit, Siang, Subansiri, Tirap	Terrestrial	June–August
424.	*Peristylus goodyeroides*	300–1000 m	Kameng, Lohit, Siang, Subansiri, Tirap	Terrestrial	April–July
425.	*Peristylus prainii*	1200–1500 m	Kameng	Terrestrial	July–September
426.	*Peristylus richardianus*	1500–1800 m	Kameng	Terrestrial	August–October
427.	*Phaius flavus*	500–1000 m	Kameng, Subansiri, Tirap	Terrestrial	April–June
428.	*Phaius longipes*	1500–2000 m	Tirap	Terrestrial	September–November
429.	*Phaius mishmensis*	1200–2000 m	Lohit, Siang, Subansiri, Tirap	Terrestrial	November–January
430.	*Phaius tancarvilleae*	1000–2000 m	Kameng, Lohit, Siang, Subansiri, Tirap	Terrestrial	April–June
431.	*Phaius woodfordii*	–	Arunachal Pradesh	Terrestrial	August–October
432.	*Phalaenopsis mannii*	200–1000 m	Kameng, Lohit, Siang, Subansiri	Epiphytic	March–July
433.	*Phalaenopsis parishii*	300–1000 m	Kameng, Subansiri	Epiphytic	March–April
434.	*Pholidota articulata*	500–1000 m	Kameng, Subansiri, Tirap	Epiphytic	July–October
435.	*Pholidota convallariae* var. *breviscapa*	Upto 1700 m	Kameng	Epiphytic	November–December
436.	*Pholidota imbricata* var. *imbricata*	1500–2000 m	Kameng, Lohit, Siang, Subansiri, Tirap	Epiphytic	June–August
	Pholidota imbricate var. *sessilis*	1200–1800 m	Kameng, Subansiri	Epiphytic	July–August
437.	*Pholidota protracta*	1500–2500 m	Kameng, Subansiri	Epiphytic	October–December
438.	*Pholidota pygmaea*	300–600 m	Subansiri	Epiphytic	October–December
439.	*Pholidota undulata*	1000–2000 m	Kameng	Epiphytic	November–January
440.	*Pholidota wattii*	500–1000 m	Subansiri	Epiphytic	April–June
441.	*Phreatia elegans*	1800–2200 m	Kameng, Lohit, Siang, Subansiri, Tirap	–	July–September
442.	*Physurus hirsutus*	500–1500 m	Siang, Tirap	Terrestrial	September–November
443.	*Platanthera bakeriana*	2500–3300 m	Lohit	Terrestrial	August–September
444.	*Platanthera clavigera*	600–1200 m	Kameng, Siang, Subansiri, Tirap	Terrestrial	May–July
445.	*Platanthera dyeriana*	500–1200 m	Kameng, Siang, Subansiri	Terrestrial	July–September

Contd...

Table 19.4–Contd...

Sl.No.	Botanical Name	Altitudes	Area of Distribution	Classification	Flowering and Fruiting
446.	*Platanthera latilabris*	900–2000 m	Kameng	Terrestrial	July–September
447.	*Platanthera leptocaulon*	900–2000 m	Kameng	Terrestrial	August–October
448.	*Platanthera stenantha*	200–3000 m	Kameng	Terrestrial	July–October
449.	*Pleione hookeriana*	2200–4000 m	Kameng, Lohit, Siang	Epiphytic	May–July
450.	*Pleione humilis*	2200–3000 m	Kameng, Siang	Epiphytic	February–April
451.	*Pleione maculata*	1000–1400 m	Tirap	Epiphytic	January–March
452.	*Pleione praecox*	1200–1800 m	Kameng, Siang, Tirap	Epiphytic	July–September
453.	*Podochilus cultratus*	500–1000 m	Tirap	Epiphytic	September–November
454.	*Podochilus khasianus*	800–1300 m	Kameng	Epiphytic	June–August
455.	*Polystachya concreta*	1000–1500 m	Tirap	Epiphytic	August–October
456.	*Pomatocalpa armigerum*	500–1000 m	Kameng	Epiphytic	August–October
457.	*Pomatocalpa undulatum*	500–1000 m	Kameng, Subansiri	Epiphytic	March–May
458.	*Pomatocalpa wendlandorum*	500–1200 m	Kameng, Siang, Tirap	Epiphytic	May–August
459.	*Porpax elwesii*	500–800 m	Kameng, Lohit, Siang	Epiphytic	September–October
460.	*Pteroceras suaveolens*	500–1000 m	Kameng, Siang, Subansiri	Epiphytic	April–August
461.	*Renanthera imschootiana*	500–1500 m	Kameng	Epiphytic	September–October
462.	*Rhynchostylis retusa*	500–1000 m	Kameng, Siang, Subansiri, Tirap	Epiphytic	June–August
463.	*Ritaia himalaica*	1550–2200 m	Kameng, Siang, Subansiri, Tirap	Epiphytic	May–July
464.	*Robiquetia spathulata*	200–500 m	Kameng, Lohit, Subansiri	Epiphytic	May–July
465.	*Robiquetia succisa*	200–700 m	Kameng	Epiphytic	May–August
466.	*Saccolabiopsis pusilla*	500–1000 m	Kameng	–	April–June
467.	*Sarcoglyphis arunachalensis*	200–1100 m	Kameng	Epiphytic	May–June
468.	*Satyrium nepalense*	200–3000 m	Kameng, Lohit, Siang, Tirap	Terrestrial	July–September
469.	*Schoenorchis gemmata*	800–2000 m	Tirap	Epiphytic	June–August
470.	*Schoenorchis roseus*	200–1000 m	Kameng, Tirap	Epiphytic	July–September
471.	*Smitinandia micrantha*	100–1000 m	Kameng, Lohit, Siang, Subansiri, Tirap	Epiphytic	July–September
472.	*Spathoglottis ixiodes*	3000–4000 m	Kameng, Siang, Subansiri, Tirap	Terrestrial	July–September

Contd...

Table 19.4– *Contd...*

Sl.No.	Botanical Name	Altitudes	Area of Distribution	Classification	Flowering and Fruiting
473.	*Spathoglottis plicata*	200–500 m	Kameng, Lohit, Subansiri	Terrestrial	April–June
474.	*Spathoglottis pubescens*	1500–3000 m	Kameng	Terrestrial	June–August
475.	*Spiranthes sinensis*	500–2000 m	Kameng, Lohit, Siang, Subansiri, Tirap	Terrestrial	April–June
476.	*Stereochilus hirtus*	1500–2000 m	Kameng, Subansiri	Epiphytic	June–August
477.	*Stereosandra javanica*	200–800 m	Kameng	Terrestrial	May–June
478.	*Sunipia andersonii*	1000–2000 m	Kameng	Epiphytic	May–June
479.	*Sunipia bicolor*	1000–2500 m	Kameng, Subansiri	Epiphytic	November–January
480.	*Sunipia candida*	1000–2000 m	Kameng, Tirap	Epiphytic	September–December
481.	*Sunipia intermedia*	1500–1800 m	Kameng	Epiphytic	July–September
482.	*Sunipia jainii*	1400–1600 m	Kameng	Epiphytic	December–January
483.	*Sunipia racemosa*	1000–1400 m	Kameng	Epiphytic	June–August
484.	*Sunipia virens*	1000–1200 m	Kameng	Epiphytic	July–September
485.	*Taeniophyllum arunachalensis*	50–100 m	Subansiri	Epiphytic	March–April
486.	*Taeniophyllum crepidiforme*	500–1000 m	Kameng	Epiphytic	September–November
487.	*Tainia latifoilia*	300–1000 m	Kameng, Siang, Subansiri, Tirap	Terrestrial	March–June
488.	*Tainia minor*	1800–2000 m	Kameng	Terrestrial	June–August
489.	*Tainia wrayana*	200–500 m	Kameng, Tirap	Terrestrial	July–August
490.	*Thelasis longifolia*	500–1000 m	Kameng, Tirap	Epiphytic	June–August
491.	*Thelasis pygmaea*	200–1000 m	Kameng	Epiphytic	July–September
492.	*Thrixspermum centipeda*	Upto 200 m	Kameng	Epiphytic	May–August
493.	*Thrixspermum muscaeflorum*	100–500 m	Kameng, Subansiri	Epiphytic	October–November
494.	*Thrixspermum pygmaeum*	100–1500 m	Kameng, Lohit, Siang, Subansiri, Tirap	Epiphytic	April–June
495.	*Thunia alba*	600–1200 m	Kameng, Lohit, Siang, Subansiri, Tirap	Terrestrial	April–June
496.	*Trias disciflora*	At 1000 m	Kameng, Subansiri	Epiphytic	December–March
497.	*Trias nasuta*	1000–1500 m	Kameng, Subansiri	Epiphytic	October–February
498.	*Trias stocksii*	500–1000 m	Changlang, Lohit, Tirap	Epiphytic	October–December
499.	*Trichotosia dasyphylla*	200–500 m	Kameng	Epiphytic	April–June

Contd...

Table 19.4—Contd...

Sl.No.	Botanical Name	Altitudes	Area of Distribution	Classification	Flowering and Fruiting
500.	*Trichotosia pulvinata*	200–1000 m	Kameng, Lohit, Siang, Subansiri, Tirap	Epiphytic	April–June
501.	*Trichotosia velutina*	200–500 m	Kameng	Epiphytic	August–September
502.	*Tropidia angulosa*	500–1000 m	Kameng	Terrestrial	September–October
503.	*Tropidia curculigoides*	500–1000 m	Kameng	Terrestrial	September–November
504.	*Tropidia pedunculata*	500–1000 m	Kameng	Terrestrial	May–July
506.	*Tylostylis discolor*	100–1200 m	Kameng, Lohit, Siang, Subansiri, Tirap	Epiphytic	January–March
507.	*Uncifera acuminata*	2000–2500 m	Subansiri, Tirap	Epiphytic	July–September
508.	*Uncifera obtusifolia*	1000–1500 m	Tirap	Epiphytic	August–October
509.	*Vanda alpina*	1500–2200 m	Kameng	Epiphytic	July–September
510.	*Vanda bicolor*	200–800 m	Siang	Epiphytic	February–April
511.	*Vanda coerulea*	1000–1500 m	Tirap	Epiphytic	July–September
512.	*Vanda coerulescens*	300–1000 m	Kameng	Epiphytic	July–September
513.	*Vanda cristata*	1000–2000 m	Kameng, Lohit, Siang, Subansiri, Tirap	Epiphytic	April–September
514.	*Vanda parishii*	1000–2000 m	Kameng	Epiphytic	May–August
515.	*Vanda pumila*	1000–2000 m	Kameng	Epiphytic	May–July
516.	*Vanda stangeana*	1000–2000 m	Kameng, Lohit	Epiphytic	May–August
517.	*Vanda testacea*	300–2000 m	Subansiri	Epiphytic	May–July
518.	*Vandopsis undulata*	1500–2000 m	Kameng, Tirap	Epiphytic	April–June
519.	*Zeuxine flava*	500–1000 m	Kameng, Lohit, Subansiri	Terrestrial	May–July
520.	*Zeuxine goodyeroides*	1000–1200 m	Subansiri	Terrestrial	October–February
521.	*Zeuxine lindleyana*	300–800 m	Kameng, Tirap	Terrestrial	March–May
522.	*Zeuxine longilabris*	200–500 m	Kameng, Tirap	Terrestrial	March–May
523.	*Zeuxine strateumatica*	300–1000 m	Kameng	Terrestrial	January–March

Figure 19.2: *Aerides multiflora*

Figure 19.3: *Bulbophyllum parviflorum*

Figure 19.4: *Calanthe masuca*

Figure 19.5: *Cymbidium elegans*

Figure 19.6: *Dendrobium sulcatum*

Figure 19.7: *Rhynchostylis retusa*

Figure 19.8: *Cymbidium gammieanum*

Figure 19.9: *Eria javanica*

Figure 19.10: *Vanda stangeana*

Figure 19.11: *Renanthera imschootiana*

Figure 19.12: *Oncidium*

Figure 19.13: *Coelogyne corymbosa*

Figure 19.14: *Gastrochilus*

Figure 19.15: *Dendrobium nobile*

Figure 19.16: *Phaius tankervilliae*

Figure 19.17: *Cymbidoun aloifolium*

Figure 19.18: *Cymbidium lowianum*

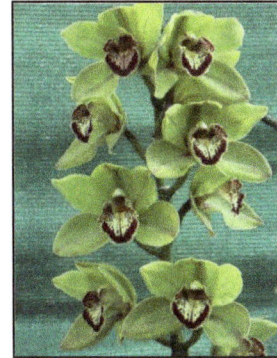

Figure 19.19: *Madrid Forest king*

educational and tourism purposes. The center also has a unique herbarium and museum depicting the various orchids of Arunachal Pradesh and neighbouring states. In addition, large number of orchid hybrids produced by the center are on display and for sale. Indeed, any tourist would have a thrilling and satisfying visit to this center. The Pakhui Wild Life Santurary across the river Kameng is an added attraction.

Sessa Orchid Sanctuary

In an effort to conserve the natural habitat of a large number of orchid species, an area of 100 sq.km has been declared as "Sessa Orchid Sanctuary " at Sessa, West Kameng district, about 20 km away from Tipi. This is a wonderful location situated between 900 to 3100m amsl encompassing tropical, sub-tropical, and temperate climatic conditions and vegetation. Such a diverse physiography favors occurrence of more than 200 orchid species with 5 new and endemic species, within this area. The Department of Environment and Forests has developed trekking routes for going through the sanctuary for visitors to enjoy the marvelous habitat of orchids in their pristine glory. There are deep gorges and valleys, high peaks and rugged terrain which make the trip of adventure tourists, as well of the nature lover, exciting.

Similarly, there are other orchid centers at Dirrang, Itanagar, Jengging and Roing which cover wide ranging habitat and diverse species of orchids. The development of farmer's orchid nurseries at Bomdila, Dirang, Hapoli and Yazali areas have further widened the scope of floriculture related tourism activities in the state. Holding of "Festival of Flowers' in various seasons at different elevation in Arunachal Pradesh would be an added attraction to the visitiors.

Promotion of Orchid Tourism

Orchid flowers have attracted the people world over for their curious shape, colour and texture. To make use of these qualities, there is vast scope for developing allied and subsidiary industries in addition to floriculture. No doubt, cut–flower production in the farm is the first and foremost economic activity that can be developed in Arunachal Pradesh. However from tourism point of view, there is a potential for developing jewelry imitating orchid flowers, carpets with the picture of orchids, bamboo and cane articles depicting the ornamental orchid flowers, moulding of paper weight, glasses etc. with orchid flowers which could be used by the traders for sale as momentos to the tourists. Further, there are orchid flowers which give fragrance (*e.g. Satyrium nepalense, Aerides odoratum,* and *Cymbidium munronianum*) which could be extracted and sold. In order to make various tourist spots in Arunachal Pradesh attractive, botanic gardens should be developed with special emphasis on orchid, which happens to be the "State Flower" in Arunachal Pradesh.

Hence, in order to develop tourism with special emphasis on orchids, such infrastructure is needed to be created without causing damage to our wild grown species and their habitat. It is of paramount importance that the tour operators are advised suitably to develop eco friendly packages for orchid tourists and trekkers so that they don't damage the habitat during the course of their visit. In fact, in USA alone, there are more than 500 Orchid Societies and a large number of enthusiasts who would like to visit the world's orchid rich habitats and "biodiversity hot spot" such as Arunachal Pradesh. It is worthy to mention here that the modern day orchid hybrids of commerce have been derived from the species of this region by contributing the useful germplasm, especially of *Cymbidium, Paphiopedilum* and *Vanda*. Therefore, it is obvious that the advance countries would like to have our unique germplasm. Thus while venturing upon specialized orchid tourism, it must be ensured that we do not lose our wild and unique germplasm, through illegal collection and trade, but this unique germplasm is conserved and used only as tourist attraction, in its natural form and habitat.

References

Anonymous. 2011. Directorate of Horticulture, Govt. of Arunachal Pradesh, Itanagar.

Bose, T.K. and Bhattacharjee, S.K., 1980. *Orchids of India,* Noya Prokash, Calcutta.

Chaddha, K.L., 2001. *Handbook of Horticulture.* ICAR, New Delhi.

Chowdhery, H.J., 1998. *Orchid Flora of Arunachal Pradesh.* Bishen Singh Mahendra Pal Singh, Dehra Dun.

Hegde, S.N., 1984. *Orchids of Arunachal Pradesh.* Forest Department, Itanagar, Arunachal Pradesh.

Mukherjee, S.K., 1983. *Orchids.* ICAR, New Delhi.

Nayar, M.P. and Sastry, 1987-88. *Red Data Book Plants of India.* Botanical Survey of India, Kolkata.

Randhawa, G.S. and Mukhopadhyay, A., 1986. *Floriculture in India.* Allied Publishers Limited, New Delhi.

2013, Biodiversity in Horticultural Crops Vol. 4
Editor: Professor K.V. Peter
Published by: DAYA PUBLISHING HOUSE, NEW DELHI

Pages 387–408

Chapter 20

Orchids of Himalayas

R.B. Ram*, Rubee Lata and M.L. Meena
Department of Applied Plant Science (Horticulture)
Babasaheb Bhimrao Ambedkar University (A Central University),
Vidya Vihar, Rae Bareli Road, Lucknow – 226 025 (U.P.), India
*E-mail: rbram@rediffmail.com

The orchid family is one of the most fascinating and attractive plants found abundantly in Himalayan region. The word Himalaya literally means 'Abode of Snow'. The Himalayan mountain system is the Earth's highest and home of the world's highest peaks. The main Himalayan range runs west to east from Indus river valley to Brahmaputra river valley forming an arc of 2,400 km long which varies in width from 400 km in the western (Kashmir-Xinjiang) region to 150 km in the eastern Tibet – Arunachal Pradesh region. Himalaya separates Indian sub continent from Tibetan Plateau. The climate of this region varies from tropical at the base of the mountain to permanent ice and snow at the highest elevation. The amount of yearly rainfall increases from west to east along the front range. The diversity of climate in terms of altitude, temperature, relative humidity, rainfall, wind velocity, sunshine and soil conditions have led to evolve a variety of distinct plant and animal species. It is estimated that nearly 10,000 species of plants are found in the Himalayan region of which about 3,160 are endemic as are 71 genera. Across the Himalayan Region there is a spectrum of orchid distribution patterns with two very recognizable endpoints: highly restricted endemics and a set of widespread species. The largest family of flowering plants in the hot spot is Orchidaceae with nearly 800 species. The family orchidaceae is one of the most diverse and most widespread groups among the angiosperms which contribute 10 per cent of all known species of flowering plants (Khumbongmayum and Das, 2006) and is regarded as advance among the angiosperms in floral complexity.

The word Orchid is derived from a Greek word *Orchis* meaning testicle because of the appearance of subterranean tubers of the genus *Orchis* (Batygina *et al.,* 2003). Orchids are perennial herbs. Orchids have been attracting botanists, naturalists and ecologists since a long time due to their incredible range of floral diversity and high economic value. Taxonomically, they represent the most highly evolved family among monocotyledons with 600-800 genera and 25,000-35,000 species. Orchids exhibit an incredible range of diversity in size, shape and colour of their flowers. They are generally available in the undisturbed forest areas. Major forest types which are identified for orchids are - tropical, temperate, alpine, tropical semi evergreen and degraded. The Greatest diversity of orchids has been observed in the tropical and sub-tropical climates where the positive factor for growth of orchids *i.e.* high humidity and thick vegetations prevail (Chakrabarti, 2010). Even though all of them are perennial herbs, they may be either terrestrial (growing on soils), epiphytes (growing on plants but not parasitizing on them), lithophytes (growing on rocks and sand grains) or saprophytes and show a great diversity in their floral structure, developed mainly as a result of their adaptation to pollination by a wide variety of insects. Of these, 657 species in 86 genera are epiphytes and 484 species in 82 genera are terrestrial. The genera *Cymbidium* and *Liparis* have both epiphytic and terrestrial species (Arora, 1980; Bhattacharya, 1969; Das and Jain, 1980; Hajra, 1983; Kothari, 1983; Issar and Uniyal, 1967, Kumar and Manial, 1994; Malhotra and Balodi, 1984; Balodi and Malhotra, 1985). The various host plants for Orchid species are *Bischofia javanica, Lagerstromia flosregine, Ficus elastic, Dillenia indica, Sterospermum chelenoides, Bombax cieba, Dysoxylum procerum, Gmelina arborea, Tectona grandis, Shorea robusta, Lagerstroemia hypholeuca, Tamarindus indica, Samanea saman,* etc. Table 20.1 shows the most preferable host plant of orchid species (Borgohain *et al.,* 2010).

They are most pampered of the plants and occupy top position among all the flowering plants valued for cut flower production and as potted plants. They are known for their longer lasting and bewitchingly beautiful flowers which fetch a very high price in the international market. The orchids also have a good market because of their medicinal property. The use of orchids in traditional medicine for treating various diseases like nervous disorder, dermal problem, dysentery, malignancy etc. were documented by Sarma *et al.* (2003). *Vanda parviflora* has antiviral and anti cancerous properties (Pangtey and Kalakoti, 1983; Rawat and Pangtey, 1983, 1985).

India alone contributed 1600 orchid species in 184 genera and many more are discovered year after year. India accounts for nearly 10 per cent of total genetic diversity of orchid flora with Himalayas as their main home (Atwood, 1986; Chadha, 1992).

Regional analyses of diversity and floristic affinities depend on historical data and per-haps reflect diversity potentials to a large degree since regional landscapes have changed radically over the centuries, largely due to the impact of human cultures. Widespread species would seem to be more resilient, not just from having broader ranges than endemic species, but also because they may have greater ecological amplitude.

The distribution pattern reveals five major phyto-geographical regions namely North Eastern Himalaya, Peninsular region, Western Himalaya, Western Ghats and Andaman & Nicobar Islands (Table 20.2).

About 110 wild indigenous species of orchids were recorded from these islands. Of these, 25 from 19 genera are endemic in nature. Important genera in this category are *Aerides, Bulbophllum, Dendrobium, Eria, Eulophia, Phalaenopsis, Pteroceras* and *Vanilla.* Nine species of orchids occurring in the islands have been placed under rare and threatened category. They are *Bulbophyllum proctractum, Dendrobium tenuicaule, Habenaria andamanica, Malleola andamanica, Malaxis andamanica, Phalaenopsis speciosa, Taeniophyllum andamanicam, Vanilla andamanica and Zeuxine andamanica.* Seventeen species of orchids existing in these islands are considered as extra-Indian species. Maximum diversity is represented in the genera like *Dendrobium, Bulbophyllum, Eulophia, Pteroceras, Eria,* etc.

Table 20.1: List of Orchid species along with their Preferred Host Plant Species

Sl.No.	Orchid Species	Hosts*									Status	Flowering
		1	2	3	4	5	6	7	8	9		
1.	Acampe regida (*Buch* Ham ex. J Smith) Hunt	–	+	–	–	–	–	–	–	–	Rare	July–Aug.
2.	A. papillosa (Lindley) Lindley	–	–	–	–	+	–	–	–	–	Rare	Nov.–Jan.
3.	Aerides odoratum Lour	+	–	–	–	–	–	–	–	–	Rare	April–May
4.	A. multiflorum Roxb	+	+	+	–	+	–	+	–	–	Common	April–May
5.	A. longicornu. Hook. Fil.	–	–	+	–	–	–	–	–	–	Rare	May–June
6.	Bulbophyllum sp.	–	+	–	–	–	–	+	–	–	Rare	–
7.	Cleisostoma appendiculatum Lindly	+	+	–	–	+	–	–	–	–	Rare	May–June
8.	Cymbidium aloifolium (L).Sw. Hook	+	+	+	+	+	+	+	–	+	Common	May–June
9.	C. bicolar Lindl.	–	–	+	–	–	–	–	–	–	Rare	March
10.	Dendrobium aphyllum (Roxb) Fisher var. Aphyllum	+	+	+	+	–	+	+	–	–	Common	April–May
11.	D. jenkinsii Wall. ex Lindl	–	+	+	–	–	–	–	–	–	Rare	May
12.	D. lituiflorum Lindl	+	+	+	–	–	–	–	–	–	Rare	April
13.	D. moschatum Sw.	+	+	–	+	–	–	+	–	–	Common	May–June
14.	D. nobile Lindl.	–	–	–	+	+	–	–	–	–	Rare	March–May
15.	D. fimbriatum Griff.	–	–	–	–	–	–	–	+	–	Rare	April
16.	D. transparans Wall	+	+	–	+	–	–	+	–	–	Rare	April–May
17.	Eria spicata (D.Don) Hook	+	+	+	–	+	–	+	–	–	Rare	July–Aug.
18.	Luisia indivisia King & Panting	+	+	–	+	+	–	–	+	–	Common	–
19.	L. trichorhiza Bl	+	–	–	+	+	+	–	+	–	Common	April–May
20.	L. inconspicua, Hook.	–	–	–	–	–	–	–	–	–	Rare	July
21.	Oberonia iridifolia (Roxb) Hook	+	+	–	–	–	+	+	–	–	Common	April
22.	Pholidota imbricata Lindl	–	+	+	–	–	–	+	+	–	Common	August
23.	Pholidota articulata Lindl.	–	–	–	–	–	–	–	–	–	Rare	July
24.	Papilionanthe teres (Roxb) Schltr	+	+	+	+	+	–	+	+	+	Common	May

Contd...

Table 20.1–*Contd...*

Sl.No.	Orchid Species	Hosts*									Status	Flowering
		1	2	3	4	5	6	7	8	9		
25.	*Rhynchostylis retusa* (Roxb) Schltr	+	+	+	–	+	+	+	–	+	Common	May–April
26.	*Sarcanthus secundus*, Griff.	–	+	–	–	–	–	–	–	–	Rare	July
27.	*Arundina graminifolia* (D.Don) Hk				Terrestrial						Common	May–Aug.
28.	*Goodyera procera* Hk.f.										Common	April–May

Abbreviations

* 1: *Lagerostroemia flos regine*, 2: *Bischofia javanica*, 3: *Ficus elastica*, 4: *Bombax cieba*, 5: *Sterospermum chelenoides*, 6: *Gmelina arborea*, 7: *Dillenia indica*, 8: *Dysoxylum procerum*, 9: *Tectona grandis*.

'+' Present; '–' Absent.

Source. Borgohain *et al.*, 2010.

Table 20.2: Distribution of Orchid Genetic Resources in Different Phyto-geographical Regions of the Country

Sl.No.	Phytogeographical Regions	No of Species
1.	Peninsular Region	250
2.	Eastern India	130
3.	Eastern Himalayas	650
4.	Western Himalayas	250
5.	Andaman & Nicobar Islands	80
6.	Central India & Gangetic Plains	60
7.	Western India	05

Source: Kumar and Manilal, 1994.

Table 20.3: Distribution of Commercially Potential Orchids in India

Eastern India and Islands	Western India	Southern and Central India	Andaman and Nicobar Islands
Aerides Lour.	*Aerides* Lour.	*Aerides* Lour.	*Aerides* Lour.
Arundina Bl.	*Bulbophyllum* Thou	*Calanthe* R.Br.	*Bulbophyllum* Thou
Bulbophyllum Thou.	*Calanthe* R.Br.	*Coelogyne* Lindl.	*Cymbidium* Sw
Calanthe R.Br.	*Coelogyne* Lindl.	*Cymbidium* Sw	*Dendrobium* Sw
Coelogyne Lindl.	*Cymbidium* Sw.	*Dendrobium* Sw	*Phalaenopsis* Bl.
Cymbidium Sw.	*Cypripedium* L.	*Rhynchostylis* Bl.	*Pholidota* Hook
Cypripedium L.	*Dactylorhiza* Neck	*Vanda* R.Br.	*Rhynchostylis* Bl.
Dendrobium Sw.	*Dendrobium* Sw	*Vanilla* Sw.	*Vanda* R.Br.
Gastrochilus D. Don	*Goodyera* R. Bl.		*Vanilla* Sw.
Goodyera R.Br.	*Rhynchostylis* Bl.		
Paphiopedilum Pfitz	*Vanda* R.Br.		
Phaius Lour.			
Phalaenopsis Bl.			
Pholidota Hook			
Rhynchostylis Bl.			
Vanda R.Br.			
Vanilla Sw.			

Source. Kumar and Manilal, 1994.

Diversity of Orchids in North-East India

The North-East India comprised of 8 states *i.e.* Arunachal Pradesh, Assam, Manipur, Meghalaya, Mizoram, Nagaland, Sikkim and Tripura has a wide range of physiography and ecoclimatic conditions and considered as the most important orchid's hot spot and harbours more than 70 per cent of the total orchid flora of our country. It is the traditional home of near about 876 orchid species belonging to 151 genera out of 1,600 species. (Jain, 1985; Tripathi & Barik, 2003; Kumar & Manilal, 1994). This region is not only rich in number of orchid species but most of the species available in this region are at the top in the list of ornamentals. Among North-Eastern states, the states - Arunachal Pradesh and Sikkim-have the highest number of orchid species and recognized as the paradise of orchids. The diverse climatic conditions ranging from humid tropical, sub-tropical, temperate and alpine zones influenced by high rainfall, varying temperature, humidity and wide ranging soil and phytogeographical situations have made these states as one of the "Biodiversity Hot-spots" in the world (Dia *et al.,* 2012).

Table 20.4: Area and Distribution of Orchids in North-Eastern States

Name of the States	Total Geographical Area (km²)	Forest Area (km²)	Forest Cover (per cent)	Orchid Species (Approx)
Arunachal Pradesh	83,743	51,540	61.5	560
Assam	78,438	30,708	39.2	193
Manipur	22,327	15,154	67.9	251
Meghalaya	22,429	15,935	75.6	352
Mizoram	21,087	8,629	52.0	244
Nagaland	16,579	9,494	42.3	241
Sikkim	7,096	6,292	60.0	525
Tripura	10,491	2,656	37.3	48

Source: Chakrabarti, 2010.

Some promising ornamental orchids of the north eastern region are *Paphiopedilum fairieanum, Paphiopedilum insigne, Paphiopedilum villosum, Paphiopedilum spicerianum, Paphiopedilum hirsutessimum, Paphiopedilum venustum, Anoectochilus sikkimensis, Aerides multiflora, Vanda coerulea, Vanda teres, Renanthera imschootiana, Rhynchostylis retusa, Pleione maculata, Pleione praecox, Pleione humilis, Cymbidium eburneum, Cymbidium devonianum,Cymbidium whiteae, Cymbidium gigantium, Dendrobium aphyllum, Dendrobium hookerianum, Dendrobium densiflorum, Dendrobium devonianum, Dendrobium thrysiflorum, Kingidium vraceanum, Thunia alba, Otochilus fuscus* and species of *Bulbophyllum, Coelogyne* and *Gastrochlihus.* (Hegde, 2001; Chowdhery, 1998).

In North – Eastern states, Arunachal Pradesh is an orchid paradise and about 142 genera and 560 species are from Arunachal Pradesh, out of which 77 genera and 384 species belong to epiphytic, 58 genera and 192 species terrestrial and 1 genus and 24 species belong to saprophytic (Rao, 1998). Arunachal Pradesh alone is representative of 16 species of genus *Cymbidium viz. Cymbidium aloifolium, C. cochleare, C. cypeifolium, C. devonianum, C. dayanum, C. hookerianum, C. iridoides, C. longifilium hook, C. lowianum C. mackinnonii duttie, C.macrorhizon and C. mastersii and* about 40 species of genus *Dendrobium* out of 20 and 48 species of genus *Cymbidium and Dendrobium* respectively known to occur in India. As per the records, the status of orchids in the state is common species (230), threatened species (31), vulnerable species (16) and endangered species (12) (Dai *et al.,* 2012). The district wise distribution of orchid species in the state is given in Table 20.5.

Sikkim Himalaya is a magnificent reservoir of diversity and is almost a rectangular piece of land, covered with extremely rugged hills and mountains, wedged in between the Himalayan kingdom of Nepal in the west and Bhutan in the east. It lies between 27°5" and 28°9" Latitude and 87°59" and 88°56" Longitude. It shares a common boundary in the whole of the North and more than half in the East with Tibetan Autonomous Region of China. The total number of orchid species in India is estimated to be around 1229, out of which 523 number of orchid species are from Sikkim alone, only next to Arunachal Pradesh having 620 species of orchids. But when land to species ratio is considered, Sikkim perhaps is the world's richest orchid diversity hot spot. Of these, 20 are monophytic orchid genera (Table 20.6) and 22 are endemic to the state (Table 20.7). Though the state is very small in size (7096 sq km), it harbours a high diversity of orchid species due to large variation in macro and micro-climatic conditions. Open canopy forests were ideal hosts for both terrestrial and epiphytic orchids (Lucksom, 2007).

Table 20.5: District-wise Distribution of Orchid Species of Arunachal Pradesh

Name of District	Number of Species	Name of District	Number of Species
Tawang	50	West Kameng	300
East Kameng	100	Papumpare	75
Lower Subansiri	150	Upper Subansiri	150
Upper Siang	200	West Siang	150
East Siang	150	Dibang Valley	200
Lohit	200	Changlang	250
Tirap	150		

Source: Dai *et al.,* 2012.

Table 20.6: Monophytic Orchid Genera in Sikkim

Sl.No.	Monophytic Orchid	Sl.No.	Monophytic Orchid
1.	*Acrochaene Lindl.*	11.	*Monomeria* (Lindl.) Bl.
2.	*Anthogonium Wall. ex Lindl.*	12.	*Ornithochilus* (Lindl.) *Wall.* ex Lindl.
3.	*Arundina Bl.*	13.	*Pennilabium* J.J. Smith.
4.	*Bulleyia Schltr.*	14.	*Risleya* King & Pantling.
5.	*Cleisocentron Bruhl.*	15.	*Ritaia* King & Pantling.
6.	*Corymborkis Thou.*	16.	*Saccolabiopsis* J.J. Smith.
7.	*Diglyphosa Bl.*	17.	*Tipularia* Nutt.
8.	*Diplopora Hook. f.*	18.	*Tylostylis* Hook. f.
9.	*Herpysma Lindl.*	19.	*Vendopsis Piftz.*
10.	*Mischobulbum Schltr.*	20.	*Didiciea cunninghamii*

Source: Lucksom, 2007.

In Assam, about 193 species of Orchids are reported in its Tropical region representing 44.39 per cent of NE and 24.42 per cent of India (Choudhury, 1988). Among these species *Rhyncostylist retusa* popularly known as "Kopaou phool", *Aerides odoratum* and *Papilionanthe teres* (*Bhatou phool*) are intimately linked with the culture of "Assamese Society". Gogoi *et al.* (2009) recorded 68 species of orchids at one of the tropical rainforest patches prevailing in Dibrugarh District of which *Anoectochilus brevilabris, Bulbophyllum spathulatum, Ceratostylis sabulata, Podochilus khasianus, Thelasis longifolia, Trichotosia velutina,* and *Tylostylis discolour* are recorded newly from the region. The species of *Paphiopedilum, Vanda, Cattleya, Cymbidium etc.* have high commercial importance as ornamental plant

(Chowlu and Das, 2007) in the global market for their extremely beautiful and long lasting flowers. Beside these, some other species of *Dendrobium, Cymbidium, Orchis, Aerides, Cypripedium, Vanilla* etc have high ethno botanical importance (Bordoloi, 2002).

Table 20.7: Endemic Orchids of Sikkim and the Altitudinal Distribution Range

Sl.No.	Name of the Species	Habitat	Altitudinal Range
1.	*Bulbophyllum trichocephalum* var. *sikkimnense* S. Z. Lucksom.	Lithophytic	800–1000 m
2.	*Calanthe anjanii* S. Z. Lucksom.	Terrestrial	2000–2500 m
3.	*Calanthe keshabii* S. Z. Lucksom.	Terrestrial	2000–2600 m
4.	*Calanthe yuksomnensis* S. Z. Lucksom.	Terrestrial	1000–2700 m
5.	*Coelogyne pantlingii* S. Z. Lucksom.	Epiphyte	2100 2500 m
6.	*Epigeneium treutleri* (Hook.f.) Ormerod	Epiphyte	Tropical valley
7.	*Cremastra appediculata* var. *sonamii* S.Z.Lucksom.	Terrestrial	920–1000 m
8.	*Cymbidium whiteae* King & Pantling.	Epiphyte	800–2000 m
9.	*Dendrobium eriiflorum* Griff.	Epiphyte	800–1000 m
10.	*Tipularia cunninghamii* (King & Prain) S.C.Chen, S.W.Gale & P.J.Cribb	Terrestrial	4000 m
11.	*Goodyera dongchenii* S.Z.Lucksom	Epiphyte	2000–2300 m
12.	*Gastrochilus sonamii* S.Z.Lucksom	Epiphyte	2300–2700 m
13.	*Neottia alternifolia* (King & Pantl.) Szlach.	Terrestrial	3000–3500 m
14.	*Liparis chungthangnensis* S.Z.Lucksom	Lithophyte	1800 2000 m
15.	*Liparis dongchenii* S.Z.Lucksom	Terrestrial	1500 2000 m
16.	*Liparis lydiae* S.Z.Lucksom	Epiphyte	1000 1300 m
17.	*Liparis platyrachis* Hook.f.	Epiphyte	1500 2000m
18.	*Liparis pygmaea* King & Pantling	Lithophyte	4350 m
19.	*Crepidium saprophytum* (King & Pantl.) A.N.Rao	Terrestrial	1800 m
20.	*Oberonia kingii* S.Z. Lucksom	Epiphyte	1000 m
21.	*Stigmatodactylus paradoxus* (Prain) Schltr.	Terrestrial	2000 m
22.	*Peristylus pseudophrys* (King & Pantling) Kranzlin	Terrestrial	1800 m

Source: Lucksom, 2007.

Manipur has an official record of only about 280 orchid species. In order to preserve the orchid of this area, the centre for orchid gene conservation of the eastern Himalayan region was set by the Science and Engineering Research Council (SERC) of the Dept. of Science and Technology (DST), Government of India. The research centre is based in the beautifull, hilly location of Hengbung, in the Senapati district of Manipur. Manipur is bordered by the Indian states of Nagaland to the north, Mizoram to the south and Assam to the west and by Burma to the east. It covers an area of 22,347 sq km (8,628 sq mi), and has a hilly topography. Forest constitutes about 78 per cent of the state. Despite its small size, Manipur forest ranges from tropical to sub alpine, supporting great floral diversity, especially orchids. In this state, the trees were found laden with thick population of *Bulbophyllum, Dendrobium species viz., D. aggregatum, D. crepidatum, D. fimbriatum, D. densiflorum, D. ochreatum, D. primulinum, Acampe ochracea, Eria pannea,* green flowered *Calanthe, Cymbidium aloifolium, C. gigantium, Vanda coerulea, Hygrochilu parishii* (syn. *Vandopsis parishii*) and *Vanda bicolor.* Other vandaceous species were also found growing in profusion. This place is also the habitat of the rare and endangered *Renanthera imschootiana* which is a monopodial orchid (small, strap shaped glossy leaves with unequal, rounded, bilobed tips). These plants were growing mostly on oak trees, 3 to 4.5m above the ground. A new record from the Manipur state on the worn – out road to Kenulu, about 22 km from Willong Khullen is *Ascocentrum himalaicum* found hanging down from the oak trees (Kishor and Nanda, 2011). This species have terete leaves and pinkish – red small flowers and are distributed from the east Himalaya to West and Southwest Yunnan in China. This species was first described in 1987 by Eric A Christenson (Borgohain, 2010). The local villagers were growing different orchid species around their homes. These include *Dendrobium, Cymbidium* and *Vanda* species (Kishor and Nanda, 2011).

Diversity of Orchids in Uttarakhand

Uttarakhand lies between 28-41' and 31-28' N latitude and 77-35' and 81-01' E longitude. The region has 53,483 sq. km. total geographical area and about 64.8 per cent of its area under forest cover (FSI, 1999). The varied topography and climatic conditions met within the state are conterminous with a very rich biodiversity. The vegetation of the State ranges from tropical deciduous to alpine vegetation and is broadly categorized into Subtropical (in the lower altitude region up to 800 m. It comprises moist tropical and dry deciduous vegetation) and temperate (found at an altitude ranging from 700-1400 m) and Sub alpine & Alpine (found above 3000 m, an area of about 1000 sq. km). The altitudinal variations and climatic changes in the state have shown the great diversity in the habit, which resulted in luxuriant and varied floristic composition. In Uttarakhand 72 genera with 236 species of orchids are recorded. Among them, 17 species are found medicinally important. Taking all the monocotyledonous families into account, Orchidaceae is the 2nd largest family after Poaceae in Uttarakhand. Genus *Habenaria* with maximum of 17 species followed by *Dendrobium-16* species and *Bulbophyllum-11* species (Joshi *et al.,* 2009).

Gori valley in Eastern Uttarakhand is regarded as orchid hotspot. The valley is situated at the junction of Western and Central Himalayas. Out of the 255 species of orchids so far recorded from Western Himalaya, 124 species are recorded from the Gori valley alone. Much of the Gori valley lies in the sub-tropical belt. It has a rich and moist riverine forest, essential for growth of orchids (Jalal and Rawat, 2009)

Chhotanagpur Plateau

Chhotanagpur lies in Deccan province in the Indian Region of Paleotropic Kingdom (Takhtajan, 1978). It occupies 2.4 per cent of India's geographic area, situated between 22°01' – 25°30' N latitude and 83°30' – 87°52' E longitude, with a total area of 79,714 km² of which 29.61 per cent is under forest cover (FSI 2005). According to the survey conducted by All India Coordinated Research Project on Taxonomy of orchids (AICOPTAX) in Chhotanagpur Plateau during 2002 – 2006, sixty three species were collected and documented from the study area with *Dendrobium* as one of the largest group of epiphytic orchids comprising of 11 species, namely, *Dendrobium aphyllum, D. bicameratum, D. cucullatum, D. crepidatum, D. formosum, D. fimbriatum, D. herbaceum, D. moschatum, D. peguanum, D. regium* and *D. transparens.* Most of the species were found in the Sal (*Shorea robusta*) dominated forests. Results show that *D. aphyllum* was the most common amongst 11 species and it was distributed through out the altitudinal gradient, whereas, rest of the orchid species were localised at comparatively higher altitudes. *D. herbaceum* was always found on the upper areas of the plateaus between 900 - 1000 m asl. *D. crepidatum* was seen in both epiphytic as well as lithophytic conditions, whereas *D. moschatum* were found growing as lithophytes along the streams. Rest of the species were epiphytic. (Kumar *et al.,* 2007 & 2011).

Table 20.8: Orchids Genera and Species in Uttarakhand

Sl.No.	Genera	Species	Sl.No.	Genera	Species
1.	*Acampe* Lindl	2	37.	*Habenaria* Willd.	17
2.	*Aerides* Lour	2	38.	*Hemivilia* Lindl	1
3.	*Anoectochilus* Blume	1	39.	*Herminium* Linn	8
4.	*Aorchis* Vermeulen	2	40.	*Kinidium* P.F. Hunt	2
5.	*Aphyllorchis* Blume	2	41.	*Livaris* L.C. Rich	10
6.	*Archineottia* Chen	1	42.	*Listera* R.Br.	4
7.	*Arundina* Blume	1	43.	*Luisia* Gaud	3
8.	*Ascocentrum* Schltr. ex J.J. Sm.	1	44.	*Malaxis* Soland ex Swartz.	7
9.	*Brachycorthis* Lindl	1	45.	*Neottia* Guettard	2
10.	*Bulbophyllum* Thouars	11	46.	*Neottianthe* (Reichb.)Schltr	2
11.	*Calanthe* Ker- Gawl	8	47.	*Nervilia* Comers.ex Gaud.	7
12.	*Cephalanthera* Rich	1	48.	*Oberonia* Lindl	9
13.	*Cheirostylis* Blume	1	49.	*Oreorchis* Lindl	3
14.	*Chiloschista* Lindl	1	50.	*Ornitho chillus* (*Lindl*) Wall ex Benth.	1
15.	*Cleisostoma* Blume	1	51.	*Otochilus* Lindl	1
16.	*Coelogyne* Lindl	5	52.	*Pachystoma* Blume	1
17.	*Corallorhiza* Gagnebin	1	53.	*Pecteilis* Rafin	2
18.	*Cryptochilus* Wall.	1	54.	*Pelatantheria* Ridl	1
19.	*Cymbidium* Swartz	9	55.	*Peristylus* Blume	9
20.	*Cypripedium* Linn.	4	56.	*Phaius* Lour	1
21.	*Dactylorhiza* Necker ex Neuski	1	57.	*Pholidota* Lindl ex Hook	2
22.	*Dendrobium* Swartz	16	58.	*Platanthera* Rich	2
23.	*Didiciea* king & Prain ex King & Pantl	1	59.	*Pleione* D.Don	4
24.	*Diphylax* Hook f.	1	60.	*Ponerorchis* Reichb.f.	
25.	*Diplomeris* D.Don	1	61.	*Pteroceras* Hasselt ex Hassk	1
26.	*Epipactis* Zinno	3	62.	*Rhvnchostylis* Blume	1
27.	*Epipogium* Gmelin ex Borkhaussen	2	63.	*Satyrium* Swartz.	1
28.	*Eria* Lindl	9	64.	*Smitinandia* Holtt.	1
29.	*Eulophia* R.Br. ex. Lindl	9	65.	*Sviranthes* Rich	2
30.	*Flickinfleria* Hawkes	2	66.	*Sunivia* Lindl	1
31.	*Galeala* Lour.	1	67.	*Thelasis* Bhime	1
32.	*Gastrochilus* D.Don	4	68.	*Thunia* Reichb.f.	1
33.	*Gastrodia* R.Br.	1	69.	*Tropidia* Lindl	1
34.	*Geodorum* G. Jackson	1	70.	*Vanda* W.Jones ex R.Br.	5
35.	*Goodyera* R.Br.	6	71.	*Vandopsis* P fitz.	1
36.	*Gymnadenia* R.Br.	1	72.	*Zeuxine* Lindl	3

| | **Total Genera – 72** | | | **Total Species - 236** | |

Source: Joshi *et al.,* 2009.

Table 20.9: Description of Various Orchids and their Occurrence

Orchids	Time of Flowering	Colour of Flower	Occurrence
Acampe longifolia Lindl.	April–May	Pale yellow, lip white	Tropical Sikkim, upper Assam
Acampe rigidia Hunt.	April–May	Pale yellow, lip white	Tropical Sikkim, upper Assam
Acanthephippium striatum Lindl.	July–August	Pale pink with bold red lines	Khashi Hills, Sikkim, Chotanagpur
Aerides fieldingii Williams	May–June	Bright rose; lip-rose purple	Khashi and Jaintia Hills, Sikkim
Aerides odoratum Lour.	June–July	Waxy while with amethyst purple	West Bengol, Orissa, Chota-Nagpur, Gharwal and Kumoau Hills, Tropical Himalayas, Sikkim, Khashi Hills
Aerides vandarum. f.	March–April	White, tip-yellow	Northern India, Sikkim, Khashi Hill, Manipur
Arachnis cathcartii J.J.Sm	November–December	Pale yellow Crossed by red streaked red White	Sikkim
Biermannia bimaculat King & Pantl.	July	White	Tista Valley, Sikkim
Bulbophyllum hirtum Lindl.	October–January	Yellow to greenish white, lip yellow with papillose margin Pale-yellowish green	Khasia Hills, Sikkim
Bulbophyllum leptanthum Hk.f.	July	Pale yellowish green	Sikkim, Khasia Hills
Bulbophyllum Odoratissimum Lindl.	July	White with Yellow tinge	Sikkim, Khasia Hills
Calanthe griffithii Linn	April–May	Brownish green tip yellow	Sikkim
Coelogyne corymbosa Lindl.	April–May	Creamy white	Khasia Hills, Sikkim
Coelogyne cristata Lindl.	March–April	Snow white	Khasia Hills, Sikkim,Kumaon Hills Dehra Dun, Sikkim
Coelogyne fimbriata Lindl.	October–December	Musk yellow, lip yellowish streaked with reddish brown	Hills
Coelogyne oculata Hk.f.	July	White	Sikkim
Coelogyne ochracea Lindl.	May–June	White	Sikkim, Kumaon,Assam
Coelogyne longipes Rolfe:	May–June	Flowers white.	Arunachal Pradesh, Nagaland, Meghalaya, Manipur, Sikkim
Coelogyne ovalis Lindl.	October–December	Lip white or yellowish streaked with purple	Sikkim, Kumaon and Meghalaya
Cymbidium dayanum Reichb. f.	March–April	Flowers ivory-white	Sikkim, Meghalaya
Cymbidium hookerianum Reichb. f.	February	Flowers bright olive-green, Lip yellowish, midlobe blotched or speckled with red	Sikkim

Contd...

Table 20.9–*Contd...*

Orchids	Time of Flowering	Colour of Flower	Occurrence
Cymbidium iridioides D. Don	October–November	Flowers light yellow-green, strip Sikkim ed longitudinally with red	Kumaon, Sikkim, Meghalaya
Cymbidium munronianum King & Pantl.	May	Flowers straw-coloured streaked or dotted purple, Lip pink with yellow recurved tip	Sikkim, Darjeeling
Dendrobium amoenum Wall	June	Flowers sepals and petals white tipped with magenta, Lip white with a purple spot near the tip, greenish yellow towards the middle	Orissa, Kumaon, Garhawal Hills, Sikkim and Meghalay
Dendrobium candidum Wall	May–June	Flowers white, Lip white with a yellow spot near the base	Kumaon, Manipur, Sikkim, Meghalaya
Dendrobium chrysotoxum Lindl.	April–May	Flowers long-lasting, golden yellow Lip yellow, deeper, in font and streaked with red in the throat	Manipur, Sikkim, Assam, Meghalaya
Dendrobium crystallinum Reichb. f.	April–June	Flowers white tipped with magenta, Lip yellow with a white border and blotch of amethyst in front	Sikkim
Dendrobium densiflorum Lindl.	April–May	Flowers butter-yellow Lip orange with pale edges	Sikkim and Meghalaya
Dendrobium gibsonii Lindl.	July–August	Flowers saffron-yellow to orange-yellow; Lip upper surface papillose, with two brownish purple spot	Assam, Sikkim, Meghalaya
Dendrobium hookerianum Lindl.	September	Flowers golden yellow; Lip apricot-yellow with or without two oblique dark purple patches	Assam, Sikkim, Meghalaya
Dendrobium lindleyi SteudRoxb. *D. aggregatum*	March–May	Flower yellow, Lip orange-yellow	Sikkim, Assam
Dendrobium longicornu Lindl. *D. hirsutum* Griff.	September–November	Flower pure white, Lip pale brown veined with orange	Sikkim, Meghalaya, Arunachal Pradesh and Nagaland
Dendrobium moschatum Buch. Ham.; *D. calceolaria* Carey ex Hk	May–July	Flower ochraceous, creamy buff, flushed with rose; sepals with reddish tips and reddish or orange veins. Lip pale yellow	Kumaon, Chota Nagpur, Assam, Manipur, Orissa, Sikkim, Meghalaya
Dendrobium nobile Lindl.	April–May	Flower white tinted with amethyst, mentum brown. Lip maroon-purple in the throat, front portion white, with a purple tip	Assam, Manipur, Orissa, Sikkim, Meghalaya
Dendrobium palpebrae Lindl.	April–May	Flowers white or rose, disc of lip orange.	Sikkim.

Contd...

Table 20.9–Contd...

Orchids	Time of Flowering	Colour of Flower	Occurrence
Dendrobium primulinum Lindle	April	Flowers white with pink lips Lip primrose-yellow, with a tinge of purple in the throat	Sikkim, Dehra Dun, Mussourie, Garhwal
Dendrobium fimbriatum Hk. f.	March–May	Flowers deep rich orange-yellow, Lip rich orange	Chota Nagpur, Orissa, Kumaon, Manipur, Western Ghats, Sikkim, Meghalaya
Dendrobium pulchellum Roxb. *D. dalhausieanum* Wall.	February–April	Flowers Cream-yellow or lemon-yellow to almost white, with a more or less definite rosy suffusion. Lip velvety, cream-white	Sikkim, Meghalaya, Arunachal Pradesh, Nagaland and Manipur
Dendrobium transparens Wall	May	Flowers white, tinged purplish rose towards tips, Lip whitish with a dark blood-red or dark purple blotch on the throat	Sikkim, Kumaon, Orissa, Chota Nagpur, Assam, Meghalaya, Manipur and Garhwal Hills
Eria coronaria Reichb. f.	November	Flowers white. Lip flushed with purple, disc yellow	Sikkim, Meghalaya
Eria javanica Bl.; *E. fragrans* Reichb. f.	April–June	Flowers white to pale yellowish	Tropical valley of Sikkim
Eria pannea Lindle.; *E. calamifolia* Hk. f.	May	Flowers: inner surface golden brown. Lip dark brown	Tropical valley of Sikkim
Flickingeria fimbriata A. D. fimbriatum Bl.	May	Flowers white or pinkish and speckled with red; the midlobe of the lip greenish yellow	Sikkim, Meghalaya and Nilgiri Hills
Goodyera procera Hk.f.	May	Flowers white	Sikkim, Assam, Meghalaya, Arunachal Pradesh, Nagaland, Chota Nagpur, and Western Ghats, Seshachalam Hills
Habenaria oitchisoni Reichb. f.	July–August	Flowers pale greenish.	Kashmir, Kumaon, Sikkim
Habenaria oitchisoni Hk. f.	July–August	Flowers green and white	Simla, Meghalaya, Sikkim
Habenaria pachycaulon Hk. f.	July–August	Flowers purple	Sikkim
Ornithochilus difformis (Wall. Ex Lindl.) Schlecher.; *O. fuscus* Wall.	February–April	Flowers greenish yellow with purplish stripe. Lip yellowish with a purple midlobe	Tropical Himalayas from Garhwal to Sikkim and Meghalaya
Pholidota articulate Lindl.	June–August	Flowers yellowish white; bracts yellow, shaded with green, part of lip with 5 low longitudinal yellow ridges	Kumaon, Sikkim Meghalaya and Manipur
Pholidota imbricate (Roxb.) Lindl.	May–August	Flowers pale pink; lip often with yellow spots	Orissa, Chota Nagpur, Deccan Peninsula, Andaman and Nicobar Islands, Assam, Kumaon, Sikkim, Meghalaya

Contd...

Table 20.9—Contd...

Orchids	Time of Flowering	Colour of Flower	Occurrence
Papilionanthe teres (Roxb.) Schltr.; *Vanda teres* Lindl.	March–April	Flowers variable in colour, generally rose-coloured, rarely white. Lip yellow within, crimson spotted	West Bengal, Assam, Meghalaya, Andaman and Nicobar Islands, tropical valley of Sikkim.
Papilionanthe vandarum (Roxb.f) Garay; *Aerides vandarum* Reichb. f	March–April	Flowers white. Lip flushed with yellow	Northern India, Sikkim, Meghalaya, Manipur
Phaius tankervilliae (Banks ex L' Hert.) Bl.; *Limodonum tankervilleae* Banks ex L' Hert.	March–April	Flowers shaded with red, white on the exterior surface, lip white with an orange-yellow base and red lines across the yellow disc.	Shimla, Nagpur, Sikkim and Meghalaya.
Phalaenopsis mannii Reichb. f.	May	Flowers yellow with brown marking.	Assam, Sikkim, Meghalaya.
Platanthera bakeriana Krzl.	July	Flowers green.	Sikkim
Pleione humilis (J. E. Sm.) D. Don; *Epidendrum humile* J.E. Sm	September–November	Flowers bluish white, lip striped with rich purplish crimson.	Sikkim
Pleione maculate (Lindl.); *Coelogyne maculata* Lindl.	October–November	Flowers white	Northern India, Sikkim, Meghalaya, Assam
Pleione praecox (J.E. Sm.) D.Don; *Coelogyne praecox* (J.E.Sm.) Lindl.	November–December	Flowers rose-purple, with a pale rose or white lip, throat yellow.	North India, Manipur, Sikkim, Meghalaya.
Satyrium nepalense D. Don.	September–October	Flowers pink or white	Kashmir, Meghalaya and hilly parts of southern India; Darjeeling, Sikkim, Manipur.
Vanda alpine Lindl.	June–July	Flowers greenish yellow	Garhwal, Kumaon, Sikkim, Meghalaya

Source. Upadhyay and Das, 2003.

Loss/Erosion of Orchid Biodiversity

The orchids in their natural habitat are highly vulnerable to loss or erosion. Protection of valuable orchid species in their natural habitats is an urgent need as orchids are very sensitive to the ecological disturbances. Their disappearance indicates a change in the quality of soil and air of the region thus they also work as ecological indicators. The vulnerability to erosion seems from two sources: the first being their highly specialized life cycle - mode of living, dependency on pollinators for pollination, lack of reserved food material in the seeds, reliance on mycorrhizal fungi for seed germination and second ornamental and therapeutic value which they possess, have made them so sought after the man (Ram, *et al.,* 2011).

Certain species of exotic orchids found in North East India, are now severely depleted due to widespread deforestation, indiscriminant collection and reckless smuggling. A recent survey found that about seventy orchid species, out of a total eight hundred seventy six are on the verge of extinction. The uncontrolled orchid export trade and illegal smuggling are major problems to conserve orchids in their natural habitats. Other factors affecting orchid loss include the improper use of land, unscientific cultivation (Jhuming) of the local people, over grazing of livestock, construction of roads, dams, bridges, natural factors like forest fires, over extraction and the general exploitation of natural resources which ultimately cause serious damage to the orchid diversity available in this region (Chakrabarti, 2010).

The changing pattern of rainfall and decrease in the forest cover have contributed to their decrease. Rainfall is decreasing every year. This affects the growth of orchids which thrive in regions with regular rainfall. With more and more wild places being opened up for tourism, the beautiful orchids are the most affected. Every tourist wants to take home these exotic beauties. Awareness among tourist with special informative posters can certainly help to check this vandalism.

Each species is adapted to live in a specialized environment because of their specialized requirements and many species are very restricted in distribution. Any destruction, degradation or defragmentation of natural habitat beyond a tolerable limit causes threat for their survival. Many of them because of their small population size and restricted distribution, require intensive care and habitat management and may survive only with human support.

Nearly 250 species of native orchids are under the threats of various categories. Certain species like *Aphyllorchis gallani, Coelogyne truetleri, and Anoectochilus rotandifolius, Paphiopedilum charlsworthii, Paphiopedilum wardii, Vanda wightiana, Pleione lagenaria, Zeuxine pulchra* probably have vanished from Indian lands.

The ubiquitous "Blue vanda" of Cheerapunjee has disappeared due to degradation of environment. Orchids are mainly shade loving; therefore, they have no chance of survival, once forests are cleared. While clearing forests for cultivation, many trees bearing orchids die (Joshi *et al.,* 2009).

Rare Endangered Orchids of Uttarakhand

Twelve taxas of rare endangered Orchids of Uttarakhand were recorded in Red Data Book (RDB) of Indian Plants (Nair and Shastry, 1987, 1988, 1990):

TAXA RDB Status

1. *Aphyllorchis gallani* Duthie Endangered
2. *Archinottia microglottis* (Duthie) Chen Rare
3. *Cyperipidium elegans* Reichb.f. Rare

4. *Cyperipidium himalaicum* Rolfe Rare

5. *Diplomeris hirsuta* (Lindl.) Lindl Vulnerable

6. *Eria occidentalis* Seid Rare

7. *Eulophia mackinnonii* Duthie Rare

8. *Flickingeria hesperis* Seid Endangered

9. *Aphyllorchis parviflora* King & Pant! Rare

10. *Calanthe alpina* Hookf.ex.Lindl. Rare

11. *Calanthe pachystalix* Reichb.f. ex. Hook f. Endangered

12. *Cypripedium cordigerum* D. Don Rare

Out of known 1600 species of Indian orchids, 352 are endemic of which 40 are "endangered" and 72 are "vulnerable". Among Indian orchid species *Cymbidium, Dendrobium* and *Vandas* are endangered. These endemic species are exclusive biological capital of the country. Once lost or became extinct it is irrecoverable loss for the country as biodiversity is sovereign right of the country as per Convention of Biodiversity. Many plants are often lost by poor cultural conditions, indifferent housing, changing and often inexperienced staff, and a shortage of funds for the care and maintenance of plants. For maintaining of live collections, provenance field data are required to establish and maintain the plants in the most suitable way. The troubles which are likely to be confronted in maintaining live collections could be conjectured by their specialized life cycle, distribution and mode of living.

There is urgent need for development of agro technique for each species to be conserved in field gene banks. Unlike other flowering plants, they have complex life cycle and specific requirement for temperature, light, nutrient etc. for proper growth and development. It would be difficult to maintain the germplasm at one place until the center is equipped with environmentally controlled glasshouses and the targeted orchids are studied for their climatic requirements and agronomic practices. For example, at Royal Botanical Garden, Kew, nearly 5000 species of orchids from different parts of the world have been conserved in green houses having eight different sets of environmental conditions. The problems, which are likely to be confronted in maintaining live collections, can be conjectured by their distribution and mode of living. Many often these collections are lost due to poor cultural conditions, indifferent housing, changing and often inexperienced staff, and a shortage of funds for the care and maintenance of plants. There are various other organizations working for *ex-situ* conservation of orchids in different agro climatic zones of the country. There is an urgent need to bring such organizations under the National Active Germplasm Sites.

Biodiversity Conservation

In situ Conservation

The *in situ* conservation of species ensures their natural growth, proliferation and perpetuation without hindering the process of evolution as part of natural ecosystem. India has an elaborate *Protective Area Network (PAN) comprising 86 National Parks, 480 Wildlife sanctuaries covering about 4.66 per cent of total geographical area of the country. There is further plan to expand this network to 160 national parks and 698 Wildlife sanctuaries to cover 5.69 per cent of total geographical area of the country.* This network automatically provides the protection to the species lie in them. Unfortunately, many important and endangered orchids *viz., Paphiopedilum druryi* in Aghasthymalai hills of Kerala, *Vanda coerulea* in Meghalaya, *Paphiopedilum wardii* and *P. specerianum* in Assam, *Renanthera imscootiana* in Arunachal

Pradesh and many more lie outside PAN. A few conscious State Governments like Arunachal Pradesh, Sikkim, Karnataka, and West Bengal have designated the orchid rich habitats as "Orchid Sanctuaries". These sanctuaries attract Wildlife Protection Act, 1972 as amended in 1992 (Ram, *et al.*, 2011).

The people of North Eastern region conserved orchids with great care. Orchids are associated with the traditional culture, religion, myth, food and folk medicines of the local people (Tribes) of North Eastern region (Arora, 1996; Dutta & Dutta, 2005; Ramakrishnan, 1992). Thus, they conserve orchids in their natural habitats in sacred groves or shrine forests or in the form of village forest reserves based on their religious beliefs. The former can be seen in Meghalaya and Manipur while the latter is common in Mizoram (Darlong & Barik, 1998). Sacred groves or shrine forests are the forest patches rich in biodiversity and represent a long tradition of environmental conservation by the tribal communities of North Eastern India. There are a large number of sacred groves in the states of Arunachal Pradesh, Meghalaya, Manipur, Sikkim and Karbi-Anglong area of Assam (Table 20.10). These are among the few least disturbed forest patches in the region serving as the original treasure house of orchid diversity. The people living near these groves/forests have vast knowledge about conservation and utilization of orchid's wealth. They completely prohibit any human interference in these sacred groves to destroy the natural habitats of various orchid species for their religious belief that the Gods and the spirits of their ancestors live in these groves (Khumbongmayum *et al.*, 2004). As a result, the sacred groves are still well protected in spite of a rapid decline of traditional value system with the advent of Christianity and other anthropogenic disturbances (Chakrabarti, 2010). In Arunachal Pradesh near about 65 sacred groves or Gumpa forests are documented where the rich orchid diversity present in that areas are conserved by the tribal people.

Table 20.10: Sacred Groves of North-Eastern Region to Conserve Orchids

States	Local Term	No. of Documented Sacred Groves
Arunachal Pradesh	Gumpa *Forests* (*Sacred Groves attached to Buddhist monasteries*)	65
Assam	Than, Madaico	40
Manipur	Gamkhap, Mauhak (*sacred bamboo reserves*)	365
Meghalaya	Law Lyngdhoh	83
Sikkim	Gumpa *Forests*	56

Source: Chakrabarti, 2010.

Ex situ Conservation

Field Gene Banks

In India, orchids have been the concerns of botanists who collected them for study and conserved them from fear of loss in their natural habitats. The Botanical Survey of India established 3 National Orchidaria at Shillong, Yurcaud and Howrah for conservation and multiplication of orchids. Similarly, States like Arunachal Pradesh, Assam, Mizoram, Karnataka, Nagaland, West Bengal, Sikkim, Himachal Pradesh and Odisha have also collected and conserved the orchids. The Tropical Botanical Garden and Research Institute (TBGRI), (Trivandrum) Kerala, National Research Centre for Orchids, Sikkim and several other organizations are also engaged in the conserving orchids. The orchids are often conserved in polyhouse/glasshouse termed as orchidarium. These structures need to be constructed keeping in view, the climatic conditions required for the species to be conserved. The germplasm conserved in such structures are at high risk of disease and pest and require proper care and monitoring of the plants (Ram, *et al.,* 2011).

Artificial Natural Habitats

Other than orchidaria, orchids can also be conserved in "artificial natural habitat" where epiphytic orchids are tied on suitable host plants and terrestrials planted in ground. The objective of this method is providing similar condition as that of nature. The conserving orchid germplasm by this method reduces cost of maintenance and incidence of diseases and pests. But its applicability is limited by availability of suitable host and adaptability of species in new environment. The species having wider adaptability or specific to that particular locality can be conserved by this method. This method has successfully been tried for conserving various epiphytic and terrestrial orchids at Darjeeling Campus of National Research Centre for Orchids (Ram, *et al.,* 2011).

Orchid Seed Banks

Orchids produce millions of seed in a single capsule but they lack in metabolic machinery and functional endosperm and therefore, require mycorrhizal association for germination in nature. Consequently, the percentage of germination is calculated to be 0.01 to 0.2. Many of orchids were germinated through asymbiotic technique where germination percentage was as high as 90. The seeds of orchids are orthodox in nature and provide a great scope for long-term storage through cryopreservation technique. Owing to their minute size, a large number of seeds can be maintained in small volume. However, long-term storage of orchid seeds would require the studies in respect of storage duration, seed viability etc. The seed storage is usually used for genome preservation.

In vitro Conservation

Maintenance of orchid germplasm in field gene bank requires huge investment and is also affected by insect pests and diseases. Therefore, *in vitro* conservation of orchid germplasm may require attention. *In vitro* conservation technique can also be used for revitalization of orchid germplasm affected by virus and virus like diseases as meristem culture technique eliminates many of viruses. Though the orchids were the first plants to be tissue cultured but *in vitro* conservation of orchids needs study in terms of genetic stability, storage duration etc. The collection of living orchids will be used in future in a variety of ways by taxonomists, cytologists and molecular biologists for genome analysis and many other researchers for improving our knowledge about the biology of this family (Ram, *et al.,* 2011).

Conclusion

☆ India is one of the primary/secondary centers of orchid biodiversity and the major regions of diversity are North eastern Himalayas, Western Ghats, and Andaman & Nicobar Islands. *India has alone contributed 1600 orchid species in 184 genera. India accounts for nearly 10 per cent of total genetic diversity of orchid flora with Himalayas as their main home.*

☆ Himalayan Region has a spectrum of orchid distribution patterns with two very recognizable endpoints; highly restricted endemics and a set of widespread species, known as "Biodiversity Hot spot" with nearly 800 species.

☆ Orchids exhibit an incredible range of diversity in size, shape and colour of their flowers. They are generally available in the undisturbed forest areas. Major forest types which are identified for orchids are - tropical, temperate, alpine, tropical semi evergreen and degraded-. Greatest diversity of orchids has been observed in the tropical and sub-tropical climates where the positive factor for growth of orchids *i.e.* high humidity and thick vegetations prevail.

☆ Successful conservation of threatened species calls for protection and management of their habitats. However in reality it is extremely difficult to realize this, because the preservation of actual habitats is generally an economic problem and is more or less limited. The existing protected territories not always coincide with local areas of distribution of rare and endangered species.

☆ Orchids are protected species under the Convention on International Trade in Endangered Species (CITES) under schedule VI of the wild life protection Act (1972), all the nine species of "Lady's slipper" can only be sold if they are grown in registered nursery.

☆ Orchids are mostly collected from the wild using non sustainable, destructive method like collecting the entire plant, rhizome, tuber and roots and other reproductive parts like fruit and seeds. Such destructive collection methods are the major factors influencing orchid population. Further, low regeneration rate and loss of habitat add to the serious threat to orchid population. Such rapid depletion from the wild requires urgent conservation measures.

☆ The indigenous species need to be evaluated for the desired attributes so that these can be used as potted plants or as parents for hybridization. The field gene banks may suffer due to disease and improper management, which would result in erosion of genetic resources. Hence, seed conservation and *in-vitro* conservation deserve attention.

☆ For promoting production of flowers (loose flowers or cut flowers), decorative plants and contract seed multiplication, sophisticated technology may be imported initially. However, the infrastructure should then be built up within the country. There is no dearth of brain and bank.

☆ If the modern scientific forest management practices could be properly integrated with the traditional wisdom and religious beliefs of the people of North Eastern India associated with the sacred groves and contributing to forest protection, these sacred groves could become a very useful model for orchid's conservation in the region. Moreover, there is urgent need to promote the concept of sacred groves of the local people and to develop some scientific technology whereby the forest departments of each state of this region could provide technical inputs to improve the canopy cover and regeneration of trees in the degraded sacred groves. This will ultimately help to conserve our valuable orchid diversity of this region.

☆ A moral awakening is need of the hour.

☆ Creation of public awareness and promotion of conservation strategies are essential.

☆ Dualism, basic research and commercialization have to be looked at with equal respect and the old socialist idea that public investment must be for social welfare not private gain may be shunned.

☆ Nevertheless, the interaction between universities and industry has to be furthered. However, shared ambition would be the vital nexus between them.

References

Arora, C.M. 1980. New record of some orchids from North Western Himalaya. VI - *Indian Journal of Forest.* 3: 78 – 79.

Arora, R. K. 1996. Role of Ethno botany in the conservation and use of plant genetic resources in India. In: Ethnobotany in human welfare (ed. SK. Jain). Deep Publication. India.

Atwood, J. T. 1986. The size of Orchidaceae and the systematic distribution of epiphytic orchids. *Selbyana* 9(1): 171-186.

Balodi, B. and Malhotra, C. L. 1985. *Herminium mackinnoni* Duthie - An overlooked species from Kumaun. *J. Eco. Taxon. Bot.* 6: 465 - 466.

Bhattacharya, U.C. 1969. New distributional records of orchids for West Himalaya. *Bull. Bot Soc.* Bengal. 23: 161-165.

Bordoloi, D. 2002. *Udvid Aswajya*. Anahita Publication, Jorhat.

Borgohain, A.; Gogoi, B. and Nath, P. C. 2010. Orchid diversity and host specificity in Deopani Reserve Forest, Sadiya, Assam. *NeBIO*. 1(3): 16 – 20.

Chadha, K. L. 1992. The Indian orchid scenario. *Journal of the Orchid Society India* 6:1- 4.

Chakrabarti, S. 2010. Conservation of orchids by the people of North Eastern India. *NeBIO*. 1(1): 48 – 52.

Choudhury, H. J. 1988. *Orchid flora of Arunachal Pradesh*. Dehradun, Bishen Singh Mahendra Pal Singh.

Chowlu, K. and Das, A. K. 2007. Orchids of floricultural importance from Arunachal Pradesh (India). *Pleione*. 1(2): 21 - 25.

Dai, O.; Nimasow, G.; Bamin, S. and Chozom, K. 2012. Floriculture prospects in Arunachal Pradesh with special reference to orchids. *Journal of Biodiversity and Environmental Science.* 2 (3): 18 – 32.

Darlong, V. T. and Barik, S. K. 1998. Traditional Practices and Recent Initiatives in Biodiversity Conservation in North east India; Lessons from Peoples Experiences. In: Abstract of the Seminar on Environmental Problems in North Eastern India, Deptt. of Ecology, Assam University, Silchar, India.

Das, S. and Jain, S.K. 1980. Orchidaceae, Genus *Coelogyne*. Fasc. *Fl. India.* 5: 1-33.

Dutta, B. K and Dutta, P. K. 2005. Potential of ethnobotanical studies in North East India: an overview. Indian Journal of Traditional Knowledge 4: 7-14.

FSI. 2005. State of forest report. FSI. Dehra Dun.

Gogoi, K.; Borah, R. L. and Sharma, G. C. 2009. Orchid flora of Joypur Reserve Forest of Dibrugarh district of Assam, India *Pleione,* 3(2): 135-147.

Hajra, P.K. 1983. New Species of Lister a from Nandadevi National Park,Chamoli, District, Uttar Pradesh. *Bull. Botanical Survey of India.* 25: 181-182.

Hegde, S.N. 2001. Prospects of Floriculture Industry in Arunachal Pradesh with special reference to Orchids. Arunachal Forest News, 19 (1&2), 172–185.

Issar, R.K. and Uniyal, M.R., 1967. Orchid of Uuttarakhand Himalayas Ind.forester 93:713-716.

Jain, S. K. 1985. Conservation of Orchids in India. In: *Progress in Orchid Research,* (Eds.) Chadha and Singh. IIHR/UNDP, Bangalore.

Jalal, J.S. and Rawat, G.S. 2009. Community Initiative Weaves New Hope For Orchid Conservation. Proc. National Conference on Orchid genetic diversity: Conservation and Commercialization. December 09-11, 2009, organised by The Orchid Society of India (TOSI),Chandigarh.

Joshi, G.C.; Tewari, L. M.; Lohani, N.; Upreti, K.; Jalal, J. S. and Tewari, G. 2009. Diversity of orchids in Uttarakhand and their conservation strategy with special reference to their medicinal importance. *Report and Opinion*, 1(3): 47 – 52.

Khumbongmayum, A. D.; Khan, M. L. and Tripathi, R. S. 2004. Sacred groves of Manipur, ideal centres for biodiversity conservation. *Current Science*. 87 (4): 430 - 433.

Khumbongmayum, A. D. and Das. 2006. *Galeola falconeri* Hook. F., an endangered giant saprophytic orchid. *Current Science*, 91(97): 871 – 873.

Kishor, R. K. and Nanda, Y. 2011. Orchid hunting in Willong, Manipur. *The Orchid Review*. September, 2011: 167 – 172.

Kothari, M. J. and Hunt, P.F., 1983. *Vandopsis undulata* (Lindl.) Smith (Orchidaceae) in Pithoragarh District. *Indian Journal of Forestry*. 6:160-161.

Kumar, C.S.; Manilal, K.S. 1994. A catalogue of Indian orchids. Dehradun, Bishen Singh Mahendra Pal Singh.

Kumar, P.; Jalal, J.S. and Rawat, G.S. 2007. Lists of species – Orchidaceae, Chotanagpur, State of Jharkhand, India. Check List, 3(4), 297–304.

Kumar, P.; Rawat, G. S. and Wood, H. P. 2011. Diversity and Ecology of *Dendrobiums* (Orchidaceae) in Chotanagpur Plateau, India. *Taiwania*. 56(1): 23-36.

Lucksom, S.Z. 2007. The Orchids of Sikkim and North-east Himalaya. S.Z. Lucksom, Sikkim, India. 984 pp.

Malhotra, C. L. and Balodi, B. 1984. A New species of *Corallarhiza gagnebin* from Gori Valley. *Bull. Botanical Survey of India*. 26 (1-2): 108-109.

Nair, M.P. and Shastry, A.R.K. (eds.), 1987–Vol-I; 1988–Vol-II; 1990–Vol-III. Red Data Book of Indian Plants, Botanical Survey of India Calcutta.

Pangtey, Y.P.S. and Kalakati, B.S. 1983. A note on the occurrence of *Cheirostylis griffithii* Lindl. (Orchidaceae) from Western Himalaya. *Ind. J. for* 6:170.

Rao, A. N. 1998. Notes on the genus *Taeniophyllum* (Orchidaceae) in Arunachal Pradesh with two new records to India. *Rheedia*, 8(2), 163–166.

Ramakrishnan, P. S. 1992. Tropical forests, exploitation, conservation and management. *Impact of Science on Society*. 42: 149 - 162.

Ram, R. B.; Lata, R. and Meena, M.L., 2011. Conservation of Floral Biodiversity of Himalayan Mountain Regions with Special Reference to Orchids. *Asian Agri-History*. 15 (3): 231 – 241.

Rawat, G.S. and Pangtey, Y.P.S. 1983. *Herminium josephii* Rchb. f. *chidaceu* - a new record for Western Himalaya. *Ind. J. For*. 6:171.

Rawat, G.S. and Pangtey, Y.P.S. 1985. *Neottianthe calcicola* (W.W. Sm.) Schlt. (Orchidaceae) New to the flora of India. *Current Science*. 54 (19): 1005 – 1006.

Sarma, C. M., Bora, R. K. and Basumatary, N. 2003. Medicinally important orchids of Northeast India. *Indian Journal of Environment & Eco-planning*.

Takhtajan, A. L. 1978. Floristic regions of the World, Nauta, Leningrad, English edn., Transl. T. J. Crovello. 1986. University of California Press, Berkeley.

Tripathi, R. S. and Barik, S. K. 2003. National Biodiversity strategy and action plan report, Northeast India. Ministry of Environment and Forest, New Delhi.

Upadhyay, R. C. and Das, S. P. 2003. Prospects and potential of orchid export from India. *Indian Horticulture*. 48(3): 22-23.

2013, Biodiversity in Horticultural Crops Vol. 4
Editor: Professor K.V. Peter
Published by: DAYA PUBLISHING HOUSE, NEW DELHI

Pages 409–446

Chapter 21

Temperate and Sub-Tropical Vegetables

D. Ram, Mathura Rai and Major Singh

Indian Institute of Vegetable Research,
Varanasi – 221 005, India
E-mail: singhvmo@gmail.com

Plant genetic resources can be described as the total genetic diversity of the cultivated species and their wild relatives (Ford-Lloyd, 2001). If we say 'Biological Diversity' it means the variability among living organisms from all sources including, *inter alia,* terrestrial, marine and other aquatic ecosystem and the ecological complexes in which they are part. This includes diversity within species, between species and ecosystems (Article 2 of the Convention on Biological Diversity) whereas, biodiversity can be defined as the quality, range or extent of differences between the biological entities in a given situation. Around 2,50,000 plant species have been described so far and a much larger number still remains to be duly recognized yet only about 3,000 out of them were picked up from the wild by human beings for their use and grown on varying scales since the beginnings of agriculture, considered to be nearly ten thousand years old. In view of mankind's increased dependence on about 30 major food crops which 'Feed the World' (Harlan, 1975). In this way, biodiversity is of tremendous importance in the daily diet of human being, who depends on the bark, young shoots, buds, flowers, fruits, leaves, roots, tubers and mushrooms, especially in the season of hunger, when other cultivated crops are unavailable. Wild plants have often played an important role in many diets due to their higher nutritional value than cultivated species. These are too directly or indirectly consumed/used as vegetables in the form of roots, tubers, bulbs, rhizomes, leaves, stems/shoots/culms, fruits and in a few cases as flowers. The prized genetic materials along with their wild and close relatives are used by plant breeders for development of improved crop varieties/hybrids. They also constitute a priceless reservoir that contains genes conferring better adaptation to stress environments and resistance to diseases and pests.

Center of Origin and Regions of Diversity

A Russian plant explorer and geneticist, Nikolai I. Vavilov, was one of the first scientists to recognize the importance of plant genetic diversity in nature. On the basis of geographical survey, he summarized that the cultivated plants originated in eight basic geographical centers of origin and two subcentres (Vavilov, 1951). His studies revealed that out of the wide range of plant diversity in the tropical and subtropical regions of the world, the major food crops have come mainly from high mountain valleys, isolated from each other to a large extent and with a great habitat range. The world's diversity in cultivated plants is distributed in twelve regions (Zeven and de Wet, 1982). Over 90 per cent of plant species for food and agriculture are located in the economically developing parts of the world, namely, the African, Asian and Latin American continents and the Far East Islands (Madeley, 1996).

For the underutilized, neglected and less known domesticated/cultivated vegetables, this diversity is mainly confined to seven regions, namely the Chinese-Japanese, Indo-Chinese-Indonesian, Hindustani, Mediterranean, European-Siberian, African and South American regions (Arora, 1985), where approximately 495 of the 540 species occur. The region possessing maximum diversity at global levels is tropical America, tropical Asia and the Mediterranean which list the diversity in major vegetables in different regions. In the tropical Asian region both India and China hold maximum diversity. Indian subcontinent is well known since long as an important center of origin and diversity of a large number of vegetable crops. India is a homeland of 167 cultivated species and 329 wild relatives of crop plants (Arora, 1991).

It is observed that people made useful selection in several vegetable crops, which were eventually domesticated and cultivated. Several weedy species were never or only temporarily domesticated, remaining as weeds but often hybridizing by chance with the cultivated ones and thus, enhancing the diversity in cultivated plants. India is 'Primary centre' for crops like eggplant, smooth gourd, ridge gourd, cucumber etc. and secondary center for cowpea, okra, chillies, pumpkin and several Cole crops (Table 21.1).

Agro-Climatic Zones of India

India is located between 8° N to 38° N latitude and 68° E to 97.5° E-longitude and exhibits extreme variations in edapho-climatic situations, agro-climatic regions and floristic diversity. The altitudinal variations are observed from below sea level to above vegetational limits in the Himalayas *i.e.* more than 3500 m above average sea level; the climates change from monsoon to temperate/alpine in the northern/Himalayan zone. Sehgal *et al.* (1992) reported occurrence of 20 agro-climatic regions in India based on physiographic, climatic and cultural features. Around 80 species of major and minor vegetables, apart from several wild/gathered kinds, occur (Choudhury, 1967 and Seshadri, 1987) in Indian condition. Concentration of genetic diversity comprising native species and landraces occur more in Western Ghats and North-Eastern Himalayas. The richness of plant diversity is largely due to ecological diversity superimposed with tribal and ethnic diversification, plant usages and religious rituals. Crops in which rich diversity occurs in India include cowpea (*Vigna unguiculata*), common bean (*Lablab purpureus*), cole crops (*Brassica* species), okra (*Abelmoschus esculentus* and related species), Brinjal (*Solanum melongena* and related species), sweet potato (*Ipomoea batatas*), taros (*Colocasia* and *Alocasia*), yams (*Dioscorea esculenta, Dioscorea alata, Dioscorea deltoidea*), sword bean (*Canavalia species*), velvet bean (*Mucuna* species), and elephant foot yam (*Amorphophalus* species). *Solanum* spp. are widely distributed in the North-Eastern region; yams in western ghats and northeastern states, chives, leeks and other wild *Allium* spp. in Kumaon and Garhwal Himalayas; cluster bean in Western arid zone;

Table 21.1: Center of Diversity of Major Vegetable Crops

Gene Centre	Primary Center	Secondary Center
Chinese-Japanese	Egg plant, wax gourd, chinese cabbage,Kangkong, welsh onion	Water melon, amaranth
Indo- Chinese	Wax gourd, Sponge gourd, ridge gourd, bitter gourd, sword bean, winged bean, taro, chayote, cucumber, bottle gourd, yam,	Chinese cabbage, bottle gourd, cucumber, yambean, amaranth, yard long bean, Kangkong
Hindustani centre	Egg plant, wax gourd, cucumber, ridge gourd, bitter gourd, sponge gourd, hyacinth bean, drumstick, okra, Kangkong	Water melon, melon, Roselle, bottle gourd, amaranth
Central Asia	Onion, garlic, carrot, spinach	Egg plant, water melon, melon, cauliflower
Near East	Onion, garlic, leek, beet	Okra
Mediterranean	White cabbage, cauliflower, water melon, broccoli, radish	Sweet pepper, garlic, okra
African	Egg plant, water melon, melon, bottle gourd, cowpea, okra, Roselle, locust bean	Onion, shallot, lima bean, White cabbage amaranth
European-Siberian	Lettuce	Onion, White cabbage, common bean, cauliflower, spinach, carrot
Central America and Mexican region	Tomato, hot pepper, pumpkin, squash, yambean, sweet potato, common bean	–
South American region	Tomato, hot pepper, cassava, pumpkin, lima bean, chayote, amaranth sweet potato	Common bean
North American	–	Tomato, Egg plant, melon, water melon, pepper, squashes, onion, lettuce, lima bean, okra, pumpkin

Lablab bean in Deccan plateau; cucurbits in Rajasthan and M.P; and leafy vegetables like *Amaranthus*, and *Fagopyrum* spp. in Western Himalayan region.

North-Western and Eastern Himalayan Region

Under Western and Eastern Himalayan region, only temperate vegetables are predominantly found. Enormous diversity occurs in *Allium* species- leek, shallot and other introductions of *Allium sativum* and *Allium cepa.* Sporadic diversity also occurs in asparagus, spinach, chenopods, amaranthus, and *Beta vulgaris.* Other vegetables like *Brassicas,* squash, *Cucurbita* spp., *Cucumis,* chillies, bell pepper, peas, faba bean, cowpea, horse radish, artichoke, potato, Colocasia, tomato, parsley, coriander, ginger, *Sechium edule* and *Cyclanthera pedata* do express a lot of variability.

North-Eastern Region including Assam

In this region also, most of the temperate vegetable crops predominate. Maximum diversity occurs in leafy vegetable crops like amaranth and *Brassica* species. Other vegetables like chillies, tomato, brinjal, okra, taros, yams and cucurbits are grown in this tract. Several kinds of beans like winged bean, French bean and lima bean are specialized due to edaphic and climatic factors. In the lower tract, rich diversity occurs in cucumber, pointed gourd, chayote, bitter gourd, spine gourd (*Momordica dioica* Roxb) and meetha karela (*Cyclanthera pedata*). Among the wild species *Abelmoschus manihot* (*pungens* form), *Alocasia macrorrhiza, Amorphophallus bulbifera, Colocasia esculenta, Cucumis hystrix, Cucumis trigonus, Dioscorea alata, Luffa graveolens, Moghania vestita, Momordica cochinchinensis, Momordica macrophylla, Momordica subangulata, Trichosanthes cucumerina, Trichosanthes dioica, Trichosanthes dicaelosperma, Trichosanthes khasiana, Trichosanthes ovata, Trichosanthes truncata* and *Solanum indicum* are prevalent.

The Northern Plains/Gangetic Plains including Tarai Region

The Northern Plains/Gangetic Plains including Tarai Region is one of the richest pockets of diversity in major vegetable crops. Due to year round cultivation, more diversity emerges and it is well fitted into the cropping patterns. Rich diversity can be observed in *Cucumis* species, *Luffa* species, *Cucurbita* species, *Benincasa hispida, Lagenaria siceraria, Momordica* species, *Trichosanthes* species, *Solanum* species, *Capsicum* species, *Abelmoschus* species, *Brassicas* and tuber crops.

North Western/Indus Plains

In this region, variability exists in *Cucumis* species, *Momordica* species, *Citrullus* species, *Solanum* species, *Amaranth, Chenopodium, Abelmoschus* species, *Capsicum* species and *Allium* species. There are certain sporadic pockets, which are rich in indigenous germplasm like Cucurbits, okra, eggplant and garlic. Specific adoptable diversity of *Caralluma* species may be spotted in this region. The wild species of *Momordica balsamina, Citrullus colocynthis* and *Cucumis prophetarum* are prevalent.

The Central Region/Plateau

More diversity occurs in *Cucurbita* species, ash gourd, round gourd, bitter gourd, pointed gourd, ridge gourd, okra, eggplant, chillies, tomato, root and bulbous crops and sporadically in leaf vegetables. As indigenous vegetables, more diversity exists in cucurbits, eggplant, okra and chillies. The wild species like *Cucumis setosus, Cucumis trigonus, Luffa acutangula var. amara, Momordica cymbalaria* etc. are more commonly prevailing in this region.

The Western and Eastern Peninsular Region

The Western and Eastern Peninsular Region is an extremely rich region for cucurbits (cucumber, bitter gourd, bottle gourd and squashes), eggplant, okra and chillies (both annual and perennial types). More landrace diversity occurs in snake gourd in the western and for *Luffa* species and eggplant in eastern region. Sporadic diversity can be spotted for leaf vegetables like *Amaranth, Brassicas, chenopods, Spinach, Beta vulgaris, Basella rubra* and *Basella alba*. Several wild species like *Abelmoschus angulosus, A. moschatus, A.manihot, A. ficuleneous, Amorphophallus campanulatus, Colocasia antiquorum, Cucumis hystrix, C. setosus, C. trigonus, Luffa acutangula var amara, Luffa graveolens, Luffa umberrata, Momordica cymbalaria, Momordica denticulata, Momordica dioica, Momordica cochinchinensis, Momordica subangulata, Solanum indicum, Solanum melongena, Trichosanthes anamalaiensis, Trichosanthes bracteata, Trichosanthes cordata, Trichosanthes cuspidata, Trichosanthes horsfieldii, Trichosanthes piniana, Trichosanthes perottitiana, Trichosanthes nerifolia, Trichosanthes himalensis, Trichosanthes multiloba* and *Trichosanthes villosa* are quite prevalent in this region.

Variability distribution of different vegetable crops to different agro-ecological regions of India are presented in Table 21.2.

**Table 21.2: Distribution of Major Vegetable Crops Variability to
Different Agro-Ecological Regions of India**

Sl.No.	Agro-Ecological Regions	Geographical Ranges	Variability in Vegetable Crops
1.	Humid Western Himalayan region	J&K, H.P. and parts of U.P.	Cucurbits, radish, carrot, turnip, peas, cowpea, chillies, brinjal, okra, spinach, fenugreek, amaranth, *Solanum khasianum, Solanum hirsutum, Sechium edule, Basella rubra*
2.	Humid Bengal/Assam basin	W.B. and Assam	Cucurbits, radish, cowpea, chillies, brinjal, okra, spinach, beet, *Abelmoschus manihot* ssp. *manihot,* amaranth, *Solanum indicum, Solanum khasianum, Solanum surattense, Cucumis sativus var. vikkimensis, Edgeria dargelingensis, Melothria assamica, Momordica cochinchinensis, Sechium edule, Tuladiantha coordifolia, Basella rubra*
3.	Humid Eastern Himalayan region and Bay Islands	Arunchal Pradesh, Nagaland, Manipur, Mizoram, Tripura, Meghalaya, Andaman & Nicobar islands	Cucurbits, radish, cowpea, pea, chillies, brinjal, okra, spinach amaranth, *Abelmoschus manihot ssp. tetraphyllus, Solanum khasianum, Solanum torvum, Solanum sisymbrifolium, Solanum ferox, Solanum verbasifolium, Cucumis hystrix, Luffa echinata, Sechium edule*
4.	Sub-humid Sutlej Ganga Alluvial plains	Punjab, U.P. and Bihar	Cucurbits, radish, peas, brinjal, okra, spinach beet, fenugreek, onion, garlic, *Abelmoschus manihot ssp. tetraphyllus var. pungens, Abelmoschus tuberculatus, Solanum indicum, Solanum khasianum, Solanum torvum, Solanum surattense, Solanum hispidum, Cucumis hardwickii, Cucumis trigonus*
5.	Humid Eastern and South Eastern uplands	East M.P., Orissa and A.P.	Cucurbits, radish, carrot, cowpea, chillies, brinjal, okra, spinach, amaranth, garlic, *Abelmoschus manihot ssp. manihot, Solanum surattense, Solanum torvum*
6.	Arid Western plains	Haryana, Rajasthan and Gujarat	Cucurbits, cauliflower, radish, carrot, peas, cowpea, chillies, brinjal, okra, spinach beet, fenugreek, onion, garlic, amaranth, *Abelmoschus tuberculatus, Abelmoschus ficuleneus, Abelmoschus manihot ssp. tetraphyllus, Solanum torvum, Solanum nigrum, Citrullus colocynthes*
7.	Semi-arid Lava Plateau and Central Highlands	Maharashtra and West M.P.	Cucurbits, cauliflower, radish, carrot, cowpea, chillies, brinjal, okra, spinach, fenugreek, amaranth, onion, *Solanum surattense, Solanum torvum, Solanum nigrum, Solanum khasianum, Cucumis setosus, Luffa acutangula var. acutangula*
8.	Humid to Semi Arid Western Ghats and Karnataka Plateau	Karnataka, Tamil Nadu, Kerala and Lakshadweep islands	Cucurbits, chillies, brinjal, okra, *Abelmoschus crinitus, Abelmoschus angulosus, Abelmoschus ficuleneus, Abelmoschus moschatus, Abelmoschus manihot var. tetraphyllus, Solanum trilobatum, Solanum indicum, Solanum incanum, Solanum pubescens, Solanum surattense, Solanum torvum, Luffa acutangula var. acutangula, Melothria angulata, Basella rubra*

Variability, Distribution and Genetic Erosion of Vegetable Crops

The valuable genetic resources of vegetable crops are vanishing rapidly. Three main processes which cause loss of genetic diversity of cultivated crop species are genetic erosion, genetic vulnerability and genetic wipe out. Several other factors like shrinking of natural resources, population pressure, urbanization; deforestation, monoculture and changing cropping pattern including use of hybrids/improved varieties are responsible for loss of genetic diversity. These are not mutually exclusive but are, in fact inter-locked by the demand of increasing population and the rising expectations. This genetic erosion is taking place at a time when new tools of biological research enable scientist to focus as much on the diversity of genes as on the diversity of genotypes. There is considerable success in protecting and preserving the agro-biodiversity albeit under *ex situ* conditions during the past one quarter century but much remain to be accomplished.

Table 21.3: Variability and Distribution Status of Vegetable Crops

Crops	Genera	CS	DS	GVS	GES	GCP
Brinjal	*Solanum*	C	W	W	M	H
Chillies	*Capsicum*	C	W	H	M	H
Tomato	*Lycoperscon*	C	W	H	M	H
Cluster bean	*Cyamopsis*	C	R	M	M	M
Cowpea	*Vigna*	C	W	H	M	H
French bean	*Phaseolus*	C	W	M	M	M
Lablab bean	*Lablab*	C	W	H	M	M
Mucuna bean	*Mucuna*	C	R	H	M	M
Pea	*Pisum*	C	R	M	M	H
Carrot	*Daucus*	C	R	M	M	H
Elephant foot yam	*Amorphophallus*	C	R	H	M	M
Giant taro	*Alocasia*	W	R	H	M	M
Radish	*Raphanus*	C	W	H	M	H
Taro	*Colocasia*	W	W	H	M	M
Yam	*Dioscoria*	W	R	H	M	M
Garlic	*Allium*	C	W	M	M	H
Onion	*Allium*	C	R	M	M	H
Okra	*Abelmoschus*	C	W	H	M	H
Amaranth	*Amaranthus*	C	W	H	M	M
Climbing spinach	*Basella*	C	W	M	M	M
Fenugreek	*Trigonella*	W	R	M	M	M
Palak	*Beta*	C	R	M	M	H
Bitter gourd	*Momordica*	C	W	H	M	H
Bottle gourd	*Lagenaria*	C	W	H	M	N
Cucumber	*Cucumis*	C	W	H	M	M
Pointed gourd	*Trichosanthes dioica*	C	L	H	H	M
Pumpkin	*Cucurbita*	C	W	H	M	H
Luffa	*Luffa*	C	W	H	H	H
Tinda	*Prae-citrullus*	C	L	M	M	M
Water melon	*Citrullus*	C	R	H	M	M
Ash gourd	*Benincasa*	C	W	H	M	M
Snake gourd	*Trichosanthes anguina*	C	W	H	M	H

CS: Crop status (C-cultivated, W-Wild); DS: Distribution status (W- wide spread distribution, R- regional distribution, L- localized distribution); GVS: Germplasm variability status (H- high, M- medium, L- low); GES: Genetic erosion status (H- high, M- medium, L- low); GCP: General Crop priorities (H- high, M- medium).

Genetic Resources of Vegetable Crops in the Indian Gene Center

Wild plants have often played an important role in many diets due to their higher nutritional value than cultivated species. Wild species and putative ancestral forms of vegetable crops contain valuable genes that are of immense need in vegetable breeding programme using conventional methods or modern biotechnology. These genetic resources can be utilized in the development of new cultivars, strains and hybrids and also in restructuring of the existing ones which lack one or the other attribute. The distributional pattern of the wild plant genetic resources in different botanical/phyto geographical regions and the areas of their concentration where rich diversity of wild species still continue to perpetuate, are of special significance for undertaking programmes on collection as well as for *in situ* conservation of biodiversity. The important families possessing wild genetic diversity are Brassicaceae (*Brassica*), Malvaceae (*Abelmoschus*), Leguminoceae (*Canavalia, Dolichas, Trigonella, Vigna*), Cucurbitaceae (*Citrullus, Coccinia, Luffa, Momordica, Neoluffa, Trichosanthes*), Solanaceae (*Solanum*), Amaranthaceae (*Amaranthus*), Dioscoreaceae (*Dioscorea*), Amaryllidaceae (*Allium*), Araceae (*Alocasia, Amorphophallus, Colocasia*). The bulb of greater galangal (*Alpinia galangal* family Zingiberaceae) can also be eaten raw. In the Garwhal areas, *Cornus capitata* and *Cornus controversa* (family Cornaceae) are eaten raw. In Rajasthan, the whole plant of *Gisekia pharnaceoide* (family Mulluginaceae) is widely consumed during food shortage but in the South and West (Deccan Region) the leaves are used as greens, as are the leaves of *Glinus trianthemoides*.

Genetic Resources of Solanaceous Vegetables

Tomato, brinjal and chillies are the most important Solanaceous vegetables, widely grown all over the country throughout the year for their edible fruits.

Lycopersicon species

Tomato is a warm season crop, sensitive to frost and usually cultivated in sub-tropical and mild cold climate. It is native of South America and Mexico, however, domestication took place in Mexico (Boswell, 1949) and now grown all over the world. The greatest variability occurs in Mexico, Central America and Coastal Peru. The wild species of *Lycopersicon* have shown to be a valuable source of resistance to various diseases and pests to quality traits in addition, which is a vital requirement in any breeding programme. *Lycopersicon pimpinellifolium* is resistant to *Fusarium* wilt, early blight and leaf curl disease and is a rich source of ascorbic acid; *Lycopersicon cheesmanii* is resistant to salinity and source of high total soluble solids (TSS); *Lycopersicon chilense* is tolerant to leaf curl disease; *Lycopersicon chmielewskii* has high content of TSS; *Lycopersicon hirsutum* is an excellent source of resistance to several pests and to leaf curl disease; *Lycopersicon peruvianum* is resistant to root knot, tobacco mosaic virus, leaf curl and has high ascorbic acid content and *Lycopersicon chilense* is a good source of drought resistance.

Solanum species

Brinjal or eggplant (*Solanum melongena* L.) being the crop of Indian Origin has developed some secondary variability in China (Vavilov, 1926). A large number of land races/traditionally grown cultivars are developed in different agro-ecological zones of India and these landraces possess valuable genes for resistance to biotic and abiotic stresses and adaptation to various environments. Maximum diversity and distribution of *Solanum* species were noted in Southern India, foothills of Himalayas and north-eastern region. The widely distributed species in the region include *Solanum torvum, Solanum indicum, Solanum insanum, Solanum surattense, Solanum pubescens* and *Solanum khasianum*. A sum of 36 accessions of cultivated eggplant and 91 other *Solanum* species from Sri Lanka were also acquired.

Some distinct species include *Solanum verbascifolium, Solanum surattense, Solanum indicum, Solanum nigrum* and *Solanum viarum* from Nepal.

Chillies and Bell Pepper

Capsicum species (Chillies or Pungent Pepper and Bell Pepper or Simla Mirch)

Chillies (*Capsicum annuum* L.) is grown in warm to hot and humid climate throughout the country. The large fruited bell pepper of today originated from tiny pungent and pointed fruited wild species *Capsicum annuum*. It is considered a cool season crop and grown in sub-tropical and temperate region. There are four other species (*viz. Capsicum frutescens, Capsicum chinense, Capsicum baccatum and Capsicum pubescens*), commonly grown in Southern and Central America except *Capsicum frutescens,* which is also grown in some parts in USA. Due to continuous selection, long history of cultivation and popularity of the crop, sufficient genetic variability was generated and now India is treated as a secondary center of diversity for chillies. In hot chillies (*Capsicum annuum* L.), rich variability in plant and fruit morphological traits occur throughout the country particularly in south peninsular region, northeastern region, in foothills of Himalaya and Gangetic plains. The variability includes plant type, fruit size/shape (long, short, pointed, smooth, wrinkled), bearing habit (fruits facing upwards, horizontal, downwards) and pungency in addition to annual and perennial types. However, in bell pepper less variability has been observed. Mostly introduced lines are grown particularly in mid hill of Himalaya and parts of hilly region of Central India including Karnataka and Maharashtra.

Genetic Resources of Cucurbitaceous Vegetable

Cucurbita species

The genus *Cucurbita* comprised of about 27 species, both wild and cultivated mostly in the tropical and sub tropical regions of Central and South America. In this genus, a few species are of commercial importance *viz.* Field pumpkin (*Cucurbita moschata* Duch. Ex. Poir.), winter squash pumpkins (*Cucurbita maxima* Duch.), Summer squash or vegetable Marrow (*Cucurbita pepo* L), winter squash pumpkins (*Cucurbita mixta* Pang. syn. *C. angyrosperma*) and Malabar gourd or Fig leaf gourd (*Cucurbita ficifolia* Boucha). The vegetable Marrow (*Cucurbita pepo* L), is an introduced crop which withstands cooler climate. *Cucurbita texa* Gray grows wild in texas (Chadha and Lal, 1993).

Cucumis species

The genus *Cucumis* comprises of about 26 species. The major crops of economic importance are cucumber (*Cucumis sativus*), muskmelon (*Cucumis melo*), snap melon (*Cucumis melo* var. *momordica*), longmelon (*Cucumis melo* var. *utililisimus*). The Indian sub-continent is the center of origin for *Cucumis sativus* and center of diversity for *Cucumis melo* (Zeven and de Wet, 1982). The wild species (*Cucumis hardwickii*) is growing in natural habitats in the foothills of Himalayas. The free hybridization with cultivated *sativus*, with no reduction of fertility in F_2 generation suggested that *Cucumis hardwickii* is likely progenator of cultivated cucumber. The cucumber *Cucumis sativus var. sativus* has 3-5 lobed leaves, ovary usually 3 placentiferous with fruits oblong, obscurely trigonus or cylindric. However, *Cucumis sativus* var. *sikkimensis* has 7-9 lobed leaves, ovary often 5 placentiferous, fruits ovoid-oblong, adapted in temperate and humid climate. Muskmelon (*Cucumis melo*) based on the distribution of diversity can be grouped into seven sub sets (Munger and Robinson, 1991).

Cucumis melo var. agrestis (Kachri)- It is a wild type with slender vines and small inedible fruits, probably synonym of *Cucumis melo var. callosus* (*Cucumis callosus.*) and *Cucumis melo var. trigonus* (*Cucumis trigonus*).

Cucumis melo var. cantaloupensis Naud. (Musk melon or cantaloupe)- Medium sized fruits with netted, warty or scaly surface, flesh usually orange but sometimes green, aromatic or musky flavour. Fruits dehiscent at maturity, usually andromonoecious.

Cucumis melo var. flexuosus Naud. (*Cucumis melo var. utilissumus*) – Snake melon, snake cucumber, Tar-kakari, fruits long and slender consumed at immature stage, monoecious.

Cucumis melo var. momordica (*Cucumis melo var. momordica*)- Snap-melon or phut is a monoecious crop grown in India and other Asian countries. It has white to pale orange, less sweet pulp. The smooth surface of the fruit starts cracking at the time of maturity.

Cucumis melo var. conomon (sweet or pickling melon)- It is generally andromonoecious in nature and bears small fruits with skin, white flesh.

Cucumis melo var. inodorous Naud. (Winter melon)- The fruit is smooth or wrinkled surface with white or green flesh, and lacking musky odor. It is also andromonoecious in nature and usually requires more time for maturity.

Cucumis melo var. dudaim Naud. (Mango melon, Pomegranate melon)- Fruits small, globular, smooth motted but not netted, flesh with acid flavor, pickling type.

Out of the seven melons *Cucumis melo var. agretis, Cucumos melo* var. *cantaloupensis* Naud. (musk melon), *Cucumis melo var. utilissumus, Cucumis melo* var. *momordica*, and *Cucumis melo var. flexuosus* Naud are available in Indian Sub- continent.

Luffa species

The Indian gene center has rich diversity in genetic resources of *luffa* species. The genus comprised of 9 species in the world and out of this 7 species (*Luffa acutangula, Luffa cylindrica, Luffa echinata, Luffa graveolens, Luffa hermaphrodita, Luffa tuberosa* and *Luffa umbellata*) are native to India. There is ambiguity with regard to *Luffa tuberosa* and *Luffa umbellate* because they are considered synonym to species of *Momordica* and *Cucurbita* respectively (Chadha and Lal, 1993). The sponge gourd (*Luffa cylindrica*) and ridge gourd (*Luffa acutangula*) have rich diversity in vine and fruit morphological characteristics throughout India particularly in North-Eastern region including Sikkim, West Bengal, Western, Central and Southern India. *Luffa hermaphrodita* is considered originated from *Luffa graveolans*, another potential species distributed in parts of North-Central India. *Luffa acutangula* var. *amara* grows in peninsular India and is a wild relative of *Luffa cylindrica* and *Luffa echinata* in natural habitats in Western Himalayas, central India and Gangetic plains.

Lagenaria species

Under genus *Lagenaria*, 6 species are reported. Out of these *Lagenaria abyssinica, Lagenaria siceraria*, and *Lagenaria leucantha* are common. Among them *L. siceraria* is generally cultivated in all tropical parts of the world, especially in India and a few African countries. Remaining species are wild, perennial and diocious in nature. Two wild species *i.e. Lagenaria abyssinica*, and *Lagenaria bravifolia* are perennial in nature. There are suggestions that *Lagenaria* occurs in wild form in South America and in India (de Candolle 1982).

Citrullus species

The genus, *Citrullus* has two species of economic importance *viz.* water melon (*Citrullus lanatus* (Thunb.) and round gourd (*Citrullus vulgaris* Schrad var. *fistulosus*). Based on the evidence (de Candolle 1982) and linguistic data (Filov and Vilenskaya, 1972), *Citrullus lanatus* originated in Africa and reached India in prehistoric times. Cultivation of large watermelon (*Citrullus lanatus* var. *citrodes*), is comparatively recent and Soviet varieties grown today have shape of their African ancestors (*Citrullus lanatus* var. *caffer*). Shimotsuma (1963) reported that *Citrullus vulgaris, Citrullus colocynthis, Citrullus ecirrhosus* and *Citrullus nandinianus* are related and cross compatible with each other. Whitaker (1933) considered *Citrullus colocynthoides* as probably an ancestor of watermelon.

Trichosanthes species

Trichosanthes has 22 species and is reported of Indian origin/parts of Tropical Asia or Indo-Malayan region. Among these, *Trichosanthes anguina L.* (snake gourd), *Trichosanthes dioica Roxb.* (pointed gourd) are cultivated throughout the country. The major zone for distribution of diversity for *Trichosanthes dioica* is north-central and north-eastern India including West Bengal whereas *Trichosanthes anguina* is distributed throughout the country. However, the rich diversity in *Trichosanthes anguina* was observed in North-Eastern states, West Bengal, Malabar Coast and in Eastern ghats in low and mid hills.

Momordica species

The genus *Momordica* has 60 species reported to occur in the world and 7 species in India. The cultivated species are *Momordica charantia* (*Bitter gourd*) grown all over the country in tropical and sub-tropical climate. *Momordica dioica* (Kartoli) grows all over West Bengal, Assam, parts of Bihar and adjoining areas; *Momordica cochinchinensis* (kakrol or sweet gourd or golkakora) is popular particularly in Tripura, Assam and West Bengal. Occurance of other species like *Momordica cymbalaria* mainly in Western Ghats, Maharashtra, Southwards and Eastern peninsular tract, *Momordica denudata* in peninsular tract, *Momordica macrophylla* in North-Eastern region and *Momordica subangulata* in North-Eastern hills, Eastern Ghats and Deccan peninsula are reported.

Benincasa species

Benincasa hispida Cogn. known as ash gourd or wax gourd, is a native of Java and Japan. It was domesticated in India during pre-historic times. Widely grown all over the country in tropical and sub-tropical climates and it possesses variability in fruit morphological characteristics and quality.

Coccinia species

Coccinia has about 35 species distributed in tropical Africa and Asia. Out of these, only one species *i.e. Coccinia grandis* is under cultivation. The related species are *Coccinia histella and Coccinia sessilifolia,* widely cultivated in several countries like Africa, Central America, China, Malaysia, Australia and other tropical Asian countries (PIER, 2001).

Sechium species

Genus *Sechium* is said originated in mountainous region of America and Mexico. It includes *Microsechium compositum, Microsechium gintonii, Sechium compositum, Sechium edule, Sechium hintoni* and *Sechium jamaicense.* Principally it was confined to Tropical and sub-tropical climate and particuarly in mid hill conditions. Maximum variability occurs in Sikkim. In Meghalaya and Mizoram, it is grown on commercial scale and is popular in Darjeeling Hills, also.

Leaf Vegetables

The Indian sub-continent is considered the place of origin for a number of leaf vegetables like *Beta vulgaris* var *bengalensis, Basella rubra, Cichorium endivia, Enhyndra fluctuant, Lactuca indica, Amaranthus* etc. Most of the leaf vegetables possess good amount of diversity, acclimatized under extreme variation in altitude with long history of its cultivation. The wild relatives of different leaf vegetables possess valuable genes particularly for resistance to biotic and abiotic stress conditions.

Beta Species

Beta species include *Beta adanensis, Beta altissima, Beta atriplicifolia, Beta bengalensis, Beta bourgaei, Beta brasiliensis, Beta campanulata, Beta chilensis, Beta cicla, Beta corolliflora, Beta lomatogona, Beta macrocarpa, Beta macrorhiza, Beta maritima, Beta maritima* subsp. *danica, Beta maritima* var. *atriplicifolia, Beta maritima* var. *erecta, Beta maritima* var. *prostrata, Beta nana, Beta orientalis, Beta palonga, Beta patellaris, Beta patula, Beta perennis, Beta procumbens, Beta trigyna, Beta trojana, Beta vulgaris, Beta vulgaris cv. conditiva, Beta vulgaris cv. saccharifera, Beta vulgaris f. rhodopleura, Beta hybrida, Beta vulgaris* subsp. *adanensis, Beta vulgaris* subsp. *cicla, Beta vulgaris* subsp. *flavescens, Beta vulgaris* subsp. *lomatogonoides, Beta vulgaris* var. *glabra, Beta vulgaris* var. *cicla, Beta vulgaris* var. *crassa, Beta vulgaris* var. *atriplicifolia, Beta vulgaris* subsp. *macrocarpa, Beta vulgaris* var. *altissima, Beta vulgaris* subsp. *vulgaris, Beta vulgaris* subsp. *provulgaris, Beta vulgaris* subsp. *patula, Beta vulgaris* subsp. *maritime, Beta vulgaris* subsp. *orientalis, Beta vulgaris* var. *erecta, Beta vulgaris* var. *flavescens, Beta vulgaris* var. *foliosa, Beta vulgaris* var. *grisea, Beta vulgaris* var. *macrocarpa, Beta vulgaris* var. *maritime, Beta vulgaris* var. *orientalis, Beta vulgaris* var. *perennis, Beta vulgaris* var. *pilosa, Beta vulgaris* var. *prostrata, Beta webbiana, Beta x intermedia* and *Beta vulgaris* var. *trojana*. The spinach beet also known as Indian spinach beet (*Beta vulgaris* var. *bengalensis* Hort.) is a close relative of beet root. It is grown in the plains and hills of India from Punjab to North Eastern Region including West Bengal, Orissa, M.P., Rajasthan, Gujarat and Maharashtra.

Amaranthus species

The origin of various species of cultivated amaranth is complicated, because the wild ancestors are pantropical cosmopolitan weeds (Mohideen and Irulappan, 1993). *Amaranthus spinosus, Amaranthus hybridus* and *Amaranthus dubius* are tropical types. *Amaranthus retroflexus, Amaranthus viridis, Amaranthus lividus* and *Amaranthus graecizans* are more temperate hot season weeds. In all probably, it might have originated in India and have spread to neighbouring countries by traders, buddies monks and others. In India a number of domesticated forms are available all over the country especially in North Eastern States, West Bengal, Orissa, Tamil Nadu, Andhra Pradesh, Karnataka and Kerala. Distribution pattern of *Amaranthus blitum* (weed throughout India), *Amaranthus caudatus* (wild throughout India), *Amaranthus gangeticus* (weed warmer parts of India, variants as *Amaranthus mangostanus*), *Amaranthus paniculatus* (warmer parts of India), and *Amaranthus spinosus* (weed throughout India) are reported.Wide ranges of variability in plant morphological characteristics are available in different parts of the country. In Orissa and North Eastern States a special type is grown where stem is consumed after full growth.

Spinacia species

Genus *Spinacia* belongs to the family Chenopodiaceae and said to originate in south west Asia. There is well description about four types of plants with reference to sex expression, which are: (i) extreme males (ii) vegetative males (iii) monoecious and (iv) females. The monoecious plants may be predominantly staminate, pistillate or pure pistillate, early but with some staminate flower late or almost equally staminate or pistillate throughout the season. The greater variability can be spotted in the plains and in foothills. Variability occurs in plant morphological characteristics, seed type and foliage yield attributing components.

Chenopodium species

Genus *Chenopodium* belongs to family Chenopodiaceae. *Chenopodium album* is found as weed in plains and *Chenopodiaceae murale* in hills upto 3000 m. In both species, maximum variability can be observed in upper Gangetic plains, extending to northern hills; elsewhere only sporadic distribution.

Basella species

Climbing spinach or Poi is grown all over the country particularly in the plains and mid-hills during hot humid season The genus *Basella* includes *Basella alba, Basella cordifolia, Basella rubra, Basella tuberosa* and *Basella vesicaria*. Rich varability in plant morphological characteristics include stem fleshiness, leaf shape and size in South India, North eastern region including Assam, West Bengal and Orissa.

Trigonella species

The genus *Trigonella* has 22 species, which are having different chromosome numbers. The 15 species having 2n=16 chromosomes include *Trigonella angeuna, Trigonella arabica, Trigonella bolansae, Trigonella calliceras, Trigonella coerulea, Trigonella corniculata, Trigonella cretica, Trigonella foenum-graecum, Trigonella gladiota, Trigonella glomerate, Trigonella gordej, Trigonella hamosa, Trigonella lypokyi, Trigonella monspeliach, Trigonella radiata:* three species have 2n=18 (*Trigonella stellata, Trigonella striata, Trigonella ornithopodioides*): two species have 2n=44, (*Trigonella geminiflora, Trigonella grandiflora*) (Fedorov, 1974) and another species has 2n= 44 (*Trigonella polycerata*). Two species of *Trigonella* are of economic importance. These are *Trigonella foenum-graecum* L., the common Methi and the other is *Trigonella corniculata* Kasuri or Champa methi. It is found wild in North Western parts of India (Bailey, 1950).

Portulaka species

Parslane (*Portulaka oleracea* L.) is indigenous to Himalaya. It occurs in several forms during hot weather, in the cultivated and wild state. The cultivated forms have broad leaves whereas in wild; both broad and small leaf types are seen growing. It is a tiny creeper with small, succulent fleshy leaves and tender stems, cooked like spinach.

Brassica species

Chinese cabbage (*Brassica pekinensis* and *Brassica chinensis*) is indigenous to China and eastern Asia where it has been in cultivation since 5th century. There are two types of Chinese cabbage-one that heading type (*Brassica pekinensis*) and other open leaf type (*Brassica chinensis*). The former forms an erect, moderately compact, usually cylindrical heads, while the later develops clusters of succulent leaves with forming head. Rich diversity in plant type occurs in foot hills of Himalaya and north eastern states including Sikkim.

Allium species

The Alliums include onion, garlic and their relatives like leeks, shallots and chives considered prized vegetables due to food, medicine or religious purposes. The genus *Allium* is represented by about 750 species widely distributed over the warm temperate and temperate zones of the northern hemisphere. Regions of high species diversity occur in Turkey, Iran, North Iraq, Afghanistan, Soviet, middle Asia and West Pakistan and a second less pronounced center of species diversity occurs in Western North America. About 40 species of *Allium* are reported in India of which seven are cultivated, the main being onion and garlic. Much diversity has built up in India in these crops as they have been cultivated from ancient times and selections have been made in different agro-ecological zones. Both these crops are known only in cultivation and while no known wild ancester is known for *Allium cepa, Allium longicuspis,* endemic to central Asia, proposed as a progenitor of *Allium sativum.* Within *Allium cepa* two horticultural varieties are recognized: the dry bulb onion and the multiplier onion (shallots). In *Allium sativum* despite it being propagated exclusively by the division of ground cloves or aerial bulbils, a wide range of variability is available in several traits. The leeks (*Allium ampeloprasum*) succeed fairly well in India thriving at elevations above 300 m and often remain fit for use until the end of rainy season. The chives (*Allium shoenoprasum*) are distributed in the Northern Himalayas from Kashmir to Kumaon. Cultivated and wild types interbreed and there is plenty of variation in cultivated forms. Attention is being given on these and other wild species, as they possess many disease resistant genes potentially useful in improvement programmes. *Allium ampeloprasum* for white rot resistance, leek yellow stripe virus, leek rust; *Allium roylei* for downey mildew; *Allium fistulosum* for pink rot, smut and onion fly, and *Allium tuberosum* for *Fusarium oxysporum* and leek moth are particularly noteworthy.

Raphanus species

Genus *Raphanus* includes *Raphanus caudatus, Raphanus landra, Raphanus lyratus, Raphanus pterocarpus, Raphanus raphanistrum, Raphanus raphanistrum subsp. landra, Raphanus sativus, Raphanus sativus var. caudatus, Raphanus sativus var. longipinnatus, Raphanus sativus var. niger, Raphanus sativus var. oleiferus, Raphanus sativus var. oleiformis, Raphanus sativus var. sativus, Raphanus violaceus* and *X Brassicaraphanus* species.

Daucus species

Genus includes several species like *Daucus aureus, Daucus bicolor, Daucus bocconei, Daucus brachiatus, Daucus broteri, Daucus capillifolius, Daucus carota,, Daucus carota cv. atrorubens, Daucus carota subsp. carota, Daucus carota subsp. commutatus, Daucus carota subsp. drepanensis, Daucus carota subsp. gadecaei, Daucus carota subsp. gummifer, Daucus carota subsp. hispanicus, Daucus carota subsp. hispidus, Daucus carota subsp. maritimus, Daucus carota subsp. maximus, Daucus carota var. atrorubens, Daucus carota var. boissieri, Daucus carota var. commutatus, Daucus carota var. sativus, Daucus crinitus, Daucus durieua, Daucus gingidium,*etc. In these genus two types *i.e.* Asiatic and European are more popular

Abelmoschus species

The center of diversity of *Abelmoschus* includes West Africa (Benin, Togo, and Guinea), India, and Southeast Asian countries, *i.e.*, Burma, Indochina, Indonesia, and Thailand (Chevalier 1940; Van Borssum Waalkes, 1966; Siemonsma, 1982a, b, Hamon 1988; Hamon and Hamon 1991; Bisht *et al.,* 1995). *Abelmoschus esculentus* is the commercially cultivated species of okra. However, in West Africa *A. caillei* is also cultivated for its fresh fruit. The geographical distribution of *A. caillei* is from Northern Guinean extending to southeast Cameroon. *Abelmoschus esculentus* is cultivated in most humid parts of Guinea, mostly in an intercropped system with *A. caillei* (Hamon and Charier 1983; Hamon 1988). The genus *Abelmoschus* comprises nine species (IBPGR, 1991) and among them *Abelmoschus moschatus* and *A. manihot* are semi wild, and show a greater diversity than the cultivated forms (Hamon *et al.,* 1991). *Abelmoschus moschatus* is morphologically and genetically the most different from other species (Hamon and Yapo, 1986) with probably the widest geographical distribution (Hamon and Charrier, 1983), which extends from the South Pacific Islands through Indochina and India and over to the central and western parts of Africa. The cultivated form (*Abelmoschus moschatus* var. *moschatus*), better known as "*musk mallow,*" "*Jew's mallow,*" or "*ambrette*" in Africa, Asia and America has fragrant seeds used for making perfume (Hamon and Charrier 1983). The subspecies *biakensis* is endemic to Papua New

Guinea (Charrier, 1984). The cultivation of *A. manihot* is mainly in Southeast Asia and is known under the vernacular name "*qibika*", also found in India, northern Australia and less frequently in the American continent and tropical West Africa. Its cultivation is exclusively for leaf consumption. The species *A. tetraphyllus* (IBPGR, 1991) *A. ficulneus, A. crinitus, A. angulosus* (Van Borssum Waalkes, 1966), and *A. tuberculatus* (Pal *et al.,* 1952) are truly wild. *Abelmoschus ficalneus* is from the Northern parts of Australia to Southeast Asia and from India to most parts of Africa. *Abelmoschus tetraphyllus, A. crinitus, A. angulosus* and *A. tuberculatus* are exclusively of Asian origin. *Abelmoschus tetraphyllus* var. *tetraphyllus* and var. *pungens* are endemic to Southeast Asia with extension of the former to Papua New Guinea and New Ireland. *Abelmoschus crinitus* is found at low altitudes in India and Southeast Asia and *A. angulosus* at high altitudes in India, Sri Lanka, Indochina, and Indonesia. *Abelmoschus tuberculatus* is endemic to northern and western parts of India (Charrier, 1984; Velayudhan and Upadhyay, 1994).

Based on different Floras, Floristic accounts and other published evidences main areas of concentration of wild species are as follows:

✰ Semi-arid tracts of north and northwestern India are for *A. ficulneus* and *A. tuberculatus.*

✰ *Tarai* range and the foothills of the Himalayas are for *A. crinitus, A. manihot* and *A. tetraphyllus* or *A. tetraphyllus* var. pungens*.*

✰ Western and Eastern Ghats and also peninsular tracts are for *A. manihot* (including *tetraphyllus* type), *A. angulosus* mainly confined to hilly tracts up to 2000 m in south India with only meager distribution elsewhere.

✰ Northeastern regions are for *A. crinitis* and *A. manihot*, mostly var. *pungens.*

In some of the wild taxa, intra-specific variation exists and has been taxonomically identified as *A. manihot* var. *tetraphyllus* and *pungens*, in *A. angulosus* var. *purpureus* and *grandiflorus*, in *A. moschatus* var. *biakensis* and *rugosus*, and in *A. tuberculatus* var. *deltoidefolius.*

Augmentation/Collection of Vegetable Germplasm

There are various ways to augment the vegetable germplasm

Introduction/Collection from Abroad as Identor

During the past, introductions of new vegetable varieties in every crop are becoming very popular among farmers or being used in selections. Examples can be cited in every crop in vegetable group where more than half the varieties grown in India are either direct introductions or selections thereof. Some of the notable introductions during recent years are bacterial wilt and fusarium wilt resistant lines of tomato from Taiwan, nematode resistant varieties Quinte and Early Rouge from Canada. Carrot variety Beta III has high carotene content. This has been used by many human nutrition programmes in India to cope up Vit A deficiencies. In watermelon, high yielding varieties tolerant to *Fusarium* wilt have been introduced from USA. In muskmelon a multiple disease resistant line AC70-54 has resistance to Gummy stem blight, powdery and downy mildew. In cucumber also gynoecious line Gy-4 was introduced and used for breeding high yielding multiple disease resistance including mildews and *Fusarium* wilt.

In Chinese cabbage, heat tolerant and black rot resistant lines were introduced from Taiwan. In cabbage, lines resistant to club rot, black rot and heat tolerant have also come from Taiwan. In capsicum, TMV tolerant, nematode tolerant and high yielding varieties World Beater and Yolo Wonder have all proved useful introductions. In chillies, hot pepper lines from AVRDC, Taiwan are very successful at several locations in India. A number of accessions of cucurbitaceous vegetable crops and their wild relatives were introduced from different countries. This includes *Citrullus lunatus* (8, all from USA), *Cucumis melo* (15, from France, Japan), *Cucumis sativus* (15 from Japan, USA), *Cucumis heptadactylus* (1 USA), *Cucumis metuliferus* (16 USA), *Cucumis anguria* (94 USA), *Cucurbita pepo* (2 from USA), *Cucurbita species* (4 from Algeria, Japan), *Luffa cylindrica* (1 Japan) and *Lagenaria siceraria* (1 Japan,).

A number of vegetable germplasm were introduced from other countries and are given in Table 21.4.

Table 21.4: Some Promising and Useful Introductions

Crop species	Accessions	Country	Promising and Useful Attributes	Indentor
Cucumis sativus	EC-399914 -37	UK	Resistant to CMV, downy mildew	PAU, Ludhiana
	EC-398030	China	Early maturing determinate type bear fruits in cluster	IIVR, Varanasi
	EC-398966-67	USA	Resistant to angular leaf spot	Sandoz India
	EC- 398968-70	USA	Resistant to Anthracnose	Sandoz India
	EC- 398971 73	USA	Resistant to fruit rot	Sandoz India
	EC- 398974-90	USA	Resistant to downy mildew	Sandoz India
	EC- 398991-9007	USA	Resistant to leaf spot	Sandoz India
Citrullus lanatus	EC-393240-43	USA	Small medium round fruits, red fleshed for breeding purpose	IIVR, Varanasi
Citrullus lanatus	EC-402549	USA	Multiple disease resistant to race-2, anthracnose, fusarium wilt and gummy stem blight with early maturity	–
Cucumis melo	EC-399866-212	UK	Resistant to downy mildew, powdery mildew and CMV	–
Cucumis melo	EC- 382726-36	France	Germplasm with male sterility genes MS1 to MS5, white fruit (gene), Fusarium wilt resistant (genes form-1, form-3)	IARI, New Delhi
Cucumis sativus	EC-382737-39	USA	Gynoecious lines	IARI, New Delhi, PAU, Ludhiana
Cucumis species	EC-382500-69	USA	Different *Cucumis* species with nematode resistance	PAU, Ludhiana
Cucurbita pepo	EC-380995	USA	Unique, small, oblong early maturing, the flesh comes out in strings, can be bakes or used in breads	IIVR, Varanasi
	EC- 380996	USA	Hull-less seeded pumpkin; seeds can be popped and eaten like snack, high in nutrition and good source of Zinc.	IIVR, Varanasi
Citrullus lanatus	EC-380989-91	USA	Yellow fleshed water melon, early maturing, sweet crisp flesh	IIVR, Varanasi
	EC-382753	USA	Breeding line SSDL resistant to Fusarium wilt and anthracnose	IIVR, Varanasi
	EC-378523-23	USA	Fusarium wilt resistant	IIVR, Varanasi

Contd...

Table 21.4–*Contd...*

Crop species	Accessions	Country	Promising and Useful Attributes	Indentor
Bitter gourd	EC-399808			
Sponge gourd	EC-305586			
Ridge gourd	EC-284347			
Bottle gourd	EC-305378			
Palak	EC-284349			
Tomato	Sioux, Fireball, Marglobe, Best of all, Roma, Money Maker			
C. pepo	EC-516625-96			
Watermelon	EC-509468, EC-477972, EC-477973, EC-477974			
C. melo	EC-468986, EC-477977, EC-477978, EC-477979, EC-477980, EC-477981			
C. sativus	EC-497645, EC-497646, EC-497647			
Bell pepper	California Wonder, Yolo Wonder, Oshkosh, Ruby king, King of North, Early Giant, Chinese Giant, World Beater			
Onion	Texas Early Grano, EC-4144	USA	Mild flavoured, large sized, yellow, suitable for fresh use	
	Bermuda Yellow, EC-109123	Philippines	Non-boltig salad onion	
	Giza-6, Giza-20, EC-200936, EC-200937	Sudan	High yielding, large sized, dark red pungent types	
	Nu Mex Br-1, EC-169372	USA	Short day, resistant to bolting and pink root disease	
	Sweet Sandwich, EC-169373	USA	Dehydrator/Salad type	
Garlic	EC-244949, EC-244858	Taiwan	Large bulbs, purple cloves	
	EC-210991	Egypt	Compact white bulbs, bold cloves	
	EC-158250	Taiwan	Large light pink bulbs, bold cloves Bold white cloves, field tolerant to purple blotch	

Contd...

Table 21.4–*Contd...*

Crop species	Accessions	Country	Promising and Useful Attributes	Indentor
Sweet potato	EC-332805	Puerto Rico	Red skin and yellow flesh	
	C-12	Peru	Deep orange, semi erect, resistant to leaf scales, drought tolerant	
Okra	Ghana Red	Ghana (Africa)		
Radish	Japanese White, China Red, Red Tail Radish			
Turnip	Snow Ball, Purple Top, White Globe			
Peas	Early Badger, Bonneville, Sylvia, Asauji, Arkel			
Cauliflower	Snowball-16, Improved Japanese			
Carrot	EC-178385,	Italy,	High Vit.A	
	EC-187201,	USA,	High Vit.A	
	EC-27883 to EC-274886	Netherlands	Male sterile and restorer line, resistant to club rot	
Sem	IC-16862, Pusa Bunch			

Collection through Exploration

Emphasis was made for collection of cucurbitaceous vegetables from different parts of the country in collaboration with NBPGR, state Agricultural Universities and ICAR Institutes. During the year, a number of crop specific explorations were undertaken to collect major and minor cucurbits. This includes bitter gourd, pointed gourd, cucumber, snap melon, kakrol, kartoli, pumpkin, sponge gourd and ridge gourd.

Table 21.5: Useful Indigenous Collections through Explorations

Tomato	HS-101, HS-102, HS-110, S-12, Punjab Kesari, Kuber, Kalyanpur T-1, KS-1, KS-2, Angurlatha, CO-1, CO-2, CO-3
	Brinjal Pusa Purple Long, Pusa Purple Round, Pusa Purple Cluster, Pusa Bhairav, Arka Sheel, Arka Kusumkar, Arka Navneet, CO-1, MDU-1, Junagarh Long.
Hot pepper	NB-34 to NB-41, NP-46A to NP-51, Jwala, Pusa Red, Kalyanpur Red, Yellow Kalyanpur, Chaman, Pant C-1, Pant C-2, G-1, G-2, G-3, G-4, G-5

Exploration within India with International Collaboration

Following explorations have been done within India with international collaboration and outside India for enhancement of vegetable germplasm.

Table 21.6 Explorations (Multiple Crops) within India with International Collaboration

Collaboration/Scheme	Areas Surveyed	Crops Collected	Collections
PL-480 scheme	Different parts of india	Cluster bean	942
		Legumes	8926
Indo-Canadian	Kumaon Himalayas	*Brassica* spp.	18
Indo-USAID collaborative explorations	Parts of Rajasthan, Madhya Pradesh and Uttarakhand	Cucumber	194
		Snapmelon	236
		Cucumis callosus	156
		Snake cucumber (*C. melo* var. *utilissimus*)	24
		Other cucurbits	48
IPGRI collaborative explorations	(i) North-eastern region (ii) Parts of India, Nepal and Sri Lanka (iii) South Asia	Okra, eggplant and their wild relatives	4665

Explorations and Collections of Germplasm Outside India with International Collaboration

Germplasm collections are also made outside India under certain programmes time to time.

Table 21.7 Explorations and collection of germplasm outside India

Country	Crops Collected
Central Asian Republics of the former USSR	Cassava, okra, winged bean, velvet bean, cucurbits, *Capsicum*, *Hibiscus* and tomato.
East Africa-Malawi and Zambia	Tuber crops, legumes, *Amaranthus* and other vegetables
Kenya, Sudan and Ethiopia, Bangladesh	Eggplant

Status of Germplasm in India

The Indian gene center has rich diversity in vegetable crops. The main aim of NBPGR is collection, evaluation and maintenance of plant genetic resources of different crops. Not only this, it plays an important role in *ex-situ* conservation of crop genetic resources (National gene bank comprising long-term seed storage, *in vitro* repository and clonal field gene banks). Conservation and sustainable management of plant genetic resources is a cooperative endeavor and requires active collaboration between NBPGR and the concerned national and international agencies. The National Bureau of Plant Genetic Resources, New Delhi from 1976 to 30-4- 2003, has a total of 26,988 accessions of various vegetable germplasm collected from Indo- gangetic plains, North-eastern region, North-western Himalayas, eastern Uttar Pradesh, Bihar, parts of Maharashtra, West Bengal, Assam, peninsular region, tarai region in Uttar Pradesh, Chattishgarh, Madhya Pradesh, Vidarbha (MS) and tribal dominated belts in central india, Orissa and adjoining Andhra Pradesh. The NBPGR New Delhi has systematically assembled over 2900 germplasm lines of *Lycopersicon* including wild species from diverse agro-climatic zones of world. The major donor countries include USA, Philippines, U.K.,

Canada, Australia, Portugal, Italy, Argentina, Cyprus, Mauritius, Israel, Netherlands, Nepal, Tanzania, Turkey, Japan, Iraq, Chile, Yugoslavia, Denmark, Switzerland, China, Trinidad, South Africa, Czechoslovakia, Ghana, Germany, Hungary, Taiwan, Poland, Bulgaria, South America, Columbia, Hong Kong, Spain, United Arab Republic, Burma, Nigeria, New Zealand and Brazil. Realizing the importance *Solanum* species, explorations were undertaken in collaboration with IPGRI covering almost all brinjal growing areas of India and neighboring countries *viz.* Bangladesh, Nepal and Sri Lanka for collection of variability. Germplasm sampled from neighboring countries includes 175 accessions of eggplant and 27 other *Solanum* species from Bangladesh.

Table 21.8: Vegetable Germplasm Holdings in National Gene Bank (as on 30-4-2005)

Vegetables	Number of Collections	Vegetables	Number of Collections
Cucumber (*Cucumis sativus*)	638	Bitter gourd (*Momordica charantia*)	973
Snap melon (*C. melo var. momordica*)	697	Kartoli (*Momordica dioica*)	18
Musk melon (*C. Melo*)	238	Kakrole (*Momordica cochinchinensis*)	12
Kachri (*Cucumis callosus*)	348	Sponge gourd (*Luffa cylindrica*)	852
Pumpkin (*Cucurbita moschata*)	1280	Ridge gourd (*Luffa acutangula*)	628
Bottle gourd (*Lagenaria siceraria*)	1113	Satputia (*Luffa hermaphrodita*)	19
Snake gourd (*Trichosanthes anguina*)	289	Ash gourd (*Benincasa hispida*)	439
Pointed gourd (*Trichosanthes dioica*)	324	Round gourd (*Citrullus fistulosus*)	64
Ivy gourd (*Coccinia indica*)	154	Water melon (*Citrullus lanatus*)	129
Musk melon	430	Kartoli	80
Tomato	1408	Brinjal	5029
Chillies	2573	French bean	1534
C. hardwickii	58	Jack bean	27
Rice bean	307	Lablab bean	904
Cluster bean	133	Canavalia	18
Broad bean	123	*Cowpea*	642
Garden pea	687	*Onion*	697
Garlic	550	Colocasia	874
Elephant foot yam	821	*Potato*	247
Radish	175	*Carrot*	32
Coriander	378	*Fenugreek* (*Trigonella foenum-graecum*)	183
Cauliflower	141	*Spinach*	133
Cabbage	20	*Sweet potato*	27
Colocasia	440	Dioscorea	335

Procedure for Genetic Resources Exploration and Collection

A number of cultivated vegetables crops and their wild relatives possess enormous variability and genes resistant to biotic and abiotic stresses. It has rich array of vegetable wealth both of indigenous and introduced types distributed in diverse ecological habitats and geographical regions. Due to long history of cultivation and continuous selection, several land races/primitive types suitable to different climatic and soil conditions have been developed in various vegetable crops. Following steps are needed during exploration and collection of genetic resources.

Survey

In vegetables, a large group of crops are annuals and propagated by seeds, while some are perennials and propagation is done by vegetative means. The principles for both groups of crops differ to some extent. Initially the survey is conducted for a particular crop in their region of diversity known as coarse grid survey and further intensive survey is undertaken from the source locality for a particular genotype having desirable gene, known as fine grid survey. In coarse grid survey, a series of collections are made at a wide interval over the whole area. Further the collection from the area, where interesting and intensive variation appear, the second expedition and intensive survey would then provide fine grid sampling by concentrating upon the gene pool in this region. Hawkes (1976) is of the view that the above sampling strategy may be attempted wherever possible depending upon the availability of time and resources. Before proceeding on expedition for gene pool sampling in vegetable crops, a curator should have sound knowledge and understanding about the vegetable crop including their primitive/landraces/local types,their wild relatives,distribution of diversity,keen observation of the variation in plants and environment, biotic and abiotic stresses and their identifying key characteristics, understanding of the gene pool concept, breeding system and population structure, coarse and fine grid sampling and propagation/seed multiplication technique.

Diversity Distribution and Variation in Adaptability Behavior

Evolution under domestication could be viewed at three different stages

1. Domestication of wild species
2. Stabilizing semi-domesticated populations for production and productivity
3. Breeding for yield increase

Plant domestication is the induction of preferred taxa to, and their subsequent evolution in, man-made habitats. Most of the popular vegetable crop species have a long history of domestication. Yet wild species are till date exploited from their natural habitats particularly in tribal belts *e.g.* ferns, potherbs, roots and tuber crops. Adoption of some lesser-known vegetable crop species is also being taken up in different agro-ecological zones *e.g. Dioscorea* spp., *Colocasia* spp. and other yams in North-East India and coastal Kerala, *Houttuynia* spp. in Himalayan belt etc.

The basic changes which occur in a species on domestication could be due to selective harvesting and re-sowing as well as specific cultivation package/s in man made habitats. Harvesting automatically favours plants with the highest yield of economic plant parts *e.g.* bigger raw fruit size, shape or colour of fruit/juice leaves in different vegetable crops, Re-sowing with seed from specific plants automatically increases the fitness of these phenotypes. Selection favours uniformity in plant stand, increase in vigour particularly at seedling stage, loss of seed dormancy etc. Above all, there is definite increase in both production and per unit area productivity due to farmer level selection in the semi-domestcated types (de Wet, 1977).

The mechanism underlying increase in adaptedness, uniformity in crop population and increase in yield are associated with development of clusters of inter-related alleles at different loci and gradual amalgamation of clusters into large synergistically interacting complexes. The adaptedness of variable forms *i.e.* types of taxon, land races of primitive cultivars etc. are manifested as their superior ability to live and reproduce in respective environments.

Thus the genetic variation and the variation for adaptability of forms within species provide unlimited resources both for sustainable production and crop improvement. Recent theories on handling of plant genetic resources, including delineation of core subsets (Rana, 1992), give due weightage to both geographical as well as genetic background of accessions for their inclusion into reprehensive subset, out of large collection.

Table 21.9: Specific Adaptability of Various Vegetable Crop Species

Crop	Area/region	Remarks
Common bean	High altitudes of Himachal Pradesh and Western Arunachal Pradesh	Large variability
Lablab bean	Tripura	Large variability
Tree bean (*Parkia roxburgah*)	Mizoram, Manipur, Nagaland	Local preference
Horse-radish	Deccan Plateau	Local preference
Potato	Lahaul Valley	High yield Av. Tuber wt. 1 kg
Solanum gilo	North Eastern Region	Introduced from Africa & naturalized
Brinjal	Bundelkhand (U.P.) and Tripura	Primitive types available (Bholanath & Sheonath types)
Tomato	NEH (Meghalaya)	Cultivated in paddy lands from February to April
Bell pepper	Mizoram	Variability for fruit colour (white, red and Black)
Sweet Potato	Tripura	Large variability and Cross-4 highly successful
Radish	Meghalaya and Arunachal Pradesh	Radish Sel.1 highly successful
Red pepper	Assam	Baghi chillies, said to have high capsicine content (only medicinal use)
Chow-chow	Mizoram, Karnataka, Maharashtra	Highly naturalized
Bitter gourd	Tamil Nadu, Kerala, U.P. Bihar, West Bengal, Maharashtra	Highly specific adaptability
Kakrol	Mizoram, Tripura, West Bengal, Bihar, Vindhya Hills of U.P.	A specialty favorite with natives
Water melon	Rajasthan, Punjab, Haryana, Western U.P., Karnataka, M.P.	Near river beds only, develop more sweetness in arid zones
Musk melon	Rajasthan, Eastern U.P., Punjab	Develop more sweetness in arid zones
Buffallo gourd	Rajasthan	Develop more starch in roots and provide foliage, fruit (flesh contains more ware) for animals

Adaptability to Biotic and Abiotic Stresses

Wilsie (1962) reported distribution of climates, soil types, migration, introduction and selection as some of the factors controlling plant distribution in the post-domestication period. For example, although cowpeas are long day plants and beans are short day plants, yet farmers' selections in different agro-climatic niche have stabilized various day neutral forms of both cowpeas and beans. Similarly, in onions long day types are available and the relatively insensitive types are grown for bulbs in lowlands (Shanmugavelu, 1989).

Gableman and Gerloff (1978) reported occurrence of genetic variation for efficiency in micronutrient utilization in vegetable crops. Tolerance to salts and acidity or alkalinity also critically affects adaptability and evolution of forms, thereby making availability of diverse germplasm. Crops like garden beet, spinach, *Atriplex* spp. and asparagus have high salt tolerance whereas radish and beans have low tolerance to salts. Similarly, cucurbits prefer a soil pH between 6.0- 7.0 and no cucurbit can successfully grow below pH 5.0. Yet among the group, muskmelon is slightly tolerant to acidity while cucumber, pumpkin and squashes fall under moderate category.

Evolution of pathogen virulence has a specific significance in relation to history of cultivation particularly mixed cropping where the host forms are likely to have polygenic complexes for horizontal resistance. Thus screening and testing of diverse germplasm, including the low yielding cultivars and landraces are important to get diverse genetic base particularly for resistance or tolerance to biotic and abiotic stresses

Planning of Exploration

Broadly two kinds of explorations are formulated based on priority of crops, the area/region to be surveyed. Both, crop specific and region specific/multi crop collections are advocated depending on the situation. The crop specific explorations are undertaken as per the needs of breeders. However the multi crop specific explorations are undertaken to cover probity areas, which hold rich crop genetic diversity in various vegetable crops, and maturity period, coincided at the same time. The exploration is carried out by explorer, crop specialist/researcher, extension worker, and by botanist. The duration of exploration varies according to the mission; within the country it is about 20 to 45 days whereas in the foreign country larger duration is required, as otherwise it will be more expensive and less remunerative.

Selection of Site

Collection sites indicate the place of sample collection. This includes cultivated fields, natural habitats (mountains, valley, river sites, sea shore, forest etc.), local village market, threshing floors and farmers seed store etc. In case of annual vegetable crops, the collection site is individual field or farm (if farmers use different seed stock) or a group of field or farms (if farmers use common seed stock). Further the collection site may be reduced if crop is self pollinated and increased if the crop is cross-pollinated and showing high genetic diversity. In case of wild and weedy or weedy forms and relatives the selection of appropriate sampling site to collect diversity is more difficult. First, such species are often continuously distributed in nature or they follow the island model or the stepping stone model (Jain, 1975). In such cases, there are no artificial boundries defining discrete, relatively homogenous populations as in cultivated crops. Marshal and Brown (1975) suggested that the plant explorer must decide where, in the continuum he/she will collect and the total size of the target area to be sampled (*e.g.* 1000, 10,000 or 100,00 m²). Gene flow through pollen and seed dispersal is extremely limited in wild vegetable species. Natural population of wild species often shows marked genetic differences over small distances (Marshal and Brown, 1975).

Specific Area of Collection

Tribal dominated region inhabited by different ethnic group provides another interesting situation for gene pool sampling of primitive types of vegetable germplasm. Such areas often possess rich diversity of various vegetable crops. The tribals hold primitive cultivars of different vegetable crops and usually do not exchange seeds between the ethnic groups. In tribal areas particularly in Northeastern parts of Madhya Pradesh, Chhatisgarh, West Bengal, Bihar and Orissa, village markets/

haat, are the most desirable place to collect diversity in a number of vegetable crops. The areas which remain to be covered include remote and less accessible areas, hilly terrain and tribal dominated belt.

Number of Sites

Number of sites depends on availability of variability and pollination nature (self and cross pollinated) of crop. Each site provides opportunity for sampling different set of alleles, if no information on the distribution of variation is available. The number of sites often depends on the length of the season, relative abundance of the target species and roughness of the terrain etc. Changes in eco-habitats/niches and farming system should also be taken into consideration for deciding number of sites. The optimum number of sites for sampling for a given species could be fifty.

Sample Size and Sampling

The optimum sample size per site would be the number of plants required to obtain, with 95 per cent certainly that all the alleles at a random locus occurring in the target population with frequency greater than 0.05. Hawkes (1976) suggested that bulked seed samples from up to 50 per cent plants (minimum) should be collected from one site. However, this becomes difficult particularly in the crops like brinjal and cucurbits where availability of this much number of fruits is very rare at a site. Further in case of cross pollinated vegetables, each sample size should be of 10,000 to 12,000 seeds and in case of self fertilized crops, sample size should be of nearly 8000 seeds.

Random Sampling

Random sampling is usually practiced to collect maximum genetic variability of species, irrespective of relative frequency or rarity of any genes or linked genetic complexes. For this, at each collection site, single spike or a few pods/fruits are randomly collected at every second or third place along with a number of transacts through the crop at least at 50 places scattered over an area or sample are taken in the clustered form. Hawkes (1976) suggested that bulked seed samples from up to 50 individuals and certainly not more than 100 should be collected randomly as multiple samples at one site. Sample should be clustered and cluster should be spread as evenly as possible throughout the collecting area. The sampling technique may be appropriate in wild and weedy species.

Biased Sampling

The random sampling is not feasible in many of the vegetables like brinjal and most of the cucurbits because farmers keep a few selected fruits only for seed purpose and many times ripened fruit is not available in the field. Under this situation the explorer sticks to biased sampling. Harlan (1975) strongly pleaded about subjective, biased sampling. For biased sampling, crop specialist category of explorer is desirable because he has to be aware of the morphological and physiological characters of the crop to be sampled. These characters include pattern, rate and habit of growth (Fordham, 1971) and seed source in relation to quality and germination requirement (Flint, 1970). Sampling of specific genotype forms a known source; also forms important part in biased sampling.

Conservation Strategy of Vegetable Genetic Resources

The rich heritage and ethnic culture have favoured to preserve the richest diversity including rare landraces/primitive types of useful vegetables like eggplant, cucumber, ridge and sponge gourd and a number of root and tuber crop species. A number of vegetable crops were brought to India from other regions by travelers, invaders like Persian, Turkey, Mughals, Portuguese, Dutch, French and British which acclimatized and developed good amount of diversity. Wild relatives/species of some of the important vegetable crops have commercial importance to a great extent. Most of these wild species grown in the natural habitats possess genes resistant to biotic and abiotic stresses. However, diversity for valuable genetic resources is threatened in recent times. Therefore, conservation of the vegetable genetic wealth particularly their wild relatives are thus essentially required for future utilization.

The conservation of PGR involves two basic strategies as per objective and priorities:

1. *In situ*
2. *Ex situ*

In situ

In situ conservation of PGR has to be carried on farm, natural habitats/wild, landraces and locally adapted materials are cultivated, utilized and conserved as part of traditional farming systems by paying due regard to natural ecosystems. The main aim of *in situ* conservation is to allow the populations to maintain/perpetuate itself within the community environment, to which it is adapted so that it has the potential for continued evolution (Anonymous, 1989 a,). Landraces/primitive cultivars developed in traditional agriculture system also need *in situ* conservation in their respective areas. These farming systems are of particular importance in maintaining local genetic diversity and providing food for local consumption and local markets. It will provide an option for further evolutionary changes likely to take place. However, for majority of the situation, *in situ* conservation is ideal method of conserving wild plant genetic resources and perennial vegetables, which either do not set or set recalcitrant seed, do not produce plants true to type. *In situ* conservation of plant genetic resources has a number of advantages as compared to *ex situ* conservation (Anonymous, 1989 a).

1. It usually allows increased probabilities of conserving a large range of potentially interesting alleles
2. It is specially adapted to species, which cannot establish or regenerated outside their natural habitats.
3. Allow natural evaluation to continue, a valuable option for conservation of diseases and pest resistant species, which can evolve with their parasites, providing breeders with dynamic source of resistance.
4. It can serve several factors at once, since genepool of value to different sectors (crop breeding, production etc.) may often overlap and so can be maintained in the same protected area.
5. It facilitates research on species in their natural habitats.
6. It assures protection of associated species of economic importance.

The 'National Man and Biosphere Committee' of the Department of Environment has already identified sites as potential areas for biosphere reserve (Table 21.10). This covers all the major biogeographic regions of the Indian subcontinent where flora and fauna can be conserved in natural habitats. These areas have been identified based on their rich genetic diversity; floristic uniqueness endemic wealth of flora/fauna and they are in totality representative of ecosystems occurring in different biogeographic regions.

Ex situ

Ex situ consevation requires collection and systematic storage of seeds/propagules outside the natural habitats of species for short, medium and long term. *Ex situ* conservation PGR in the form of seeds in domesticated plant species for use in their genetic improvement was initiated with the

Table 21.10: List of Biosphere Reserves, their Location and Demarcation of Area

Biosphere Reserve	State(s) Location	Area (sq.km)
Kerala	Tamil Nadu, Karnataka, Nilgiri	5,520
Namdapha	Arunachal Pradesh	4,500
Nanda Devi	Uttar Pradesh	2,000
Uttarkhand	Valley of Uttar Pradesh and Uttaranchal	3940
North Island of Andamans	The Andaman and Nicobar Island	1,375
Gulf of Mannar	Tamil Nadu	10,500
Kaziranga	Assam	7,600
Sunderbans	West Bengal	*
Thar Desert	Rajasthan	*
Manas	Assam	600
Kanha	Madhya Pradesh and Chattishgarh	*
Nokrek (Tura range)	Meghalaya	80
Rann of Kutch	Gujrat	5,000

*: In the process of demarcation and delineation.

Source: Anonymous, 1987.

establishment of Imperial Agricultural Research Institute (IARI) in 1905 in village Pusa, Darbhanga district, Bengal (now in Bihar). In 1936, this institute was shifted to New Delhi. The conservation efforts of important PGR were initially started in the Division of Botany and later in the Division of Plant Introduction through frequent multiplication of seed material. In 1976, the Division of Plant Introduction was developed into National Bureau of Plant Genetic Introduction, which in 1977 was re-named as National Bureau of Plant Genetic Resources (NBPGR). NBPGR, New Delhi, is the nodal agency for *ex situ* conservation of PGR for food and agriculture and is maintaining active and base collections of various crop species and their wild relatives including vegetables in a network of gene banks in the country. The various possible approaches can be grouped into:

Plant Conservation

It includes:

Botanical Gardens

India has more than 100 botanical gardens under different management systems located in different bio-geographic regions. Central and State Government manage 33 botanical gardens which maintain the diversity in the form of plants or plant populations (MoEF, 1998).

Arboreta

An arboretum generally refers to a place established for conservation of tree species or vegetatively propagated crops. The Regional Plant Resource Centre at Bhubaneshwar has established an arboretum with 1,430 species of tree, a palmeratum of 100 different types of palms, a bombastum with 61 collections of bamboo and an orchidarium housing 220 species of orchids. At the national level, the interest in establishment of arboreta is very weak probably because of the cost involved.

Herbal Garden

This generally refers to the gardens which predominantly conserve herbs and shrubs of medicinal and aromatic values. The concept of herbal garden has been picked up by the non-governmental organization (NGOs)in India. Several non-governmental organizations in different parts of the country, particularly in tribal areas in Gujarat, Karnataka, Maharashtra, Madhya Pradesh, Chhattishgarh and Uttrakhandhave established herbal gardens with the objective of conservation of local biodiversity in medicinal and aromatic plants and other economically important species. At community level, effort for such herbal gardens may increase income of local communities.

Field Genebank

Most institutions dealing with perennial or vegetatively propagated domesticated plants have field repositories for conservation of PGR.

Clonal Repositories

It includes clonally propagated crops.

Seed Conservation

It includes

Low Temperature Storage of Orthodox Seeds (Seed Gene Bank)

The seed genebank is responsible for conservation of seed accessions on long-term basis as base collections for posterity. It has 12 storage modules with a capacity to hold about one million accessions including vegetable seeds at –20°C. The present base collection holding of vegetable seeds in the National Gene Bank (NGB) is 15032.

Cryopreservation: Storage of Orthodox Seeds and Recalcitrant (Embryonic Axes) Seeds

Cryopreservation is storage of biological samples in viable conditions at ultra low temperature of liquid nitrogen at –150 to –196 °C. A total of 5328 accessions belonging to 502 species representing all major crops, have been cryopreserved at moisture content of 5-8 per cent in the vapour phase of liquid nitrogen. Further, investigations are on to develop protocols for cryopreservation of more recalcitrant seed species and establish their base collections. As an alternative complementary method, attempts are also being made to cryopreserve pollen in trees or vegetatively propagated species.

In vitro Genebank Conservation

Includes (a) conservation of cells, tissues, organs in glass or plastic containers under aseptic conditions through slow growth of cultures, (b) Cryopreservation of cultures (tissues, organs, pollen or cultures in liquid nitrogen at –150 to –196 °C.). In India, to undertake the conservation of vegetatively propagated species, a National Facility for Tissue Culture Repository was established at NBPGR in 1986 with the financial assistance from Department of Biotechnology and since then the infrastructure has been further strengthened. The *in vitro* repository holds1,396 accessions belonging to 124 vegetatively propagated plant species maintained for short term and medium term storage.

DNA Conservation

The basic objective in conservation of PGR is conservation of genetic diversity existing in the form of a functional unit called 'gene'. The whole genome in the form of genomic library or a sequence of DNA in the form of DNA library may be conserved following the appropriate DNA conservation method.

Each technology is selected based on its merits in terms of utility, security, complementary and the advantages over the others.

National Active Germplasm Sites

The 40 National Active Germplasm Sites (NAGS) are holding active collections of relevant crop species. Eleven of them have the medium term seed storage facilities in the form of cold storage modules maintained at 4°C and 35-40 per cent relative humidity. Active collections are basically for sustainable use in various crop improvement programmes. The NAGS are based at premier crop or crop group-based institutes. The crop-based institutes have multi-disciplinary team of scientists, better equipped for evaluation of germplasm for agronomic potential and various yield reducing factors to identify accessions with desirable traits. This helps to generate information on potential value of various accessions to provide the needed support for their utilization in breeding programmes. The National Active Germplasm Site (NAGS) for vegetable crops is Indian Institute of Vegetable Research, Varanasi (U.P.).

Germplasm Registration

The Indian Council of Agricultural Research (ICAR) at NBPGR, New Delhi has provided a mechanism for registering potentially valuable germplasm under 'Registration of Plant Germplasm'. A number of vegetable germplasm are being registered for their unique traits. The validity of registered germplasm is fixed in term of duration *i.e.* 15 and 18 years for clonally and seed multiplied vegetable crops, respectively.

Exchange of Germplasm

The ultimate goal of plant introduction to any region of the world is the use of germplasm to its fullest potential. In earlier days, the agencies for plant introduction activities were travelers, pilgrims, invaders, explorers and naturalists. The movement of plants within the old world countries was possible much earlier because of geographic contact compared to the exchange of plants between the new world and the old world. Owing to these factors, the old world crops, *viz.* wheat and soybean were unknown in America about 400 years ago of which America is now a major exporter. Similarily, new world crops like chillies, and sweet potato introduced during the seventeeth century in India, show much more diversity due to better adaptation to Indian conditions.

After the monumental work of N.I. Vavilov (1928), a Russian Botanist and explorer, the concept of 'centres of origin' was analyzed, understood and recognized. Organized plant explorations and collections were undertaken jointly by different countries to these centers and the material collected was brought to their areas of collection and evaluated for various attributes. The spread of such findings created interest in plant breeders all over the world to acquire such materials for their utilization as primary or secondary introductions which further helped in the movements of crop plants (Dadlani *et al.*, 1981). The classical example of introduction of new crops that helped other countries more than the place of origin of such plants are the introduction of sunflower to USSR from central Mexico, USA, Chinese soybean to North America, Ethiopian coffee to central and south America, Bahian cocoa to West Africa, Amazonian rubber to Malaysia, Asiatic yam into tropical America and Africa, wheat from Near East to USA; maize, tomato, potato and sweet potato from South America to Europe and Asia.

Table 21.11: Registered Germplasm of Vegetables

Crop	Line		Unique Traits
Pointed gourd	IIVR PG-105	INGR-03035	Seedless fruits, obligate parthenocarpic, long duration fruiting (IIVR, Varanasi)
Bitter gourd	GY-63	INGR-03037	Gynoecious line with high yielding ability and attractive fruits (IIVR, Varanasi)
Water melon	RW-187-2	INGR-01037	High high yielding ability with yellow coloured flesh (ARS, Durgapura, Jaipur)
Water melon	RW-177-2	INGR-01038	Leaf mutant with simple unlobed leaves (ARS, Durgapura, Jaipur)
Bottle gourd	Androman-6	INGR-99009	Andromonoecious sex form of bottle gourd (NDUA&T, Fazabad)
	PBOG-54	INGR-99022	Segmented leaf genotype
	Cut Leaf		GB Pant Univ. of Agril. & Tech. Pantnagar
Potato	MP/99-322	INGR 04109	High starch & dry matter, low amylose and resistant to late blight
Chillies	MS1 & B line	INGR-04052	CGMS line with good GCA (IIHR, Bangalore)
	MS3 & B line	INGR-04053	CGMS line with good GCA (IIHR, Bangalore)
	IHR14 (PMR-14)	INGR-04054	Fertlity restorer line (IIHR, Bangalore)
	IHR-11-2 (IHR11#2)	INGR-04055	Fertlity restorer line (IIHR, Bangalore)
	MS-2A & 2B	INGR-03077	CMS line
	Capsicum Selection-2 (Nishat-1)	INGR-00001	High yielding with early maturity, export quality, superior shelf life and high nutritive value
	DRLT-1107	INGR-03092	High pungency
Brinjal	Hisar Jamuni (H-9)	INGR-99037	Intermediate, semi spreading, medium tall thornless, better ratoon for spring crop, fruit colour is retained for a long time less seed content, serves purpose of long and round types.
	IIHR-3 (96-2-1)	INGR-03074	Resistance to bacterial wilt
Cucunber	AHC-2	INGR-98017	High yielding and long fruited types
	AHC-13	INGR-98018	High yielding, unique, small fruited types, drought hardy, bear fruits even under high temperature in arid rergion with limited irrigation
Kachari	Kachari AHK-119	INGR-98013	High yielding and drought hardy
Fenu greek (Methi)	Hissar Methi-346	INGR-01012	Downy mildew resistant genotype with quick germination, faster initial growth, long pod, bold seed with green tan seed coat colour and light green narrow leaves
Fenu greek (Methi)	Hissar Methi-350	INGR-01012	Powdery mildew resistant and Downy mildew tolerant genotype. Plant is tall erected; multi branched with whitish green stems and dark green narrow leaves.
Peas	PMR-27	INGR-03078	Tendril type with resistance to powdery mildew
	DPP-62	INGR-03079	Yellow wrinkled seeds with resistannce to powdery mildew
Round melon	HT-10	INGR-99038	Intermediate semi spreading vine type roundmelon, tolerant to downy mildew and root rot wilt complex.
Snapmelon	AHS-10	INGR-98015	High yielding and drought hardy
Snapmelon	AHS-82	INGR-98016	High yielding and drought hardy
Tomato	Hisar Lalit	INGR-03036	Resistant to root knot nematode
	IIHR-2195	INGR-03076	Resistant to tomato leaf curl virus and bacterial wilt
Soybean	P-1366	INGR-01035	Vegetable type

Procedure for Exchange of Germplasm

The NBPGR has brought out a brochure 'Guidelines for the exchange of seed/planting material' which has been widely circulated amongst scientists in India. The guidelines for import of seed/planting material for research purpose have been revised (Anon, 1989) in view of the enactment of new Seed Development Policy by Government of India, which has also been circulated among scientists. The Govt has made the issuance of import permit mandatory for import of all seed/planting material for research purposes.

Import of Germplasm

All requests for indenting germplasm from abroad are to be made to NBPGR in prescribed application proforma giving specific details of the required materials stating the sources/country as well as address of the organization, so that the 'Import permit' is issued and sent to concerned scientist (s) for sending the same to the Bureau.

Concerned scientist/organization abroad is advised to take into consideration the following requirements for mailing the materials to India:

Only healthy, viable and clean seed material (free from soil, pests, pathogens and weeds) is to be forwarded without any seed treatment so as to facilitate proper quarantine examination. It may, however, be fumigated, if considered necessary.

The material is required to be accompanied with the 'Import permit' (which is to be sent to them alongwith our request letter) and phytosanitary certificate with additional declaration, if any, based on crop inspection certifying that the material is free from particular pathogen(s)/insects (s). It is to be further ensured that the package of seed/planting material must be addressed to the Director, NBPGR New Delhi who has to take delivery of the seed/planting material and conduct quarantine examination.

Export of Germplasm

Exchange of germplasm involves not only introductions but also the supply of seed and other materials to collaborating scientist/organizations abroad. Export of vegetable germplasm is also made on the basis of request received by Bureau/ICAR Institutes/Agricultural Universities in India under various protocols/work plans/memorandum of understanding with different countries/CGIAR Institutions. This includes *Cucumis melo* (1), *Cucurbita pepo* (4), *Cucumis sativus* (1), *Luffa cylindrica* (1) to Italy; *Momordica charantia* (1) to Taiwan; *Sechium edule* (2) and *Triochosanthes dioica* (1) both to USA. During the year 2001-2002 a few cucurbitaceous germplasm were exported under phytosanitary certificate to other countries cucumber (1), bitter gourd (1) ridge gourd (1) and sponge gourd (1) to Iraq.

Table 21.12: Export of Germplasm

Crop	Number of Samples	Country
Cucumis melo	6	DPR Korea
Citrullus lanatus	1	DPR Korea
Cucumis sativus	1	DPR Korea
Momordica charantia	1	DPR Korea
Citrullus lanatus	1	Bangladesh
Cucumis sativus	1	Iraq
Lagenaria siceraria	1	Bangladesh, Iraq, Taiwan

The following guidelines are to be observed while responding to such requests:

1. Request for seed/planting materials received from concerned organizations/agencies are to be forwarded to the NBPGR with relevant information so that prompt action on the supply of desired materials could be taken.

2. Dispatch of the seed/planting material is to be channelised through the Bureau so that prompt inspection of the material could be done from quarantine angle and phytosanitary certificate issued.

3. No seed dressing with insecticides or fungicides is given while dispatching the seed to the bureau.

Plant Quarantine

During introduction, promising germplasm possesses a great potential for transfer of associated pests and diseases from one region/country to another as all types of planting materials are affected by diseases or injuries caused by one or more organisms *viz.* insects, nematodes, fungi, bacteria, viruses etc. It is also resulting in distribution of several important pests and diseases too from one country to another. Thus, the movement of germplasm has exposed the world to greater risks of introducing exotic pests and pathogens in 'clean areas'. If unhealthy looking plants are left out, latent or hidden infestation is likely to be collected, specially in case of vegetatively propagated materials like bulbs, tubers, rhizomes etc. Another danger is that some of the exotic pests and pathogens often find more favorable conditions or more suitable host plants for their growth and multiplication in the new environment than were available in their original habitat. The problems of dissemination of pests and diseases through affected plant material are further aggravated because the material can be sent across the world in a matter of few hours, thus helping them to cover the long distance in viable conditions.

Plant quarantine can be defined as "legal enforcement of measures aimed to prevent pests and pathogens from spreading, or to prevent them from multiplying further in case they have already found entry and have established in a new area". Though Plant quarantine measures may not guarantee an everlasting protection against the entry of exotic species but will certainly check or delay the introduction of these unwanted organisms and their subsequent establishment in hitherto clean areas.

Indian Plant Quarantine Regulations and Set-up

Realising the importance of plant quarantine, Govt. of India also legislated an Act in 1914, known as 'Destructive Insects and Pests Act of 1914' (DIP Act) under which various notifications have been issued from time to time, restricting plants and plant materials from other countries as well as from one state to another state within the country (domestic quarantine). The latest regulation enacted under the DIP Act is the 'Plants, fruits and seeds Order, 1989 (PFS). This was necessitated to cater to the needs of 'The New Policy on Seed Development' (NPSD) of Govt. of India which came into force on 1st October, 1988 with the objective to make avaible to the Indian farmers the best genetic materials in the world to increase our agricultural productivity and to encourage the private sector seed industry in India not only to fulfill domestic requirements but also to develop export potential. While liberalizing import, care has been taken that there is absolutely no compromise on plant quarantine requirements. Although there are several requirements under the FPS Order, 1989, but for our purpose the most important requirements are:

1. No consignment shall be imported into India without a valid import permit issued by the competent authority. For importing germplasm Director, NBPGR has been authorized by the Govt. of India to issue import permits, both for Govt. institutions as well as private seed companies. For bulk consignments, the import permit is issued by the Plant Protection Adviser to the Govt. of India.

2. No consignment shall be imported unless accompanied by an official phytosanitary certificate by an official of the exporting country.

3. Seeds/planting materials requiring isolation growing under detention shall be grown in an approved post-entry quarantine facility.

   Import of soil, earth, sand, compost, and plant debris accompanying seeds/planting materials shall not be permitted.

4. Hay, straw or any other material of plant origins shall not be used as packing material.

Safe Movement of Vegetable Germplasm

There are glaring example of losses caused by the insect-pests and diseases introduced with import of planting materials like seed, propagules, fruits, plant etc. The dreaded golden nematode of potato introduced from U.K. has already infested the entire Nilgiri hills. Of the several introduced exotic insect pests the spiraling whitefly, *Aleurodicus disperses* is quite recent (Russel, 1995) which has established successfully in Karnataka and has been found infesting 253 plant species. The introduction of exotic weeds such as *Lantana camara*, in the early 19th century from Central America, Parthenium hysterophorus from Central and South America and *Phalaris minor* from Mexico in mid 20th century into India have become endemic sources of threat to our crop production and environment.

Table 21.13: Introduction of Insects/Nematodes

Insects/Nematodes	In Association with	From
Acanthoscelides obtectus	Cajanus cajan	Brazil and Columbia
	Phaseolus vulgaris	Italy and Nigeria
Anthonomus grandis	Gossypium and Vigna species	USA
Ditylenchus dipsaci (nematode)	Allium cepa	UK
Heterodera schachtii (nematode)	Beta vulgaris	Germany, Denmark, Italy
Globoera rostochiensis (Golden nematode)	Potato	-

Table 21.14: Important Vegetable Diseases Believed to have been Introduced to India

Host	Diseases	Pathogen	First record in India
Beans	Fuscans blight	Xanthomonas phaseoli var. fuscans	Pantnagar, Shimla, Solan, Delhi, 1972
	Common blight	Xanthomonas phaseoli	Poona, 1972
Cabbage and cauliflower	Black rot	Xanthomonas campestris	Delhi, 1968
Chillies	Leaf spot	Xanthomonas vasicatoria	Maharashtra, 1950
Onion	Downy mildew	Peronospora destructor	Jammu & Kashmir, 1977
Potato	Late blight	Phytophthora infestans	1983
	Wart	Synchytrium endobioticum	West Bengal, 1953

Evaluation, Promotion and Utilization of Vegetable Germplasm

Precise evaluation and documentation of germplasm are pre-requisites for its utilization. To promote the use of promising germplasm/genetic stocks identified for evaluation, germplasm field days should be organized besides publishing research articles, crop catalogues (printed and electronic), etc. The development of core set of collection particularly in the crops having large germplasm can be a powerful tool for promoting utilization of germplasm. It is also important to accord recognition to those associated with the development of improved/unique germplasm and genetic stocks, such as, plant breeders, farmer-breeders or other developers/innovators.

Characterization, preliminary evaluation and further evaluation of germplasm are prerequisite for utilization of plant genetic resources. Until characterization is not done, and attributes/traits are not known, the germplasm has little practical utility. Germplasm evaluation, in broader sense, provides the description of the materials in a collection. Characterization and evaluation are normally the responsibility of crop curator, while further characterization and evaluation should be normally be carried out by plant breeder. The evaluation of genetic resources is a multi-dimensional endeavor; involving various disciplines like as cytology, agronomy, genetics and biochemistry etc. The multidisciplinary participation to generate wide spectrum information of the germplasm collection would lead to its meaningful utilization. Various steps in plant genetic resource handling beginning from augmentation to seed increase, characterization, preliminary evaluation, detailed evaluation, utilization, regeneration and documentation need descriptions for practical purpose, as given below:

Characterization and Preliminary Evaluation (CPE)

The CPE consists of recording a limited number of additional traits, thought desirable by the consensus of specialists/users of the particular crops/species, which help in identifying useful germplasm. Evolution data refer to environmentally influenced characters. The important ones include site data; data on plant, leaf, flower, fruits, seeds and reaction to pests and diseases. Passport data-Data collected by curator including origin of the sample or it's known history. This is very important for identification and help in designating core collection, identifying duplicates and planning future exploration. Preliminary evaluation consists of recording data on a limited number of agronomic traits thought desirable by a consensus of research workers of the particular crop. The traits, in general, have quantitative inheritance and, are influenced environmentally. Thus, to better assess their perfotrmance, the germplasm should be evaluated in the agro-ecological region from where the accessions are collected or in a similar environment. There is no prescribed limit for the number of accessions in a trial planted for preliminary evaluation. Some trials may have many hundreds of accessions, each in small and perhaps single line plots. To minimize the inter-genotype competition, accessions should be arranged according to their height, growth habit, maturity and other such traits.

Large number of accessions in a trial usually need a compromise between statistical and practical consideration. If these are ever in conflict, then practical consideration should take precedence over the statistical ones. In general, the preliminary evaluation trials undertaken by the curators have two features in common that set them apart from other field experiments (IPGRI, 2001).

Advanced Evaluation (AE)

The AE consists of recording a number of additional descriptors useful in crop improvement. These include important agronomic traits, stress tolerance, disease and pest resistance and quality characters etc.

Table 21.15: Evaluation and Characterization of some Vegetables

Crops	Evaluating Center	Promising Accession
Bitter gourd	NBPGR, New Delhi	High yielder-UB-22, UB-66, IC-74158, C-1621, K-3374
	IIVR, Varanasi	High yielder-IC-44428B, NIC-72285, IC-85604A, IC-85608B, IC-85611, IC-85636, V-89/0-103C, VRBT-29, VRBT-50
		High number of fruits/plant-IC-45364A, VRBT-78, Ic-44426D, IC-44428B, IC-45350, NIC-72285, IC-85611
		CMV Resistance- IC-44428B, IC-72285, IC-85608B, V-89/0-103C, IC-85611, IC-85636, VRBT-50
		Fruit fly Resistance- IC-85641 and V-89/0-103C
Bottle gourd	NBPGR,New Delhi	High yielder (Summer) - U-10-315, U-11-10, U-8-217, U-8-46, U-8-198, BDJ-864
		High yielder (Kharif) – U-11-216A, U-11-50, BDJ-53, BDJ-783, BDJ-896, EC-187247, BDJ-616
	IIVR, Varanasi	High yielder-U-10-316, IC-92363, VRBG-2, VRBG-4
		Red pumpkin beetle resistance- U-10-316, NIC-1225
		Leaf miner resistance- U-10-316, IC-92363 IC-92418
		Downy mildew resistance- U-10-316, IC-92362
Ivy gourd	IGKVV, Raipur	High yielder-(3.20-21.21kg/plant)- Ac-59, 35, 54, 16, 6, 48, 10, 11, 14, 34, 52, 24, 17, 36, 2, 4, 31, 29
		Number of fruit/plant (209.33-1292.7)- Ac-35, 59, 11, 5, 16, 54, 24, 17, 10, 51, 61, 22, 58, 60, 53, 32, 31, 63, 37, 23
		Length of internodes (7.79-11.43cm)-Ac-17, 61, 5,14, 51, 6, 11, 15, 37, 1, 4, 29, 32, 36, 22, 2, 9, 55
		Diameter of stem (3.30-4.96)-Ac-4, 51, 63, 15, 36, 6, 11, 59, 61, 24, 62, 17, 16, 7, 32, 52, 58, 32, 2, 54, 5, 23, 4, 22, 9, 60, 29
		Length of fruit (3.97-7.60)- Ac-6, 1, 62, 11, 33, 54, 61, 31, 59
		Diameter of fruit (1.63-3.20)- Ac-59, 52, 31, 14, 35, 37
	IIVR, Varanasi	High yielder-VRK-05, VRK-10, VRK-20, VRK-35
		Diameter of stem-U-325/DA/DR/20, 325/DA/DR/36
		Length of internodes (CM)-U-35/DA/DR/48
		Resistance to mosaic virus-VRK-05, VRK-10, VRK-20, VRK-35
		Resistance to leaf miner- vrk-01, vrk-06, vrk-22, vrk-04, vrk-31, vrk-33 and vrk-55
Pointed gourd	IIVR, Varanasi	High yielder- VRPG-13, VRPG-44, VRPG-70, VRPG-72, VRPG-93
		Number of fruit/plant- VRPG-13, VRPG-44, VRPG-70, VRPG-72, VRPG-88, VRPG-93, VRPG-113
		Heat tolerance- VRPG-72, VRPG-99, VRPG-19, U-34/BP/DR/07, VRPG-110, VRPG-110A, VRPG-110B
		Resistance to gall forming nematode- U-34/BP/DR/44, VRPG-72, VRPG-18, VRPG-96
Ridge gourd	NBPGR, New Delhi	Promising- IC-93399, U-324-138, U-20-202A, U-24-134, Nic-10216
Sponge gourd	NBPGR, New Delhi	Promising- IC-92779, IC-92745 (very early), D-24-163, Nic-10237

Contd...

Table 21.5–*Contd...*

Crops	Evaluating Center	Promising Accession
Satputia	NBPGR, New Delhi	Promising- DCB-1363
Brinjal	NBPGR, New Delhi	IC-126879, NIC-5938, EC-304992 (Long deep purple); IC-90764, IC-112312, IC-90975, IC-126869, IC-144084 (Purple Long); IC- 74209A, IC- 136142, IC- 144083, NIC- 4573 (Green Long), IC-14405, NIC-13009 (Deep Purple Oblong); IC-90957, IC-127239, IC-113802, IC-112821 (Purple oblong); IC-127163 (white with purple stripe oblong), IC- 137748, IC- 144078 IC- 136328 (Creamy white oblong)
Chillies	NBPGR, New Delhi	IC-119676, IC-119639, PSR-1775, JBT-12/83 MNCH-64, BD-119, BDS-2573
Tomato	NBPGR, New Delhi	EC-106285, EC-118292, EC-163683, EC-170662, EC-110578 (High yield); EC-41824, EC-164855, EC-169308, EC-122063, EC-126955 (Heat tolerant); EC-315487, EC-320578 (Moderate heat tolerant lines); EC-260639, EC-232429, EC-267729, EC-267726 (Suitable for fresh market); EC-161252, EC-100845, EC-110268 (Suitable gor transportation); EC-54628, EC-52062 (Paste type); EC-357833, EC-357827, EC-362956, EC-362940, EC-362944, EC-362947 (early maturing type)

Observations are recorded on qualitative and quantitative traits. Qualitative traits include morphological, physiological and biochemical characters related to survival and scored using a number of checks to determine variation within and among the traits. Quantitative traits are subjected to environmental factors and are responsible for adaptation and productivity. These include productivity, quality components, resistance to diseases and pests as well as tolerance to adverse conditions or stress.

Much emphasis is presently given on multi-locational and multi-disciplinary approach of germplasm evaluation. For preliminary evaluation, locally adapted cultivars should be used as check and screening for specific diseases under controlled conditions or at hot spots should be carried out. Augmented design is invariably used due to large number of germplasm holdings under evaluation. Care should be taken to minimize natural cross-pollinated contamination and erroneous labeling. While, regenerating care should be taken to preserve original structure and productivity of accessions. Main dangers are due to inappropriate handling of materials during sowing/planting, harvesting, threshing, cleaning, sub-sampling, packing and labeling. Frequent regeneration should also be avoided by producing sufficient seeds during initial seed increase and conservation in medium term storage in working collection and long term gene bank as base collection.

Future Thrust

In order to meet the needs of vegetable breeders in the country and elsewhere, some more elite germplasm will be introduced and indigenous variability will have to be collected. Considering the introductions already made and supplied to different centers in the last 40-45 years and crop diversity build-up since 1940, following priorities need emphasis:

1. To assemble and inventorise all the existing germplasm with different centers in various crops and characterization of the same at multi-location using standard descriptors (already developed by the NBPGR in collaboration with crop specialists) including diseases/pests resistance and quality attributes. Thus, status of field gene bank can be assessed.

2. Multiplication, seed increase and conservation of germplasm in the National Gene Bank at NBPGR with necessary passport and evaluation data.

3. To collect germplasm materials in different crops as per identified gaps and areas of germplasm availability including hot spot locations. Under NATP, collection of vegetable germplasm was given top priority. A massive programme through NBPGR Headquarters and 10 Regional Stations (acting as Zonal centers) in collaboration with several CCPIs under NATP on Plant Biodiversity project is under operation. In this task, IIVR, Varanasi, IIHR, Bangalore, other ICAR institutes, SAUs, NGOs etc. are helping a great deal.

4. The NBPGR imparts training in Exploration/collection of germplasm. It has planned to conduct training on wild relatives of crop plants, and plant identification, taxonomy/ biosystematics etc.

5. The wild germplasm assumes high priority for collection. It should be handled carefully for maintenance as these are the sources of biotic/abiotic stresses, diseases/pest resistance etc.

6. Accessioning of all existing germplasm with different partners and registration of elite germplasm have become a necessary in the light of recent global development after signing of CBD and WTO.

7. Re-survey/collection in those areas from where useful genotypes have already been identified based on evaluation studies and biochemical/molecular linked desirable traits

References

Anonymous, 1987. Biosphere reserves. Proc. 1st Nat. Symp. Sep.1986. Ministry of Environment and Forests, New Delhi-258 p.

Anonymous, 1989 a. Plant Genetic Resources: their conservation *in situ* for human use. IUCN-UNESCO/ FAO,38 p.

Ariyo, O.J., 1993. Genetic diversity in West African okra (*Abelmoschus caillei*) Multivariate analysis of morphological and agronomical characteristics. *Gen. Resources Crop Evol.*, **40**: 25-32.

Arora, R.K. 1985. Genetic resources of less known cultivated food plants. *NBPGR Sci. Monogr.* No. 9:125.

Baswell, V.R. 1949. Our vegetable travelers. *Nat. Geogr. Mag.* **6**: 145.

Bailey, L.H. 1950. The standard Encyclopedia oh Horticulture. The Mac Millan Co., New York.

Bisht, I.S., Mahajan, R.K. and Rana, R.S., 1995. Genetic diversity in South Asian okra (*Abelmoschus esculentus*) germplasm collection. *Ann. Apl. Biol.*, **126**: 539-550.

Bos,I.1983. the optimum number of replication when testing lines or families on a fixed number of plots. *Euphytica* **32**: 311-318.

Chadha, M.L. and Tarsem, Lal, 1993. Improvement of cucurbits. *In* Advances in Horticulture, Vol. 5 Vegetable Crops. (Eds.) K.L. Chadha and G.Kalloo, Malhotra Publishing house, New Delhi

Choudhury, B. 1967. Vegetables: National Book Trust, India

Charrier, A., 1984. Genetic Resources of the Genus *Abelmoschus Med.* (okra, English translation). IBPGR, Rome.

Chevalier, A., 1940. L'origine, la culture et les usages de cinq Hibiscus de la section *Abelmoschus. Rev. Bot. Appl.*, **20**: 319-328 and 419.

Dean, A. and Voss, D.1999. *Design and Analysis of experiments.* Springer Texts in statistics Springer, New York, USA, 67-97 pp.

De Candolle, A. 1982. Origine des plantes cultivers. Paris (English Translation, 1886, Kegan Paul).

Dadlani, S.A., Singh, B.P. and Singh, R.V. 1981. System of national and International exchange of germplasm and Methods of recording followed at NBPGR. *In* Mehra, K.L. Arora. R.K. and Wadhi, S.R. (Eds.) Plant Exploration and collections, Sci. Monog. No. 3. NBPGR, New Delhi.

De Wet, J.M.J. 1977. Increasing cereal yields: Evolution under domestication pp. 111-118. In D.S. Seigler (Eds) *Crop Resources*. Academic Press. New York.

Fedorov.1974. Chromosome numbers of flowering plants. Atto Koetlz. Science Publishers, 624, Koring Stein, West Germany.

Federer, W.T. and Raghavarao, D. 1975. On Augmented Design. *Biometrics* 31: 29-35.

Federer, W.T. Nair, R.C and Raghavarao, D. 1975. Some augmented row-column designs. *Biometrics* 37: 361-374.

Fatokun, C.A. 1987. Wide hybridization in okra. *Theor. Appl. Gen.*, 74: 483-486.

Filov, A.I. and Vilmskaya, G.M. 1972. Cultivated cucurbits in different languages of the world. *Trudypo Prikladnoi Botanike Genetike L Selektsii*, 47: 138-150.

Flint, H.L. 1970. Importance of seed source ton propagation. Combined Proc. Int. Plant Propgator's Society, Wooster, Ohio 20: 470-475.

Fordham, A.J. 1971. Canadian hemlock variants and their propagation. Combined Proc. Int. *Plant Propagator's Society*, Wooster, Ohio 20: 171-179.

Ford-Lloyd, B.V. 2001. Genotyping in plant genetic resources. In: Henry RJ (ed) Plant genotyping-The DNA Fingerprinting of Plants. CABI Publishing, UK, 59-82, pp.

Gableman, and Gerloff (1978). Isolating plant germplasm with altered efficiency in mineral nutrition. *Hort. Sci.* 13: 682-684.

Hamon, S., 1988. Organisation evolutive du genre *Abelmoschus* (Gombo): co-adaptation et evolution de deux species de gombo cultivees en afrique de l'ouest *Abelmoschus esculentus* et *Abelmoschus caillei*. (Eds.) ORSTOM, T.D.M. 46.

Hamon, S. and Hamon, P., 1991. Future prospects of the genetic integrity of two species of okra (*Abelmoschus esculentus* and *Abelmoschus caillei*) cultivated in West Africa. *Euphytica*, 58: 101-111.

Hamon, S. and Charrier, A., 1983. Large variation of okra collected in Benin and Togo. *Pl. Genet. Resources Newsletter*, 56: 52-58.

Hamon, S., Charrier, A., Koechlin, J. and van-Sloten, D.H., 1991. Potential improvement of okra (*Abelmoschus* spp.) through the study of its genetic resources (in French). *Pl. Genet. Resources Newsletter*, 86: 9-15.

Hamon, S. and Yapo, A., 1986. Perturbation induced within the genus *Abelmoschus* by the discovery of a second edible okra species in West Africa. *Acta Horticulturae*, 182: 133-143.

Harlan, J.R. 1975. Practical problem in exploration of seed crops. *In* Crop genetic resources for today and tomorrow. (eds.) O.H. Frankel and J.G. Hawkes. Cambridge Univ. Press, London. 111-115 pp.

Hawkes, J.G. 1976. Sampling gene pools. Proc. NATO Conf. Conservation of threatened plants. Sec.I. *Ecology*. Plenum London.

IBPGR, 1991. Report on international workshop on okra genetic resources. Oct. 1990, IBPGR, Rome, pp. 2-3 and 8-12.

IPGRI, 2001. The Design and Analysis of evaluation Trials of Genetic Resources Collections. A guide for gene bank managers. IPGRI, Technical Bull. No. 4.International Plant genetic Resources Institute, Rome, Italy.

Jambhale, N.D. and Nerkar, Y.S., 1986. Parbhani Kranti- yellow vein mosaic resistant okra. *HortScience,* **21**: 1470-1471.

Jambhale, N.D. and Nerkar, Y.S., 1990. Okra. *In: Vegetable Growing Handbook: Organic and Traditional Methods* (Eds. Splittstoesser, W.E.). Melbourne, Thomas Nelson Australia,. 589-607 pp.

Jain, S.K. 1975. Population structure and the effect of breeding system. *In* Crop Genetic Resources for today and tomorrow. (eds.) O.H. Frankel and J.G. Hawkes. Cambridge Univ. Press, London. 15-36 pp.

Kempton R.A. and Fox, P.N.1997. Statistical Methods for Plant Variety Evaluation. Chapman and Hall, U.K.

Koechlin, J., 1991. African okras (*Abelmoschus* spp.): study of the diversity with regard to breeding (in French). In: Travaux et Documents Microfiches, 72. *Thesis,* Institut National Agronomiques Paris-Grignon, 180 pp.

Lin, C.S. and poushinsky, G. 1983. A modified augmented design for an early stage of plant selection involving a large number of test lines without replication. *Biometrics* **39**: 553-561.

Madeley, J. 1996. Yours for Food: Plant Genetic Resources and Food Security. Section 1: Plant Genetic Resources. People's Plan. BUKO Agrar, Nernstweg,Hamburg, Germany.

Marshal, D.R. and A.H.D. Brown. 1975. Optimum sampling strategies in genetic conservation. *In* Crop Genetic Resources for today and tomorrow. (eds.) O.H. Frankel and J.G. Hawkes. Cambridge Univ. Press, London. 37-52 pp.

Ministry of Environment and Forest (MoEF). 1998. Implementation of article 6 of the Convention on Biological Diversity in India. National Report (interim), Ministry of Environment and Forestry, Government of India.

Mohideen, M. K. and Irulappan, I. 1993. Improvement of Amaranthus. *In* Advances in Horticulture. Vol.5, Vegetable crops. 305-323. (eds) K.L. Chadha and G.Kalloo, Malhotra Printing House, New Delhi.

Munger, H.M. and Robinson, R.W. 1991. Nomenclature of *Cucumis melo* L. *Cucurbit Genet. Coop. Rep.* **14**:33.

Pal, B.P., Singh, H.B. and Swarup, V., 1952. Taxonomic relationship and breeding possibilities of species of *Abelmoschus* related to okra (*Abelmoschus esculentus* L.). *Bot. Gaz.,* **113**: 455-464.

PIER, 2001. " Invasive plant species: *Coccinia grandis*" Pacific Island Ecosystems at risk. United States Department of Agriculture, Natural Resources Conservation Services. The Maui Weekly Wrap. Maui News, p 10-18.

Rana, R.S. 1992. Core collection and the priorities of national programmes: Indian Perspective. Paper presented at the Intl. Workshop on Core Collection of plant Germplasm. Held at CENARGEN, Brasilia, Brazil. Sept.

Rana, R.S., Saxena, R.K. and Kochhar, Sudhir 1994. Descriptors and their descriptors status in relation to PGR conservation. *In* R.S.Rana *et al.* Conservation and Management of Plant Genetic resources. NBPGR, New Delhi. pp. 192-201

Seshadri, V.S. 1987. Genetic resources and their utilization in vegetable crops, pp. 335-343. in Plant Genetic Resources: Indian Perspective (Eds. R.S.Paroda, R.K. Arora and K.P.S. Chandel), NBPGR, New Delhi.

Sehgal, J.L., D.K. Mandal, C. Mandal and S. Vadivelun 1992. Agro-ecological regions of India. Tech. Bull. NBSS Pub. No.24. NBSS & LUP (ICAR) N.Delhi.

Shanmugavelu, K.G. 1989. Production Technology of vegetable crops. Oxford and IBH Fransisco. 448 p.

Siemonsma, J.S., 1982a. La culture du gombo (*Abelmoschus* spp.) legume fruit tropical avec reference speciale a la Cote d'Ivoire. Thesis Univ. Wageningen, the Netherlands.

Siemonsma, J.S., 1982b. West African Okra, morphological and cytological indications for the existence of a natural amphidiploid of [*Abelmoschus esculentus* (L.) Moench. and *Abelmoschus* manihot (L.) Medikus. *Euphytica*, 31:241-252.

Shimotsuma. 1963. Cytogenetics and evolutionary studies in genus *Citrullus*, Seiken Zin Rept. Kihara Inst. Biol. Res. 15: 24-34.

Van Borssum Waalkes, J., 1966. Malesian *Malvaceae* revised. *Blumea*, 1: 1-251.

Vavilov, N.I.1926. Studies on the origin of cultivated Plants. Bull. Appl. Bot. 26 (2):1-248.

Vavilov, N.I.1951. Phyto-geographical basis of plant breeding. In: (Selected Writings of N.I Vavilov and translated by KS Chester) *The Origin, Variation, Immunity and Breeding of cultivated plants, Chronica Botanica* 13:364. Waltham Mass., USA.

Velayudhan, K.C., and Upadhyay, M.P., 1994. Collecting okra, eggplant and their wild relatives in Nepal. *Pl. Genet. Resources Newsletter*, **97**: 55-57.

Whitaker, T.W. 1933. Bot. Gaz. 94: 780-790.

Wilsie, C.P. 1962. Crop adaptation and distribution. W.H. Freeman and Co. San Fransisco. 448 pp.

Zeven, A.C. and J.M.J. de Wet, 1982. Dictionary of cultivated plants and regions of diversity. Wageningen. 259 pp.

Zeven, A.C. and Zhukovsky, P.M. 1975. Dictionary of cultivated plants and their centers of diversity: excluding ornamentals, forest trees and lower plants. PUDOC, Wageningen. 219 pp.

Previous Volumes–Contents

Biodiversity in Horticultural Crops Vol 1/*Peter, K V & Z Abraham eds*

2007, xviii+364p., col. plts., figs., tabls., ind., 25 cm   Rs. 1800

ISBN 978-81-7035-490-1

Biodiversity in Horticultural Crops Vol 2/*Peter, K V ed*

2008, xix+320p., col. plts., figs., tabls., ind., 25 cm   Rs. 1500

ISBN 81-7035-562-5

Part I: General

Biodiversity in Horticultural Crops Vol 3/*Peter, K V ed*

2011, xxiv+336p., col. plts., tabls., figs., ind., 25 cm   Rs. 1600

ISBN 978-81-7035-672-1

Index